ISBN 978-1-332-30571-1
PIBN 10311788

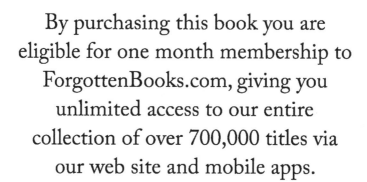

English
Français
Deutsche
Italiano
Español
Português

www.forgottenbooks.com

Mythology Photography **Fiction**
Fishing Christianity **Art** Cooking
Essays Buddhism Freemasonry
Medicine **Biology** Music **Ancient**
Egypt Evolution Carpentry Physics
Dance Geology **Mathematics** Fitness
Shakespeare **Folklore** Yoga Marketing
Confidence Immortality Biographies
Poetry **Psychology** Witchcraft
Electronics Chemistry History **Law**
Accounting **Philosophy** Anthropology
Alchemy Drama Quantum Mechanics
Atheism Sexual Health **Ancient History**
Entrepreneurship Languages Sport
Paleontology Needlework Islam
Metaphysics Investment Archaeology
Parenting Statistics Criminology
Motivational

PHYTOLOGIA

Designed to expedite botanical publication

Vol 36	June 1977	No 1

CONTENTS

Published by Harold N Moldenke and Alma L Moldenke

303 Parkside Road
Plainfield, New Jersey 07060
U S A

Price of this number $2 50, per volume, $9.75 in advance or $10 50 after close of the volume, 75 cents extra to all foreign addresses, 512 pages constitute a volume

NOTES ON THE SPECIES OF ERYTHRINA. IX.

B. A. Krukoff[1]

Contents

Introduction

Eight papers covering the morphology, distribution, chromosomes, palynology, alkaloids, and amino acids of _Erythrina_ were published in the September and December 1974 issues of _Lloydia_. A supplement to the symposium is planned for publication, also in _Lloydia_, sometime in 1977. I have written a brief paper for this supplement to the symposium, but this present paper includes information which is best published separately.

Four hundred eighty-nine new collections were examined in connection with the preparation of these two papers. For a list of species reduced to synonymy, extensions of ranges, etc., see my paper which is to be published in the supplement to the symposium.

1. __Erythrina fusca__ Loureiro, Fl. Cochinch. 427. 1790.

I have seen the following collections from the New World: Belize   L. Dieckman 196 (MO). Panamá: Colón· R. L. Wilbur 11169 (MO). Venezuela: Mérida: Luis Ruíz Terán 493 (MO). Colombia   Chocó AI. Gentry 9286, Cundinamarca: E. Forero et al. 417. Ecuador: Guayas: L. Holm-Nielsen et al. 7238 (AAU). Surinam. Landsbosh 136 (MO). Brazil: Amazonas: Prance et al. 23520, Bento S. Pena s.n. (March 21, 1975) (MO).

The following collections from the Old World were available Tonga   M. Hotta 5105 (TI), 5466 (TI).

Two other collections from Guayas, Ecuador, were examined but are not cited here.

[1] Consulting Botanist of Merck Sharp & Dohme Research Laboratories, Rahway, New Jersey, and Honorary Curator of the New York Botanical Garden.

2. <u>Erythrina crista-galli</u> L. Mant. 99. 1767.

Argentina: Buenos Aires· <u>O. Boelcke et al.</u> <u>14406</u> (MO).
Paraguay: <u>Jím Conrad</u> <u>2191</u>.

Two other collections, one from Brazil (Paraná) and the
other from Argentina (Corrientes) were examined but are not
cited here; also one from Belize, from cult. plant.

3. <u>Erythrina falcata</u> Bentham in Mart. Fl. Bras. 15(1): 172.
1859.

Brazil: Rio de Janeiro: <u>G. Pabst</u> <u>7369</u>; Santa Catarina:
<u>L. B. Smith et al.</u> <u>12800</u> (MO).

One collection from Brazil (Rio de Janeiro) was examined
but is not cited here.

4. <u>Erythrina dominguezii</u> Hassler, Physis 6: 123. 1922.

Argentina. Jujuy: <u>A. Krapovickas et al.</u> <u>26580</u> (MO).

5. <u>Erythrina ulei</u> Harms, Verh. Bot. Ver. Brand. 48: 172. 1907.

Ecuador· Zamora Chinchipe: <u>Elbert L. Little et al.</u> <u>275</u>.
Perú: Cuzco: <u>C. Vargas C.</u> <u>15537</u> (MO).

This is the first record of this species from Zamora
Chinchipe.

6. <u>Erythrina verna</u> Velloso, Fl. Flum. 304. 1825.

One collection from Brazil (Rio de Janeiro) was examined
but is not cited here.

7. <u>Erythrina poeppigiana</u> (Walpers) O. F. Cook, Bull. U. S.
Dept. Agr. Bot. 25. 57. 1901.

Venezuela: Mérida· <u>Luis Ruíz Terán</u> <u>494</u> (MO). Colombia:
Chocó: Rio Tolo, alt. 50 m, <u>E. Forero et al.</u> <u>1021</u>. Ecuador:
Morona Santiago· <u>Elbert L. Little et al.</u> <u>366</u>. Perú· San
Martín: <u>Schunke</u> <u>8231</u>.

This is the first record of this species from Chocó and
Morona Santiago. The collector states on the label of Chocó
specimen: "árbol muy común en toda el región."

Two collections, one from Ecuador (Esmeralda) and the other
from Bolivia (Bení) were examined but are not cited here.

8. Erythrina suberosa Roxburgh, Fl. Ind. 3. 253. 1832.

   Four collections from NW India were examined but are not cited here.

10. Erythrina stricta Roxburgh, Fl. Ind. 3 251. 1832.

   Thailand: C. F. van Beusekom 421 (AAU).

12. Erythrina arborescens Roxburgh, Pl. Coromandel 3 14, pl. 219. 1819.

   China: Sikang: C. Y. Chiao 1772 (S). India: Darjeeling, alt. ± 2200 m, H. Ohashi et al. s.n./1972.

13. Erythrina subumbrans (Hasskarl) Merrill, Philipp. Jour. Sci. Bot. 5· 113. 1910.

   India: G. Thanikaimoni 1191 (PO); Mysore: C. J. Saldanha 2182 (MO), 15834 (MO), 16454 (MO). Ceylon· Kostermans 25272 (AAU), 25327 (AAU), Kokawa & Hotta 5206 (AAU). Philippines: Quezon Justo P. Rojo 71 (MO).

   Seven collections deposited at S were examined but are not cited here. They are from the Philippines, Sumatra, Java, and Celebes.

14. Erythrina breviflora Alph. DeCandolle, Prodr. 2 413. 1825.

   México· Jalisco· 12 km NW of Los Volcanos, alt. 1900 m, Breedlove 35846.

15. Erythrina edulis Triana, M. Micheli, Jour. de Bot. 6 145. 1892.

   Ecuador Cotopaxi L. Holm-Nielsen 1140 (AAU); Tungurahua: L. Holm-Nielsen et al. 241 (AAU), 278 (AAU), Los Rios· Al. Gentry 9650 (S); Loja· A. Paredes 108 (MO); Napo Al. Gentry 12407, Morona Santiago Alberto T. Ortega U. 207. Peru· Huanuco: J. Schunke 8313.

   This is the first record of this species from Morona Santiago.

   Four collections from Ecuador (Pichincha, Loja and Napo) were examined but are not cited here.

18. Erythrina schimpffii Diels, Bibl. Bot. 116 96. 1937.

   Ecuador El Oro· alt. ± 360 m, Plowman 3457 (ECON).

19. Erythrina montana Rose & Standley, Contr. U. S. Nat. Herb.
    20· 179. 1919.

    México: Durango: alt. ± 2200 m, Davidse 10023.

22a. Erythrina herbacea L. subsp. herbacea. Erythrina herbacea
     L. Sp. Pl. 706. 1/53 sens. str.

     U. S. A.: Georgia: Rick Volosen s.n. (Apr. 12, 1970) (MO);
     Mississippi: Fleet N. Lee 30 (MO). Texas· D. S. Correll 23431.

22b. Erythrina herbacea L. subsp. nigrorosea Krukoff & Barneby
     Phytologia 25(1): 6. 1972.

     México Jalisco: C. Johnson 1973/325 (MO).

28b. Erythrina lanata Rose subsp. occidentalis (Standley)
     Krukoff & Barneby, Phytologia 27: 117. 1973.

     México: Jalisco: L. A. Perez 360.

29. Erythrina goldmanii Standley, Contr. U. S. Nat. Herb. 20:
     181. 1919.

     México: Chiapas: D. E. Breedlove 20397 (MO), 20593
     (Cintalapa, alt. ± 1000 m), 36747 (Cintalapa, alt. ± 900 m),
     23492 and 30362 (Ocozocoautla de Espinosa, alt. 800-1000 m),
     23561 (Tuxtla Gutiérrez, alt. 530 m), 23748 (Tonalá, alt. ± 60 m),
     26917 and 36829 (Arriaga, alt. ± 250 m), 23838 (MO) (Chiapa de
     Corzo), 37639 (Villa Corzo, alt. ± 900 m), 38655 (Angel Albino
     Corzo, alt. ± 900 m).

     The above-cited and previous collections of D. E. Breedlove
     from the State of Chiapas cover practically all municipalities in
     this State.

30. Erythrina caribaea Krukoff & Barneby, Phytologia 25: 9. 1972.

     México: Chiapas: mun. Ocozocoautla, alt. ± 700 m, lower
     montane rain forest, Breedlove 38698; mun. Ocosingo, alt. ± 300 m,
     tropical rain forest, Breedlove 33907.

31. Erythrina folkersii Krukoff & Moldenke, Phytologia 1: 286.
     1938.

     México: Chiapas: Las Margaritas, alt. ± 300 m, Breedlove
     et al. 34268. Belize: J. Dwyer et al. 88 (MO), 12121 (MO),
     12507 (MO), 12714 (MO).

36. <u>Erythrina chiapasana</u> Krukoff, Brittonia 3 304. 1939.

    México   Chiapas   Las Margaritas, alt. ± 1700 m, <u>Breedlove</u>
    <u>33398</u>.

    Berlin, Breedlove, and Raven (Principles of Tzeltal Plant
    Classification, p. 14, 1973) state that this species is common
    in Pine-Oak-Liquidambar Forest in mun. Tenejapa, Chiapas.

41. <u>Erythrina chiriquensis</u> Krukoff in Brittonia 3· 222. 1939.

    Panamá   <u>Croat 15639</u>, Chiriqui· <u>E. A. Lao 354</u> (MO) (alt.
    ± 2000 m).

45. <u>Erythrina steyermarkii</u> Krukoff & Barneby, Mem. N. Y. Bot.
    Gard. 20(2) 175. 1970.

    Costa Rica: Cartago· along road from Turrialba to Moravia
    de Chirripo, <u>Burger, W. et al. 10032</u>.

49. <u>Erythrina lanceolata</u> Standley, Contr. U. S. Nat. Herb. 17:
    432. 1914.

    Costa Rica· Alajuela   alt. ± 500 m, <u>W. Burger et al. 10020</u>.

50. <u>Erythrina costaricensis</u> M. Micheli, Bull. Herb. Boiss. 2·445.
    1894.

    Panamá   <u>J. A. Duke 14503</u>, Cocle· <u>Croat 26126</u> (MO); Canal
    Zone. <u>Croat 8667</u> (MO), <u>AI. Gentry et al. 8657</u> (MO), <u>Mireya D.</u>
    <u>Correa A. 484</u> (MO); Colón <u>Davidse et al. 10068</u> (MO); San Blas:
    <u>Duke 14857</u>. Colombia. Chocó· Rio Tolo, <u>Forero et al. 973</u>;
    Antioquia <u>AI. Gentry 9222</u> (MO).

    This is the first record of this species from San Blas.

52. <u>Erythrina americana</u> Miller, Gard. Dict. ed. 8, No. 5. 1768.

    México: Veracruz Orizaba, <u>M. Souza 4438</u>; Chiapas: mun.
    Bochil, along road to Simojovel, alt. 1250 m.

    This is the first record of this species from the State of
    Chiapas.

53. <u>Erythrina berteroana</u> Urban, Symb. Ant. 5· 370. 1908.

    Panamá   Cocle: El Valle de Antón, <u>M. Nee 9219</u>, Canal
    Zone <u>AI. Gentry 8629</u> (MO), <u>8746</u> (MO), Colón· <u>Davidse 10068</u> (MO).

    This is the first record of this species from Colón.

54. <u>Erythrina rubrinervia</u> H. B. K. Nov. Gen. & Sp. 6: 434. 1824.

   Colombia: Cundinamarca: <u>E. Forero 300</u> (AAU).

58. <u>Erythrina gibbosa</u> Cufodontis, Arch. Bot. Sist. Fitog. & Genet. 10: 34. 1934.

   Panamá. Cocle: <u>Duke 13232</u>; Veraguas· <u>Mori & Kallunki 2537</u>.

61. <u>Erythrina peruviana</u> Krukoff, Brittonia 3: 262. 1939.

   Ecuador: Zamora Chinchipe, alt. $\pm$ 1000 m, <u>Elbert L. Little et al. 276</u>.

   This is the first record of this species from Zamora Chinchipe.

62. <u>Erythrina mitis</u> Jacquin, Hort. Schoenb. 2: 47. 1797.

   Venezuela: <u>Gentry et al. 14823</u>; Miranda: <u>Davidse 4114</u> (MO).

64a. <u>Erythrina corallodendrum</u> L. var. <u>corallodendrum</u>
    <u>Erythrina corallodendrum</u> L. Sp. Pl. 706. 1763.

   Jamaica· St. Ann, <u>Dulcie Powell 979</u> (MO).

70. <u>Erythrina oliviae</u> Krukoff, Phytologia 19(3): 128. 1969.

   México: Puebla: km 230-231 of the México-Oaxaca highway, bank of dry stream, <u>J. Mejicanos 1977/1</u> (Feb. 16, 1977), <u>1977/2</u>.

   These two collections were made from the same two trees from which the type and all other collections of this species were made. The trees were leafless (the leaves fell in December) and had mature fruits.

71. <u>Erythrina caffra</u> Thunberg, Prodr. Pl. Cap. 121. 1800.

   Eight collections from South Africa (Natal) were examined but are not cited here.

   For illustration of this species, see: Palmer, E. & Pitman, N. Trees of South Africa 2: 955-957. 1972, and Killick, Fl. Pl. of Africa 43 (3--4): t. 1709, 1710. 1976.

72. <u>Erythrina lysistemon</u> Hutchinson, Kew Bull. 1933: 422. 1933.

   Tanzania: <u>C. F. Paget-Wilkes 211</u> (MO). South Africa: near Port St. John, <u>R. D. A. Bayliss 7046</u> (MO).

Eleven collections from South Africa, S. Rhodesia, Tanzania, and from plants in cultivation in Australia (Brisbane) and Hong Kong were examined but are not cited here.

For illustration of this species, see   Palmer E. & Pitman, N. Trees of South Africa 2: 957-959. 1972.

73. Erythrina humeana Sprengel, Syst. 3 243. 1826.

Six specimens from South Africa (Cape Province and Transvaal) were examined but are not cited here.

For illustration of this species, see. Palmer, E. & Pitman, N. Trees of South Africa 2: 961-962. 1972.

74. Erythrina zeyheri Harvey, Fl. Cap. 2. 236. 1862.

Three collections of this species from South Africa (Transvaal) were examined but are not cited here.

75. Erythrina acanthocarpa E. Meyer, Comm. Pl. Afr. Austr. 1: 151. 1836.

Six collections from South Africa (Cape Province) were examined but are not cited here.

77. Erythrina brucei Schweinfurth, Verhand. Zoo.-Bot. Gesell. Wien 18: 653. 1868. et auct. plur., pro majore parte, leguminibus seminibusque exceptis; emend. Gillett, Kew Bull. 15. 428. 1962.

Two collections from Ethiopia, collected at altitudes of ± 2250 m and 2800 m, were examined but are not cited here.

79. Erythrina senegalensis Alph. DeCandolle, Prodr. 2: 413. 1825.

Guinée   Jacques-Georges 5348. Nigeria: western state, Roy C. Brown 921 (MO).

Three collections from Senegal and Sierra Leone were examined but are not cited here.

81. Erythrina mildbraedii Harms in Mildbr. Deutsch. Zentr.-Afr. Exp. 190//1908, 2: 264, tab. 30. 1911.

Nigeria   Ibadan Div., F. N. Hepper 2290 (S).

85. Erythrina decora Harms in Engl. Jahrb. 49 441. 1913.

South West Africa   Hakasgabirga, E. Busch 7957 (S).

For illustration of this species, see· Palmer, E. & Pitman, N. Trees of South Africa 2: 952. 1972.

88.   Erythrina addisoniae Hutchinson & Dalziel, Kew Bull. 1929: 17. 1929.

Guinée· Jacques-Georges 4745 (MO), 5523 (MO).

89.   Erythrina droogmansiana DeWildeman & Th. Durand, Bull. Soc. Roy. Bot. Belg. 40 19. 1901.

Zaire   Orientale. J. Louis 9464 (S).

For the occurrence of this species in Uganda, see the paper by B. Verdcourt & T. J. Synnott in Kew Bulletin 30(3): 471-473. 1975.

91.   Erythrina sacleuxii Hua, Bull. Soc. Linn. Paris n.s. 1: 54. 1898.

Kenya. Kwale Dist., P. J. Greenway 9646 (S). Tanzania: Bezirk Lindi, H. J. Schlieben 5634 (S), 6193 (S).

93.   Erythrina sigmoidea Hua, Bull. Mus. Hist. Nat. Par. 3: 327. 1897.

Nigeria: Latilo et al. 69374 (MO).

94.   Erythrina latissima E. Meyer, Comm. Pl. Afr. Austr. 1: 151. 1836.

Four collections from Rhodesia and South Africa (Natal) were examined but are not cited here.

For illustrations of this species, see Killick, Fl. Pl. of Africa 43(3-4)· t. 1709, 1710. 1976, also Palmer, E. & Pitman, N., Trees of South Africa 2: 959-961. 1972.

95.   Erythrina abyssinica Lamarck, Encycl. Bot. 2: 392. 1788; DC Prodr. 2: 413. 1825.

Ruanda· P. Auquier 2652 (MO). Tanzania: G. W. Frame 529 (MO), C. F. Paget-Wilkes 210 (MO). Malawi: Jean Pawek 6347 (MO), 7748 (MO). Rhodesia   Adele Lewis Grant s.n. (July 1928) (MO).

Twenty-three collections from Ethiopia, Eritrea, Kenya, Tanzania, Mozambique, S. Rhodesia, and Angola were examined but are not cited here.

96. Erythrina variegata L. Herb. Amboin. 10. 1754, Amoen. Acad.
    4 122. 1759.

India   G. Thanikaimoni 1181 (Pondichery). Taiwan. C.
Owatari s.n. (March 1, 1898) (TI), L. Sasaki s.n. (Sept. 8, 1965
(TI). Ryukyu Islands   R. J. Alvis 73 (TI). Okinawa Islands:
S. Hatusima 17497 (TI). Jaluit Islands   G. Koidzumi s.n. (Jan.
1915) (TI). Palau· T. Tuyama s.n. (Aug. 28, 1939) (TI).

Seven collections from Java, Celebes, Solomon Islands, and
Fiji were examined but are not cited here; also three collections
from Virgin Islands (St. Croix and Tortola) which are from
cultivated plants.

102. Erythrina velutina Willdenow, Gest. Nat. Freunde Berlin Neue
     Schr. 3: 426. 1801.

Venezuela: Anzoátegui. Luis Ruíz Terán 350 (MO). Ecuador:
Manabi: C. A. Dodson & L. B. Thien 1009, Al. Gentry 12204,
Guayas: Al. Gentry 10077; Galápagos Islands: F. R. Fosberg 44917
(MO), 44944 (MO). Perú: Plowman 5437 (Tumbes, alt. ± 550 m).

104. Erythrina burttii Baker f., Jour. Bot. 70: 254. 1932.

Kenya: Kajiado Dist.· Greenway 9579 (S).

105. Erythrina burana R. Chiovenda, Att. R. Accad. Ital., Mem.
     Sc. Fis. Mat. & Nat. 11: 27. 1940.

Ethiopia: Harar: students of Imper. Ethiop. College of
Agr. & Mech. Arts s.n. (S).

The collectors state that this species is very common in
Harar and is planted on campus.

106. Erythrina perrieri R. Viguier, Not. Syst. 14. 175. 1952.

Mauritius: Ile aux Cerfs, D. Lorence 7 (1975) (MO).

The collector states on the label: "flowers scarlet,
occasional locally."

This is the first record of this species from Mauritius.

Hybrids

1. Erythrina x bidwillii Lindley, Bot. Reg. 33 pl. 9. 1849.

Japan: Tokyo, Shinjiku-gyon, M. Togashi s.n. (July 6, 1968)
(TI) (cult.).

7. Erythrina x sykesii Barneby & Krukoff, Lloydia 37· 447. 1974.

Western Australia: Perth.   Univ. Campus, David Ladd s.n. (June 9, 1970) (MO).

The collector states on the label: "common Perth street tree."

Bibliography

(In order to conserve space, we are citing here only the papers which are not cited in Supplements III-VIII.)

1. Barton, Derek H. R. et al. Phenol oxidation and biosynthesis, Part XXII. The alkaloids of E. lysistemon, E. abyssinica, E. poeppigiana, E. fusca, and E. subumbrans (as "E. lithosperma Blume"). J. Chem. Soc. (C.) 652-654. 1971.

2. Bhakuni, D. S. & N. M. Khanna. Chemical examination of the bark of Erythrina variegata (as "indica Lam."). J. Sci. Ind. Res. Sect. B 18: 494. 1959.

3. Bocquet, G. & J. O. Derron. Les Erythrina de la République de Sao Tomé et Príncipe. Ber. Schweiz. Bot. Ges. 85(4): 298-302. 1975.

4. Chopra, R. N., S. Ghosh & B. N. Sen. Some common indigenous remedies--Erythrina variegata (as "indica"). Indian J. Med. Res. 22: 265. 1934.

5. Ghosal, S., S. K. Dutta & S. K. Bhattacharya. Erythrina-- Chemical and pharmacological evaluation II: Alkaloids of Erythrina variegata L. J. Pharm. Sci. 61: 1274. 1972.

6. Krukoff, B. A. Notes on the species of Erythrina. VIII. Phytologia 33: 342-355. 1976.

7. Letcher, R. M. Alkaloids of Erythrina lysistemon. 11-- Methoxyerythraline, a new alkaloid. J. Chem. Soc. (C.) 652. 1971.

8. Singh, H. & A. S. Chawla. Isolation of erysodine, erysotrine and hypaphorine from Erythrina suberosa Roxb. seeds. Experientia 25: 785. 1969.

9. Singh, H., A. S. Chawla, J. W. Rowe & J. K. Toda. Waxes and sterols of Erythrina suberosa bark. Phytochemistry 9: 1673. 1970.

10. Singh, H. & A. S. Chawla. Chemical constituents of Erythrina suberosa Roxb. seeds. J. Pharm. Sci. 59 1179. 1970.

11. Singh, H. & A. S. Chawla. Study of the chemical constituents of seeds of Erythrina variegata (as E. variegata var. orientalis"). Planta Med. 19. 71. 1970.

12. Singh, H. & A. S. Chawla. Study of Erythrina suberosa leaves. Planta Med. 19: 378. 1971.

13. Singh, H., A. S. Chawla, A. K. Jindal, M. R. Subbaram & K. T. Achaya. Seed oils of Erythrina arborescens and E. stricta. Indian J. Technol. 10 115. 1972.

14. Singh, H. et al. Investigation of Erythrina spp. VII. Chemical constituents of Erythrina variegata (as "E. variegata var. orientalis"). Lloydia 38· 97-100. 1975.

15. Gustafson, R. Erythrina in Southern California. Hortulus Aliquando 1. 5-11. 1975-6.

SUPPLEMENTARY NOTES ON AMERICAN MENISPERMACEAE. XII.
NEOTROPICAL TRICLISIEAE AND ANOMOSPERMEAE

B. A. Krukoff[1]

Contents

Introduction

Since the latest paper in this series (Supplement XI) was
published, 47 new collections were examined, adding to our knowledge of several species. Extensions of ranges were noted for
four species, but no new species were described. Two new collections of poorly collected and poorly understood Anomospermum
matogrossense were of particular interest.

The chemical work on neotropical Triclisieae and Anomospermeae by Prof. Michael P. Cava and his associates continues, as
may be seen from the bibliography. Abuta splendida, Sciadotenia
toxifera and Abuta grisebachii were studied, and four new ben-
zylisoquinoline alkaloids (krukovine, sciadenine, grisabine, and
grisabutine) were isolated and characterized.

Of the 16 known New World genera of Menispermaceae, chromosomes were studied of only four genera (Calycocarpum, Menispermum,
Cocculus, and Cissampelos). Dr. William L. Theobald, director of
Pacific Tropical Botanical Garden in Hawaii, is now receiving
seeds, largely from the collectors of MO and NY. It is interesting that the first two batches of seeds were sent by Dr. J.-J.
de Granville from French Guiana. They were of the recently
described remarkable Elephantomene eburnea and Anomospermum sp.;
seeds of these germinated four to five months after planting.

[1]Consulting Botanist of Merck Sharp & Dohme Research
Laboratories, Rahway, New Jersey, and Honorary Curator of the
New York Botanical Garden.

SUPPLEMENTARY NOTES ON THE AMERICAN SPECIES

OF STRYCHNOS. XV.

B. A. Krukoff[1]

Contents

Introduction

Since the previous paper in this series was submitted for publication in 1976, 75 new collections were examined. The newly examined collections added to our knowledge of several species, and extensions of range were noted for ten. The extensions of S. nigricans to the well-collected State of Paraná, Brazil, and of S. poeppigii to Panama, Panama, are most interesting. No new species were described. It was particularly interesting to see fruits (unfortunately, not completely mature) of S. tabascana for the first time. The shells of these fruits are thicker than those of the closely related S. panamensis. The corolla of S. tabascana is pubescent externally, whereas that of S. panamensis is glabrous.

The chemical work of Professor Marini-Bettolo and his associates is continuing, as may be seen from the bibliography.

In a previous paper, I mentioned the very timely and interesting contributions made by Dr. Ghillean T. Prance concerning the botanical ingredients of Curare as prepared by four Indian tribes. Three species of Strychnos were identified previously: S. bredemeyeri used by Mayongong and Sanama Indians of Roraima, and S. cogens and S. solimoesana used by Jamamadi Indians of the basin of the Rio Purus. Other specimens, sterile and from comparatively young plants, are still under study. Details on the preparation of Curare will be published elsewhere.

S. peckii is reported by Alberto T. Ortega as an ingredient of Curare in Morona Santiago, Ecuador.

[1]Consulting Botanist of Merck Sharp & Dohme Research Laboratories, Rahway, New Jersey, and Honorary Curator of New York Botanical Garden.

17

6. Strychnos rondeletioides Spruce ex Bentham, Jour. Linn. Soc. 1: 104. 1856.

Venezuela: Amazonas. Paul E. Berry 638. Brazil: Amazonas: basin of Rio Purus, Prance et al. P21206, P21207, P3403. Perú: Loreto· Juan Revilla 174, Al. Gentry et al. 16677, s.n. (Jan. 1976).

Specimens from the basin of the Rio Purus are from plants which are used as a fish poison by Paumari Indians. Their name for this plant is "Jadadakaikapihai."

10. Strychnos brachiata Ruiz & Pavón, Fl. Per. 2: 30. 1799.

Colombia· Boyaca· C. Sastre 766. Perú: San Martín: Mariscal Cáceres, Tocache Nuevo, J. Schunke 2310.

This is the first record of this species from Boyaca.

11. Strychnos trinervis (Velloso) Martius, Syst. Mat. Med. Bras. 121. 1843.

Brazil: Minas Gerais: Serra de Cipó, JBR 114491 (MO), Guanabara: JBR 109209, Paraná: mun. Antonina, Hatschbach 33409.

12. Strychnos panamensis Seemann, Bot. Voy. Herald, 166. 1854.

Mexico: Chiapas: Mapastepec, alt. ± 180 m, Breedlove & Thorne 30708. Panama· Chiriqui· Croat 21935 (MO); Canal Zone: Croat 10097. Venezuela: Zulia: F. D. Chitti & Benkowski 3146. Colombia: Chocó: Al. Gentry 9430.

13. Strychnos tabascana Sprague & Sandwith, Kew Bull. 1927: 128. 1927.

Mexico. Chiapas: mun. La Trinitaria, alt. ± 1300 m, montane rain forest, Breedlove 38882; mun. Las Margaritas, alt. ± 350 m, tropical rain forest, Breedlove 33158.

Breedlove 38882 is a very valuable specimen, as it is the first specimen seen with immature fruits. The shells of these fruits are thicker than those of S. panamansis.

18. Strychnos medeola Sagot ex Progel in Mart. Fl. Bras. 6(1): 282. 1868.

Brazil: Pará: Nilo T. Silva 3352 (IAN).

19. Strychnos toxifera Robert Schomburgk ex Bentham, Jour. Bot.
Hook. 3 240. 1841.

Guyana   Kanuku Mtns., R. Goodland & Maycock 461. Colombia:
Chocó Al. Gentry & Aguirre 15196 (MO). Brazil: Manaus
Prance et al. 23569.

These are the first records of this species from Chocó as
well as from the lower Rio Negro.

21. Strychnos diaboli Sandwith, Kew Bull. 1931: 486. 1931.

Venezuela· Amazonas· San Carlos de Rio Negro, Paul E.
Berry 1544.

This is the first record of this species from Venezuela.

23. Strychnos sandwithiana Krukoff & Barneby, Mem. N. Y. Bot.
Gard. 20(1): 36. 1969.

Perú: San Martín: Mariscal Cáceres, Tocache Nuevo,
J. Schunke 43, 56.

This is the first record of this species from Peru.

24. Strychnos jobertiana Baillon, Adansonia. 12: 367. 1879.

Venezuela   Amazonas   San Carlos de Rio Negro, Paul E.
Berry 1405, 1447. Brazil: Amazonas   Manaus-Itacoatiara road,
km 13, W. Rodrigues & A. Loureiro 9519.

25. Strychnos pseudo-quina A. St. Hilaire, Mém. Mus. Paris
9: 340. 1822.

Brazil. Mato Grosso Cuiabá, Hatschbach 34010, 36072;
Minas Gerais. J.B.R. 130177 (MO), Sao Paulo: Oswaldo Handro
439 (S).

28. Strychnos solimoesana Krukoff, Brittonia 4: 280. 1942.

Brazil   Amazonas· basin of Rio Purus, Prance et al.
P21254; Manaus-Porto Velho road, Prance et al. 22884.

Prance et al. P21254 is an ingredient of Jamamadi arrow
poison and Prance 22884 is the first record of this species
from the basin of the Rio Negro.

31. Strychnos peckii B. L. Robinson, Proc. Amer. Acad. 49: 504.
1913.

Perú: San Martín: Mariscal Cáceres, Tocache Nuevo,
J. Schunke 39, 42. Ecuador Morona Santiago, alt. ± 300 m,
Alberto T. Ortega U. 401.

Ortega's label reads: "usado para envenenar las flechas."
Schunke's specimens are the first record of this species from
San Martín.

32. Strychnos erichsonii Richard Schomburgk, Reisen 3: 1082.
     1848. nomen: ex Progel in Mart. Fl. Bras. 6(1): 274. 1868.

Surinam: J. C. Lindeman 522 (SW plateau covered by ferrobauxite, 550-710 m alt.), LBB 15301 (dist. Para). French Guiana:
Maroni River, Sastre & Moretti 4024. Brazil: Amazonas· Prance
24541 (basin of Rio Içá, ± 5 km above mouth), Mori & Prance 9121,
9126 (Rio Jandiatuba, ± 10 km downstream from São Paulo de
Olivença). Perú: lower Anpiyacu, north of Rio Marañon, Prance
24699.

The two collections from Rio Jandiatuba are from white-water
varzea forest, both collected on February 26 in flower. They are
the best collections I have ever seen in flower; I refer particularly to the well-preserved creamy-yellow papillose corolla tubes.

32a. Strychnos croatii Krukoff & Barneby.

Panama. Panama· Croat s.n. (transect. #114).

35. Strychnos bredemeyeri (Schultes) Sprague & Sandwith, Kew
     Bull. 1927: 128. 1927.

Venezuela. Amazonas Paul E. Berry 1607. Brazil:
Roraima: vicinity of Auaris, Prance et al. 21502.

Prance writes on the label· "used as an ingredient of
Mayongong and Sanama Curare."

"Cumuduá" or "Cumarua" (Mayongong Indian dialect), "Mogoli"
(Sanama Indian dialect).

36a. Strychnos mitscherlichii Richard Schomburgk, Reisen 2· 451.
     1848, var. mitscherlichii.

Surinam· Lely Mts., SW plateau covered by ferrobauxite,
550-710 m, Lindeman et al. 234, 734. Brazil: Amazonas: Prance
et al. 23445 (near Manaus), 24426 (Rio Solimões, Ilha Jurupari
and vicinity). Perú: San Martín, Mariscal Cáceres, Tocache
Nuevo, J. Schunke 40.

37. Strychnos solerederi Gilg in Engler, Bot. Jahrb. 25 (Beibl.
     60): 40. 1898.

     Perú San Martín· Mariscal Caceres, Tocache Nuevo,
J. Schunke 8144.

39. Strychnos guianensis (Aublet) Martius, Syst. Mart. Med.
     Bras. 121. 1843.

     Brazil Amazonas   upper Rio Solimões, Mori & Prance 9036
(Paraná de Tonantins), 2206 (Igarapé Preto, near Belem). Perú:
Loreto AI. Gentry et al. 16675, Juan Revilla 172.

43. Strychnos panurensis Sprague & Sandwith, Kew Bull. 1927:
     132. 1927.

     Colombia. Chocó Duke 13331 (MO) (Rio Truando), 15800
(MO). Perú· San Martín Mariscal Cáceres, Tocache Nuevo,
J. Schunke 44.

47. Strychnos cogens Bentham, Jour. Bot. Hook, 3. 241. 1841.

     Brazil: Amazonas: basin of Rio Purus, Prance et al. 23438.

     This is the main ingredient of Jamamadi arrow poison, "Iha"
(Jamamadi Indian dialect).

     This is the first record of this species from the basin of the
Rio Purus.

48. Strychnos melinoniana Baillon, Bull. Soc. Linn. Paris 1:
     256. 1880.

     Surinam: Lely Mountains, alt. ± 650 m, Mori & Bolten 8493.

53. Strychnos fendleri Sprague & Sandwith, Kew Bull. 1927: 129.
     1927.

     Marini-Bettolo and collaborators investigated the alkaloids
found in this species (Gazzetta Chimica Italiana 106 773-777.
1976). The stem bark was found to contain seven tertiary alkaloids. The structure of four of these is reported.

56. Strychnos parvifolia DC., Prodr. 9. 16. 1845.

     Brazil: Guanabara. JBR 114845.

59. Strychnos brasiliensis (Sprengel) Martius, Flora 24 (Beibl.
     2): 84. 1841.

Brazil: Rio de Janeiro: JBR 141312 (MO), 141313 (MO);
Guanabara: JBR 55680; Paraná: Hatschbach 35135, 35625.

63. Strychnos brachistantha Standley, Field Mus. Publ. Bot. 12:
     412. 1936.

Belize: Corozal: alt. ± 33 m, Croat 24961.

64. Strychnos nigricans Progel in Mart. Fl. Bras. 6(1): 280.
     1868.

Brazil: Paraná: Rio Putuña, Hatschbach 35591.

This is the first record of this species from the State of
Paraná.

69. Strychnos poeppigii Progel in Mart. Fl. Bras. 6(1): 282.
     1868.

Panama: Panama: natural bridge along Madden Lake, Croat
12403.

This is the first record of this species from Panama.

70. Strychnos tarapotensis Sprague & Sandw., Kew Bull. 1927:
     131. 1927.

Perú: San Martín: T. Plowman & H. Kennedy 3812 (ECON),
J. Schunke 8150 (Mariscal Cáceres, Tocache Nuevo), Madre de Dios:
Plowman & Davis 5067.

This is the first record of this species from Madre de Dios.

Bibliography

(In order to conserve space, I am citing here only the papers
which are not cited in Suppl. VII-XIV.)

71. Krukoff, B. A. Supplementary notes on the American
     species of Strychnos. XIV. Phytologia 33: 305-322. 1976.

109r. Marini-Bettolo, G. B., et al. XXIX. New indole alkaloids
     from Strychnos fendleri Sprague & Sandwith. Gazzetta
     Chimica Italiana 106: 773-777. 1976,

109s. Marini-Bettolo, G. B., et al. Sul curaro Yanoáma. Un
     nuevo tipo di curaro indigeno: "Curare di torrefazione
     e percolazione." Lincei. Rend. Sc. fis. mat. e nat. 38:
     34-38. 1965.

A new species of Bidens (Asteraceae) from Brazil

B. L. Turner

The University of Texas, Austin, Texas 78712

Field work in Brazil during 1974 with the entomologist, Dr. D. Otte of the Philadelphia Academy of Sciences, resulted in the discovery of the following species belonging to the Section Selvorngia. The section was previously thought to be monotypic with the single species B. graveolens Gard.

Bidens goiana Turner, Sp. nov. Fig. 1

Herbae erectae ad 1 m altae caulibus debilibus, B. graveolens valde similes sed foliis minoribus planissime ellipticis, capitulis floribusque minoribus, floribus perspicue flavis.

Erect weak-stemmed herb up to 1 m tall. Much resembling B. graveolens but with smaller, more elliptic leaves, smaller heads and floral parts, and decidedly yellow flowers. Chromosome number, $\underline{n}$ = 22 pairs.

HOLOTYPE (LL)· Brazil. Goias. 40 km ENE of Brasilia. In burned-over, short, open forests. Sandy sterile soils. 5 Feb. 1974. B. L. Turner 9125

The species is found as a populational unit in the same region in which B. graveolens occurs. In addition to its smaller habit and more delicate, flexuous, inflorescence, B. goiana can be recognized by the dill-like smell of its crushed foliage and decidedly yellow flowers. The crushed foliage of B. graveolens has a lemon smell and the flowers are variously purplish- to brown-yellow.

The chromosome number of Bidens goiana is tetraploid on a base of $\underline{x}$ = 11, while B. graveolens is tetraploid on a base of $\underline{x}$ = 12 (Turner, et al., in press), although it is possible that tetraploid populations ancestral to the latter gave rise to B goiana by aneuploid loss at the higher level.

I am grateful to M. C. Johnston for the Latin diagnosis. Supported, in part, by N S. F. Grant 1013950.

Fig. 1. Habit sketch of _Bidens goiana_ (X 1/2)

CLUSIA SECTION COCHLANTHERA - - AGAIN

Bassett Maguire
The New York Botanical Garden

Twice I have reviewed the content of the section Cochlanthera
(Choisy) Engler of the genus Clusia Linnaeus of the Clusiaceae. As
the last paper (1977) was about to be released from the press, there
came into my hands a small collection made in 1962 on the Cambridge
Calima Valley Expedition of that year. Immediately thereafter a
duplicate of the same collection came to me from the U. S. National
Herbarium.

It proved to represent still another, and undescribed, Clusia
of the section Cochlanthera. It is necessary to place the new
species on record. This is the eighth now known for the section,
and the fifth apparently endemic to Colombia, having been collected
there in the Pacific Department of Valle.

The new plant is placed in the subsection Cochlanthera, its
closest relative being Clusia centricupula Cuatrecasas, also of
Valle, Colombia. Both species, amply distinct as shown by the key,
are insufficiently known, our species only by the type and that of
Cuatrecasas by only two collections. It is not known whether the
two species are closely sympatric.

Clusia calimae Maguire, sp nov

Frutex vel arbor mediocris; ramulis plus minusve 4-angulatis,
4-costatis, internodiis 3-4 cm longis; foliis mediocribus, laminis subcoriaceis, obovatis, (4)6-8 cm longis, (2.5)3.0-4.5 cm
latis, costa prominenti, venis lateralibus, prominulis, angulo
45° adscendentibus, apice late obtuso vel rotundato, basi acutius-
cula, brevi-decurrenti; petiolis crassis, 4-8 mm longis, 5-8 mm
latis, subamplectantibus; inflorescentia 12-15-flore, bracteis
parvis; floribus masculinis: sepalis 9, duobus inferioribus jugis
decussatis, late semiorbicularibus, ca 6 mm longis, 5 mm latis,
minute marginatis; superioribus imbricatis, subchartaceis, valde
scarioso-marginatis, late semiorbicularibus, 10-12 mm latis, 8-
10 mm longis; petalis 8, obovato-oblanceolatis, aliquantum pandu-
riformibus, 14-16 mm latis, 22-25 mm longis; staminibus introrsis,
paucis, 14-16, 2-seriatis, liberis vel minute ad basim connatis;
annulo deficienti; filamentis 3-4 mm longis, ad basim 1.2-1.6 mm
latis; antheris 1.2-1.5 mm longis, valde recurvatis; staminodiis
in massa centrali 6-8 mm diam, 4-5 mm alta; ovario deficienti; nec
floribus foemineis nec fructibus visis.

Type. Shrub 12-15 ft, in hedge, pink flowers and buds, edge of thick forest, near Las Delicias, at 5000 ft alt, NW of Restrepo, Valle, Colombia, 1 Aug 1962, <u>J. W. Robinson 201</u> (holotype K).

Distribution. COLOMBIA. Valle: shrub 12-15 ft, in hedge, pink flowers and buds, edge of thick forest near Las Delicias, NW of Restrepo, 5000 ft alt, 1 Aug 1962, <u>Robinson 201</u> (holotype K, isotype US, fragment NY).

<u>Clusia calimae</u> would be placed in the key (Maguire, 1977, p 133) <u>immediately after</u> <u>C</u>. <u>centricupula</u> in the following manner:

1. Receptacle of ♂ flowers shallowly discoid, forming a glutinous mass in the center of the disc (subsect <u>Cochlanthera</u>).

   2. Stamens fewer than 50.

      3. Leaf petiole slender, not winged; coastal mountains of Venezuela.   1. <u>Clusia</u> <u>cochlanthera</u> Vesque.

      3. Leaf petiole broadly winged, 1.0-1.5 cm long; Pacific Colombia.

         4. Stamens ca 35, 3-cyclic; leaf blades broadly elliptic or elliptic-obovate, 4-7 mm broad, 10-15 cm long, the apex acute or acutish.
            2. <u>Clusia</u> <u>centricupula</u> Cuatrecasas.

         4. Stamens 14-16, unicyclic; leaf blades obovate, 3.0-4.5 cm broad, 6-8 mm long, the apex broadly rounded.
            3. <u>Clusia</u> <u>calimae</u> Maguire.

   2. Stamens ± 100.   4. <u>Clusia</u> <u>lunanthera</u> Maguire.
                        5. <u>Clusia</u> <u>cochlitheca</u> Maguire.

1. Receptacle of ♂ flowers provided with a prominent coroniform androphore (subsect <u>Orthoneura</u>).
                6. <u>Clusia</u> <u>orthoneura</u> Standley.
                7. <u>Clusia</u> <u>celiae</u> Maguire.
                8. <u>Clusia</u> <u>cochliformis</u> Maguire.

[*]Maguire, B. Mem. N. Y. Bot. Gard. 10(1): 58-61. 1958.
Maguire, B. Caldasia 11(55): 129-146. 1977.

<u>Clusia cochlanthera</u> Vesque

Dr. Steyermark has obtained a fine suite of flowering material from the type locality of <u>Clusia cochlanthera</u>, the second and apparently only specimen so obtained since the original collections made some 125 years ago. These new materials now permit confirmation of my interpretation and circumscription made in the earlier papers.

VENEZUELA. Estado Carabobo: staminate epiphytic tree, leaves coriaceous, flowers with cream-white to pale yellow petals, selva siempre verde en las laderas arriba de las cabeceras de río Gián, este de Los Tanques, al sur de Borburata, 750-1100 m alt, 31 Mar 1966, <u>Julian A. Steyermark & Cora Steyermark</u> <u>95390</u> (NY, U, VEN).

A FIFTH SUMMARY OF THE VERBENACEAE, AVICENNIACEAE, STILBACEAE, DICRASTYLIDACEAE, SYMPHOREMACEAE, NYCTANTHACEAE, AND ERIOCAULACEAE OF THE WORLD AS TO VALID TAXA, GEOGRAPHIC DISTRIBUTION, AND SYNONYMY

Supplement 7

Harold N. Moldenke

Since the publication of the 6th supplement to the above work in PHYTOLOGIA, volume 34, number 3, on October 1, 1976, 2,009 new herbarium specimens have come to me from 15 institutional and private herbaria on five continents. This new material, in addition to a vast amount of new literature which has been examined by my wife and/or myself, has brought to light numerous new taxa, new geographic records, new invalid names, spellings, and accreditions, as well as numerous additional emendations and corrections of former entries. These are presented herewith (as promised on page 974 of the original work).

Herbarium specimen or literature citations substantiating these records are presented, as usual, in my monographs of the genera involved or in their periodic supplements, mostly published in PHYTOLOGIA. Citation to place of publication of the names listed in Part II are also given in detail in these monographs or their supplements.

Addenda and errata to Part I: The known geographic distribution of the accepted taxa:

CANADA:
  Québec:
  Verbena hastata L. [Sternes Island]
UNITED STATES OF AMERICA:
  Vermont:
  Verbena hastata L. [Grand Isle County]
  Virginia:
  Lachnocaulon anceps (Walt.) Morong [Dinwiddie County]
  Verbena urticifolia L. [Nelson County]
  North Carolina:
  Callicarpa americana L. [Hattaras Island]
  Eriocaulon decangulare f. parviceps Moldenke [Carteret County]
  Verbena bonariensis L. [Tyrrell County]
  South Carolina:
  Lachnocaulon anceps f. glabrescens Moldenke [Kershaw County]
  Florida:
  Clerodendrum kaempferi (Jacq.) Sieb. [Dade County]
  Duranta repens L. [Hillsborough County; Sanibel Island]
  Lachnocaulon anceps (Walt.) Morong [Calhoun & Lee Counties]

Florida [continued]:
Lachnocaulon anceps f. glabrescens Moldenke [Highlands County]
Lachnocaulon beyrichianum Sporleder [Martin & Orange Counties]
Lachnocaulon eciliatum Small [Highlands County]
Lachnocaulon engleri f. abludens Moldenke [Pasco County]*
Lachnocaulon minus (Chapm.) Small [Bay & Madison Counties]
Lantana camara var. aculeata (L.) Moldenke [Broward County;
    [Sanibel Island]
Lantana camara var. mista (L.) L. H. Bailey [Key Largo]
Lantana camara var. ternata Moldenke [Highland County]
Phyla strigulosa (Mart. & Gal.) Moldenke [Dade County]
Stachytarpheta jamaicensis (L.) Vahl [Egmont Key]
Verbena tenuisecta Briq. [Madison County]
Alabama:
Eriocaulon lineare Small [Escambia & Geneva Counties]
Eriocaulon texense Körn. [Escambia, Mobile, & Washington Coun-
    ties]
Lachnocaulon beyrichianum Sporleder [Mobile County]
Lachnocaulon minus (Chapm.) Small [Mobile County]
Verbena bipinnatifida Nutt. [Pickens County]
Mississippi:
Lachnocaulon anceps (Walt.) Morong [Covington County]
Ohio:
Phyla lanceolata (Michx.) Greene [Auglaize County]
Iowa:
Verbena bracteata Lag. & Rodr. [Clinton, Jackson, & Jones
    Counties]
xVerbena engelmannii Moldenke [Louisa & Van Buren Counties]
xVerbena moechina Moldenke [Louisa County]
xVerbena rydbergii Moldenke [Louisa County]
Verbena simplex Lehm. [Jones County]
Verbena stricta Vent. [Jones County]
Verbena urticifolia L. [Clinton & Jackson Counties]
Michigan:
Verbena bracteata Lag. & Rodr. [Oakland County]
Verbena hastata L. [Leelanau County]
Verbena stricta Vent. [Leelanau County]
Verbena urticifolia var. leiocarpa Perry & Fernald [Oakland
    County]
Kansas:
Verbena hastata L. [Lyon County]
Missouri:
Phyla lanceolata (Michx.) Greene [Cass County]
Verbena urticifolia var. leiocarpa Perry & Fernald [Shannon
    County]
Verbena xutha Lehm. [Saint Louis]

Arkansas:
  Callicarpa americana L. [Independence & union Counties]
  Eriocaulon körnickianum Van Heurck & Muell.-Arg. [Saline County]
Louisiana:
  Lachnocaulon anceps f. glabrescens Moldenke [Vernon Parish]
  Lantana camara var. mista (L.) L. H. Bailey [Plaquemines &
    Tangipahoa Parishes]
  Phyla nodiflora var. reptans (Spreng.) Moldenke [Cameron &
    Terrebonne Parishes]
 Texas:
  Lantana camara f. parvifolia Moldenke [Cameron County]
  Phyla nodiflora var. texensis Moldenke [Caldwell County]
 New Mexico:
  Verbena bracteata Lag. & Rodr. [McKinley County]
MEXICO:
  Eriocaulon microcephalum H.B.K. [Durango]
  Lantana camara var. mista (L.) L. H. Bailey [Yucatán]
  Lantana frutilla var. obtusifolia Moldenke [Sinaloa]
  Lantana hirta var. pubescens Moldenke — delete the asterisk
  Lantana hispida H.B.K. [Tamaulipas]
  Lantana kingi Moldenke [Tamaulipas]
  Lippia alba var. globiflora (L'Hér.) Moldenke [Tamaulipas]
  Phyla nodiflora var. texensis Moldenke [Michoacán]
  Tonina fluviatilis Aubl. [Veracruz]
  Verbena elegans H.B.K. [Durango]
GUATEMALA:
  Aegiphila laxicupulis Moldenke [Jutiapa]
  Lantana hirta Grah. [Guatemala]
  Lantana hirta var. pubescens Moldenke [Sacatepéquez]
  Lippia myriocephala Schlecht. & Cham. [Huehuetenango]
BELIZE:
  Priva lappulacea f. albiflora Moldenke
HONDURAS:
  Lippia oxyphyllaria (Donn. Sm.) Standl. [Choluteca & Morazán]
NICARAGUA:
  Aegiphila laxicupulis Moldenke [Chontales]
  Clerodendrum ligustrinum var. nicaraguense Moldenke [Corn Island]
  Lantana glandulosissima Hayek [Esteli]
  Lantana hirta Grah. [Esteli & Matagalpa]
  Lantana trifolia L. [Jinotega]
  Phyla betulaefolia (H.B.K.) Greene [Rio San Juan]
COSTA RICA:
  Aegiphila magnifica var. pubescens Moldenke [Puntarenas]
  Lippia controversa Moldenke [Alajuela]
  Paepalanthus costaricensis Moldenke [Alajuela]
  Verbena parvula Hayek [Heredia]

BAHAMA ISLANDS:
Citharexylum fruticosum f. bahamense (Millsp.) Moldenke [Acklin]
Lantana arida Britton [Eleuthera]
Lantana tiliaefolia Cham. [Great Inagua]
Phyla strigulosa var. sericea (Kuntze) Moldenke — to be deleted
CUBA:
Paepalanthus seslerioides Griseb. [delete "Oriente"]
Paepalanthus seslerioides var. wilsonii Moldenke [Pinar del Rio]
ISLA DE PINOS:
Lachnocaulon anceps (Walt.) Morong — to be deleted
Lachnocaulon anceps f. glabrescens Moldenke
Paepalanthus seslerioides Griseb. — to be deleted
Paepalanthus seslerioides var. carabiae Moldenke*
Paepalanthus seslerioides var. wilsonii Moldenke
CAYMAN ISLANDS:
Lantana camara var. aculeata (L.) Moldenke [Little Cayman]
HISPANIOLA:
Lantana trifolia var. quadriverticillata Jiménez [Dominican Re-
public]*
VIRGIN ISLANDS:
Lantana reticulata Pers. [St. Croix]
WINDWARD ISLANDS:
Avicennia germinans var. guayaquilensis (H.B.K.) Moldenke [St..
Vincent]
Lantana involucrata L. [St. Lucia]
TRINIDAD AND TOBAGO:
Avicennia schaueriana f. candicans Moldenke [Trinidad]
NORTHERN SOUTH AMERICAN ISLANDS:
Duranta repens L. [Margarita]
COLOMBIAN CARIBBEAN ISLANDS:
Verbena litoralis H.B.K. [San Andrés]
LESSER ANTILLES:
Stachytarpheta gibberosa Reichenb. — to be deleted
COLOMBIA:
Aegiphila grandis Moldenke [Cauca, Huila, & Valle del Cauca]
Aegiphila grandis var. cuatrecasasi (Moldenke) López-Palacios
[Magdalena]*
Aegiphila grandis var. sessiliflora (Moldenke) Moldenke [Antio-
quia, Cauca, Cundinamarca, Huila, & Valle del Cauca]*
Aegiphila longifolia Turcz. — to be deleted
Aegiphila mollis var. longifolia (Turcz.) López-Palacios [Meta
& Santander]
Asgiphila mollis var. puberulenta (Moldenke) López-Palacios
[Antioquia]
Aegiphila novogranatensis Moldenke [Cundinamarca & Tolima] —
delete the asterisk
Aegiphila sessiliflora Moldenke — to be deleted

COLOMBIA [continued]:
  Aegiphila sessiliflora var. cuatrecasasi Moldenke — to be
    deleted
  Aegiphila sufflava Moldenke [Amazonas]
  Aegiphila truncata Moldenke — to be deleted
  Bouchea boyacana Moldenke [Antioquia]
  Duranta sprucei var. breviracemosa Moldenke — delete the asterisk

  Lantana camara L. [Antioquia]
  Lantana fiebrigii var. puberulenta Moldenke [Cundinamarca]*
  Lantana glutinosa Poepp. [Nariño]
  Lantana maxima Hayek [Caldas]
  Lantana trifolia L. [Córdoba]
  Lippia americana f. hyptoides (Benth.) Moldenke [Antioquia]
  Lippia schlimii var. glabrescens (Moldenke) Moldenke [Quindiu]
  Stachytarpheta angustifolia f. elatior (Schrad.) López-Palacios
    [Córdoba]
  Stachytarpheta cayennensis (L. C. Rich.) Vahl [Caldas]

  Vitex orinocensis var. multiflora (Miq.) Huber [Arauca & Cór-
    doba]
VENEZUELA:
  Avicennia schaueriana Stapf & Leechman [Delta Amacuro]

  Duranta coriacea Hayek [Mérida]
  Duranta repens L. [Amazonas]
  Duranta repens var. alba (Masters) L. H. Bailey [Bolívar]
  Duranta repens var. canescens Moldenke [Falcón]

  Lippia americana f. pilosa Moldenke [Lara]
  Lippia hirsuta var. moritzii (Turcz.) López-Palacios [Táchira]
  Paepalanthus meseticola Moldenke & Steyerm. [Bolívar]*
  Petrea aspera Turcz. [Amazonas]
  Syngonanthus yapacanensis var. hirsutus Moldenke [Amazonas]*
  Tonina fluviatilis f. obtusifolia Moldenke — to be deleted
GUYANA:
  Avicennia schaueriana f. candicans Moldenke
  Syngonanthus tenuis (H.B.K.) Ruhl.
  Tonina fluviatilis f. obtusifolia Moldenke — add an asterisk
SURINAM:
  Avicennia elliptica var. martii Moldenke — to be deleted
  Avicennia schaueriana f. candicans Moldenke
ECUADOR:
  Aegiphila grandis Moldenke [El Oro]
  Aegiphila integrifolia var. lopez-palacii Moldenke [Napo]*
  Aegiphila lopez-palacii Moldenke [Pichincha]*
  Aegiphila lopez-palacii var. pubescens Moldenke [Pichincha]*
  Aegiphila novogranatensis Moldenke [Pichincha]
  Aegiphila rimbachii Moldenke [Pichincha]
  Aloysia scorodonioides (H.B.K.) Cham. [Guayas & Imbabura]

ECUADOR [continued]:
  Aloysia triphylla (L'Hér.) Britton [Pichincha]
  Citharexylum gentryi Moldenke [Los Ríos]*
  Citharexylum macrophyllum Poir. [Napo]
  Citharexylum montanum Moldenke [Imbabura]
  Duranta repens L. [El Oro]
  Duranta sprucei var. breviracemosa Moldenke [Pichincha]
  Duranta triacantha A. L. Juss. [Cotopaxi]
  Lantana camara var. moritziana (Otto & Dietr.) López-Palacios
    [Imbabura]
  Lantana maxima Hayek [El Oro]
  Lantana reptans Hayek [Azuay]
  Lippia alba (Mill.) N. E. Br. [Napo]
  Lippia americana f. hyptoides (Benth.) Moldenke [Manabí]
  Petrea volubilis L. [Guayas]
  Phyla strigulosa var. sericea (Kuntze) Moldenke [El Oro, Guayas,
    Loja, & Manabí]
  Priva lappulacea (L.) Pers. [Manabí]
  Priva lappulacea f. albiflora Moldenke [El Oro]
  Stachytarpheta straminea Moldenke [El Oro & Napo]
  Verbena demissa Moldenke [Loja]
  Verbena demissa f. alba Moldenke [Pichincha]*
  Verbena hispida Ruíz & Pav. [Loja]
  Verbena litoralis H.B.K. [Carchi & Imbabura]
  Verbena litoralis f. magnifolia Moldenke [Napo]*
  Verbena parvula var. gigas Moldenke [Loja]
  Verbena parvula var. obovata Moldenke [Pichincha]*
  Vitex gigantea H.B.K. [Napo]
PUNA ISLAND:
  Lippia americana L.
GALAPAGOS ISLANDS:
  Stachytarpheta cayennensis (L. C. Rich.) Vahl [Narborough]
  Verbena litoralis H.B.K. [Narborough]
  Verbena stewartii Moldenke [Narborough]
PERU:
  Citharexylum dentatum D. Don [Junín]
  Clerodendrum tessmanni Moldenke [San Martín]
  Hierobotana inflata (H.B.K.) Briq. [Ica]
  Junellia hayekii Moldenke [Ayacucho]
  Lantana reptans Hayek — delete the asterisk
  Syngonanthus caulescens var. angustifolius Moldenke [Amazonas]
  Syngonanthus compactus Ruhl. [San Martín]
  Verbena fasciculata Benth. [Ica]
  Vitex excelsa var. petiolata Moldenke [Loreto]*
BRAZIL:
  Aegiphila candelabrum Briq. [Minas Gerais]
  Aegiphila glandulifera var. paraënsis Moldenke [Amazônas]

BRAZIL [continued]:

Aegiphila lanceolata Moldenke [Amazônas]

Aegiphila longifolia Turcz. — to be deleted

Aegiphila mollis var. longifolia (Turcz.) López-Palacios [Amazônas]

Aegiphila sellowiana Cham. [Amazônas]

Avicennia elliptica var. martii Moldenke [delete "Pará"], add
asterisk

Avicennia germinans var. guayaquilensis (H.B.K.) Moldenke [Cea-
rá & Pará]

Avicennia schaueriana Stapf & Leechman [Graguatá Island; delete
Maranhão & Florianopolis Island]

Avicennia schaueriana f. candicans Moldenke [Bahia, Ceará, Guanabara, Maranhão, Paraíba, Paraná, Pernambuco, Rio de Janeiro,
Santa Catarina, & São Paulo; Florianopolis, Gobernador, Pinheiros, & Santa Catarina Islands]

Casselia confertiflora (Moldenke) Moldenke [Bahia]

Citharexylum pernambucense Moldenke [Maranhão]

Clerodendrum philippinum Schau. [Amazônas]

Duranta repens L. [Espirito Santo]

Eriocaulon megapotamicum Malme [Paraná]

Eriocaulon sellowianum var. longifolium Moldenke [Mato Grosso]

Eriocaulon spruceanum f. viviparum Moldenke [Roraima]

Lantana camara L. [Bahia]

Lantana camara var. aculeata (L.) Moldenke [Espirito Santo]

Lantana camara var. moritziana (Otto & Dietr.) López-Palacios
[Amazônas]

Lantana canescens H.B.K. [Paraíba]

Lantana fucata var. longipes Moldenke [Rio Grande do Sul &
Santa Catarina]

Lantana lippioides Spreng. — to be deleted

Lantana procurrens Schau. [Minas Gerais]

Lantana trifolia L. [Minas Gerais & Rio de Janeiro]

Lantana trifolia f. hirsuta Moldenke [Rio de Janeiro]

Lantana triplinervia Turcz. [Rio de Janeiro]

Lantana triplinervia f. armata Moldenke [Guanabara & São Paulo]*

Lantana triplinervia var. puberulenta (Moldenke) Moldenke [Gua-
nabara]

Lantana viscosa Pohl [Guanabara; Fundão Island]

Leiothrix rufula var. brevipes Moldenke [Rio de Janeiro]*

Lippia alba var. globiflora (L'Hér.) Moldenke [Amazônas, Bahia,
& Rio de Janeiro]

Lippia balansae Briq. [Mato Grosso]

Lippia campestris Moldenke [Rio Grande do Sul]

Lippia elliptica Schau. [delete "Minas Gerais"]

Lippia gardneriana Schau. [Minas Gerais]

Lippia gracilis Schau. [Goiás, Maranhão, & Mato Grosso]

Lippia lepida Moldenke [Distrito Federal]

BRAZIL [continued]:
  Lippia lorentzii Moldenke [Roraima]
  Lippia mattogrossensis Moldenke [delete "Distrito Federal"]
  Lippia nepetacea Schau. [Minas Gerais]*
  Lippia obscura Briq. [Distrito Federal & Goiás]
  Paepalanthus argillicola Alv. Silv. [Rio de Janeiro]
  Paepalanthus ciliatus (Bong.) Kunth [Guanabara]
  Paepalanthus elongatus f. graminifolius Herzog [Minas Gerais]
  Paepalanthus erectifolius Alv. Silv. [Rondonia]
  Paepalanthus filifolius Moldenke — to be deleted
  Paepalanthus formosus Moldenke [Mato Grosso]
  Paepalanthus glabrifolius Ruhl. [Guanabara]
  Paepalanthus guaraiensis Moldenke [Goiás]*
  Paepalanthus polytrichoides Kunth [Rondônia]
  Paepalanthus saxicola var. pilosus Moldenke [Goiás]*
  Paepalanthus tortilis var. glaberrimus Mart. & Moldenke [Guana-
      bara & Rio de Janeiro]*
  Paepalanthus tortilis var. minor Moldenke [Espirito Santo]*
  Petrea nitidula Moldenke [Mato Grosso]
  Stachytarpheta dichotoma f. albiflora Moldenke [Mato Grosso]
  Stachytarpheta sessilis Moldenke [Minas Gerais]
  Syngonanthus baldwini Moldenke [Pará]
  Syngonanthus bisumbellatus (Körn.) Ruhl. [Roraima]
  Syngonanthus caulescens var. angustifolius Moldenke [Minas Ger-
      ais, Paraná, Rio Grande do Sul, & Rondônia]
  Syngonanthus caulescens var. discretifolius Moldenke [Pará]*
  Syngonanthus densus (Körn.) Ruhl. [Pará]
  Syngonanthus elegantulus Ruhl. [Guanabara]
  Syngonanthus elegantulus var. glaziovii Moldenke [Minas Gerais]*
  Syngonanthus fertilis (Körn.) Ruhl. [Amazônas]
  Syngonanthus glandulosus Gleason [Mato Grosso]
  Syngonanthus humboldtii var. glandulosus Gleason [Mato Grosso &
      Rondônia]
  Syngonanthus longipes Gleason [Amazônas]
  Syngonanthus niveus var. rosulatus (Körn.) Moldenke [Bahia]
  Syngonanthus reclinatus (Körn.) Ruhl. [Rio de Janeiro]
  Syngonanthus reflexus Gleason [Amazônas]
  Syngonanthus ruprechtianus (Körn.) Ruhl. [Minas Gerais]* — this
      is the corrected entry
  Syngonanthus umbellatus (Lam.) Ruhl. [delete "Goiás"]
  Syngonanthus widgrenianus (Körn.) Ruhl. [delete "Paraná"]
  Syngonanthus xeranthemoides var. confusus (Körn.) Moldenke [de-
      lete "Mato Grosso"]
  Syngonanthus xeranthemoides var. hirsutus Moldenke [Mato Grosso]
  Verbena balansae Briq. [Mato Grosso]
  Verbena campestris Moldenke [Paraná]
  Vitex sprucei Briq. [Roraima]
  Vitex triflora var. floribunda Huber [Rondônia]

MARACÁ ISLAND:
  Syngonanthus tenuis (H.B.K.) Ruhl. — to be deleted
BOLIVIA:
  Aegiphila saltensis Legname [Tarija]
  Junellia minima (Meyen) Moldenke [Santa Cruz]
  Junellia seriphioides (Gill. & Hook.) Moldenke [Oruro]
  Lantana brachypoda Hayek [Tarija]
  Lippia alba var. globiflora L'Hér.) Moldenke [Santa Cruz]
  Lippia chacensis Moldenke — to be deleted
  Lippia lorentzii Moldenke [Santa Cruz]
  Vitex triflora Vahl [Pando]
PARAGUAY:
  Lantana aristata var. longipedunculata Moldenke*
  Lippia intermedia Cham.
  Syngonanthus glandulosus var. epapillosus Moldenke
ARGENTINA:
  Aegiphila saltensis Legname — delete the asterisk
  Citharexylum jörgensenii (Lillo) Moldenke [Salta]
  Junellia digitata (R. A. Phil.) Moldenke [La Rioja]
  Lantana camara var. aculeata (L.) Moldenke [Corrientes]
  Lantana fucata var. longipes Moldenke [Salta]
  Neosparton aphyllum (Gill. & Hook.) Kuntze [La Rioja]
  Verbena minutiflora Briq. [Toledo Island]
MACARONESIA:
  Phyla nodiflora var. canescens (H.B.K.) Moldenke [Gran Canaria]
FRANCE:
  Phyla nodiflora var. canescens (H.B.K.) Moldenke
  Verbena supina f. erecta Moldenke
BALEARIC ISLANDS:
  Phyla nodiflora (L.) Greene [Minorca]
GERMANY:
  Verbena urticifolia L.
AEGEAN ISLANDS:
  Vitex agnus-castus L. [Psara]
CRETE:
  Phyla nodiflora var. reptans (Spreng.) Moldenke
ALGERIA:
  Lantana camara L.
LIBYA:
  Verbena supina L.
CENTRAL AFRICAN REPUBLIC:
  Syngonanthus chevalieri H. Lecomte — to be deleted
  Syngonanthus wahlbergii (Wikstr.) Ruhl.
TANGANYIKA:
  Vitex bunguensis Moldenke*
KENYA:
  Holmskioldia n. sp. ined.

ZAMBIA:
  Kalaharia uncinata var. hirsuta (Moldenke) Moldenke
RHODESIA:
  Lantana petitiana A. Rich.
MOZAMBIQUE:
  Clerodendrum ternatum Schinz [Zambezia]
NAMIBIA:
  Lantana camara L.
  Lippia javanica (Burm. f.) Spreng.
SOUTH AFRICA:
  Holmskioldia tettensis (Klotzsch) Vatke [Natal]
  Lantana camara L. [Transvaal]
SEYCHELLES ISLANDS:
  Phyla nodiflora var. reptans (Spreng.) Moldenke [Praslin]
ARABIA:
  Phyla nodiflora var. reptans (Spreng.) Moldenke [Yemen]
AFGHANISTAN:
  Phyla nodiflora (L.) Greene
  Phyla nodiflora var. canescens (H.B.K.) Moldenke
PAKISTAN:
  Lantana rugosa Thunb. [Kohat & Quetta]
INDIA:
  Clerodendrum glaucum Wall. -- to be deleted
  Clerodendrum serratum var. wallichii C. B. Clarke [Assam, Mani-
    pur, Uttar Pradesh, & West Bengal] — delete the asterisk
  Eriocaulon breviscapum Körn. [Meghalaya]
  Eriocaulon brownianum var. latifolium Moldenke [Meghalaya]
  Eriocaulon infirmum Steud. [Meghalaya]
  Lantana camara var. angustifolia Moldenke — to be deleted
  Lantana camara f. parvifolia Moldenke [Kerala]
  Lantana rugosa Thunb. [West Bengal]
  Lantana triplinervia Turcz. [Andhra Pradesh, Bihar, Madras, &
    Mysore]
  Symphorema polyandrum Wight [Union Territory]
SRI LANKA:
  Citharexylum spinosum L.
  Clerodendrum kaempferi (Jacq.) Sieb.
  Lantana camara var. mista (L.) L. H. Bailey
  Lantana tiliaefolia Cham. -- to be deleted
BANGLADESH:
  Phyla nodiflora var. reptans (Spreng.) Moldenke
CHINESE COASTAL ISLANDS:
  Phyla nodiflora var. reptans (Spreng.) Moldenke [Hainan]
  Sphenodesme ferruginea (W. Griff.) Briq. [Hainan]
THAILAND:
  Clerodendrum serratum var. obovatum Moldenke*
  Clerodendrum serratum var. pilosum Moldenke*
  Clerodendrum serratum var. wallichii C. B. Clarke

THAILAND [continued]:
  Eriocaulon laosense var. maxwellii Moldenke*
  Glossocarya mollis var. maxwellii Moldenke*
  Gmelina arborea var. canescens Haines
  Premna macrophylla var. thailandica Moldenke*
INDOCHINA:
  Clerodendrum serratum var. wallichii C. B. Clarke [Cambodia &
    Tonkin]
MALAYA:
  Clerodendrum serratum var. wallichii C. B. Clarke [Perak &
    Selangor]
  Eriocaulon australe R. Br. [Penang]
  Eriocaulon sexangulare L. [Penang]
  Geunsia pentandra (Roxb.) Merr. [Selangor]
  Sphenodesme pentandra var. wallichiana (Schau.) Merr. [Penang]
  Vitex gamosepala W. Griff [Pahang]
  Vitex gamosepala var. kunstleri King & Gamble [Penang]
MALAYAN ISLANDS:
  Vitex siamica F. N. Will. [Bumbon Besar]
  Vitex trifolia var. bicolor (Willd.) Moldenke [Bumbon Besar]
JAPAN:
  Callicarpa australis Koidz. [Kyushu & Shikoku] delete the aster-
    isk
RYUKYU ISLAND ARCHIPELAGO:
  Callicarpa australis Koidz.
FORMOSA:
  Callicarpa longifolia Lam.

PHILIPPINE ISLANDS:
  Tectona philippinensis Benth. [Iling]
BONIN ISLANDS:
  Callicarpa parvifolia Hook. & Arn. [Anijima]
MARIANAS ISLANDS:
  Premna mariannarum f. dentata Moldenke [Guguan]*
  Vitex trifolia var. bicolor (Willd.) Moldenke [Maug & Pagan]
GREATER SUNDA ISLANDS:
  Eriocaulon heterolepis var. nigricans Körn. [Sabah]
  Eriocaulon merrillii Ruhl. — to be deleted
  Eriocaulon sollyamum var. sumatramum Van Royen [Sumatra]*
PHOENIX ISLANDS:
  Lantana camara L. [Canton]
BISMARK ARCHIPELAGO:
  Avicennia alba Blume [Manus]
SOLOMON ISLANDS:
  Lantana camara var. mista (L.) L. H. Bailey [Guadalcanal]
  Lantana montevidensis (Spreng.) Briq. [Guadalcanal]
AUSTRALIA:
  Clerodendrum heterophyllum (Poir.) R. Br. [Queensland]
  Lantana camara var. aculeata (L.) Moldenke [New South Wales]

GREAT BARRIER REEF:
    Clerodendrum heterophyllum f. angustifolium Moldenke [Strad-
      broke]
    Duranta repens L. [Stradbroke]
    Eriocaulon australe R. Br. [Stradbroke]
    Eriocaulon scariosum J. E. Sm. [Stradbroke]
    Verbena bonariensis L. [Stradbroke]
NEW ZEALAND:
    Avicennia marina var. resinifera (Forst. f.) Bakh. [Pollen]
HAWAIIAN ISLANDS:
    Holmskioldia sanguinea Retz. [Hawaii]
SAMOAN ISLANDS:
    Premna taitensis Schau. [Upolu]
    Stachytarpheta jamaicensis (L.) Vahl [Upolu]
 CULTIVATED:
    Aegiphila elata Sw. [England]
    Aegiphila foetida Sw. [England]
    Aegiphila laevis (Aubl.) Gmel. [England]
    Aegiphila trifida Sw. [England]
    Aloysia scorodonioides (H.B.K.) Cham. [Ecuador]
    Avicennia officinalis L. [England]
    Callicarpa arborea Roxb. [England]
    Callicarpa dichotoma (Lour.) K. Koch [England]
    Callicarpa longifolia Lam. [England]
    Callicarpa macrophylla Vahl [England]
    Callicarpa reticulata Sw. [England]
    Callicarpa tomentosa (L.) Murr. [England]
    Caryopteris mongholica Bunge [Germany]
    Citharexylum caudatum L. [England]
    Citharexylum fruticosum L. [England]
    Citharexylum fruticosum var. subserratum (Sw.) Moldenke [England]
    Citharexylum fruticosum var. villosum (Jacq.) O. E. Schulz [Eng-
      land]
    Citharexylum sericeum Lodd. [England]*
    Clerodendrum colebrokianum Walp. [Florida]
    Clerodendrum floribundum R. Br. [Australia]
    Clerodendrum glabrum E. Mey. [Australia]
    Clerodendrum heterophyllum f. angustifolium Moldenke [Australia
      & Great Barrier Reef]
    Clerodendrum serratum var. wallichii C. B. Clarke [Singapore]
    Clerodendrum umbellatum var. speciosum (Dombrain) Moldenke [Ecu-
      ador]
    Congea tomentosa Roxb. [Tobago]
    Cornutia latifolia (H.B.K.) Moldenke [England]
    Duranta repens var. alba (Masters) L. H. Bailey [Ecuador]
    Eriocaulon australe R. Br. [England]
    Eriocaulon decangulare L. [England]

CULTIVATED [continued]:
  Eriocaulon sexangulare L. [England]
  Ghinia spicata (Aubl.) Moldenke [Germany]
  Gmelina philippensis Cham. [Tobago]
  Holmskioldia sanguinea Retz. [Germany & Jamaica]
  Lantana achyranthifolia f. grandifolia Moldenke [Germany]
  Lantana annua L. [France]
  Lantana aristata var. longipedunculata Moldenke [Germany]
  Lantana camara var. mutabilis (Hook.) L. H. Bailey [Colombia]
  Lantana camara f. parvifolia Moldenke [Germany & Italy]
  Lantana fiebrigii Hayek [Germany]
  Lantana horrida H.B.K. [Germany]
  Lantana jamaicensis Britton [Germany]
  Lantana montevidensis (Spreng.) Briq. [Michigan]
  Lippia alba var. globiflora (L'Hér.) Moldenke [India]
  Lippia callicarpaefolia H.B.K. [Germany]
  Paepalanthus lamarckii Kunth [England]
  Petrea rugosa var. casta Moldenke [Colombia]
  Phyla nodiflora var. reptans (Spreng.) Moldenke [Italy]
  Phyla scaberrima (A. L. Juss.) Moldenke [Venezuela]
  Phyla strigulosa (Mart. & Gal.) Moldenke [France]
  Syngonanthus niveus (Bong.) Ruhl. [Germany]
  xVerbena hybrida Voss [Ecuador]
  Verbena monacensis Moldenke [Venezuela]
  Verbena officinalis L. [Mexico]
  Verbena sulphurea D. Don [Germany]
  Verbena supina f. erecta Moldenke [Germany]
  Verbena tenuisecta Briq. [St. Croix]
  Vitex trifolia var. subtrisecta (Kuntze) Moldenke [Australia]
FOSSILIZED:
  Tectona grandis L. f. [Pleistocene of India]

Additions and emendations to Part II: An alphabetic list of re-
jected scientific names proposed in these groups, including
misspellings and variations in accreditation:

Aegiphila arborescens ♀ breviflora Schau. = A. integrifolia
  (Jacq.) Jacq.
Aegiphila arborescens ♂ longiflora Schau. = A. bracteolosa Mol-
  denke
Aegiphila brachiata Cham. & Schl. = A. deppeana Steud.
Aegiphila cuatrecasasii Moldenke = A. cuatrecasasi Moldenke
Aegiphila gloriosa var. paraensis Hock. = A. gloriosa var.
  paraensis Moldenke
Aegiphila grandis var. cuatrecasasii (Moldenke) López-Palacios =
  A. grandis var. cuatrecasasi (Moldenke) López-Palacios

Aegiphila longifolia Turcz. = A. mollis var. longifolia (Turcz.)
    López-Palacios
Aegiphila sesiliflora Moldenke = A. grandis var. sessiliflora
    (Moldenke) Moldenke — this is the corrected entry
Asgiphila sessiliflora Moldenke = A. grandis var. sessiliflora
    (Moldenke) Moldenke
Aegiphila sessiliflora var. cuatrecasasi Moldenke = A. grandis
    var. cuatrecasasi (Moldenke) López-Palacios
Aegiphila truncata Moldenke = A. grandis Moldenke
Aloysia attennata Walp. = Lippia vernonioides Cham.
Aloysia gratissima (Gill. ex Hook.) Tronc. = A. gratissima (Gill.
    & Hook.) Troncoso
Avicennia officinalis var. lanceolata Kuntze = A. germinans var.
    guayaquilensis (H.B.K.) Moldenke
Avicennia schaeereana Stapf & Lehm. = A. schaueriana Stapf &
    Leechman
Avicennia schaueriana f. glabrescens Moldenke = A. schaueriana
    Stapf & Leechman
Bontia P. Br. ex Airy Shaw in Willis = Avicennia L.
Bontia L. ex Loefl. = Avicennia L.
Bouchea dichotoma Mohr [in part] = Stachytarpheta dichotoma (Ruíz
    & Pav.) Vahl — this is the corrected entry
Bouhea Moldenke = Bouchea Cham.
Bouhea fluminensis Moldenke = Bouchea fluminensis (Vell.) Moldenke
Bouhea fluminensis var. pilosa Moldenke = Bouchea fluminensis var.
    pilosa Moldenke
Burchardia Heist ex Duham. = Callicarpa L.
Callicarpa acuminata L. = C. acuminata H.B.K.
Callicarpa maingayi King = C. maingayi King & Gamble
Callicarpa pendulata R. Br. = C. pedunculata R. Br.
Calymega Poit. ex Mold. = Vitex Tourn.
Camara lamii folio &c. Dill. = Lantana camara var. mista (L.)
    L. H. Bailey
Camara melissae folio &c. Dill. = Lantana camara L.
Camara trifolia &c. Plum. = Lantana trifolia L.
Citharexylon sericeum Lodd. = Citharexylum sericeum Lodd.
Citharexylon subflavescens Moldenke = Citharexylum subflavescens
    Blake
Citharexylum cinereum ♀ Lam. = C. spinosum L.
Citharexylum cyanocarpum Hook. & Arn. = Rhaphithamnus spinosus (A.
    L. Juss.) Moldenke
Citharexylum flexuosum D. Don = C. flexuosum (Ruíz & Pav.) D. Don
Citharexylum serrectum Griseb. = C. caudatum L.
Clerodendron floribundum Lindl. = Clerodendrum emirnense Bojer
Clerodendron hirsutum D. Don = Clerodendrum umbellatum Poir.
Clerodendron obovatum Walp. = Clerodendrum obovatum (Roxb.) Walp.

Clerodendron ovatum α R. Br. = Clerodendrum floribundum R. Br.
Clerodendron ovatum β R. Br. = Clerodendrum floribundum R. Br.
Clerodendron speciosissimum Hort. Angl. = Clerodendrum kaempferi
    (Jacq.) Sieb.
Clerodendron spinosum Spreng. = Clerodendrum spinosum (L.) Spreng.
Clerodendron ternuifolium H.B.K. = Clerodendrum ternifolium H.B.K.
Clerodendrum divaricatum Jack = C. serratum var. wallichii C. B.
    Clarke — this is the corrected entry
Clerodendrum kaempferi (Jack.) Sieb. = C. kaempferi (Jacq.) Sieb.
Clerodendrum teruatum Schinz = C. ternatum Schinz
Clerodendrum teruatum var. lanceolatum (Guerke) Moldenke = C.
    ternatum var. lanceolatum (Gürke) Moldenke
Congea azurea Vahl = C. tomentosa Roxb.
Congea tomentosa var. β Schau. = C. tomentosa Roxb.
Congea tomentosa α latifolia Schau. = C. tomentosa Roxb.
Congea tomentosa β oblongifolia Schau. = C. tomentosa Roxb.
Cornutia microcalicina Pav. & Mold. = C. microcalycina Pavon &
    Moldenke
Cornutia pyramidata var. albida Anon. = C. pyramidata var. isthm-
    mica Moldenke
Cryptocaly Benth. = Phyla Lour.
Dupatia Griseb. = Paepalanthus Mart.
Dupatia seslerioides Griseb. = Paepalanthus seslerioides Griseb.
Dupatya alsinoides Wr. & Sauv. = Paepalanthus alsinoides C. Wright
Dupatya alsinoides (Wright & Sauv.) Britton = Paepalanthus alsin-
    oides C. Wright
Dupatya ruprechtiana (Körn.) Kuntze = Syngonanthus ruprechtianus
    (Körn.) Ruhl. — this is the corrected entry
Dupatya ruprechtiana Kuntze = Syngonanthus ruprechtianus (Körn.)
    Ruhl. — this is the corrected entry
Duranta repens γ mutisii 3. acuminata a. glabrifolia Kuntze = D.
    coriacea Hayek — this is the corrected entry
Eriocaulon alpestre var. alpestre [Hook. f. & Thoms.] ex Van Royen
    = E. alpestre Hook. f. & Thoms.
Eriocaulon brevipedunculatum Suesseng. & Heine = E. kinabaluense
    Van Royen
Eriocaulon ovoideum var. ulei Knuth = Dioscorea amarantoides var.
    ulei Knuth, Dioscoreaceae
Eriocaulon umbellatum Humb. = Syngonanthus umbellatus (Lam.) Ruhl.
Eriocaulon umbellatum Kunth = Syngonanthus umbellatus (Lam.) Ruhl.
Eriocaulon vernonioides (Kunth) D. Dietr. = Syngonanthus xeranthe-
    moides var. vernonioides (Kunth) Moldenke
Ghinia verbenacea Sw. = G. boxiana Moldenke
Glandularia Schau. = Verbena [Dorst.] L.
Glandularia disecta (Willd.) Schnack & Covas = Verbena dissecta
    Willd.
Glossocarya mollis Wall. ex Griff. = G. mollis Wall.

Gmelina elliptica J. C. Sm. = G. elliptica J. E. Sm.
Goniostachyum graveolens Small = Lippia graveolens H.B.K.
Junelia Mold. = Junellia Moldenke
Lachnocaulon anceps (Walt.) DC. = L. anceps (Walt.) Morong
Lachnocaulon digynum Sporl. = L. digynum Körn.
Lantana aculeata Auct. = L. camara var. mista (L.) L. H. Bailey
Lantana alba var. trifoliata Benth. = L. indica Roxb.
Lantana asperata Hort. Paris. = L. camara var. nivea (Vent.) L. H.
     Bailey
Lantana bellowsiana Hort. = L. montevidensis (Spreng.) Briq.
Lantana camarra L. = L. camara L.
Lantana capensis Thunb. = Spielmannia jasminum Medic., Myoporaceae
Lantana chiapensis Moldenke = L. chiapasensis Moldenke
Lantana cinerea Otto & Dietr. = L. brasiliensis Link
Lantana cremulata Otto & Dietr. = L. camara var. splendens (Medic.)
     Moldenke — this is the corrected entry

Lantana crispa Thunb. = Spielmannia jasminum Medic., Myoporaceae
Lantana crocea ♂ rugosa Otto & Dietr. = L. urticaefolia Mill.
Lantana crocea ♀ planifolia Otto & Dietr. = L. urticaefolia Mill.
Lantana cuneifolia Mart. = L. chamissonis (D. Dietr.) Benth.
Lantana emvolutrata Kummer = L. annua L.
Lantana geminata (H.B.K.) Spreng. = Lippia alba var. globiflora
     (L'Hér.) Moldenke — this is the corrected entry
Lantana geminata (Kunth) Spreng. = Lippia alba var. globiflora
     L'Hér.) Moldenke — this is the corrected entry
Lantana geminata Spreng. = Lippia alba var. globiflora (L'Hér.)
     Moldenke
Lantana glutinosa var. albiflora Mold. = L. glutinosa f. albiflora
     Moldenke
Lantana graveoleus Crutchfield & Johnston = Lippia graveolens
     H.B.K.
Lantana involucrata f. candida Fosb. = L. involucrata L.
Lantana lippioides Spreng. = L. canescens H.B.K.
Lantana montevidensis Spreng. = L. montevidensis (Spreng.) Briq.
Lantana nivea ♀ Vent. = L. camara var. mutabilis (Hook.) L. H.
     Bailey
Lantana salviaefolia Cham. = L. fucata Lindl.
Lantana salviaefolia L. = L. fucata Lindl.
Lantana splendens Medic. = L. camara var. splendens (Medic.) Mol-
     denke
Lantana svennsonii Mold. = L. svensonii Moldenke
Lantana svennsonii f. albiflora Mold. = L. svensonii f. albiflora
     Moldenke
Lantana variegata Otto & Dietr. = L. purpurea Hornem.
Lantana youngii Kummer = L. tiliaefolia Cham.
Lippia chacensis Moldenke = L. lorentzii Moldenke
Lippia citrata Willd. = L. alba var. globiflora (L'Hér.) Moldenke
     — this is the corrected entry

Lippia citrosa (Small) = Lantana microcephala A. Rich.
Lippia corimbosa Troncoso = L. corymbosa Cham.
Lippia dictamnus Mart. = L. francensis Moldenke
Lippia discolor Hort. = Phyla scaberrima (A. L. Juss.) Moldenke
Lippia floribunda H.B.K. = L. schlimii var. glabrescens (Moldenke)
    Moldenke — this is the corrected entry
Lippia floribunda Humb. & Bonpl. = L. schlimii var. glabrescens
    (Moldenke) Moldenke -- this is the corrected entry
Lippia floribunda Humb. & Kunth = L. schlimii var. glabrescens
    (Moldenke) Moldenke -- this is the corrected entry
Lippia floribunda Kunth = L. schlimii var. glabrescens (Moldenke)
    Moldenke — this is the corrected entry
Lippia globiflora f. pubescens Kuntze = L. alba var. globiflora
    (L'Hér.) Moldenke — this is the corrected entry
Lippia globiflora albiflora Kuntze = L. alba var. globiflora
    (L'Hér.) Moldenke — this is the corrected entry
Lippia hatschbachi Moldenke = L. hatschbachii Moldenke
Lippia hemisphaeria Jacq. = L. americana L.
Lippia hispida Jacq. = Lantana achyranthifolia f. grandifolia
    Moldenke
Lippia imundata Mart. = L. lorentzii Moldenke
Lippia lacunosa var. ovatifolia Moldenke = L. lacunosa var. acuti-
    folia Moldenke
Lippia latoovata Mart. = L. lupulina Cham.
Lippia lavandulaefolia Schwaegr. = L. javanica (Burm. f.) Spreng.
Lippia lopezzi Mold. = L. lopezii Moldenke
Lippia macrocalyx Mart. = L. microcephala Cham.
Lippia mcvaughii Moldenke = L. mcvaughi Moldenke
Lippia myriocephala H.B.K. = L. myriocephala Schlecht. & Cham.
Lippia nepetacea Schau. — to be deleted
Lippia nodiflora var. repens (Spreng.) Ross = Phyla nodiflora (L.)
    Greene
Lippia nodiflora ♂ sarmentosa DC. = Phyla nodiflora (L.) Greene
Lippia pumila Cham. & Sohl. = L. pumila Cham.
Lippia pycnocephala Seem. = L. myriocephala var. hypoleia (Briq.)
    Moldenke
Lippia rhodocnemis Mart. = L. rhodocnemis Mart. & Schau.
Lippia rubra Hort. = Lantana achyranthifolia f. grandifolia Mol-
    denke
Lippia saturaeaefolia Mart. = L. satureiaefolia Mart. & Schau.
Lippia sericea Schau. = L. sericea Cham.
Lippia stöchas Mart. = L. sericea Cham.
Lippia strigulosa f. parvifolia (Mold.) Fosberg = Phyla strigulosa
    var. sericea (Kuntze) Moldenke
Lippia substrigosus Turcz. = L. substrigosa Turcz.
Lippia tergulifera Briq. = L. tegulifera Briq.
Lippia thymoides Mart. = L. thymoides Mart. & Schau.

Macrostegia Nees in DC. = <u>Vitex</u> Tourn.
Mallelou Rheede ex Adans. = <u>Vitex</u> Tourn.
Melasanthus triphyllus Pohl = <u>Stachytarpheta</u> rhomboidalis (Pohl)
     Walp.
Paepalanthus alsinoides Wright & Sauvalle = <u>P.</u> alsinoides C.
     Wright
Paepalanthus colombianus Cleef = <u>P.</u> columbiensis Ruhl. — this is
     the corrected entry
Paepalanthus drouetii L. B. Sm. = <u>Syngonanthus</u> <u>drouetii</u> L. B. Sm.
     — this is the corrected entry
Paepalanthus filifolius Moldenke = <u>P.</u> <u>capillifolius</u> Moldenke
Paepalanthus phlepsae Moldenke = <u>P.</u> phelpsae Moldenke
Paepalanthus rupprechtianus Körn. = <u>Syngonanthus</u> <u>ruprechtianus</u>
     (Körn.) Ruhl. — this is the corrected entry
Paepalanthus ruprechtianus Körn. = Syngonanthus <u>ruprechtianus</u>
     (Körn.) Ruhl. — this is the corrected entry
Paepalanthus vernonioides var. a Kunth = Syngonanthus xeranthe-
     moides var. vernonioides (Kunth) Moldenke — this is the
     corrected entry
Paepalanthus vernonioides var. ♂ Kunth = Syngonanthus xeranthe-
     moides var. vernonioides (Kunth) Moldenke — this is the
     corrected entry
Petrea mexicana Humb. & Bonpl. = <u>P.</u> volubilis var. pubescens Mol-
     denke — this is the corrected entry
Petrea pubeseens Turcz. = <u>P.</u> pubescens Turcz.
Petrea pubeseens var. klugii Mold. = <u>P.</u> pubescens var. klugii
     Moldenke
Phyla betulacea (H.B.K.) Greene = <u>P.</u> betulaefolia (H.B.K.) Greene
Phyla betulaefolia H.B.K. = <u>P.</u> betulaefolia (H.B.K.) Greene
Pitraea cuncato-ovata (Cav.) Caso = <u>P.</u> cuneato-ovata (Cav.) Caro
Priva lappula Andrews = <u>P.</u> lappulacea (L.) Pers.
Pygmaopremna Nayar, Yogan., & Subram. = Pygmaeopremna Merr.
Pygmaopremna herbacea Nayar, Yogan., & Subram. = <u>Pygmaeopremna</u>
     herbacea (Roxb.) Moldenke
Sép'halicá W. Jones = Nyctanthes L. & <u>N.</u> arbor-tristis L.
Siphonanthus Schau. = Clerodendrum Burm.
Stachytarpha aristata Vahl = Stachytarpheta orubica (L.) Vahl
Stachytarpha cajanensis Vahl = Stachytarpheta cayennensis (L. C.
     Rich.) Vahl
Stachytarpha ciliata Kunze = Stachytarpheta indica (L.) Vahl
Stachytarpha dichotoma Vahl = Stachytarpheta dichotoma (Ruíz &
     Pav.) Vahl — this is the corrected entry
Stachytarpha gibberosa Reichenb. = Stachytarpheta dichotoma (Ruíz
     & Pav.) Vahl
Stachytarpha glauca Walp. = Stachytarpheta glauca (Pohl) Schau.
Stachytarpha glauca ♂ Schau. = Stachytarpheta glauca (Pohl) Schau.
Stachytarpha glauca ♀ subintegrifolia Schau. = Stachytarpheta
     glauca var. subintegrifolia Schau.

Stachytarpha hirsuta Jacq. f. = Stachytarpheta canescens H.B.K.

Stachytarpha hirsutissima Link = Stachytarpheta canescens H.B.K.

Stachytarpha hirta Kunth = Stachytarpheta dichotoma (Ruíz & Pav.)
      Vahl

Stachytarpha hirta H.B.K. = Stachytarpheta dichotoma (Ruíz & Pav.)
      Vahl

Stachytarpha jamaicensis Gardn. = Stachytarpheta dichotoma (Ruíz
      & Pav.) Vahl

Stachytarpha marginata Vahl = Stachytarpheta jamaicensis (L.) Vahl

Stachytarpha maximiliani ♀ glabrata Schau. = Stachytarpheta maxi-
      miliani var. glabrata Schau.

Stachytarpha microphylla Walp. = Stachytarpheta sanguinea Mart.

Stachytarpha palustris Schott = Stachytarpheta angustifolia f.
      elatior (Schrad.) López-Palacios

Stachytarpha pilosiuscula H.B.K. = Stachytarpheta jamaicensis (L.)
      Vahl

Stachytarpha pilosiuscula Kunth = Stachytarpheta jamaicensis (L.)
      Vahl

Stachytarpha quadrangula Nees & Mart. = Stachytarpheta quadrangula
      Nees & Mart.

Stachytarpha rhomboidalis Walp. = Stachytarpheta rhomboidalis Schau.

Stachytarpha triphylla Walp. = Stachytarpheta rhomboidalis Schau.

Stachytarpha umbrosa H.B.K. = Stachytarpheta dichotoma (Ruíz &
      Pav.) Vahl

Stachytarpha umbrosa Kunth = Stachytarpheta dichotoma (Ruíz & Pav.)
      Vahl

Stachytarpha urticifolia Sims = Stachytarpheta urticaefolia
      (Salisb.) Sims

Stachytarpha veronicaefolia Cham. = Stachytarpheta cayennensis (L.
      C. Rich.) Vahl

Stachytarpha villosa Schau. = Stachytarpheta villosa Cham. — this
      is the corrected entry

Stachytarpha zuccagni Roem. & Schult. = Stachytarpheta mutabilis
      var. violacea Moldenke

Stachytarpheta gibberosa Reichenb. = S. dichotoma (Ruíz & Pav.)
      Vahl

Stachytarpheta glauca Walp. = S. glauca (Pohl) Schau.

Stachytarpheta hirta H.B.K. = S. dichotoma (Ruíz & Pav.) Vahl --
      this is the corrected entry

Stachytarpheta jamaicensis Gardn. = S. dichotoma (Ruíz & Pav.)
      Vahl — this is the corrected entry

Stachytarpheta microphylla Walp. = S. sanguinea Mart.

Stachytarpheta reticulata Mart. ex Schau. = S. reticulata Mart.

Stachytarpheta rhomboidalis Walp. = S. rhomboidalis Schau.

Stachytarpheta triphylla Walp. = S. rhomboidalis Schau.

Stachytarpheta umbrosa H.B.K. = S. dichotoma (Ruíz & Pav.) Vahl —
      this is the corrected entry

Stigmatococca Mart. ex Mold. = Aegiphila Jacq.
Stigmatococca Willd. = Ardisia Sw., Myrsinaceae
Syngonanthus Ruhl. = Syngonanthus Ruhl.
Syngonanthus rufoalbus Alv. Silv. = S. rufo-albus Alv. Silv.
Syngonanthus rupprechtianus (Körn.) Ruhl. = S. ruprechtianus
 (Körn.) Ruhl. — this is the corrected entry
Syngonanthus umbellatus var. brachyphylla Huber = S. umbellatus
 f. brachyphyllus (Huber) Moldenke — this is the corrected
 entry
Syngonanthus vernonioides Ruhl. = S. xeranthemoides var. vernon-
 ioides (Kunth) Moldenke
Syngonanthus wahlbergii (Koern.) Ruhl. = S. wahlbergii (Wikstr.)
 Ruhl.
Tomonea verbenacea Sw. = Ghinia boxiana Moldenke
Tonina guianensis Samuels = T. fluviatilis Aubl.
Upata Rheede ex Adans. = Avicennia L.
Verbena a.Stachytarpheta Endl. = Stachytarpheta Vahl
Verbena b. Bouchea Endl. = Bouchea Cham.
Verbena bornariensis L. = V. bonariensis L.
Verbena carolinensis etc. Dill. = V. carolina L.
Verbena carolinensis &c. Dill. = V. carolina L.
Verbena dermani Mold. = xV. dermeni Moldenke
Verbena disecta Willd. = V. dissecta Willd.
Verbena gobiflora Ruiz & Pav. = Lippia alba var. globiflora
 (L'Hér.) Moldenke
Verbena halii Small = V. halei Small
Verbena jamaicensis St.-Hil. = Stachytarpheta dichotoma (Ruíz &
 Pav.) Vahl
Verbena littoralis var. albiflora Moldenke = V. litoralis var.
 albiflora Moldenke
Verbena longifolia H.B.K. = V. longifolia Mart. & Gal.
Verbena mexicana trachelii fol. &c. Dill. = Priva mexicana (L.)
 Pers.
Verbena xmoenchina Moldenke = xV. moechina Moldenke
Verbena moranensis H.B.K. = V. elegans H.B.K.
Verbena odorata J[uss.] = Lippia javanica (Burm. f.) Spreng.
Verbena odorata Pers. = Lippia a₁ba var. globiflora (L'Hér.)
 Moldenke — this is the corrected entry
Verbena paniculata β pinnatifida Lam. = V. hastata L.
Verbena paniculata β pinnatifida Schau. = V. hastata L.
Verbena phlogiflora α vulgaris Cham. = V. phlogiflora Cham.
Verbena phlogiflora α vulgaris Schau. = V. phlogiflora Cham.
Verbena phlogiflora β macilenta Cham. = V. megapotamica Spreng.
Verbena rigida var. glandulos Moldenke = V. rigida var. glandu-
 lifera Moldenke — this is the corrected entry
Verbena rigida var. glandulosa Moldenke = V. rigida var. glandu-
 lifera Moldenke

Verbena stellarioides ♂ decurrens Cham. = V. stellarioides Cham.
Verbenaceae Auct. = Verbenaceae J. St.-Hil.
Verbeneae Reich. = Verbenaceae J. St.-Hil.
Verbeneae Schau. = Verbenaceae J. St.-Hil.
Verbeneae [Labiatarum sectio] ex parte Reichenb. = Verbenaceae J.
     St.-Hil.
Verbera Bert. = Verbena [Dorst.] L.
Verbera diceras Bert. = Verbena sulphurea D. Don
Vitex caribaea Hook. & Arn. = Vitis californica Benth., Vitaceae
Vitex montevidensis ♀ parviflora Cham. = V. schaueriana Moldenke
Vitex montevidensis ♀ parviflora Schau. = V. schaueriana Moldenke
Vitex montevidensis multinervis Cham. = V. megapotamica (Spreng.)
     Moldenke
Vitex negundo Roxb. = V. negundo L.
Vitex sellowiana ♀ parviflora Cham. = V. mexiae Moldenke
Vitex sellowiana ♀ parviflora Schau. = V. mexiae Moldenke
Vitex triflora Moldenke = V. triflora Vahl
Vitex trifolia &c. Pluk. = V. trifolia L.
Viticastrum ramosum Presl = Sphenodesme racemosa (Presl) Moldenke
Viticeae Schau. = Viticoideae Briq.
Volkameria buchananii Roxb. = Clerodendrum buchanani (Roxb.) Walp.
Volkameria foetida Hamilt. = Clerodendrum bungei Steud.
Volkameria inermis L. f. = Clerodendrum inerme (L.) Gaertn.
Volkameria inermis ♀ Ait. = Clerodendrum ligustrinum (Jacq.) R.Br.
Zapania Juss. ex Steud. = Lippia Houst.
Zapania Schau. = Lippia Houst.
Zapania geminata (H.B.K.) Gibert = Lippia alba var. globiflora
     (L'Hér.) Moldenke -- this is the corrected entry
Zapania hispida Zuccagni = Priva mexicana (L.) Pers.
Zapania mexicana Lam. = Priva mexicana (L.) Pers.
Zapania odorata Pers. = Lippia alba var. globiflora (L'Hér.)
     Moldenke -- this is the corrected entry
Zapania prismatica Poir. = Bouchea prismatica (L.) Kuntze
Zappania odorata Pers. = Lippia alba var. globiflora (L'Hér.) Mol-
     denke

* A few copies of the original (1971) work, 974 pp., are still
available for $25 plus postage from Mrs. Alma L. Moldenke, 303
Parkside Road, Plainfield, New Jersey 07060, U.S.A.

Harold N. Moldenke

LANTANA TRIPLINERVIA f. ARMATA Moldenke, f. nov.

Haec forma a forma typica speciei ramis crasse armatis spinis duris magnis arcte recurvatis et corollis aureo-flavis vel lilacinis recedit.

This form differs from the typical form of the species in its stems and branches being very coarsely and viciously armed with stout, thick, strongly recurved thorns in great profusion and in the corollas being either golden-yellow, the throat ringed with light-orange, or the tube and limb lilac, the throat ringed with vermillion.

The type of this form was collected by George Eiten and W. D. Clayton (no. 6194) in tall grass of low secondary forest between the road and a nearby creek 1.3 km. southeast of the center of the city of Pariquera-açu on the road to Iguape, 24°43' S., 47°52 1/2' W., at 50 meters altitude, in the Municipio de Pariquera-açu, São Paulo, Brazil, on February 18, 1965, and is deposited in the United States National Herbarium in Washington. The collectors report the common name, "ribeirão turvo".

LEIOTHRIX RUFULA var. BREVIPES Moldenke, var. nov.

Haec varietas a forma typica speciei pedunculis maturis ca. 3 cm. longis recedit.

This variety differs from the typical form of the species in having its mature peduncles during anthesis and fruit only about 3 cm. long.

The type of the variety was collected by Alberto Castellanos (no. 25666) at Abr. Rebouças, Itatiaia, Rio de Janeiro, Brazil, at 2350 meters altitude, on December 3, 1964, and is deposited in my personal herbarium.

PAEPALANTHUS GUARAIENSIS Moldenke, sp. nov.

Herba parva caulescens; caule ca. 2 cm. longo dense folioso; foliis parvis linearibus 5—8 mm. longis recurvatis dense villosulis acutis; pedunculis filiformibus 4—6 cm. longis obscure pilosulo-puberulis 3-sulcatis solitariis vel paucis; capitulis hemisphaericis vel in maturitate globosis griseis ca. 2 mm. latis.

Small herb; stems short, erect, ca. 2 cm. long, densely foliose; lower leaves densely appressed to the stems and closely imbricate, about 5 mm. long and 1 mm. wide, pilose, the upper ones spreading-recurved, linear, 5—8 mm. long, acute, densely whitish-villosulous; peduncles 1 or 2 at the apex of the stem, erect, filiform, 4—6 mm. long, pilosulous-puberulent, very obscurely so in age, 3-sulcate, twisted; heads small, at first hemispheric, later globose, grayish, about 2 mm. wide; involucral bractlets oblanceolate-elliptic, stramineous, 0.5—0.8 mm. long, subacute apically, densely barbate at the

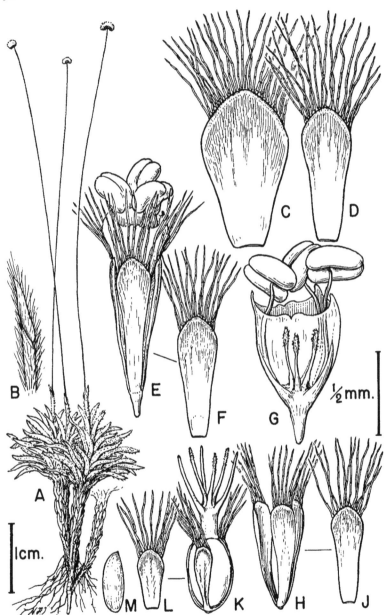

Fig. 1. Paepalanthus guaraiensis Moldenke

apex; receptacular bractlets blackish, rather broadly obovate or oblanceolate, ca. 1 mm. long and 0.6 mm. wide, densely white-barbate at the apex; for floral characters see illustration, Fig. 1: A - Habit; B - Sheath; C - Receptacular bractlet; D - Involucral bractlet; E - Staminate floret; F - Staminate sepal; G - Staminate floret with sepals removed; H - Pistillate calyx; J - Pistillate sepal; K - Pistillate floret with one petal removed; L - Pistillate petal; M - Ovary. Drawing by Haruto Fukuda.

The type of the species was collected by Gert Hatschbach and R. Kummrow (no. 38508) in "dos rochedões de arenito, nas anfractuosidades, Rod. Belem - Brasilia, mun. Guarai", Goiás, Brazil, on March 30, 1976, and is deposited in my personal herbarium.

PAEPALANTHUS TORTILIS var. GLABERRIMUS Mart. & Moldenke, var.nov.

Haec varietas a forma typica speciei foliis vaginisque glaberrimis recedit.

This variety differs from the typical form of the species in having its leaves and sheaths completely glabrous.

The type of the variety is Herb. Martius 551 from the Serra da Brocca, Rio de Janeiro, Brazil, collected in July, 1839, and deposited in the herbarium of Columbia University. It was originally determined and distributed by Martius as Eriocaulon tortile var. glaberrimum, but this name appears to be a cheironym.

SYNGONANTHUS YAPACANENSIS var. HIRSUTUS Moldenke, var. nov.

Haec varietas a forma typica speciei pedunculis densissime longeque hirsutis recedit.

This variety differs from the typical form of the species in its peduncles during anthesis being very densely long-hirsute with ascending, shaggy, grayish hairs.

The variety is based on A. Gentry & S. Tillett 10869, collected in thickets and forest, mostly on white sand, at 110 m. altitude, on the road from San Fernando de Atabapo to Santa Barbara 12--40 km. from San Fernando, Amazonas, Venezuela, on March 24, 1974, and is deposited in my personal herbarium.

VERBENA DEMISSA f. ALBA Moldenke, f. nov.

Haec forma a forma typica speciei corollis albis recedit.

This form differs from the typical form of the species in having white corollas.

The type of the form was collected by Santiago López-Palacios (no. 4200) in the Cráter del Pululagua, La Reventazón, Pichincha, Ecuador, at 2100 meters altitude, on January 23, 1977, and is deposited in my personal herbarium. The collector notes: "Hierba rastrera. Espigas por lo general simples. Flores blancas."

VERBENA LITORALIS f. MAGNIFOLIA Moldenke, f. nov.

Haec forma a forma typica speciei foliis multo maioribus laminis usque 15 cm. longis 8 cm. latis recedit.

This form differs from the typical form of the species in its much larger leaves, which may be up to 15 cm. long and 8 cm. wide, coarsely and somewhat irregularly serrate-dentate along the margins, some of the larger teeth being bidentate.

The type of the form was collected by Santiago López-Palacios (no. 4188) at Tena, Napo, Ecuador, at 500 meters altitude, on January 11, 1977, and is deposited in my personal herbarium. The collector notes: "Hierba de 1—1.2 m. Fls. blanco-morado muy pequeñas, espigas muy delgadas y alargadas".

VERBENA PARVULA var. OBOVATA Moldenke, var. nov.

Haec varietas a forma typical speciei laminis foliorum obovatis recedit.

This variety differs from the typical form of the species in having its leaf-blades mostly more or less obovate.

The type of the variety was collected by Santiago López-Palacios (no. 4250) at Quito, Pichincha, Ecuador, at an altitude of 2800 meters, on February 6, 1977, and is deposited in my personal herbarium. The collector notes: "Hierba decumbente, espigas cilíndricas, en su mayoría de a 3, fls. moradas. Hojas algo obovadas."

AEGIPHILA LOPEZ-PALACII Moldenke, sp. nov.

Arbor ramis ramulisque crassis valde medullosis obtuse tetragonis dense pulverulento-puberulis valde lenticillatis; foliis permagnis oppositis obovatis ca. 35—40 cm. longis 14—19 cm. latis subcoriaceous in siccitate brunneis ad apicem rotundatis ad basin acutis integris utrinque minuteque pulverulento-puberulis; venis crassis perspicuis, secundariis multis rectis; inflorescentiis axillaribus glomeratis multifloris perspicue bracteolatis; bracteolis linearibus elongatis.

Tree, 4—10 m. tall or even taller; branches and branchlets very stout and medullose, obtusely tetragonal, somewhat decussately flattened, conspicuously elevated-lenticellate, densely pulverulent-puberulent; leaves decussate-opposite, very large, coriaceous, brunnescent in drying; petioles very stout, 3—4 cm. long, densely pulverulent-puberulent, flattened above; leaf-blades obovate, very large, 35—40 cm. long, 14—19 cm. wide, rounded apically, entire-margined, rather abruptly acute basally, minutely pulverulent-puberulent on both surfaces under a handlens, the venation coarse and conspicuous; midrib very stout, rounded-elevated beneath, densely pulverulent-puberulent; secondaries very numerous, 15 or more per side, mostly rather straight and parallel, extending almost to the margins at right angles from the midrib, prominent beneath, flattened above; veinlet reticulation rather abundant, prominulous beneath, inconspicuous above; inflorescence axillary, glomerate, sessile, many-flowered, conspicuously many-bracteolate; bractlets linear, about 2 cm. long and 1 mm. wide, twisted, pulverulent-puberulent; calyx campanulate-obconic, about 9 mm. long, 6 mm. wide at the apex, densely puberulent, the rim mostly irregularly 2-lipped; corolla not well preserved, apparently hypocrateriform, the tube very slender, included by the calyx, the lobes spreading, ca. 3 mm. long; stamens exserted from the corolla-tube, apparently equaling the lobes in length.

The type of this most remarkable species was collected by my good friend and colleague, Santiago López-Palacios (no. 4237) — in whose honor I am pleased to name it in small recognition of the very important and careful field work which he is conducting in this most difficult and perplexing group of plants — at Campamento San José, beyond Los Bancos, at 2200 meters altitude, Pichincha, Ecuador, on February 4, 1977, and is deposited in my personal herbarium. The collector notes: "Árbol de unos 4 m (existen ejemplares mayores de 10 m), con cálices secos en inflorescencias glomeradas sésiles."

AEGIPHILA LOPEZ-PALACII var. PUBESCENS Moldenke, var. nov.

Haec varietas a forma typica speciei laminis foliorum subtus dense subtomentello-pubescentibus recedit.

This variety differs from the typical form of the species in having the lower surface of the leaf-blades densely subtomentellous-pubescent.

The type of the variety was collected by Santiago López-Palacios (no. 4201) in the Cráter del Pululagua, La Reventazón, at 2100 meters altitude, Pichincha, Ecuador, on January 23, 1977, and is deposited in my personal herbarium. The collector notes: "Arbolito en crecimiento de 2—4 m. en lugares húmedos y sombreados; esteril".

AEGIPHILA INTEGRIFOLIA var. LOPEZ-PALACII Moldenke, var. nov.

Haec varietas a forma typica speciei laminis foliorum distincte serrulatis recedit.

This variety differs from the typical form of the species in having the margins of its leaf-blades distinctly serrulate.

The type of the variety was collected by Santiago López-Palacios (no. 4257) at Tena, Río Uglo, Napo, Ecuador, at 500 m. altitude, on February 10, 1977, and is deposited in my personal herbarium. The collector notes: "Arbolito 2-5 m. Hojas anchamente elípticas, claramente aserradas, cimas axilares multifloras. Cáliz verde, 4-lobado, de unas 7 mm. Corola blanca, tubo delgado de 1 cm. de largo, lobulos extendidos de unos 12 mm. de diámetro."

ADDITIONAL NOTES ON THE ERIOCAULACEAE. LXXI

Harold N. Moldenke

SYNGONANTHUS FLAVIDULUS (Michx.) Ruhl.
  Additional bibliography: Britton & Br., Illustr. Fl., ed. 2,
imp. 3, 1: 455 & 680, fig. 1144. 1936; Moldenke, N. Am. Fl. 19: 43
& 44. 1937; Moldenke, Phytologia 1: 336 & 343--344. 1939; Durand &
Jacks., Ind. Kew. Suppl. 1, imp. 2, 145. 1941; Worsdell, Ind. Lond.
Suppl. 2: 426. 1941; Britton & Br., Illustr. Fl., ed. 2, imp. 4, 1:
455 & 680, fig. 1144. 1943; Jacks. in Hook. f. & Jacks., Ind. Kew.,
imp. 2, 1: 878 & 879 (1946) and imp. 2, 2: 402. 1946; Moldenke,
Alph. List Cit. 1: 17, 31, 35, 38, 42, 43, 45, 63, 90, 98, 99, 138-
140, 152, 153, 164, 169, 191, 221, 234, 257, 275, 276, 283, 286,
290, & 292--295. 1946; Moldenke, Known Geogr. Distrib. Erioc. 2, 3,
29, 34, 40, 48, & 58. 1946; Britton & Br., Illustr. Fl., ed. 2,
imp. 5, 1: 455 & 680, fig. 1144. 1947; Moldenke, Phytologia 2: 350
(1947) and 2: 496. 1948; Moldenke, Alph. List Cit. 2: 377, 413,
460, 470, 480, 504, 507, 508, 511--513, 524, 545, 554, 572, 583,
617, 630, 639, & 641 (1948), 3: 660, 675, 697, 721, 725, 736, 741,
742, 756, 759, 760, 772, 774, 776--778, 787, 790, 806, 813, 822,
835, 841, 842, 850, 851, 895, 899, 917, 931, 937, 940, 943, 946, &
958 (1949), and 4: 1001, 1003, 1112, 1118, 1132, 1164, 1176, 1177,
1181, 1191, 1192, 1201, 1204, 1216, 1221, 1222, 1227, 1241, 1252,
1288, 1289, & 1292. 1949; E. D. Merr., Ind. Rafin. 82. 1949; Moldenke, Known Geogr. Distrib. Verbenac., [ed. 2], 7, 8, 10, 11, &
213. 1949; Gleason, New Britton & Br. Illustr. Fl., imp. 1, 1: 372
& 481 (1952) and imp. 1, 3: 585 & 591. 1952; Moldenke, Phytologia
4: 313--316. 1953; Thorne, Am. Midl. Nat. 52: 282. 1954; Core, Pl.
Tax. 268. 1955; Gleason, New Britt. & Br. Illustr. Fl., imp. 2, 1:
372 & 418 (1958) and imp. 2, 3: 585 & 591. 1958; Durand & Jacks.,
Ind. Kew. Suppl. 1, imp. 3, 145. 1959; Moldenke, Résumé 10, 11, 13,
14, 280, 282, 288, 292, 302, 325, 414, & 491. 1959; Moldenke, Résumé Suppl. 1: 2 & 16. 1959; Jacks. in Hook. f. & Jacks., Ind.
Kew., imp. 3, 1: 878 & 879 (1960) and imp. 3, 2: 402. 1960; Moldenke, Résumé Suppl. 3: 3, 31, & 35 (1962), 4: [1]--3 (1962), 5: 2
(1962), and 6: 10. 1963; Gleason, New Britt. & Br. Illustr. Fl.,
imp. 3, 1: 372 & 481 (1963) and imp. 3, 3: 585 & 591. 1963; Radford, Ahles, & Bell, Guide Vasc. Fl. Carol. 106 & 107. 1964;
Thanikaimoni, Pollen & Spores 7: 183 & 187, tab. 1. 1965; Kral,
Sida 2: 327--332. 1966; Shinners, Sida 2: 441 & 447. 1966; Grimm,
Recog. Flow. Wild Pl. 36. 1968; Moldenke, Résumé Suppl. 16: [1] &
25 (1968) and 17: [1] & 9. 1968; Rickett, Wild Fls. U. S. 2 (1):
[85] & 135, pl. 27 (1968) and 2 (2): 674. 1968; Moldenke, Résumé
Suppl. 18: [1] & 13. 1969; Moldenke, Phytologia 18: 80, 369, 370,
379, & 380 (1969) and 19: 28 & 75. 1969; Tomlinson in C. R. Metcalfe, Anat. Monocot. 3: 149, 156, 157, 161, 162, 168, 169, 172,
175, 182--186, 190, & 191, fig. 33 H, I, & K, 35 I, & 39 A--D.
1969; Britton & Br., Illustr. Fl., ed. 2, imp. 6, 1: 455 & 680,
fig. 1144. 1970; Moldenke, Phytologia 20: 41, 42, 52, & 424. 1970;

S. Ell., Sketch Bot., imp. 3, 2: 566—567 & 728. 1971; Long & Lakela, Fl. Trop. Fla., ed. 1, 259, 262, 930, & 958. 1971; Moldenke,
Fifth Summ. 1: 23, 25, 26, 30, 32, 481, & 487 (1971) and 2: 496,
500, 513, 534, 578, 583, 593, 636, 764, & 962. 1971; Moldenke,
Phytologia 25: 125 & 225 (1973), 26: 17, 27, & 179 (1973), and 29:
204. 1974; Michx., Fl. Bor.-Am., imp. 2, 2 [Ewan, Class. Bot. Am.
3]: 166. 1974; Moldenke, Phytologia 31: 375 (1975) and 34: 248,
277, & 486. 1976; Lakela, Long, Fleming, & Genelle, Pl. Tampa Bay,
ed. 3 [Bot. Lab. Univ. S. Fla. Contrib. 73:] 39 & 180. 1976; Long
& Lakela, Fl. Trop. Fla., ed. 2, 259, 262, 930, & 958. 1976; Moldenke, Phytologia 35: 304, 313, 346, 347, & 457—458. 1977.

Additional & emended illustrations: Britton & Br., Illustr. Fl.,
ed. 1, 1: 373, fig. 902 (1896), ed. 2, imp. 1, 1: 455, fig. 1144
(1913), and ed. 2, imp. 2, 1: 455, fig. 1144. 1923; M. F. Baker,
Fla. Wild Fls. 122. 1926; J. K. Small, Man. Southeast. Fl. 257.
1933; Britton & Br., Illustr. Fl., ed. 2, imp. 3, 1: 455, fig.
1144 (1936), ed. 2, imp. 4, 1: 455, fig. 1144 (1943), and ed. 2,
imp. 5, 1: 455, fig. 1144. 1947; Thanikaimoni, Pollen & Spores 7:
183, tab. 1. 1965; Kral, Sida 2: 328. 1966; Rickett, Wild Fls. U.
S. 2 (1): [85] (in color). 1968; Tomlinson in C. R. Metcalfe,
Anat. Monocot. 3: 156, 168, & 182, fig. 32 H, I, & K, 35 I, & 39
A—D. 1969; Britton & Br., Illustr. Fl., ed. 2, imp. 6, 1: 455,
fig. 1144. 1970.

Recent collectors describe this plant as a low herb, clump-
forming or solitary, the leaves dense, recurved, rosette-forming,
flattened against the substratum, with hairs tending to be pustu-
late-based, at least some of those on the upper portions of the
peduncles clavate or gland-tipped, the flowers white, and the
bracts straw-colored and shiny. D'Arcy refers to the heads as
"bright white", but to me in the field they have usually had a
yellowish cast.

Collectors have found the species growing in bogs, roadside
swales, longleaf pine sandhill bogs and bog margins, clearings in
longleaf pine - saw palmetto flats, and sandy arid pinelands, in
open pine-palmetto forests, hillside bogs, cleared pinelands,
railroad ditches, and low areas in sandhills, in sandy prairies,
clearings in pine flatwoods, moist ground of pinelands, white
sand scrub, roadside ditches, and moist broad shallow sandy-peaty
ditches, under Taxodium in moist sedge associations, at pond
edges, in wet sandy peat in bogs in longleaf pine - saw palmetto
flatwoods, pine flatwoods ditches, moist areas in pineland bogs,
and slash pine - saw palmetto flatwoods, in damp white sand along
the borders of shallow ponds in sand barrens, in low and moist
pinelands, low scrubland, sandy openings in scrub, boggy areas,
and the sandy open scrub-covered edge of pinelands, at the edges
of cutover pinewoods, in sandy or sandy-peaty soil, in the high
pine borders of swamps, on the shores of sinkhole lakes, at the
edge of cypress ponds and ponds with surrounding shrubbery, in the
sandhills bordering bogs, in clearings in shrub bogs, and in sandy
peat of seepage from hillside bogs, by ponds in pond cypress flatwoods, and in hammocks with Asimina, Blechnum, and Nephrolepis.

D'Arcy reports it "frequent in wet grassy ditches" in Florida,
Wentz found it "common at edge of beach of 5-acre pond" and Tomlinson found it "abundant in sandy prairie and cutover pineland,
forming dense tufts in drier areas" in the same state, but Myint
refers to it as "occasional along streams and in grassy pinelands".
It has been found in flower from March to July and in fruit in
March, April, June, and July. Lakela and her associates (1976)
aver that it flowers in the "summer".

Thorne (1954) refers to the species as "rare". Radford and his
associates (1964) also report it as "rare" [in the Carolinas] in
bogs, savannas, and low pinelands in Bladen, Brunswick, New Hanover, and Sampson Counties, North Carolina. Harper (1906) records
it from Appling, Berrien, Coffee, Colquitt, Decatur, Dodge, Dooly,
Emanuel, Irwin, Montgomery, Tattnall, Telfair, and Wilcox Counties,
Georgia. The Masseys found it "abundant on low roadsides with
Eriocaulon" in North Carolina. In some recent floras it is listed
as occurring in Virginia, but as yet I have seen no material to
substantiate this claim. Ruhland (1930) makes the remarkable as-
sertion that its natural distribution is "an Flussufern von Pennsylvania bis Karolina" -- what the basis is of his Pennsylvania
"record" is unknown to me. There is nothing in the Berlin herbarium of this species from Pennsylvania or Virginia. Certain
species of Eriocaulon have also been reported from Pennsylvania,
perhaps on the basis of unlabeled specimens in the Schweinitz
herbarium.

Common names for Syngonanthus flavidulus are "bantum buttons",
"bog-buttons", "dupatya", "shoe buttons", "shoe-buttons", and
"yellow pipewort".

The specific initial letter is sometimes uppercased for no
valid reason. The Eriocaulon caespitosum of Cabanis, listed in
the synonymy, is based on a specimen in the Berlin herbarium from
Ebenezer, Mississippi, inscribed "Eriocaulon caespitosum mihi,
Restiaceae". The E. caespitosum of Poeppig, however, is a synonym
of Paepalanthus bifidus (Schrad.) Kunth, while E. caespitosum
Wikstr. is now known as Syngonanthus caespitosus (Wikstr.) Ruhl.
Rafinesque (1840) described his Eriocaulon flavidulum var. cinereum from "Florida, Alabama, leaves broadly subulate 2 inches,
scape 3 to 4, fls. dark gray, bracts greenish". Kunth (1841) drew
up his description of S. flavidulus from a specimen in the Berlin
herbarium labeled as from Palisot de Beauvois and originally from
North America. He says "Descr. juxta specimen a Belvisio sub
nomine Eriocauli setacei acceptum". He also asserts that his
Paepalanthus nardifolius (now known as Syngonanthus fischerianus)
of Brazil is "P. flavidulo proxima affinis, differt foliis angustioribus et rigidioribus, vaginis longioribus, sepalis masculis
exterioribus angustato-acutatis, glabris".

Morong (1891) speaks of Körnicke's critical examination of the
plants originally called Eriocaulon flavidulum by early writers on
the American flora: "Körnicke (Linnaea, 27, 590) under the name
Eriocaulon flavidulum, Mx., following Pursh (El. 1, 92) and Elli-

ott (Bot. ii, 566), states that two plants have been sent from
North America under this name and that he regards Kunth's P.
flavidulus as something distinct from the plant of Michaux. That
which he describes is undoubtedly something distinct and is clearly an Eriocaulon, but, so far as I can judge, it corresponds very
nearly, if not quite, to E. articulatum [now known as E. pelluci-
dum]. The plant of Elliott is also, I think, that species.
Michaux distinctly calls his species puberulent and the scapes aggregated and five striate, while his other characters correspond
very well with our plant. There is not, so far as ascertained,
any other in the habitat given by him, 'Carolina', that bears
such characters."

The Müller (1860) work listed in the bibliography of S. flavidulus is sometimes cited as "1858", but actually was not published
until 1860 -- pages 1--160 were issued in 1858, pages 161--640 in
1859, and pages 641--966 in 1860. The Holm (1901) work is sometimes erroneously cited as "1904". The right-hand color illustration given by Rickett (1968) is most misleading because the
flower-heads seem to be bluish-tinted when actually they are yellow
ish straw-color.

Grimm (1968) describes S. flavidulus: "Its straw-colored flower
heads are on naked stalks to 12 inches tall and arise from a cluste
of short, awl-like leaves which are woolly at the base. It grows
wet pinelands and bogs of the coastal plain from N. C. south to Fla
and Ala., blooming May to October". Kral (1966) tells us that
"Characteristics which distinguish this species from other Eriocaul
aceous plants of the United States and Canada are as follows: 1.
Roots unbranched, spongy-thickened, non-septate....this in contrast
to roots branched and slender-fibrous in Lachnocaulon and roots
thickened-septate in Eriocaulon. 2. Leaves of the rosette very
copious, very narrowly linear, and definitely recurved to flatten
against the substratum.....this in contrast to ascending-spreading
leaf habit of sympatric Eriocaulaceae. 3. Trichomes of the leaves
tending to be pustular based, a characteristic not found on sympatric Eriocaulaceae; at least some of the trichomes of the upper
scape clavate or glandular-tipped. 4. Both sets of perianth parts
present, the flowers seemingly actinomorphic....this in contrast
to Lachnocaulon, in which only one set of parts is present or Eriocaulon, in which zygomorphy is apparent."

Material of Syngonanthus flavidulus has been misidentified and
distributed in some herbaria as Eriocaulon sp., E. lineare Small,
E. parkeri B. L. Robinson, E. septangulare With., E. setaceum L.,
Lachnocaulon sp., L. anceps (Walt.) Morong, L. glabrum Körn., and
L. michauxii Kunth.

On the other hand, the A. Ruth s.n. [Jesup, June 1893], distributed as S. flavidulus, is actually Eriocaulon compressum Lam.,
while the Bernhardi s.n. [Philadelphia] is E. parkeri B. L. Robinson [as is also the "E. flavidulum Michx." recognized by Ruhland
in his monograph (1903) as distinct from Syngonanthus flavidulus

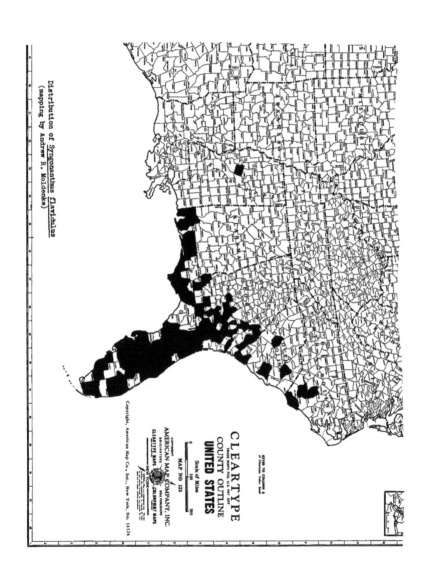

Distribution of *Syngonanthus flavidulus*
(mapping by Andrew R. Moldenke)

(Michx.) Ruhl.], G. L. Fisher s.n. [Mobile, May 12, 1928], Herb.
Umbach 10992, Lighthipe 173, and Perdue 1765 are Lachnocaulon
anceps (Walt.) Morong, and J. Kohlmeyer 2034 is L. minus (Chapm.)
Small. R. M. Harper 1608 is a mixture of S. flavidulus and
Eriocaulon lineare Small, Schallert 16912 is a mixture with
Lachnocaulon anceps, Meebold 28099 is a mixture with Lachnocaulon
glabrum, and Dress & Read 7495 is a mixture with Xyris sp.

    Additional citations: NORTH CAROLINA: Brunswick Co.: Massey &
Massey 3284 (Mi, N). Sampson Co.: Ahles & Laing 24651 (Hi—
97139). County undetermined: W. Bennett 417/73 [Jampon Beach]
(Hm). SOUTH CAROLINA: Berkeley Co.: Ravenel s.n. [Santee Canal,
Auh.] (Ms—15488). GEORGIA: Berrien Co.: R. Kral 24254 (N).
Brantley Co.: Kuns 99 (Ws). Brooks Co.: R. Kral 28693 (W—
2673941). Clinch Co.: R. Kral 24288 (N); A. R. Moldenke 332
(Fg). Early Co.: R. F. Thorne 4963 (Ca—906387, N). Effingham
Co.: R. Kral 24103 (N). Irwin Co.: R. Kral 27121 (W—2673951).
Jeff Davis Co.: A. R. Moldenke 350 (Z). Lanier Co.: R. Kral
24265 (N). Liberty Co.: R. Kral 24217 (N). Lowndes Co.: R. M.
Harper 1608, in part (W—431916); A. R. Moldenke 314 (Fg, S).
Miller Co.: R. F. Thorne 4194 (Vi). Pierce Co.: R. Kral 24149
(N). Screven Co.: R. Kral 24028 (N), 24051 (N). Tattnall Co.:
Ahles & Mueller 54172 (Hi—202837). Ware Co.: R. Kral 25307 (N);
Kuns 4 (Ws, Ws); A. R. Moldenke 340 (Fg). Wayne Co.: R. Kral
24184 (N), 24198 (N); A. E. Radford 7968 (Hi—129168), 7971 (Hi—
57247). FLORIDA: Baker Co.: West & Arnold s.n. [Sapp, 25 Apr.
1940] (Ca—841820). Bay Co.: Moldenke & Moldenke 26700 (Ac). Clay
Co.: W. M. Canby s.n. [Hibernia, March 1869] (Ca—405217, Dt).
Collier Co.: Atwater M.210 (Hi—182373); Gillis 10425 (Ld); Mee-
bold 28105 (Mu); H. E. Moore Jr. 7116 (Ba); Tomlinson 31-3-63 B
(Ft—276, Ft, Ft). Duval Co.: Clausen & Trapido 3290 (Ca—841822);
Curtiss 3020 (Ms—15491, Mu, S), 4140 (Ca—58580), 4786 (Ca—
115159, Mm—7948), s.n. [May 1875] (Ms—15489, N); Faxon s.n.
[Jacksonville, Mch. 1873] (Ws); Moldenke & Moldenke 26434 (Ld).
Franklin Co.: Hunnewell 13208 (Ws); Moldenke & Moldenke 26627 (Ac),
26646 (Ld, Ws); A. Wood s.n. [Apalachicola] (Ws). Gilchrist Co.:
D'Arcy 1508 (Sd—86713); A. E. Radford 8324 (Hi—129146). Gulf
Co.: Godfrey & Triplett 59789 (Hi—156875). Highlands Co.: McFar-
lin 4340 (Mi); Small & DeWinkeler 9966 (S). Hillsborough Co.:
Dress & Hansen 991 (Ba); Pollard s.n. [Tampa, March 7, 1898] (W—
328233). Lake Co.: Moldenke & Moldenke 26492 (Ba); G. V. Nash 143
(Ca—115160, Mm—7947). Lee Co.: Craighead s.n. [28 April 1967]
(Ft—13147); H. N. Moldenke 688 (S); Seibert 1371 (Ca—26154).
Leon Co.: Godfrey 62901 (Bl—199118, Go, N). Levy Co.: Cooley,
Wood, & Wilson 5984 (Hi—193977, N). Liberty Co.: A. R. Moldenke
282 (Fg), 284 (Fg). Manatee Co.: Friell s.n. [April 9, 1969] (Lc);

R. W. Hill s.n. [4/9/1969] (Lc); S. M. Tracy 6643 (Ca--181779, Mi,
S). Marion Co.: Dress & Hansen 2013 (Ba). Nassau Co.: A. Ruth
s.n. [March 1893] (Se--96013). Okaloosa Co.: Godfrey 56719 (Ca--
112565, N). Orange Co.: F. S. Blanton 6491 (Mi, N); Moldenke &
Moldenke 26548 (Ac); P. C. Schallert 6116 (Go); Wentz 624 (Mi).
Osceola Co.: A. A. Eaton 1060 (Ld); Myint 964 (N); P. O. Schallert
16312 (S), 16912, in part (Ut--89890b). Palm Beach Co.: W. B. Fox
s.n. [Delray Beach, April 2, 1945] (Ws). Pinellas Co.: M. S. Bebb
s.n. [Clearwater, 1894] (Ok); Genelle & Fleming 143 (N). Polk Co.:
Goodale s.n. [Conine, 9 April 1933] (Ms--69826); Meebold 28099, in
part (Mu); Milligan s.n. [May 1890] (W--503998); P. O. Schallert
6116, in part (Ok), s.n. [May 2, 1941] (Ca--841821); Topping 2609
(Mi). Putnam Co.: Moldenke & Moldenke 29829 (Ac, Ld). Saint Johns
Co.: Hunnewell 8656 (Ws); Owen s.n. [St. Augustine, May 1878] (Ca--
67949); M. C. Reynolds s.n. [Mar.--July 1875] (Ca--2426). Santa
Rosa Co.: A. R. Moldenke 267 (Fg). Sarasota Co.: R. Kral 2121 (Ms--
44937). Seminole Co.: Cooley, Eaton, & Ray 7407 (Hi—204702);
Foster, Smith, & Smith s.n. [Pl. Exsicc. Gray. 1334] (B, Ba, Bl--
72361, Ca--717066, Gg--333524, Hi, N, Ok, S, St, Ut--889b, Vi, Ws);
P. O. Schallert 6116, in part (Je--8761, Mu, Ws). Volusia Co.: H.
C. Beardslee s.n. [New Smyrna, March 1925] (Ca--841824). Wakulla
Co.: Godfrey 53293 (Hi--157562, N); N. C. Henderson 64-244 (Bl--
208900); Moldenke & Moldenke 29392 (Ac, Gz, Kh, Ld, Tu). Walton
Co.: A. R. Moldenke 269 (Fg). County undetermined: A. W. Chapman
s.n. (Ws); Herb. Amherst Coll. s.n. [East Florida] (Ms--15490);
Herb. Chapman s.n. [Fla.] (Ok). Marco Isl.: Silverstone 24 (Ws).
Pine Isl.: Lakela, Long, & Broome 30560 (N); H. N. Moldenke 940
(S). ALABAMA: Baldwin Co.: Dress & Read 7495, in part (Ba, Ld,
Mu); Iltis & Univ. Wisc. Pl. Geogr. Field Trip 25234 (Ws); S. B.
Jones s.n. [8 May 1960] (Hi--210889); C. Mohr s.n. [July 1881]
(Hi), s.n. [July 1882] (Hi); W. Wolf s.n. [Elberta, Aug. 21, 1925]
(Ca--841823). Mobile Co.: F. W. Pennell 4509 [Herb. Dreisbach
1940] (Mi). MISSISSIPPI: Holmes Co.: Cabanis s.n. [Ebenezer] (B).
NORTH AMERICA: Locality undetermined: Palisot de Beauvois s.n.
(B). LOCALITY OF COLLECTION UNDETERMINED: Curtiss s.n. [Southern
States, 1875] (Ws); Sprengel s.n. (B). MOUNTED ILLUSTRATIONS:
floral diagrams by Körnicke & Kunth (B).

SYNGONANTHUS FLAVIPES Moldenke, Mem. N. Y. Bot. Gard. 8: 100—
    101. 1953.
    Bibliography: Moldenke, Mem. N. Y. Bot. Gard. 8: 100—101.
1953; Moldenke, Phytologia 4: 316. 1963; Moldenke, Résumé 73 & 491.
1959; G. Taylor, Ind. Kew. Suppl. 12: 138. 1959; Moldenke, Fifth
Summ. 1: 127 (1971) and 2: 952. 1971.
    The type of this species was collected by B. Maguire, R. S.
Cowan, & J. J. Wurdack (no. 30465) in wet places on Savanna No.

III, at 125 meters altitude, Cerro Yapacana on the Río Orinoco, .
Amazonas, Venezuela, on December 31, 1950, and is deposited in the
Britton Herbarium at the New York Botanical Garden. Other collectors report it "locally frequent" or "locally abundant" at altitudes of 100--125 meters, referring to its "shining leaves", and
found it in flower in December and in fruit in June.

Additional citations: VENEZUELA: Amazonas: Wurdack & Adderley
42860 (N, S). Bolívar: Wurdack & Monachino 39934 (N, S).

SYNGONANTHUS FLEXUOSUS Alv. Silv., Fl. Mont. 1: 393—395, pl. 252.
    1928.

Bibliography: Alv. Silv., Fl. Mont. 1: 393—395 & 417, pl. 202.
1928; Wangerin in Just, Bot. Jahresber. 57 (1): 478. 1937; Fedde
in Just, Bot. Jahresber. 57 (2): 895. 1938; A. W. Hill, Ind. Kew.
Suppl. 9: 271. 1938; Worsdell, Ind. Lond. Suppl. 2: 426. 1941;
Moldenke, Known Geogr. Distrib. Erioc. 18 & 58. 1946; Moldenke,
Known Geogr. Distrib. Verbenac., [ed. 2], 91 & 213. 1949; Moldenke,
Résumé 107 & 491. 1959; Moldenke, Fifth Summ. 1: 173 (1971) and 2:
962. 1971; Moldenke, Phytologia 35: 350. 1977.

Illustrations: Alv. Silv., Fl. Mont. 1: pl. 202. 1928.

The type of this species was collected by Dr. Joaquim Gomes
Michaeli [Herb. A. Silveira 655] "In campis prope Barauna", Minas
Gerais, Brazil, in April, 1918, and is deposited in the Silveira
Herbarium. On page 417 of his work (1928) Silveira gives "Barau-
nas" as the type locality. Elsewhere he comments that the "Species S. squarroso Ruhl. proxima, sed foliis pubescentibus facile
distinguitur". It also closely resembles S. glaber Alv. Silv. in
general habital aspect.

Silveira, in his text, refers to "Tabula CCLIII" as illustrating S. flexuosus, but the actual illustration is labeled "TABULA
CCLII". Thus far the species is known only from the original collection.

SYNGONANTHUS FUSCESCENS Ruhl. in Engl., Pflanzenreich 13 (4-30):
    249. 1903.

Bibliography: Ruhl. in Engl., Pflanzenreich 13 (4-30): 244,
249, & 293. 1903; Prain, Ind. Kew. Suppl. 3: 175. 1908; Alv. Silv.,
Fl. Mont. 1: 417. 1928; Moldenke, Known Geogr. Distrib. Erioc. 18
& 58. 1946; Moldenke, Known Geogr. Distrib. Verbenac., [ed. 2],
91 & 213. 1949; Moldenke, Phytologia 4: 316. 1963; Moldenke, Résumé 107 & 491. 1959; Moldenke, Fifth Summ. 1: 173 (1971) and 2:
962. 1971.

This species is based on Sena s.n. [Herb. Schwacke 14569] from
the Serra do Cipó, Minas Gerais, Brazil, deposited in the Berlin
herbarium where it was photographed by Macbride as his type photograph number 10685. Hatschbach encountered the species on rocky
campos and in "solo arenoso do campo, junta a afloramentos rochos-
os", at 1200 meters altitude, flowering in August and September
and fruiting in September. Silveira (1928) cites A. Silveira 517
from the Serra do Riacho do Vento, collected in 1908.

Additional citations: BRAZIL: Minas Gerais: Hatschbach 27428 (Z),

30214 (Ld); <u>Sena s.n.</u> [Herb. Schwacke 14569; Macbride photos 10685] (B—type, N—photo of type, W—photo of type, Z—isotype).

SYNGONANTHUS GARIMPENSIS Alv. Silv., Fl. Mont. 1: 317—319, pl. 201 & 202. 1928.

Bibliography: Alv. Silv., Fl. Mont. 1: 317—319 & 417, pl. 201 & 202. 1928; Wangerin in Just, Bot. Jahresber. 57 (1): 478. 1937; Fedde in Just, Bot. Jahresber. 57 (2): 895. 1938; A. W. Hill, Ind. Kew. Suppl. 9: 271. 1938; Worsdell, Ind. Lond. Suppl. 2: 426. 1941; Moldenke, Known Geogr. Distrib. Erioc. 18 & 58. 1946; Moldenke, Known Geogr. Distrib. Verbenac., [ed. 2], 91 & 213. 1949; Moldenke, Résumé 107 & 491. 1959; Moldenke, Fifth Summ. 1: 173 (1971) and 2: 962. 1971.

Illustrations: Alv. Silv., Fl. Mont. 1: pl. 201 & 202. 1928.

This species is based on <u>A. Silveira 543</u> from "In campis arenosis in Serra do Garimpo, inter Caeté et Santa Barbara", Minas Gerais, Brazil, collected in April, 1909, and deposited in the Silveira Herbarium. Silveira (1928) notes that the species "Ab affinibus indumento foliorum facile distinguitur". Thus far it is known only from the original collection.

SYNGONANTHUS GLABER Alv. Silv., Fl. Mont. 1: 388—390, pl. 248. 1928.

Bibliography: Alv. Silv., Fl. Mont. 1: 388—390 & 417, pl. 248. 1928; Wangerin in Just, Bot. Jahresber. 57 (1): 478. 1937; Fedde in Just, Bot. Jahresber. 57 (2): 895. 1938; A. W. Hill, Ind. Kew. Suppl. 9: 271. 1938; Worsdell, Ind. Lond. Suppl. 2: 426. 1941; Moldenke, Known Geogr. Distrib. Erioc. 18 & 58. 1946; Moldenke, Known Geogr. Distrib. Verbenac., [ed. 2], 91 & 213. 1949; Moldenke, Résumé 107 & 491. 1959; Moldenke, Fifth Summ. 1: 173 (1971) and 2: 962. 1971.

Illustrations: Alv. Silv., Fl. Mont. 1: pl. 248. 1928.

This species is based on <u>A. Silveira 788</u> from "In campis prope Milho Verde, inter Serro et Diamantina, in Serra Geral", Minas Gerais, Brazil, collected in 1925, and deposited in the Silveira Herbarium. Silveira (1928) comments that the "Species ob magnitudinem foliorum pedunculorumque a <u>S. squarroso</u> Ruhl. proximo praecipue differt". It should be noted that in his text he refers to "Tabula CCXLIX" as illustrative of this species, but the actual plate is labeled "TABULA CCXLVIII". The species also resembles <u>S. flexuosus</u> Alv. Silv. in general habit. Thus far it is known only from the original collection.

SYNGONANTHUS GLANDULIFER Alv. Silv., Fl. Mont. 1: 321—322, pl. 204. 1928.

Bibliography: Alv. Silv., Fl. Mont. 1: 321—322, pl. 204. 1928; Wangerin in Just, Bot. Jahresber. 57 (1): 478. 1937; Fedde in Just, Bot. Jahresber. 57 (2): 895. 1938; A. W. Hill, Ind. Kew. Suppl. 9: 271. 1938; Worsdell, Ind. Lond. Suppl. 2: 426. 1941; Moldenke, Known Geogr. Distrib. Erioc. 18 & 58. 1946; Moldenke, Alph. List Cit. 2: 412 (1948) and 3: 935. 1949; Moldenke, Known

Geogr. Distrib. Verbenac., [ed. 2], 91 & 213. 1949; Moldenke, Phy-
tologia 4: 316. 1953; Mendes Magalhãies, Anais V Reun. Anual Soc.
Bot. Bras. 236--237. 1956; Moldenke, Résumé 107 & 491. 1959; Rennó, Levant. Herb. Inst. Agron. Minas 71. 1960; Moldenke, Fifth
Summ. 1: 173 (1971) and 2: 962. 1971.
    Illustrations: Alv. Silv., Fl. Mont. 1: pl. 204. 1928.
    This species is based on A. Silveira 549 from "In campis in
Serra do Cipó, locis arenosis....Apr. 1909, in campis prope Itambé
do Serro.....Apr. 1918", Minas Gerais, Brazil, deposited in the
Silveira Herbarium. Mendes Magalhãies (1956) also reports it collected in anthesis in March. Silveira (1928) comments that the
species "Ab affinibus (S. anthemidifloro et aliis) praecipue differt forma indumentoque bractearum involucrantium, foliorum,
pedunculorum vaginarumque pilositate et sepalorum petalorumque
colore". Thus far it is known only from these three collections.

SYNGONANTHUS GLANDULOSUS Gleason, Bull. Torrey Bot. Club 56: 394—
    395. 1929.
    Synonymy: Syngonanthus oblongus f. abbreviata Herzog ex Lützelb.,
Estud. Bot. Nordést. 149 & 151. 1923. Syngonanthus oblongus f.
abbreviatus Herzog ex Moldenke, Phytologia 4: 328. 1953.
    Bibliography: Lützelb., Estud. Bot. Nordést. 3: 149 & 151. 1923;
Gleason, Bull. Torrey Bot. Club 56: 394—395. 1929; A. W. Hill,
Ind. Kew. Suppl. 8: 231. 1933; Fedde & Schust. in Just, Bot. Jahresber. 57 (2): 16. 1937; Moldenke, Known Geogr. Distrib. Erioc. 6 &
58. 1946; Moldenke, Phytologia 2: 352. 1947; Moldenke, Alph. List
Cit. 3: 975. 1949; Moldenke, Known Geogr. Distrib. Verbenac., [ed.
2], 65, 67, & 213. 1949; Moldenke, Phytologia 4: 316 & 328. 1953;
Hocking, Dict. Terms Pharmacog. 284. 1955; Moldenke, Résumé 73, 76,
77, 107, 108, & 492. 1959; Moldenke, Résumé Suppl. 1: 6 (1959) and
12: 3. 1965; Lindeman & Görts-van Rijn in Pulle & Lanjouw, Fl. Surin. 1 [Meded. Konink. Inst. Trop. 30, Afd. Trop. Prod. 11]: 335 &
339. 1968; Van Donselaar, Meded. Bot. Mus. Rijksuniv. Utrecht 306:
397 & 402. 1968; Moldenke, Résumé Suppl. 18: 4. 1969; Teunissen &
Wildschut, Verh. Konink. Nederl. Akad. Wet. Natuurk. 59 (2): 23.
1970; Koyama & Oldenburger, Rhodora 73: 159. 1971; Moldenke, Fifth
Summ. 1: 120, 127, 131, 133, & 173 (1971) and 2: 962 & 968. 1971;
Teunissen & Wildschut, Meded. Bot. Mus. Utr. 341: 23. 1971; Anon.,
Biol. Abstr. 56 (10): B.A.S.I.C. S.265. 1973; Moldenke, Biol. Abstr. 56: 5366. 1973; Moldenke, Phytologia 26: 177. 1973; Hocking,
Excerpt. Bot. A.23: 293. 1974; Moldenke, Phytologia 28: 437 & 440
(1974), 30: 35 & 106 (1975), 31: 386 & 408 (1975), 34: 259 (1976),
and 35: 112, 291, 306—308, 354, & 359. 1977.
    This puzzling species is based on G. H. H. Tate 345, collected
in Philipp Swamp in the Roraima district of Guyana, at 5100—5200
feet altitude, on November 11, 1927, and deposited in the Britton
Herbarium at the New York Botanical Garden. Gleason (1929) comments that "The plant consists of a number of short erect stems
which are densely leafy and send out numerous peduncles from the
upper axils. In general habit it resembles S. simplex, gracilis,
and biformis, and differs from the first in its appendaged style,

from the second in its narrow acute bracts, from the last in its
symmetrical sepals, and from all three in the rounded sinuses of
its peduncular sheaths." To me, it much more closely resembles
very young forms of S. caulescens (Poir.) Ruhl. Lindeman &
Görts-van Rijn key out these perplexingly similar taxa as follows:
1. Peduncles glandular-pubescent, 7—11 cm. long; leaves 1.5 mm.
   wide, about 1 cm. long; peduncular sheaths with rounded sinus..
                                    S. glandulosus Gleason.
   1a. Peduncles pubescent to glabrous, their sheaths obliquely
       split; leaves 1—5 cm. long.
     2. Stems floating, up to 3 dm. long; leaves fenestrate, 3—4.5
        cm. long, 2 mm. wide; peduncles 2—4 together at the end of
        the stem, 3—6 cm. long; petals of the female florets slight-
        ly longer than the sepals; style without appendages.........
                                    S. macrocaulon Ruhl.
     2a. Stems to 8 dm. long, simple; leaves 1.5—4.5 cm. long, 1.5—
        4.5 mm. wide; peduncles 5—30 cm. long, in a terminal fas-
        cicle; petals of the female florets shorter than the sepals;
        style appendaged ....S. caulescens (Poir.) Ruhl.
Gleason, in his unpublished Flora of British Guiana, keys the related taxa as follows:
1. Petals of the pistillate florets shorter than the sepals.
   2. Lateral sepals of the staminate florets strongly falcate and
      inequilateral.
     3. Pistillate and staminate florets, including the pedicels,
        about equal in length.................S. simplex (Miq.) Ruhl.
     3a. Pistillate florets about twice as long as the staminate.
                                    S. biformis (N. E. Br.) Gleason
   2a. Lateral sepals of the staminate florets not falcate, equi-
      lateral.
     4. Bracts obovate, broadly rounded at the summit..............
                                    S. gracilis (Bong.) Ruhl.
     4a. Bracts oblong, acute to obtuse at the apex.
        5. Leaves rosulate; peduncles not glandular; sinus of the
           sheaths, opposite the lamina, acute..S. eriophyllus
           (Mart.) Ruhl. [now regarded as S. gracilis (Bong.) Ruhl.]
        5a. Leaves crowded on a very short stem; peduncles glandu-
           lar; sinus of the sheaths broadly rounded................
                                    S. glandulosus Gleason
He describes S. glandulosus as having "Leaves densely cespitose,
spreading or recurved, 1 cm. long, conspicuously pubescent; peduncles 7—11 cm. long, numerous, 3-costate, glandular-pubescent;
sheaths twisted, about 1 cm. long, the lamina acuminate above a
rounded sinus; heads about 5 mm. wide, white; bracts imbricate,
scarious, lanceolate to elliptic, sharply acute, the longest 3 mm.
long." He regarded it as endemic to the Mt. Roraima region.
    Recent collectors refer to the flower-heads as "light-gray",
"pale-gray", or "dull-white" and the flowers as white. They have
encountered it in cerrado and Sphagnum bogs at altitudes of 115—
2085 meters, flowering from July to April, fruiting in February,

April, and July to September. Wurdack and his associates found
it "locally abundant on moist riverbanks". Goodland found it "in
wet sandy open savanna grasslands with scattered trees, Curatella,
Byrsonima, Trachypogon, and Fimbristylis dominant". Koyama & Oldenburger (1971) report _it_ growing in association with Philodice
hoffmannseggii, Diplacrum africanum, Syngonanthus gracilis var.
koernickeanus, Baccpa monierioides, Centunculus pentander, Poly-
ga₁a paludosa, Utricularia adpressa, and Eleocharis nana. Donse-
laãr encountered it "in wet valley floor with hummocks and channels ('hog-wallow structure')".

Syngonanthus oblongus f. abbreviatus of Herzog is based on
Lützelburg 338 from Bahia, Brazil, in the Munich herbarium and
seems to be conspecific with Gleason's plant.

Hocking (1955) reports the vernacular name, "guanak", and says
that the entire plant is used in decoction form in the treatment
of dentalgia in Venezuela. Lindeman & Görts-van Rijn (1968) cite
from Surinam: B.W. 7133, Rombouts 556, and Wessels Boer 800.

Material of what appears to be S. glandulosus has been misidentified and distributed in some herbaria as S. caulescens (Poir.)
Ruhl., S. gracilis (Bong.) Ruhl., S. simplex var. appendiculifera
Ruhl., S. xeranthemoides (Bong.) Ruhl., and Paepalanthus subtilis
Miq. On the other hand, the Irwin, Grear, Souza, & Reis dos Santos 14410 and G. H. H. Tate 246, distributed as S. glandulosus actually seem to be immature S. caulescens (Poir.) Ruhl., Lützelburg
21036 is S. gracilis var. amazonicus Ruhl., and Cowan & Soderstrom
is S. huberi Ruhl. Goodland 254 is a mixture of S. glandulosus
and Paepalanthus lamarckii Kunth, while W. A. Egler 47650 is a mixture with Paepalanthus oyapockensis Herzog and Cordeiro 30 is a
mixture with S. humboldtii var. glandulosus Gleason.

Additional citations: COLOMBIA: Santander: Barkley & Bouthil-
lette 38C168 (Ld). VENEZUELA: Aragua: Pittier 5841 (W--601553).
Bolívar: Bogner 1086 (Mu); Wurdack & Monachino 41048 (Mu, N, S).
GUYANA: Goodland 254, in part (W--2546169); Irwin 501 (W--2212839);
G. H. H. Tate 345 (N--type). SURINAM: Donselaar 3605 (Ut--320379);
Stahel 7133 [574] (Ut--44056A); Rombouts 214 (Ut--44055A); Wildschut & Teunissen 11572 (Ld). BRAZIL: Amapá: W. A. Egler 47650, in
part (N). Amazônas: Lützelburg 21036, in part (Mu); Prance, Maas,
Atchley, Steward, Woolcott, Coêlho, Monteiro, Pinheiro, & Ramos
13822 (Ac, N). Bahia: Lützelburg 338 [N. Y. Bot. Gard. Type Photo
new ser. 8832] (Mu, N--photo, Z--photo). Mato Grosso: Cordeiro 31,
in part (Ld). Roraima: Ule 7929 [Herb. Mus. Goeldi 13021] (K, Z).

SYNGONANTHUS GLANDULOSUS var. EPAPILLOSUS Moldenke, Phytologia 26:
    177--178. 1973.
    Bibliography: Anon., Biol. Abstr. 56 (10): B.A.S.I.C. S.265.
1973; Moldenke, Biol. Abstr. 56: 5366. 1973; Moldenke, Phytologia
26: 177--178. 1973; Hocking, Excerpt. Bot. A.23: 293. 1974; Molden-

ke, Phytologia 28: 437 & 440 (1974) and 35: 359. 1977.

Recent collectors refer to the flower-heads of this plant as "white", "off-white", "creamish-white", or "light-gray" and the flowers as white. They have found the plant growing in very wet ground, in cerrado, on wet campos, in marshes in gallery forests, among rocks at streamsides, in swamps, and in wet sand in sedge-eriocaul savannas, at altitudes of 230—1100 meters, flowering and fruiting from November to June and in September. Ratter and his associates found it "in a stream, the leaves submerged, the flower-heads held above the surface of the water". Anderson found it at the "edge of brejo in an area of gallery forest, adjacent brejo, and nearby cerrado and campo limpo"; Cowan & Soderstrom refer to it as a "locally common herb in boggy patches atop rocks in constant mist of [water]falls", while Steyermark & Wurdack found it a "locally abundant depressed form near water level on rocky edge of river". The Eitens encountered it "at water level", "in soaking soil at brookside in light shade of narrow gallery scrub", and "in lower part of a natural grassy campo at valley head next to its border with swampy gallery forest, ground soaking, with grass clumps and puddles between clumps, soil black humusy-clay; the plant rooted in soaking soil, the base in air or covered with 1 cm. of water". Dombrowski reports it frequent in "banhado".

Anderson 9564 has very much the general appearance of a variety of S. gracilis (Bong.) Ruhl. Material of S. glandulosus var. epapillosus has been widely misidentified and distributed in herbaria as S. caulescens (Poir.) Ruhl. or as typical S. glandulosus Gleason. Philcox, Fereira, & Bertoldo 3431 is a mixture with S. nitens (Bong.) Ruhl.; Cowan & Soderstrom 2154 is S. huberi Ruhl., while Ratter, Santos, Souza, & Ferreira R.1723 is a mixture of S. huberi and S. huberi f. viviparus Moldenke.

Citations: VENEZUELA: Bolívar: Hamann 2896 (Hm); Koyama & Agostini 7285 (N); Steyermark & Wurdack 45a (N—tyoe); Vareschi & Foldats 4743 (N). SURINAM: Rombouts 556 (N, Ut--44057A). BRAZIL: Amapá: Black 49-8256 (N). Goiás: W. R. Anderson 9564 (N); Hatschbach 36947 (Ld). Mato Grosso: Eiten & Eiten 8579 (W—2757729), 8626 (W—2757731), 9145 (W—2757737); Philcox & Fereira 3412 (K), 3431, in part (K), 3505 (K). Minas Gerais: Mello Barreto 25682 (N). Paraná: Dombrowski 6764 (Z); Hatschbach 33470 (Ld). PARAGUAY: Pedersen 9399 (N), 10095 (N).

SYNGONANTHUS GLAUCUS Alv. Silv., Fl. Mont. 1: 373—374, pl. 237. 1928.

Bibliography: Alv. Silv., Fl. Mont. 1: 373—374 & 418, pl. 237, 1928; Wangerin in Just, Bot. Jahresber. 57 (1): 478. 1937; Fedde in Just, Bot. Jahresber. 57 (2): 895. 1938; A. W. Hill, Ind. Kew. Suppl. 9: 271. 1938; Worsdell, Ind. Lond. Suppl. 2: 426. 1941; Moldenke, Known Geogr. Distrib. Erioc. 18 & 58. 1946; Moldenke, Known Geogr. Distrib. Verbenac., [ed. 2], 92 & 213. 1949; Moldenke, Résumé 107 & 492. 1959; Moldenke, Fifth Summ. 1: 173

(1971) and 2: 962. 1971; Moldenke, Phytologia 35: 431. 1977.
Illustrations: Alv. Silv., Fl. Mont. 1: pl. 237. 1928.
This species is based on A. Silveira 669 from "In campis prope
Diamantina", Minas Gerais, Brazil, collected in April, 1918, and
deposited in the Silveira herbarium. In his text Silveira (1928)
refers to "Tabula CCXXXVIII" as illustrating this species, but the
plate that actually does so is labeled "TABULA CCXXXVII". Thus
far the species is known only from the original collection.

SYNGONANTHUS GOYAZENSIS (Körn.) Ruhl. in Engl., Pflanzenreich 13
(4-30): 255. 1903.
Synonymy: Paepalanthus goyazensis Körn. in Mart., Fl. Bras. 3
(1): 453. 1863. Dupatya goyazensis (Körn.) Kuntze, Rev. Gen. Pl.
2: 745. 1891. Dupatya goyazensis Kuntze apud Durand & Jacks.,
Ind. Kew. Suppl. 1, imp. 1, 145. 1902. Syngonanthus goyazensis
Ruhl. apud Prain, Ind. Kew. Suppl. 3: 175. 1908. Syngonanthus
goyazensis (Bong.) Ruhl. ex Moldenke, Résumé 361, in syn. 1959.
Bibliography: Körn. in Mart., Fl. Bras. 3 (1): 453 & 507. 1863;
Kuntze, Rev. Gen. Pl. 2: 745. 1891; Jacks. in Hook. f. & Jacks.,
Ind. Kew., imp. 1, 2: 402. 1894; Durand & Jacks., Ind. Kew. Suppl.
1, imp. 1, 145. 1902; Ruhl. in Engl., Pflanzenreich 13 (4-30):
215, 245, 255, 290, & 293. 1903; Prain, Ind. Kew. Suppl. 3: 175.
1908; Alv. Silv., Fl. Mont. 1: 418. 1928; Durand & Jacks., Ind.
kew. Suppl. 1, imp. 2, 145. 1941; Jacks. in Hook. f. & Jacks.,
Ind. Kew., imp. 2, 2: 402. 1946; Moldenke, Known Geogr. Distrib.
Erioc. 18, 30, 49, & 58. 1946; Moldenke, Phytologia 2: 498. 1948;
Moldenke, Known Geogr. Distrib. Verbenac., [ed. 2], 92 & 213.
1949; Moldenke, Phytologia 4: 316. 1953; Durand & Jacks., Ind.
Kew. Suppl. 1, imp. 3, 145. 1959; Moldenke. Résumé 107, 280, 325,
351, 419, & 492. 1959; Jacks. in Hook. f. & Jacks., Ind. Kew., imp.
3, 2: 402. 1960; Rennó, Levant. Herb. Inst. Agron. Minas 71. 1960;
Moldenke, Fifth Summ. 1: 173 & 481 (1971) and 2: 583, 636, 778, &
962. 1971.
The type of this species was collected by George Gardner (no.
4384) in Goiás, Brazil, deposited in the Berlin herbarium where
Macbride photographed it as his type photograph number 10696.
Ruhland (1903) cites only the type collection and Glaziou 22310,
both from Goiás. He suggests that S. sclerophyllus Alv. Silv.
may actually only be a variety of S. goyazensis. Silveira (1928)
cites A. Silveira 740, also from Goiás.
Hunt & Ramos refer to S. goyazensis as having white inflorescences and found it growing in waterlogged ground by a small
stream in campo cerrado, at 600—1000 meters altitude, in flower
and fruit in June. Glaziou collected it in anthesis in October.
Additional citations: BRAZIL: Goiás: G. Gardner 4384 [Macbride
photos 10696] (B—type, N—isotype, N—photo of type, W—photo of
type); Glaziou 22310 (B, W—1124171); Hunt & Ramos 6275 (N). Minas Gerais: Héringer 7057 (Z); Santos & Castellanos 24182 (Bd—
28328). MOUNTED ILLUSTRATIONS: drawings & notes by Körnicke (B).

SYNGONANTHUS GRACILIS (Bong.) Ruhl. in Rngl., Pflanzenreich 13
(4-30): 249. 1903; Uittien & Heyn in Pulle, Fl. Surin. 1:
220. 1938 [not S. gracilis Molfino, 1945].

   Synonymy: Eriocaulon gracile Bong., Mém. Acad. Imp. Sci. St.
Pétersb., ser. 6, 1: 634, pl. 46. 1831 [not E. gracile Heyne, 1946,
nor Mart., 1832, nor Mart. & Wall., 1852]. Eriocaulon glabrum
Steud., Syn. Pl. Glum. 2: [Cyp.] 281. 1855 [not E. glabrum Pennell,
1959, nor Salzm., 1959]. Paepalanthus eriophyllus Mart. ex Körn.
in Mart., Fl. Bras. 3 (1): 463. 1863. Paepalanthus glanduliferus
Mart. ex Körn. in Mart., Fl. Bras. 3 (1): 464 & 560, in syn. 1863.
Paepalanthus gracilis (Bong.) Körn. in Mart., Fl. Bras. 3 (1):
460, pl. 59, fig. 1. 1863; Malme, Svensk. Vet. Akad. Handl. 27 (3):
no. 11: 31. 1901. Paepalanthus gracilis Körn. in Mart., Fl. Bras.
3 (1): 460. 1863. Paepalanthus gracilis var. α subvar. ϼ Körn.
in Mart., Fl. Bras. 3 (1): 460, 461, & 463. 1863. Paepalanthus
gracilis var. c Körn. in Mart., Fl. Bras. 3 (1): 460, in part.
1863. Paepalanthus eriophyllus var. α Körn. in Mart., Fl. Bras.
3 (1): 463—464. 1863. Paepalanthus eriophyllus var. ϼ Körn. in
Mart., Fl. Bras. 3 (1): 464. 1863. Dupatya eriophylla (Mart.)
Kuntze, Rev. Gen. Pl. 2: 745. 1891. Dupatya gracilis ([Bong.]
Körn.) Kuntze, Rev. Gen. Pl. 2: 745. 1891. Dupatya eriophylla
Kuntze apud Durand & Jacks., Ind. Kew. Suppl. 1, imp. 1, 145.
1902. Dupatya gracilis Kuntze apud Durand & Jacks., Ind. Kew.
Suppl 1, imp. 1, 145. 1902. Paepalanthus glandulifer Mart. apud
Ruhl. in Engl., Pflanzenreich 13 (4-30): 249, in syn. 1903. Paepalanthus pohlianus Mart. ex Ruhl. in Engl., Pflanzenreich 13 (4-
30): 250, in syn. 1903. Syngonanthus eriophyllus var. calvescens
Ruhl. in Engl., Pflanzenreich 13 (4-30): 249. 1903. Syngonanthus
gracilis var. olivacea Ruhl. in Engl., Pflanzenreich 13 (4-30):
250. 1903. Syngonanthus eriophyllus (Mart.) Ruhl. in Engl., Pflanzenreich 13 (4-30): 249. 1903. Syngonanthus eriophyllus Ruhl.
apud Prain, Ind. Kew. Suppl. 3: 175. 1908. Syngonanthus gracilis Ruhl. apud Prain, Ind. Kew. Suppl. 3: 175. 1908. Syngonanthus gracilis var. a (Kunth) Ruhl. ex Alv. Silv., Fl. Mont. 1: 418.
1928. Syngonanthus gracilis var. microphylla Alv. Silv., Fl. Mont.
1: 418, nom. nud. 1928. Syngonanthus gracilis var. olivaceus
Ruhl. ex Moldenke, Known Geogr. Distrib. Erioc. 18 & 58. 1946.
Syngonatnhus gracilis (Körn.) Ruhl. ex Reitz, Sellowia 7: 125,
sphalm. 1956. Paepalanthus hirtellus Körn. ex Moldenke, Résumé
Suppl. 1: 21, in syn. 1959. Paepalanthus hirtellus var. α Körn.
ex Moldenke, Résumé Suppl. 1: 21, in syn. 1959. Paepalanthus
olivaceus Körn. ex Moldenke, Résumé Suppl. 1: 21, in syn. 1959.
Syngonanthus gracilis (Körn.) Ruhl. ex Reitz, Sellowia 11: 31,
sphalm. 1959. Syngonanthus gracilis Körn. ex Moldenke, Résumé
Suppl. 1: 23, in syn. 1959. Dupatya gracilis (Körn.) Kuntze ex
Moldenke, Fifth Summ. 1: 481, in syn. 1971. Syngonanthus gracilis

(Bong.) Ruhl ex J. A. Steyerm., Biotropica 6: 7 & 10, sphalm.
1974. Paepalanthus eriophyllus "Mart. ex Körn." apud Moldenke &
Sm. in Reitz, Fl. Ilust. Catar. I Erio: 101, in syn. 1976. Paepalanthus hirtellus "Körn. ex Moldenke" apud Moldenke & Sm. in
Reitz, Fl. Ilust. Catar. I Erio: 101, in syn. 1976. Paepalanthus
hirtellus var. α "Körn. ex Moldenke" apud Moldenke & Sm. in
Reitz, Fl. Ilust. Catar. I Erio: 101, in syn. 1976. Paepalanthus
olivaceus "ex Moldenke" apud Moldenke & Sm. in Reitz, Fl. Ilust.
Catar. I Erio: 102, in syn. 1976. Paepalanthus pohlianus "Mart.
ex Ruhl." apud Moldenke & Sm. in Reitz, Fl. Ilust. Catar. I Erio:
102, in syn. 1976. Syngonanthus eriophyllus "(Mart. ex Körn.)
Ruhl." apud Moldenke & Sm. in Reitz, Fl. Ilust. Catar. I Erio:
103, in syn. 1976. Syngonanthus gracilis var. a "Ruhl. ex Moldenke" apud Moldenke & Sm. in Reitz, Fl. Ilust. Catar. I Erio:
103, in syn. 1976. Syngonanthus gracilis var. microplylla Alv.
Silv. ex Moldenke & Sm. in Reitz, Fl. Ilust. Catar. I Erio: 103,
in syn. 1976.

Bibliography: Bong., Mém. Acad. Imp. Sci. St. Péters., ser. 6,
1: 634. 1831; Bong., Ess. Monog. Erioc. 34—35. 1831; Steud.,
Nom. Bot., ed. 2, 1: 585. 1840; Kunth, Enum. Pl. 3: 534, 578,
613, & 624. 1841; D. Dietr., Syn. Pl. 5: 268. 1852; Steud., Syn.
Pl. Glum. 2: [Cyp.] 280, 281, & 334. 1855; C. Müll. in Walp.,
Ann. Bot. Syst. 5: 926 & 941—942. 1860; Körn. in Mart., Fl.
Bras. 3 (1): 460—464, 500, 507, & 560, pl. 59, fig. 1. 1863;
Kuntze, Rev. Gen. Pl. 2: 745. 1891; Jacks. in Hook. f. & Jacks.,
Ind. Kew., imp. 1, 1: 878 (1893) and imp. 1, 2: 401 & 402. 1894;
Huber, Bol. Mus. Para. 2: 499. 1898; Malme, Bih. Svensk. Vet.
Akad. Handl. 27 (3), no. 11: 31—32. 1901; Durand & Jacks., Ind.
Kew. Suppl. 1, imp. 1, 145. 1902; Chod. & Hassl., Bull. Herb.
Boiss., ser. 2, 3: 1033 & 1034. 1903; Chod. & Hassl., Pl. Hassler.
2: 255 & 256. 1903; Ruhl. in Engl., Pflanzenreich 13 (4-30): 28,
244, 249—253, 257, 285, 289, 290, & 293. 1903; Prain, Ind. Kew.
Suppl. 3: 175. 1908; Molfino, Physis 6: 363. 1923; Alv. Silv., Fl.
Mont. 1: 418. 1928; Ruhl. in Engl. & Prantl, Nat. Pflanzenfam.,
ed. 2, 15a: 56. 1930; Stapf, Ind. Lond. 4: 518. 1930; Gleason,
Bull. Torrey Bot. Club 58: 327 & 331. 1931; Herzog in Fedde,
Repert. Spec. Nov. 29: 212. 1931; Uittien & Heyn in Pulle, Fl.
Surin. 1 [Meded. Konink. Ver. Ind. Inst. 30, Afd. Handelsmus.
11]: 220—221. 1938; Durand & Jacks., Ind. Kew. Suppl. 1, imp.
2, 145. 1941; Herter, Revist. Sudam. Bot. 7: 199. 1943; Castell.
in Descole, Gen. & Sp. Pl. Argent. 3: 83. 1945; Abbiatti, Rev.
Mus. La Plata Bot., ser. 2, 6: [311] & 312. 1946; Jacks. in
Hook. f. & Jacks., Ind. Kew., imp. 2, 1: 878 (1946) and imp. 2,
2: 401 & 402. 1946; Moldenke, Alph. List Cit. 1: 132 & 223.
1946; Moldenke, Known Geogr. Distrib. Erioc. 5—7, 18—20, 29, 30,
33, 35, 44, 45, 48, 49, 52, 57, 58, 60, & 61. 1946; Moldenke, Lilloa 12: 173. 1946; Moldenke, Phytologia 2: 352, 373, 374, 377, &
381. 1947; Moldenke in Maguire & al., Bull. Torrey Bot. Club 75:
200—201. 1948; Moldenke, Phytologia 2: 492 & 498. 1948; Moldenke,
Alph. List Cit. 3: 601, 702, 704, 710, 744, 874, 975, & 976 (1949)

and 4: 985, 1072, 1076, & 1301. 1949; Moldenke, Known Geogr. Dis-
trib. Verbenac., [ed. 2], 61, 65, 67, 68, 92, 95, 97, 100, & 213.
1949; Moldenke, Phytologia 4: 316—320. 1953; Moldenke in Maguire
& al., Mem. N. Y. Bot. Gard. 8: 101. 1953; Herter, Revist. Sudam.
Bot. 9: 188. 1954; Mendes Magalhães, Anais V Reun. Anual Soc. Bot.
Bras. 266—267. 1956; Reitz, Sellowia 7: 125. 1956; Moldenke in
J. A. Steyerm., Fieldiana Bot. 28: 825 & 826. 1957; Cuatrecasas,
Revist. Acad. Colomb. Cienc. 10: 254. 1958; R. C. Foster, Contrib.
Gray Herb. 184: 39. 1958; Durand & Jacks., Ind. Kew. Suppl. 1,
imp. 3, 145. 1959; Reitz, Sellowia 11: 31 & 131. 1959; Moldenke,
Résumé 69, 73, 76, 77, 107, 112, 115, 119, 280, 286, 288, 289,
310, 323—325, 327, 351, 352, & 492. 1959; Moldenke, Résumé Suppl.
1: 5, 6, & 20—23. 1959; Van Royen, Nov. Guin., ser. 2, 10: 39 &
44. 1959; Jacks. in Hook. f. & Jacks., Ind. Kew., imp. 3, 1: 878
(1960) and imp. 3, 2: 401 & 402. 1960; Rennó, Levant. Herb. Inst.
Agron. Minas 71. 1960; Moldenke, Résumé Suppl. 2: 5 (1960), 3: 12
& 14 (1962), 4: 4 (1962), 10: 6 & 7 (1964), and 12: 3, 4, & 12.
1965; Angely, Fl. Anal. Paran., ed. 1, 201. 1965; Van Donselaar,
Wentia 14: 40. 1965; Huinink, Wentia 17: 140—141. 1966; J. A.
Steyerm., Act. Bot. Venez. 1: 60, 122, 135, 148, 155, & 247. 1966;
Moldenke, Résumé Suppl. 15: 5 (1967) and 16: 6. 1968; Aristeguie-
ta, Act. Bot. Venez. 3: 25. 1968; Lindeman & Görts-van Rijn in
Pulle & Lanjouw, Fl. Surin. 1 [Meded. Konink. Inst. Trop. 30, Afd.
Trop. Prod. 11]: 336 & 337. 1968; Van Donselaar, Meded. Bot. Mus.
Rijksuniv. Utrecht 306: 402. 1968; Moldenke, Phytologia 18: 100,
102, 260, 261, & 388 (1969), 19: 8 (1969), 19: 339 (1970), and 20:
101. 1970; Angely, Fl. Anal. & Fitogeogr. Est. S. Paulo, ed. 1, 6:
1162 & Ind. 28. 1970; Reitz, Sellowia 22: 137. 1970; Teunissen &
Wildschut, Verh. Konink. Nederl. Akad. Wet. Natuurk. 59 (2): 23,
36, & table 1. 1970; Koyama & Oldenburger, Rhodora 73: 159. 1971;
Moldenke, Fifth Summ. 1: 120, 127, 131, 133, 134, 174, 180, 184,
189, 480, & 481 (1971) and 2: 495, 501, 502, 549, 578, 582—584,
587, 588, 591, 636—638, 962, 963, 968, & 973. 1971; Moldenke,
Phytologia 21: 418 (1971) and 22: 6. 1971; Teunissen & Wildschut,
Meded. Bot. Mus. Utr. 341: 23, 36, & table 1. 1971; Hocking, Biol.
Abstr. A.21: 30. 1972; Moldenke, Biol. Abstr. 53: 5252 (1972) and
54: 6295. 1972; Moldenke, Phytologia 25: 230 (1973), 26: 27 & 45
(1973), 28: 440 (1974), and 29: 211, 311, 319, & 323. 1974; Rodriguez M., Mem. 11 Congres. Venez. Bot. 95. 1974; J. A. Steyerm.,
Biotropica 6: 7 & 10. 1974; Cárdenas de Guevara, Act. Bot. Venez.
10: 39. 1975; J. A. Steyerm., Act. Bot. Venez. 10: 232. 1975;
Moldenke, Phytologia 30: 37, 52, & 318 (1975), 31: 383, 386, &
408 (1975), 34: 259, 260, 273, 275—277, 392, & 487 (1976), and
35: 18 & 28. 1976; Moldenke & Sm. in Reitz, Fl. Ilust. Catar. I
Eric: 62, 63, 77—80, & 98—103, pl. 8, fig. 27—31. 1976; Moldenke, Phytologia 35: 112, 125, 291, 306, 308, 338, 340, 341, 427,
440, & 442. 1977.

     Illustrations: Körn. in Mart., Fl. Bras. 3 (1): pl. 59, fig. 1.
1863; Moldenke & Sm. in Reitz, Fl. Ilust. Catar. I Erio: 63, pl.
8, fig. 27—31. 1976.
     This is a very widespread and extremely variable species. No

less than 16 rather poorly defined subspecific taxa have been proposed. The typical form apparently is found from Colombia and Venezuela, through the Guianas, to Brazil and Uruguay. Bongard's original (1831) description is "Acaule; foliis vaginas subaequantibus, confertis, linearibus, acutis, pilosiusculis; pedunculis caespitosis, filiformibus, pubescentibus; vaginis pilosiusculis. In umbrosis siccis montis Itacolumi." It is probably based on a Riedel collection in the Leningrad herbarium. Kunth (1841) adds: "A specie homonyma Martiana longe diversum".

Recent collectors describe the plant as growing 6 inches tall, with dark-green leaves, white or gray-white flower-heads, and white flowers. Tutin says: "bracts round the flowers paleaceous", while Huinink (1966) calls it a "hemixerophyte, scleromorphic, nanophyll", with a "hemispherical-shaped root-system" and inhabiting the Xyrido-Paspaletum ecologic association. Other collectors have encountered it on savannas, wet-sand savannas, white-sand savannas, savannas with a quartzite base, and dry sandy uplands, in swamps, sandy swamps, and "swamps on open level portions of plateaus", on large mesas, in moist depressions in llanos, along railroad tracks, in sand and white sand, and swampy ground by streamlets, in damp seepage patches on white-sand campinas, and in dry sandy or gravelly places in general, at altitudes of 30—2000 meters, flowering in April, May, and July to November, and fruiting in January, July to September, and November. In southern Brazil and Uruguay it is said to flower mostly in January and February.

McKee encountered S. gracilis "in dry sand in area of sandhills with low forest or scattered shrubs". Murça Pires & Cavalcante report it "common on savannas"; Goodland & Persaud "in grassland with scattered trees, Curatella, Brysonima, Trachypogon, and Fimbristylis dominant". The Maguires aver that it is "locally frequent in moist sand among rocks", "frequent in wet places along brooks", and "locally frequent annual in marshy places along streamsides". Malme (1901) reports it from "supra saxa tempore hiemali irrigata" and "in loco aperto, arenoso, humido, parce graminoso". Ruiz-Terán & López-Palacios describe it as an "Hierba rosulada, cespitosa, a la sombra de laja, en suelo húmedo. Roseto de unos 2 cm. de alto. Hojas finas, flexibles, verde subintensas, no espenescentes en al ápice. Escapos de 12--15 cm. Capítulos hemisféricos. Flores blancas."

Aristeguieta (1968) records this species from Guárico, Venezuela; Herter (1954) gives its distribution as "Sudamérica cálida". Malme (1901) cites Regnell III.1266 & 111.1801 from Minas Gerais and Mato Grosso, Brazil. Silveira (1928) cites A. Silveira 216 from Serra do Lenheiro, Minas Gerais, collected in 1896, as typical S. gracilis, but A. Silveira 227, from the Serra de Ibitipoca, collected in the same year, as his var. a.

Vernacular names reported for the species are "capim manso", "capipoatinga", "capipoatinga-mimosa", "gravatá manso", "semprevivas do campo", and "sempre-viva-do-campo".

Bongard's plate 46, often cited as illustrative of this species,

apparently was never actually published and probably is available
only in the Leningrad herbarium or library. Bongard's discussion
of the species is sometimes cited to various dates, but was actually published in 1831. The Malme (1901) work is sometimes incorrectly cited as "1903".

Paepalanthus eriophyllus Mart. and Syngonanthus eriophyllus
var. calvescens Ruhl. seem to be based, in part, at least, on
Kegel 231 in the Berlin herbarium. Uittien & Heyn (1938) aver
that Paepalanthus eriophyllus Mart. and P. glanduliferus Mart.
are typified, respectively, by Wullschlägel 763 and 762 from the
"Pará distr., plant. Berlijn", Surinam. These latter workers include S. biformis (N. E. Br.) Gleason and S. simplex (Miq.) Ruhl.
in the synonymy of S. gracilis. My disposition of the extraneous
synonyms which they list is as follows: Paepalanthus biformis N.
E. Br. is Syngonanthus biformis (N. E. Br.) Gleason, a valid species; Paepalanthus gracilis var. c Körn. is in part typical Syngonanthus gracilis (Bong.) Ruhl. and in part S. gracilis var. hirtellus Ruhl.; Eriocaulon brizoides (Kunth) Steud. is Syngonanthus
gracilis var. koernickeanus Ruhl.; Paepalanthus brizoides Kunth
is in part S. gracilis and in part S. gracilis var. koernickeanus;
and Dupatya simplex Kuntze, Eriocaulon hostmanni Steud., E. simplex (Miq.) Steud., Paepalanthus hispidus Klotzsch, and P. simplex
Miq. are all Syngonanthus simplex (Miq.) Ruhl.

The Eriocaulon gracile credited to Heyne, to Martius, and to
Martius & Wallich, referred to in the synonymy of S. gracilis (a-
bove), are all synonyms of Eriocaulon infirmum Steud.; E. glabrum
Pennell is E. peruvianum Ruhl.; E. glabrum Salzm. is Syngonanthus
gracilis var. glabriusculus Ruhl.; and Syngonanthus gracilis Molfino is a synonym of Eriocaulon argentinum Castell.

The name, Syngonanthus gracilis, as applied to the present
taxon, is very widely credited to "(Körn.) Ruhl.", but Ruhland
(1903) was in error when he wrote it thus because Körnicke (1863)
plainly cites Eriocaulon gracile Bong. as the name-bringing synonym, even though Ruhland does so only in the discussion of his
S. gracilis var. olivacea Ruhl. This so-called var. olivacea is,
therefore, the actual typical variety of the species and the names
given in its synonymy therefore belong in the synonymy of this
typical form of S. gracilis. Ruhland cites for it only Pohl s.n.
and Widgren s.n. from Minas Gerais, Brazil. Silveira (1928)
cites A. Silveira 651 from Diamantina (also in Minas Gerais), collected in 1908, as this "var. olivacea".

Macbride (1936) distinguishes S. gracilis from the closely related S. nitens (Bong.) Ruhl. by stating that in the former the
heads are smaller, only 3--5 mm. in diameter, while in the latter
they are 5--8.5 mm. thick. This, in general, is a quite valid
distinction.

Ruhland's other varieties are tentatively maintained by me and will be discussed hereinafter separately although they are, at best, rather difficult to distinguish in all cases.

Körnicke's (1863) varieties and subvarieties are being treated by me as follows: Paepalanthus eriophyllus var. α and var. β, P. gracilis var. α subvar. β, and P. gracilis var. c are typical Syngonanthus gracilis (Bong.) Ruhl.; Paepalanthus gracilis var. α is S. gracilis var. glabriusculus Ruhl.; P. gracilis var. α subvar. α is S. gracilis var. subinflatus Ruhl.; P. gracilis var. β and var. β subvar. α are S. gracilis var. koernickeanus Ruhl.; and P. gracilis var. β subvar. β is S. gracilis var. setaceus Ruhl. Ruhland's (1903) Syngonanthus eriophyllus var. glanduliferus is S. gracilis var. koernickeanus Ruhl. Paepalanthus brizoides Kunth is apparently in part typical S. gracilis and in part var. koernickeanus. It is based on two Sellow collections from "Brasilia meridionalis, inter Rio Janeiro at Campos et inter Vittoria et Bania" and one by Luschnath from "Campos prope St. Joao". He describes it as "Acaulis; caespitosus; foliis setaceo-linearibus, obtusiusculis, rigidis, glabris, recurvatis; vaginis glanduloso-pilosis, folia superantibus; pedunculis subcapillaceis, trisulcatis, vix puberulis; bracteis involucrantibus ellipticis, obtusis, aridis, stramineo-albidis, glabris, flores superantibus; sepalis exterioribus masculis et femineis angustato-acutis, glabris." He comments that it is "Affinis P. tenui".

Körnicke's var. α subvar. β is based on Clausen 68, 164, & s.n. [Cachoeira do Campo], Houllet s.n., Martius 1083 & s.n. [Itambé], and Riedel s.n. from Minas Gerais; his var. c is based on Salzmann s.n. from Bahia, Spruce s.n. from Amazônas, Pohl s.n. and Weddell 2136 from Goiás, Gardner s.n., Martius s.n., and Widgren s.n. from Minas Gerais, Martius s.n. and Vauthier s.n. from Rio de Janeiro, and Riedel 2304 from São Paulo. His Paepalanthus eriophyllus var. α is based on Kegel s.n. and Wullschlägel 762 from Surinam, while P. eriophyllus var. β is based on Wullschlägel 763, also from Surinam.

Gleason, usually most conservative in his treatment of species, still maintains S. eriophyllus as distinct from S. gracilis in his unpublished "Flora of British Guiana", citing for the former only Jenman 3768 and giving its overall distribution as only Surinam and Guiana. For S. gracilis he cites Appun 1526, Gleason 652, Linder 40, Lloyd s.n., Loyed s.n., and Parker s.n., giving its overall distribution as "Venezuela to French Guiana and Uruguay". He describes S. eriophyllus as "Leaves densely rosulate, narrowly linear, recurved, densely and persistently white-lanate, 1—2 cm. long; peduncles few, slender, 4—6 cm. high, sparsely hirtellous; sheaths glandular-hirtellous, prominently striate and twisted, the lamina acuminate; heads 3--5 mm. wide, subglobose; bracts oblong, obtuse, soon glabrous." He describes S. gracilis as "Leaves densely cespitose, often more or less recurved, narrowly linear, hirsute, 10—25 mm. long; peduncles several or many, 8—15 cm. high,

glabrous or nearly so; sheaths equaling the leaves; heads hemispheric, 3—5 mm. wide, white; bracts obovate, broadly rounded,
glabrous, silvery and scarious." In his key he distinguishes
these and some other taxa as follows:
1. Lateral sepals of the staminate florets strongly falcate and
   inequilateral.
   2. Pistillate and staminate florets, including the pedicels, a-
      bout equal in length. . . . . S. simplex
   2a. Pistillate florets about twice as long as the staminate. .
                                                    S. biformis
1a. Lateral sepals of the staminate florets not falcate, equilat-
    eral.
    3. Bracts obovate, broadly rounded at the summit. . S. gracilis
    3a. Bracts oblong, acute to obtuse at the apex.
        4. Leaves rosulate; peduncles not glandular; sinus of the
           sheaths, opposite the lamina, acute. . . S. eriophyllus
        4a. Leaves crowded on a very short stem; peduncles glandular;
            sinus of the sheaths broadly rounded. . . S. glandulosus
    Ruhland (1903) also keeps the two taxa separate, distinguishing
them as follows:
1. Folia dense rosulata, anguste linearia, rigida. Plantae in
   Guyana collectae. . . . . S. eriophyllus
1a. Folia plusminusve caespitosa, rarius rosulato-caespitosa.
    Species brasilienses. . . . . .S. gracilis
    Lindeman & Görts-van Rijn (1968) key out some of these thus:
1. Male and female flowers not very unequal in size or shape. Involucral bracts about the same length as the flowers.
   2. Leaves about 5 mm. long, densely rosulate, white-villous and
      pilose, later glabrous. Peduncles 5—7 cm. long. Involu-
      cral bracts glabrous, the inner ones ciliate. Style with-
      out appendages. . . . . .S. simplex
   2a. Leaves 1—3 cm. long, cespitose, glabrous or slightly puber-
       ulcus. Peduncles 6—30 cm. long. Involucral bracts longer
       than or equaling the flowers. Sepals at first puberulous
       in the middle, later glabrous. . . . . S. gracilis
    Also quite similar in habit, at least, to S. gracilis are S.
llanorum Ruhl., S. pauciflorus Alv. Silv., and S. planus Ruhl.
    Material of S. gracilis has been misidentified and distributed
in some herbaria as Paepalanthus exiguus (Bong.) Körn. and P. subtilis Miq. On the other hand, the Robertson & Austin 268, distributed as S. gracilis, is actually Comanthera kegeliana (Körn.)
Moldenke; Alston & Lutz 33 and B. Lutz 668 are Leiothrix dielsii
Ruhl.; Mexia 5882 is Leiothrix fulgida Ruhl.; Alston & Lutz 133
is Paepalanthus tortilis (Bong.) Mart.; Williams & Assis 6885 is
Syngonanthus biformis (N. E. Br.) Gleason; A. R. Schultz 324 is S.
chrysanthus (Bong.) Ruhl.; Donselaar 3605 is S. glandulosus Gleason; Hunt & Ramos 6140, Malme 1653, and Swallen 4912 are S. gracilis var. aureus Ruhl.; Hermann 11054 and Tutin 619 are S. gracilis

var. koernickeamus Ruhl.; B. Lutz 602 is S. gracilis var. setaceus
Ruhl.; Hassler 9430 is S. nitens var. hirtulus Ruhl.; Hassler
9436, 9436a, & 9436b are S. nitens var. koernickei Ruhl.; Brade
6578 and N. A. Rosa 477 [Herb. IFEAN 149907] are S. nitens f. pilosus Moldenke; and G. A. Black 54-16734 is S. tenuis (H.B.K.) Ruhl.
Lockhart s.n. [Caracas] is a mixture of S. gracilis with Comanthera kegeliana (Körn.) Moldenke; J. A. Steyermark 57804 is a mixture
with Paepalanthus lamarckii Kunth; Black 48-3050 is a mixture with
P. fasciculatus (Rottb.) Kunth and P. fasciculatus f. sphaeroceph-
alus Herzog; Mexia 5756 is a mixture with P. tortilis (Bong.) Mart.;
Phelps & Hitchcock s.n. [February 12, 1949] is a mixture with S.
gracilis var. glabriusculus Ruhl.; Lanjouw & Lindeman 860 is a mixture with S. gracilis var. koernickeamus Ruhl.; F. Lima s.n. [Herb.
Mus. Goeldi 12173] is a mixture with S. umbellatus (Lam.) Ruhl.;
and Vareschi & Maegdefrau 6613 is a mixture with S. yapacanensis
Moldenke.

Additional citations: COLOMBIA: Magdalena: C. Allen 527 (E—
1014370, F—1391643, F—1391775). Meta: F. W. Pennell 1427 (N, W—
1041737); Smith & Idrobo 1395 (Ca—1147411). Santander: Fassett
25068 (W—2166142, Ws). Vaupés: Schultes, Baker, & Cabrera 18093
(W—2172073), 18114 (Ss), 18539 (S, Ss); Schultes & Cabrera 14238
(Z), 14337 (Ss), 14376 (Ss), 18390 (Ss, W—2198900), 19178 (Ss),
19749b (Ss), 19918c (Ss). VENEZUELA: Amazonas: Maguire & Maguire
35022 (N); Phelps & Hitchcock s.n. [February 12, 1949] (N); J. A.
Steyermark 105141a (Ft), 57804, in part (N); G. H. H. Tate 216 (N),
259 (N); Vareschi & Maegdefrau 6613, in part (Ve—42532). Bolívar: Bernardi 6608 (N); Merxmüller 22955 (Mu); Pannier & Schwabe
s.n. [Auyantepui] (Ve); Ruiz-Terán & López-Palacios 11336 (Mi); J.
A. Steyermark 89672 (Mi); G. H. H. Tate 813 (N); Wurdack & Guppy 9
(Mu, N). Federal District: Lockhart s.n. [Caracas] (K). Guárico:
Aristeguieta 4492 (N). GUYANA: C. W. Anderson 512 (K); Carrick
973 (Kl—3973); H. A. Gleason 652 (N, W—1191105); Goodland & Persaud 791 (W—2546171); Linder 40 (N); McKee 10681 (Ws); Tutin 483
(Ut—3964A, W—1743597); Whitton 213 (K). SURINAM: Donselaar
3661 (Ut—320401); Florschütz & Florschütz 616 (Ut—80218B); Kegel
231 (B); Lanjouw & Lindeman 128 (N), 860, in part (Ut—17894B),
1792 (Ut—17893B), 1855 (Ut—17895B), 3013 (N); Wullschlägel 762
(E—photo, F—photo, N—photo, Z—photo). BRAZIL: Amapá: Irwin &
Westra 47259a (N); Murça Pires & Cavalcante 52274 (N). Amazônas:
G. A. Black 48-3050, in part (N, W—2655155); Prance 23528 (Ld);
Prance, Maas, Atchley, Steward, Woolcott, Coêlho, Monteiro, Pinheiro, & Ramos 13836 (Ac, N); Prance, Pena, Forero, Ramos, & Monteiro 4790 (N); Spruce 1502 (P). Ceará: Herb. Mus. Goeldi 49 (Gl).
Minas Gerais: P. Clausen 1831 (N); Martius s.n. [in udis irreguis
prov. Rio de Janeiro et Minarum passim] (Mu); Mexia 5756, in part

[Herb. Leonard 7656] (B). Pará: Black & Ledoux 50-10407 (Z),
50-10631 (Ca--28245, Z); W. A. Egler 160 (Bs); F. Lima s.n. [Herb.
Mus. Goeldi 12173] (Bs); E. Pereira 5022 (Bd--12468). Rio de Janeiro: Martius s.n. [in udis irreguis prov. Rio de Janeiro et Minarum passim] (Mu). Roraima: G. A. Black s.n. [Herb. Inst. Agron.
Norte 77605] (Z); Maguire & Maguire 40101 (N, Sm). Santa Catarina:
Reitz 4735 [Herb. Reitz 4737] (S); Smith & Klein 10679 (Ok).
MOUNTED ILLUSTRATIONS: Körn. in Mart., Fl. Bras. 3 (1): pl. 59, fig.
1. 1863 (B, N, Z); drawings & notes by Körnicke (B, B, B, B).

SYNGONANTHUS GRACILIS var. AMAZONICUS Ruhl. in Engl., Pflanzen-
    reich 13 (4-30): 250 [as "amazonica"]. 1903; Moldenke, Known
    Geogr. Distrib. Erioc. 18 & 58. 1946.
    Synonymy: Syngonanthus gracilis var. amazonica Ruhl. in Engl.,
Pflanzenreich 13 (4-30): 250. 1903.
    Bibliography: Ruhl. in Engl., Pflanzenreich 13 (4-30): 250 &
293. 1903; Alv. Silv., Fl. Mont. 1: 418. 1928; Herzog in Fedde,
Repert. Spec. Nov. 29: 212. 1931; Moldenke, Known Geogr. Distrib.
Erioc. 18 & 58. 1946; Moldenke, Phytologia 2: 493. 1948; Moldenke,
Known Geogr. Distrib. Verbenac., [ed. 2], 92, 95, & 213. 1949;
Moldenke, Phytologia 4: 317--318. 1953; Moldenke, Résumé 107, 112,
351, & 492. 1959; Moldenke, Fifth Summ. 1: 174 & 180 (1971) and 2:
636 & 962. 1971; Moldenke, Phytologia 25: 230 (1973) and 34: 259.
1976.
    This variety is based on Huber 351 from the "Mündungsgebiet
des Amazonas, Marajó, in einem Campos-wäldchen" and Burchell 8911
from "zwischen Junil und São João am Tocantins", Pará, Brazil, in
the Berlin herbarium.
    Ruhland's original (1903) description is "Differt foliis sub-
erecto-recurvatis, 2 cm. longis, basi vix vel non ampliatis,
juventute leviter puberulis, cito glaberrimis; vaginis laxiusculis, striatis, patenti-puberulis, calvescentibus; pedunculis
erectis, valde tortis, fusco-stramineis, 3-costatis, apice breviter rigido-pilosa excepta glabriusculis; capitulis pallide
cinereo-stramineis, duriusculis, 15--16 cm longis; bracteis involucrantibus obovatis vel ovatis, obtusis, capitulum 3--4 mm
latum vix aequantibus, glabris."
    Recent collectors describe the plant as an herb, 15 cm. tall,
the inflorescences grayish-white, and have found it growing on
moist or marshy campos, and in coarse white sand on disturbed
white-sand savannas, flowering in February, May, August, and September, and fruiting in September. Silveira (1928) cites Huber
437 from Marajó island, collected in 1896, deposited in the Silveira herbarium.
    Material of this variety has been misidentified and distribu-
ted in some herbaria as Paepalanthus nitens Körn. and Syngonanthus elegans (Bong.) Ruhl. Lützelburg 21036 is a mixture with
S. glandulosus Gleason.
    Additional citations: BRAZIL: Amazônas: Maas & Maas 454 (Ld, N);

Ule 6177 (B). Maranhão: Murça Pires & Black 2537 (Z). Mato Gros-
so: Irwin & Soderstrom 6477 (N). Minas Gerais: Lützelburg 21150
(Mu). Pará: G. A. Black 55-18613 (N); Ducke 5503 (Bs), 11656 (Bs),
s.n. [Herb. Mus. Goeldi 10676] (Bs); Spruce 610 (Mu). Rio de Jan-
eiro: Jobert 1227 (P). Roraima: M. Silva 122 [Herb. Brad. 47002]
(Ld). State undetermined: Lützelburg 20548a [Igarapé] (Mu), 21003
[Maruay] (Mu), 21004 [Rio Cotim Contá] (Mu), 21036, in part [Vera
Cruz] (Mu), 21132 [Maruay], 21156 [Maruay], 21289 [Serra da Lua]
(Mu).

SYNGONANTHUS GRACILIS var. ARAXAENSIS Alv. Silv., Fl. Mont. 1: 347.
      1928.
      Bibliography: Alv. Silv., Fl. Mont. 1: 347 & 418. 1928; Molden-
ke, Known Geogr. Distrib, Erioc. 18 & 58. 1946; Moldenke, Known
Geogr. Distrib. Verbenac., [ed. 2], 92 & 213. 1949; Moldenke, Résumé 107 & 492. 1959; Moldenke, Fifth Summ. 1: 174 (1971) and 2: 962.
1971.
      This variety is based on an unnumbered specimen collected by Dr.
J. Michaeli "In campis prope Araxá", Minas Gerais, Brazil, in April,
1919, and is no. 715 in the Silveira herbarium. Silveira's original
(1928) description is "Folia glabra, rigidula, 1 cm. longa. Pedun-
culi glabri, 20—30 cm alti. Vaginae arctae oblique fissae, glabrae
folia duplo superantes striatae, lamina erecta instructae, 2—2,5
elatae. Capitula albido-flavida, 5 mm lata. Bracteae involucrantes
obovato-rotundatae, glabrae. Bracteae flores stipantes nullae.
Sepala floris masculi utrinque pilosa, cito calvescentia. Sepala
floris feminei illis floris masculi similia." Thus far it is known
only from the original collection.

SYNGONANTHUS GRACILIS var. AUREUS Ruhl. in Engl., Pflanzenreich 13
      (4-30): 251 [as "aurea"]. 1903; Moldenke, Known Geogr. Distrib.
      Erioc. 18 & 58. 1946.
      Synonymy: Syngonanthus gracilis var. aurea Ruhl. in Engl., Pflan-
zenreich 13 (4-30): 251. 1903.
      Bibliography: Chod. & Hassl., Bull. Herb. Boiss., ser. 2, 3:
1034. 1903; Chod. & Hassl., Pl. Hassler. 2: 256. 1903; Ruhl. in
Engl., Pflanzenreich 13 (4-30): 251 & 293. 1903; Moldenke, Known
Geogr. Distrib. Erioc. 18 & 58. 1946; Moldenke, Alph. List Cit. 4:
1301. 1949; Moldenke, Known Geogr. Distrib. Verbenac., [ed. 2], 92
& 213. 1949; Moldenke, Phytologia 4: 318. 1953; Moldenke, Résumé
73, 107, 351, & 492. 1959; Moldenke, Résumé Suppl. 12: 4 (1965)
and 15: 5. 1967; Moldenke, Fifth Summ. 1: 127, 174, 180, & 184
(1971) and 2: 636 & 962. 1971; Angely, Fl. Anal. & Fitogeogr. Est.
S. Paulo, ed. 1, 6: 1162 & Ind. 28. 1972; Moldenke, Phytologia 31:
386 (1975), 34: 260 (1976), and 35: 456. 1977.
      This variety is based on Burchell 7177 from Goiás and G. Gardner
5270, Glaziou 15680, and Sena s.n. [Herb. Schwacke 14556] from Minas Gerais, Brazil, all deposited in the Berlin herbarium. The original description is: "Differt foliis caespitosis, erecto-patentibus,
rigidulis, obtusiusculis, saepe olivaceo-viridibus, puberulis, mox

calvescentibus, ad 2 cm longis, medio usque 2/3 mm latis; vaginis
folia paullo modo superantibus, olivaceis, arctis, vix striatulis,
patentissimo-puberulis; pedunculis profunde 3 sulcatis, saepe
tortis, stramineo-flavidis, pilis brevibus, sparsis, vix puberulis, cito glabriusculis, 16—18 cm altis; capitulis globosis,
duriusculis, majusculis, latitudine demum interdum 5 mm exceden-
tibus; bracteis involucrantibus florum discum vix aequantibus,
concavis, glabris, ovatis, aureo-flavis, exterioribus saepe sub-
acutis. [An fortasse species distincta?].....Varietas habitu S.
nitenti similis, valde insignis."

Recent collectors describe this plant as to 20 cm. tall, the
inflorescences 10—15 cm. tall, gray or grayish, the flower-heads
white or grayish, and the flowers white. They have found it growing in sandy soil, in cerrado, at gallery margins, in moist open
ground and boggy ground near streamlets, on savannas and savanna-
margins, on sandy open ground, in wet sand close to streams, and
on periodically flooded campos, at altitudes of 200—1700 meters,
flowering in January, March to August, and October, and in fruit
in January, May, June, and August. Irwin and his associates refer to it as "locally common in cerrado"; Murça Pires & Cavalcante found it "frequent in wet sandy savannas"; Argent encountered
it "on open ground between grass tussocks". Ratter and his asoc-
iates found it "in damp cerrado between tufts of tall grasses and
sedges which shade it", while Anderson found it in grass of "gallery forest, adjacent brejo, and nearby cerrado and campo limpo"
and in similar areas but the "higher drier slopes with grassy
campo or rocky cerrado". Hatschbach reports it it "cerrado pequeño brejo".

The Angely (1972) work cited in the bibliography is often referred to as having been published in 1970, the title-page date,
but was not actually issued until 1972.

Material of this variety has been misidentified and distributed in some herbaria as Paepalanthus hirtellus Körn., Syngonanthus fischerianus (Bong.) Ruhl., typical S. gracilis (Bong.)
Ruhl. and its var. setacea Ruhl. On the other hand, the Hassler
4671, distributed as S. gracilis var. aureus, actually is S.
nitens var. koernickei Ruhl.

Additional citations: BRAZIL: Alagoas: Mendes Magalhães 162
[Herb. Jard. Bot. Belo Horiz. 32594] (N). Amapá: Murça Pires &
Cavalcante 52380 (N). Amazônas: Fromm 1455 [E. Santos 1477;
Sacco 1712; Trinta 381] (Bd—25617). Bahia: Lützelburg 241a (Mu).
Goiás: W. R. Anderson 9567 (Ld, N), 10387(Ld, N); Fróes 30116
(Hk, Z), 30183 (Be—79494); Hunt & Ramos 6140 (N); Irwin, Souza,
Grear, & Reis dos Santos 17600 (Ld, N); Macedo 3333 (S), 3355 (S);
Ule 233 (P). Maranhão: Lisboa 2333 (Bs). Mato Grosso: Argent in
Richards 6454 (Ld, N); M. A. Chase 11905 (W—1470136); Hatschbach
24343 (Ld), 38622 (Ld); Hunt & Ramos 5711 (N); Irwin, Grear, Sou-
za, & Reis dos Santos 16285 (Ac, N), 16346 (Ld, N, W—2769018);
Irwin, Souza, Grear, & Reis dos Santos 16976 (Ac, N, W—2759017),

16978 (Ac, N, W--2759031), 17365 (Ac, N); Kuntze s.n. [200 m.,
VII.92] (N); Maguire, Murça Pires, Maguire, & Silva 56231 (N);
Malme 1653 (W--1483435); Ratter, Santos, Souza, & Ferreira R.1686a
(K); Santos, Souza, Ferreira, & Andrelinho R.1783 (Ac, N). Minas
Gerais: Glaziou 15680 (W--1124144--cotype); Occhioni 5615 [Herb.
Fac. Nac. Farmao. 14309] (Ld); J. E. Oliveira 1317 [Herb. Jard.
Bot. Belo Horiz. 45187] (N). Pará: Spruce s.n. [In vicinibus San-
tarem, Aug. 1850] (N). Piauí: Lützelburg 233 (Mu). Rondônia:
Prance, Forero, Coêlho, Ramos, & Farias 5765 (Ac, N, S). Roraima:
Ule 7665 [Herb. Mus. Goeldi 12774] (Bs, K, N). São Paulo: Eiten,
Eiten, Felippe, & Freitas Campos 3028 (N). State undetermined: G.
Gardner 2748 bis (W--936284); Glaziou s.n. (P); J. E. Pohl s.n.
[in Brasilia] (Mu). MARAJÓ ISLAND: Swallen 4912 (W--1592046).

SYNGONANTHUS GRACILIS var. BOLIVIANUS Ruhl. in Engl., Pflanzen-
        reich 13 (4-30): 252 [as "boliviana"]. 1903; Moldenke, Known
        Geogr. Distrib. Erioc. 19, 29, & 58. 1946.
    Synonymy: Syngonanthus gracilis var. boliviana Ruhl. in Engl.,
Pflanzenreich 13 (4-30): 252. 1903. Dupatya fischeriana Kuntze ex
Ruhl. in Engl., Pflanzenreich 13 (4-30): 252, in syn. 1903 [not D.
fischeriana Kuntze, 1902]. Paepalanthus gracilis var. boliviana
Ruhl. ex Moldenke, Résumé Suppl. 1: 20, in syn. 1959. Syngonanthus gracilis boliviana Ruhl. ex Moldenke, Résumé Suppl. 12: 13, in
syn. 1965. Syngonanthus gracilis bolivianus Ruhl. ex Moldenke,
Fifth Summ. 2: 637, in syn. 1971.
    Bibliography: Ruhl. in Engl., Pflanzenreich 13 (4-30): 252 &
293. 1903; Moldenke, Known Geogr. Distrib. Erioc. 19, 29, & 58.
1964; Moldenke, Known Geogr. Distrib. Verbenac., [ed. 2], 97 &
213. 1949; R. C. Foster, Contrib. Gray Herb. 184: 39. 1958; Moldenke, Résumé 115, 280, 351, & 492. 1959; Moldenke, Résumé Suppl.
1: 20 (1959) and 12: 12. 1965; Moldenke, Fifth Summ. 1: 184 & 480
(1971) and 2: 583, 636, 637, & 962. 1971; Moldenke, Phytologia 35:
454. 1977.
    This variety is based on Otto Kuntze 455 from an altitude of
200 meters in east Velasco, Santa Cruz, Bolivia, collected in July
of 1892 and deposited in the Berlin herbarium. The unnumbered
specimen in the New York Botanical Garden herbarium is probably an
isotype. Ruhland's original (1903) description is "Differt foliis
diffuso-caespitosis, in sicco pallide stramineis, vix puberulis,
obtusis, setaceo-linearibus, supra concaviusculis, subtus nervo
uno valde prominente quasi bisulcatis, 1--1,5 cm longis; vaginis
folia superantibus, oblique fissis, arctis, striatis, lamina paullo recurva, acuta instructis; capitulis bracteisque involucrantibus ut in varietate antecedente [var. recurvifolius Ruhl.]; pedunculis gracilibus, costatis, tortis, subflexuosis, 12 cm longis."
    The variety has been encountered on "campos cienagosas" at 200--
500 meters altitude, flowering in April and July. The Dupatya
fischeriana Kuntze (1902), referred to in the synonymy above, is a

synonym of Syngonanthus fischerianus (Bong.) Ruhl. and is based
on a misidentification by Kuntze of the type specimen in the Berlin herbarium. Steinbach describes the flowers of S. gracilis
var. bolivianus as "flor paposa blanca".

Citations: BOLIVIA: Santa Cruz: Kuntze 455 (B—type), s.n.
[Ost Velasco, VII.92] (N, N); Perrottet 766 (V—143602); J.
Steinbach 5507 (N, W—1472861).

SYNGONANTHUS GRACILIS var. GLABRIUSCULUS Ruhl. in Engl., Pflan-
    zenreich 13 (4-30): 251 [as "glabriuscula"]. 1903; Molden-
    ke, Known Geogr. Distrib. Erioc. 18 & 58. 1946.
    Synonymy: Eriocaulon glabrum Salzm. ex Steud., Syn. Pl. Glum.
2: [Cyp.] 281 & 334. 1855. Paepalanthus gracilis var. a Körn.
in Mart., Fl. Bras. 3 (1): 460—463. 1863. Eriocaulon glabrum
Steud. apud Jacks. in Hook. f. & Jacks., Ind. Kew., imp. 1, 1:
878, in syn. 1893. Limnoxeranthemum glabrum Salzm. ex Jacks. in
Hook. f. & Jacks., Ind. Kew., imp. 1, 2: 84. 1894. Syngonanthus
gracilis var. glabriuscula Ruhl. in Engl., Pflanzenreich 13 (4-
30): 241. 1903. Limnoxeranthemum (Eriocaulon) glabrum Salzm. ex
Ruhl. in Engl., Pflanzenreich 13 (4-30): 251, in syn. 1903. Paepalanthus gracilis var. α Körn. apud Ruhl. in Engl., Pflanzenreich 13 (4-30): 251, in syn. 1903. Paepalanthus glaber Körn. ex
Moldenke, Résumé Suppl. 1: 20, in syn. 1959. Syngonanthus gracilis var. glabriusculis Ruhl. ex Moldenke, Fifth Summ. 2: 962,
sphalm. 1971.
    Bibliography: Steud., Syn. Pl. Glum. 2: [Cyp.] 280, 281, &
334. 1855; Körn. in Mart., Fl. Bras. 3 (1): 460—463. 1863;
Jacks. in Hook. f. & Jacks., Ind. Kew., imp. 1, 1: 878 (1893) and
imp. 1, 2: 84. 1894; Ruhl. in Engl., Pflanzenreich 13 (4-30): 251,
285, 290, & 293. 1903; Jacks. in Hook. f. & Jacks., Ind. Kew.,
imp. 2, 1: 878 (1946) and imp. 2, 2: 84. 1946; Moldenke, Known
Geogr. Distrib. Erioc. 18, 35, 44, 49, & 58. 1946; Moldenke, Phytologia 2: 352 & 374. 1947; Moldenke, Alph. List Cit. 3: 710 &
975. 1949; Moldenke, Known Geogr. Distrib. Verbenac., [ed. 2],
65, 92, & 213. 1949; Moldenke in Maguire, Mem. N. Y. Bot. Gard.
9: 101. 1953; Moldenke, Phytologia 4: 318. 1953; Moldenke in J.
A. Steyerm., Fieldiana Bot. 28: 824. 1957; Moldenke, Résumé 73,
107, 288, 310, 325, 351, & 492. 1959; Moldenke, Résumé Suppl. 1:
20 (1959) and 2: 5. 1960; Jacks. in Hook. f. & Jacks., Ind. Kew.,
imp. 3, 1: 878 (1960) and imp. 3, 2: 84. 1960; Moldenke, Résumé
Suppl. 3: 14 (1962) and 12: 3. 1965; J. A. Steyerm., Act. Bot.
Venez. 1: 247. 1966; Moldenke, Résumé Suppl. 16: 6. 1968; Moldenke, Phytologia 18: 388. 1969; Moldenke, Fifth Summ. 1: 127, 133,
& 174 (1971) and 2: 501, 549, 583, 636, & 962. 1971; Angely, Fl.
Anal. & Fitogeogr. Est. S. Paulo, ed. 1, 6: 1162 & Ind. 28. 1972;
Moldenke, Phytologia 28: 440 (1974), 30: 52 (1975), 31: 383 &
408 (1975), 34: 487 (1976), and 35: 18 & 125. 1976.
    This variety is apparently based on the Paepalanthus gracilis
var. a of Körnicke (1863) and that, in turn, is presumably based

on Salzmann s.n. from "auf feuchten Wiesen" in Bahia, Brazil, and
P. Clausen 68 and Martius 1083 from Minas Gerais, Brazil, all deposited in the Berlin herbarium. Ruhland (1903) describes it as
follows: "Differt foliis irregulariter caespitosis, fere latius-
cule linearibus, chartaceo-membranaceis, pallide olivaceo- vel
glaucescenti-viridibus, obtusis, glabris, vel supra dense hirto-
puberulis, dein glabriusculis et albo-punctulatis, 1—1,5 cm longis, interdum 2/3 mm latis; vaginis arctis, leviter striatulis,
vix puberulis vel glabriusculis; pedunculis plerumque valde tortis, erectis, 3-costatis, in sulcis arcte appresso-incanis, ceter-
um sparse et longiuscule pilosis, mox glabris, 20 cm saepe exce-
dentibus; bracteis involucrantibus rotundato-obtusis, aureo-
flavidis, glabris, capitulum plus minus concavo-includentibus."
He cites from Bahia Salzmann s.n. and from Minas Gerais P. Clausen
68, Glaziou 11845 & 17308, Martius 1083, Sena s.n. [Herb. Schwacke
12828], and A. Silveira 1415.
    Recent collectors refer to this plant as a rosette herb "bear-
ing a single head", the inflorescences to 15 cm. tall, the flower-
heads grayish, and the flowers white. They have found it growing
in sandy soil, on wet campos, in moist meadows, on sandy dry or
white-sand savannas, and in Vellozia associations, at altitudes of
200—1980 meters, flowering from April to August and in November,
in fruit in May, August, and November. Irwin and his associates
found it "locally common in wet places on creek margins"; Maguire
and his associates found it "frequent on open banks" and "common
on savannas"; Murça Pires & Cavalcante refer to it as "frequent on
wet sandy savannas".
    Ruiz-Terán and López-Palacios describe it as "Hierba rosulada.
Hojas hasta 10 mm. de largo. Escapos hasta de 8 cm. de longitud,
cilíndricos. Capítulos globosos [or "hemisféricos"] 2—6 mm. de
diámetro. Flores blanquecinas". They have encountered it on "orillas de la carretera". Mori 829 is placed here only tentatively --
it is obviously very immature, the flower-heads are very small and
too pointed to be typical. It was found growing in "open barrows
of white sand with many lichens and Eriocaulaceae along ponds in
open sand."
    The Angely (1972) reference in the bibliography is often cited
as "1970", the title-page date, but was not actually issued until
1972.
    Material of this variety has been misidentified and distributed
in some herbaria under the names, Eriocaulon repens Lam., Paepalanthus sp., and Syngonanthus nitens Kunth. On the other hand,
Glaziou 15680, distributed as S. gracilis var. glabriusculus, is
actually var. aureus Ruhl. Phelps & Hitchcock s.n. [February 12,
1949] is a mixture with typical S. gracilis (Bong.) Ruhl., while
Hallé 512 is a mixture with something not eriocaulaceous.
    Additional citations: VENEZUELA: Amazonas: Foldats 3545 (Ve);
Phelps & Hitchcock s.n. [February 12, 1949], in part (N, N). Bolívar: Ruiz-Terán & López-Palacios 10979 (Mi), 11421 (Ld). SURINAM:
Maguire, Schulz, Soderstrom, & Holmgren 54216 (N). FRENCH GUIANA:

Hallé 512, in part (P). BRAZIL: Amapá: W. A. Egler 1453 [Herb.
Mus. Goeldi 24609] (Mi); Irwin & Westra 47259 (N); Murça Pires &
Cavalcante 52405 (N). Amazônas: Mori 829 (Ws). Goiás: Irwin,
Souza, & Reis dos Santos 9756 (N). Mato Grosso: Irwin, Grear,
Souza, & Reis dos Santos 15954 (Ld, N). Minas Gerais: P. Clausen
68 (B--cotype), 164 (B), s.n. (P); Martius 1083 (B--cotype, Mu--
cotype), s.n. [In uvidis altis herbaceis et turfosis prope Itambé]
(Mu). Paraíba: Coêlho de Moraes 2210 (Z), 2212a (Mm). Roraima:
G. A. Black 51-13126 (Be--70882); Prance, Forero, Pena, & Ramos
4490 (Ld, N, S). State undetermined: J. F. T. Müller 90 (P).
MOUNTED ILLUSTRATIONS: drawings & notes by Körnicke (B).

SYNGONANTHUS GRACILIS var. GRISEUS Ruhl. in Engl., Pflanzenreich
    13 (4-30): 251 [as "grisea"]. 1903; Moldenke, known Geogr.
    Distrib. Erioc. 18 & 58. 1946.
    Synonymy: Syngonanthus gracilis var. grisea Ruhl. in Engl.,
Pflanzenreich 13 (4-30): 251. 1903.
    Bibliography: Ruhl. in Engl., Pflanzenreich 13 (4-30): 251 &
293. 1903; Alv. Silv., Fl. Mont. 1: 418. 1928; Moldenke, Known
Geogr. Distrib. Erioc. 18 & 58. 1946; Moldenke, Known Geogr. Distrib. Verbenac., [ed. 2], 92 & 213. 1949; Angely, Fl. Paran. 10:
12 & 15. 1957; Moldenke, Résumé 107, 352, & 492. 1959; Angely, Fl.
Anal. Paran., ed. 1, 201. 1965; Moldenke, Fifth Summ. 1: 174
(1971) and 2: 636 & 962. 1971.
    This variety is based on three collections in the Berlin herb-
arium: Herb. Bernhardi s.n. from Bahia, Burchell 5764 from Minas
Gerais, and Schwacke 2483 from Santa Catarina, Brazil. The last-
mentioned is cited by Ruhland (1903) as from Paraná, but according
to a letter to me from Dr. Angely, dated December 3, 1957, the
locality of collection was, indeed, in the state of Paraná in 1903,
but in 1919 the boundaries of Paraná and Santa Catarina were officially changed and the locality in question is now definitely in
Santa Catarina.
    Ruhland's original (1903) description of the variety is: "Dif-
fert foliis linearibus, obtusiusculis, subtus plerumque plurinerv-
iis, glabris vel vilis longis hinc inde conspersis deinque glabriusculis, apice semper recurvatis; vaginis laxiusculis, folia
vix vel non superantibus, glabriusculis, striatulis; pedunculis
erectis, interdum acutangulo-3-costatis, pilis brevissimis sparse
instructis, cito omnino glabris, tortis, brunneo-fuscis; capitulis globosis, griseis, pallidis, densifloris; bracteis involucrantibus obovatis, obtusiusculis, glabris, concavis, pallidis."
Silveira (1928) cites A. Silveira 847 from Itacambira, Minas Gerais, collected in 1926.

SYNGONANTHUS GRACILIS var. HIRTELLUS (Steud.) Ruhl. in Engl., Pfl-
    anzenreich 13 (4-30): 249 [as "hirtella"]. 1903; Moldenke,
    Known Geogr. Distrib. Erioc. 18 & 58. 1946.
    Synonymy: Eriocaulon hirtellum Steud., Syn. Pl. Glum. 2: [Cyp.]

280. 1855. Limnoxeranthemum pubescens Salzm. ex Steud., Syn. Pl.
Glum. 2: [Cyp.] 280, in syn. 1855. Paepalanthus gracilis var. c
Körn. in Mart., Fl. Bras. 3 (1): 461, in part. 1863. Syngonan-
thus gracilis var. hirtella (Steud.) Ruhl. in Engl., Pflanzen-
reich 13 (4-30): 249. 1903. Eriocaulon hirtellus Steud. apud
Ruhl. in Engl., Pflanzenreich 13 (4-30): 249, in syn. 1903. Pae-
palanthus hirtellus var. ♀ Körn. ex Moldenke, Résumé Suppl. 1: 21,
in syn. 1959. Paepalanthus pohlianus var. ♀ Mart. ex Moldenke,
Résumé Suppl. 1: 22, in syn. 1959. Paepalanthus tristis Körn. ex
Moldenke, Résumé Suppl. 1: 22, in syn. 1959.

Bibliography: Steud., Syn. Pl. Glum. 2: [Cyp.] 280 & 334. 1855;
Körn. in Mart., Fl. Bras. 3 (1): 461. 1863; Jacks. in Hook. f. &
Jacks., Ind. Kew., imp. 1, 1: 878. 1893; Ruhl. in Engl., Pflanzen-
reich 13 (4-30): 249--251, 286, 290, & 293. 1903; Herzog in Fedde,
Repert. Spec. Nov. 29: 212. 1931; Jacks. in Hook. f. & Jacks., Ind.
Kew., imp. 2, 1: 878. 1946; Moldenke, Known Geogr. Distrib. Erioc.
18, 35, 49, & 58. 1946; Moldenke, Phytologia 2: 352 & 374. 1947;
Moldenke, Alph. List Cit. 3: 975. 1949; Moldenke, Known Geogr.
Distrib. Verbenac., [ed. 2], 65, 92, & 213. 1949; Moldenke, Phyto-
logia 4: 318. 1953; Moldenke in J. A. Steyerm., Fieldiana Bot. 28:
824. 1957; Moldenke, Résumé 73, 107, 289, 325, 352, & 492. 1959;
Moldenke, Résumé Suppl. 1: 21 & 22. 1959; Jacks. in Hook. f. &
Jacks., Ind. Kew., imp. 3, 1: 878. 1960; J. A. Steyerm., Act. Bot.
Venez. 1: 247. 1966; Angely, Fl. Anal. & Fitogeogr. Est. S. Paulo,
ed. 1, 6: 1162 & Ind. 28. 1972; Moldenke, Phytologia 30: 318
(1975), 31: 383 (1975), 34: 275 (1976), and 35: 338. 1977.

Steudel's original (1855) description of this taxon is: "Caespitosum subacaule pusillum; foliis angustissimis linearibus scab-
riusculis brevibus (vix 1/2" longis, 1/3''' latis); vaginis quam
folia duplo longioribus scapisque patenti glanduloso-pilosis;
scapis solitariis 1--4-pollicaribus; capitulo hemisphaerico glab-
ro; bracteis involucrantibus ovatis obtusis (piso parum majoribus);
flosculis ipsis basi nudis; receptaculo piloso. Limnoxeranthemum
pubescens Salzm. Bahia." Limnoxeranthemum pubescens Salzm. is regarded by Ruhland as a synonym of Syngonanthus gracilis var.
koernickeanus Ruhl., but it certainly has to go wherever the name,
Eriocaulon hirtellum Steud., goes and that is the name-bringing
synonym of Syngonanthus gracilis var. hirtellus (Steud.) Ruhl.

Ruhland (1903) cites no specimens for this variety, but implies
that it is very widely distributed in the states of Amazônas, Bahia, Goiás, Minas Gerais, Rio de Janeiro, and São Paulo, Brazil.
He comments that "Varietas latissime divulgata et habitu constan-
te est. Sed species non habenda est, quod indumentum foliorum et
vaginae valde variabilis transitum ad sequentum [vars. tenuissimus,
olivaceus, subinflatus, pallidus, amazonicus, koernickeanus, glabriusculus, aureus, griseus, setaceus, recurvifolius and bolivianus
Ruhl.] faciunt." He describes it as "Differt styli appendicibus
nullis; bracteis involucrantibus fuscescenti-flavidis, capitulum
vix includentibus; pedunculis apicem versus glandulifero-pilosis."

Recent collectors refer to the plant as having inflorescences to 8 cm. tall and the flower-heads yellow-brown. Davidse and his associates speak of it having "spikelets white", but there are no spikelets — the flowers are in heads. Collectors have encountered it on "campo cerrado", among rocks, in large swampy savannas, "in sand along streamlet at top of waterfall", and in wet places in rocky campo, at 125—1250 meters altitude, flowering from March to May and in July, August, and December, in fruit in March and December. Wurdack & Adderley refer to it as "occasional", while Anderson found it "on grassy campo with scattered trees on crystal sand, wet in places". Ruiz-Terán & López-Palacios describe the plant as "Hierba mínima, en suelo húmedo e musgoso, a la sombra de rocas de arenitica. Roseta de 7—10 mm. de largo. Escapos e-rectos, 5—6 cm." and found it growing on the "orillas de la carretera".

Paepalanthus hirtellus var. β is based on Widgren s.n. [1845] in the Berlin herbarium. Körnicke's P. gracilis var. c was based by him on Spruce s.n. from Amazônas, Salzmann s.n. from Bahia, Pohl s.n. and Weddell 2136 from Goiás, Gardner 2748 from Piaui, Gardner s.n., Martius s.n., and Widgren s.n. from Minas Gerais, Martius s.n. and Vauthier s.n. from Rio de Janeiro, and Riedel 2304 from São Paulo.

It should be noted here again that the Angely (1972) reference in the bibliography of this taxon is often cited as "1970", the title-page date, but the work was not actually issued until 1972. On the other hand, the Steyermark (1966) reference is sometimes cited as "1967", but actually was published in 1966.

Prance, Pennington, & Murça Pires 1283 & 1284 are mixtures with Paepalanthus polytrichoides Kunth and Syngonanthus bellus Moldenke.

Additional citations: VENEZUELA: Amazonas: Wurdack & Adderley 43697 (N, S). Bolívar: Davidse, Ramia, & Montes 4846 (Ld); Merx-müller 22955 (Mu); Ruiz-Terán & López-Palacios 11288 (Mi). BRAZIL: Goiás: W. R. Anderson 8065 (Ld, N); Irwin, Grear, Souza, & Reis dos Santos 13341 (N), 13388a (N). Maranhão: Murça Pires & Black 2251 (Ss), 2266 (Z). Minas Gerais: Widgren s.n. [1845] (B). Pará: Ducke s.n. [Herb. Mus. Goeldi 16257] (Bs); Murça Pires & Silva 4204 (N), 4718 (Ca--28212, N); E. Pereira 5109 [Herb. Brad. 12471] (Lw); Prance, Pennington, & Murça Pires 1283, in part (N), 1284, in part (N, S). State undetermined: G. Gardner s.n. (B). MOUNTED IL-LUSTRATIONS: drawings & notes by Körnicke (B, B, B).

SYNGONANTHUS GRACILIS var. KOERNICKEANUS Ruh. in Engl., Pflanzen-reich 13 (4-30): 250—251 [as "koernickeana"]. 1903; Moldenke, Known Geogr. Distrib. Erioc. 18 & 58. 1946.
Synonymy: Paepalanthus brizoides Kunth, Enum. Pl. 3: 534, in part. 1841. Eriocaulon brizoides Kunth ex D. Dietr., Syn. Pl. 5: 262. 1852. Eriocaulon brizoides (Kunth) Steud., Syn. Pl. Glum. 2: [Cyp.] 281. 1855; Moldenke, Résumé 286, in syn. 1959. Paepalanthus gracilis var. β Körn. in Mart., Fl. Bras. 3 (1): 460—463. 1863.

Paepalanthus gracilis var. β subvar. α Körn. in Mart., Fl. Bras.
3 (1): 460, 461, & 463. 1863. Paepalanthus glandulifer Mart. ex
Körn. in Mart., Fl. Bras. 3 (1): 464, in syn. 1863. Eriocaulon
brizoides Steud. apud Jacks. in Hook. f. & Jacks., Ind. Kew., imp.
1, 1: 877, in syn. 1893. Paepalanthus glanduliferus Mart. apud
Jacks. in Hook. f. & Jacks., Ind. Kew., imp. 1, 2: 402, in syn.
1894. Paepalanthus gracilis var. b subvar. α Körn. ex Ruhl. in
Engl., Pflanzenreich 13 (4-30): 250, in syn. 1903. Syngonanthus
gracilis var. koernickeana Ruhl. in Engl., Pflanzenreich 13 (4-30):
250. 1903. Paepalanthus gracilis var. b var. α Körn. ex Ruhl. in
Engl., Pflanzenreich 13 (4-30): 290, sphalm. 1903. Syngonanthus
eriophyllus var. glanduliferus Ruhl. ex Moldenke in Gleason & Killip, Brittonia 3: 159. 1939. Syngonanthus eriophyllus var. glandulifer Ruhl. ex Moldenke, Known Geogr. Distrib. Erioc. 57, in syn.
1946. Paepalanthus filiformis Mart. ex Moldenke, Résumé 325, in
syn. 1959. Paepalanthus filiformis var. minor Mart. ex Moldenke,
Résumé 325, in syn. 1959. Paepalanthus gracilis var. g Körn. ex
Moldenke, Résumé Suppl. 1: 20, in syn. 1959. Syngonanthus gracilis var. koernickieanus Ruhl. ex Moldenke, Fifth Summ. 2: 636, in
syn. 1971.

Bibliography: Kunth, Enum. Pl. 3: 534 & 624. 1841; D. Dietr.,
Syn. Pl. 5: 262. 1852; Steud., Syn. Pl. Glum. 2: [Cyp.] 281 & 333.
1855; Körn. in Mart., Fl. Bras. 3 (1): 460--464. 1863; Jacks. in
Hook. f. & Jacks., Ind. Kew., imp. 1, 2: 84, 401, & 402. 1894;
Ruhl. in Engl., Pflanzenreich 13 (4-30): 249--251, 285, 290, &
293. 1903; Alv. Silv., Fl. Mont. 1: 418. 1928; Moldenke in Gleason
& Killip, Brittonia 3: 159. 1939; Jacks. in Hook. f. & Jacks.,
Ind. Kew., imp. 2, 1: 877 (1946) and imp. 2, 2: 84, 401, & 402.
1946; Moldenke, Known Geogr. Distrib. Erioc. 7, 18, 33, 44, 45,
48, 49, 57, 58, & 61. 1946; Moldenke, Phytologia 2: 352, 373, &
377. 1947; Moldenke, Alph. List Cit. 4: 985. 1949; Moldenke, Known
Geogr. Distrib. Verbenac., [ed. 2], 92 & 213. 1949; Moldenke, Phytologia 4: 319. 1953; Moldenke, Résumé 69, 73, 77, 107, 286, 310,
323, 325, 351, 352, & 492. 1959; Moldenke, Résumé Suppl. 1: 5 &
20. 1959; Jacks. in Hook. f. & Jacks., Ind. Kew., imp. 3, 1: 877
(1960) and imp. 3, 2: 84, 401, & 402. 1960; Moldenke, Résumé Suppl.
3: 12 & 14 (1962) and 4: 4. 1962; Van Donselaar, Wentia 14: 70.
1965; Kramer & Van Donselaar, Meded. Bot. Mus. Herb. Rijksuniv.
Utrecht 309: opp. 500 & 509, tab. 1 & 2. 1968; Lindeman & Görts-
van Rijn in Pulle & Lanjouw, Fl. Surin. 1 [Meded. Konink. Inst.
Trop. 30, Afd. Trop. Prod. 11]: 336. 1968; Koyama & Oldenburger,
Rhodora 73: 159. 1971; Moldenke, Fifth Summ. 1: 120, 127, 131,
133, 134, & 174 (1971) and 2: 495, 549, 578, 582, 583, 636, & 962.
1971; Angely, Fl. Anal. & Fitogeogr. Est. S. Paulo, ed. 1, 6:
1162 & Ind. 28. 1972; Moldenke & Sm. in Reitz, Fl. Ilust. Catar.
I Erio: 78 & 101. 1976; Moldenke, Phytologia 34: 276 (1976) and
35: 291. 1977.

[to be continued]

BOOK REVIEWS

Alma L. Moldenke

"FACES OF THE WILDERNESS" by Harvey Brooms, xiii & 271 pp., illus.,
Mountain Press Publishing Co., Missoula, Montana 59801.
1972. $7.95.

This book carries a FOREWORD by William O. Douglas who rated
Harvey Broome as a "joyous companion" in backpacking and camping,
an "advocate extraordinary" for the preservation of unique
wilderness areas, and "rated along with Henry Thoreau and John
Muir" in outdoor literature. Harvey Broome was one of the
founders of The Wilderness Society in 1935 and one of its leaders
until his death in 1968. The Wilderness Society has sponsored
this publication.
The council of The Wilderness Society holds its annual meet-
ings not at its highly urbanized headquarters in Washington, D.C.,
but in places of wilderness concern such as the Okefinokee Swamp,
Grand Tetons, Bitterroot Forest, Olympics and Olympia Beach, or
the Great Smokies. The text comes from accounts of such trips
in the author's journal and is well garnished by a baker's dozen
of fine black/white photographs. In the Sun River country of
Montana with its "magnificent beauty and peace people like these
tiny figures around this campfire had fought for this beautiful
range. Others, in greater numbers, had burned our forests, had
denuded our plains, had extinguished noble species of game, had
muddied our rivers, and were sweeping the richness of our country
to the oceans and the Gulf....What the wilderness movement needs
to do more than anything else is to.....bring to people generally
......reverence for the natural scene."   Well worth reading!

"ECOLOGICAL CRISIS – Readings for Survival" edited by Glen A. &
Rhoda M. Love, ix & 342 pp., illus., Harcourt, Brace, &
Jovanovich, Inc., New York, N. Y. 10001. 1971. $5.95
paperbound.

This thoughtfully compelling collection of 22 papers authored
in direct and interesting language as contributions previously
published in about as many other journals and/or books certainly
should achieve the Loves' intent "to provoke discussion – and
action – toward an enhanced quality of life for all people on the
earth". The results of alternatively continuing to pollute our
air, our water, our land and ourselves cannot be dismissed as
"scare tactics" but as inexorably "eco-catastrophic".

"A FLORA OF TROPICAL FLORIDA — A Manual of the Seed Plants and
Ferns of Southern Peninsular Florida", 2nd Edition, by Robert
W. Long and Olga Lakela, xvii & 962 pp., illus., Banyan Books,
Miami, Florida 33143. 1976. $29.50.

Besides being a very well prepared, worthwhile and workable
manual, this book must also have been a highly successful seller
to need another printing within only a five year span!
In this new edition the pagination and illustrations are basically the same. On about 140 pages small text, key, nomenclature
and distribution revisions have been made. There has also been a
change in publishers from the University of Miami Press (itself
reorganized) to Banyan Books. The same "head count" of 1647 species along with 190 subspecific taxa in 762 genera in 179 plant
families has been maintained.
Botany students at area colleges and universities, many skilled
technicians in ecological and agricultural programs, botanically
trained tourists, and the increasing number of literate retiree-
migrants from the snowbelt to the sunbelt will find this book
helpful in distinguishing the individual floral members of this
fascinating tropical area of "scrub forests, hammock and tree islands, freshwater swamps, dry pineland, wet or low pineland, the
mangroves, salt marshes, wet prairies, dry prairies, coastal
strands and dunes, pond and river margins, marine communities, and
ruderal communities."
This book serves as an excellent monument to the research and
teaching careers of the authors.

"PLANTS OF THE TAMPA BAY AREA, Revised Edition (3rd) by Olga Lakela, Robert W. Long, Glenn Fleming & Pierre Genelle, 198
pp., illus., Banyan Books, Miami, Florida 33143. 1976.
$7.95 paperbound.

This contribution is No. 73 from the Botanical Laboratories,
University of South Florida at Tampa. The only illustration is
a needed page map of the area. "The purpose of this book is to
present a listing of the native, naturalized, and commonly cultivated plants of the area bordering Tampa Bay. For taxonomic
descriptions and diagnostic keys reference is made to J. K. Small's
"Manual" and to Long & Lakela's "Flora of Tropical Florida". In
all 1306 species in 582 genera in 167 families are included.
At the end of the book are corrections and late additions. The
authority for Verbena tenuisecta is Briquet; the "Vitex trifolia"
L., planted and escaping, is var. subtrisecta; the "Verbena officinalis" L. in wooded lots and berms is more likely to be V. halei
which is common in such areas. The European V. officinalis may be
grown in some gardens or occur on ballast, but is certainly not a
common escapee. Lantana aculeata is given as a synonym under L.
camara, but it is better treated as a valid variety, as is done by
the senior authors in their Manual. The same treatment should ap-

ply to L. odorata listed under L. involucrata. Avicennia germinans (L.) Stearn is an illegitimate name that must be relegated to
synonymy under A. germinans (L.) L.

This book has been and can continue to be of considerable use
in the area.

"GROWING FOOD IN SOUTH FLORIDA" by Felice Dickson, 128 pp., illus.,
    Banyan Books, Miami, Florida 33143. 1975. $5.95.

For 8 years the author has been the Farm and Garden Editor of
"The Miami Herald" and for even a decade earlier she has been an
enthusiast for tropical horticulture after having met our mutual
greatly admired friend, Dr. Edwin A. Menninger, the Flowering Tree
Man, of Stuart, Florida.

This easily readable, interesting, accurate, attractive, simple
book offers guidance for experienced or engenue, chemical or organic, large acreage or small plot, potted or hydroponic gardeners
wishing to raise appropriate temperate and newer tropical vegetable
crops. She pays tribute to the valuable directions available free
through trained county agents in the Cooperative Extension Service.

"WILD FLOWERS OF FLORIDA" by Glenn Fleming, Pierre Genelle & Robert
    W. Long, 96 pp., illus., Banyan Books, Inc., Miami, Florida.
    1976. $3.95 paperback.

This is a souvenir type book made particularly attractive by 156
fine legended color photographs taken mainly by the first two authors and made botanically valuable by checking with the last author.
On page 82, however, it is Verbena tenuisecta Briq. which is pic-
tured, not V. canadensis as there stated. In the common name and
scientific name listing on p. 15 Avicennia germinans should be
credited to "(L.) L." and Verbena canadensis to "(L.) Britton".

Out of the vast total of the Floridian flora those most commonly
encountered in fields and along roadsides have here been selected
and arranged in four color groupings — whitish, yellowish, pink-
reddish, and purple-bluish.

PHYTOLOGIA

Designed to expedite botanical publication

CONTENTS

Published by Harold N. Moldenke and Alma L Moldenke

303 Parkside Road
Plainfield, New Jersey 07060
U S A

Price of this number, $2, per volume, $9 75 in advance or $10 50 after close of the volume, 75 cents extra to all foreign addresses, 512 pages constitute a volume

PRELIMINARY TAXONOMIC STUDIES IN THE PALM GENUS ORBIGNYA MART.*

S. F. Glassman

Professor of Biological Sciences, University of Illinois, Chicago
Circle and Research Associate in Palms, Field Museum, Chicago.

Martius first erected the genus Orbignya in 1837. No species
were described in this article, but in 1844 he delineated
O. phalerata and O. humilis as the first two taxa. At later
dates, other species of Orbignya were described or transferred
from other genera (mainly Attalea) by Drude (1881), Barbosa
Rodrigues (1879,1888, 1891, 1898, 1903), Burret (1929, 1930,
1932, 1940), Bondar (1954) and several other authors.

Perhaps the most detailed treatment of Orbignya was by Burret
(1929). He recognized a total of 19 species and at the same
time divided the genus into three sections: Distichanthus
Burret, Pleiostichanthus Burret and Spirostachys Burret.   In
the first two sections the male flowers are arranged in two rows
along the rachillae of the male spadix (and in turn they are
distinguished from each other by whether the fibers in the
fruit endocarp are abundant or scarce to absent), whereas in the
section Spirostachys, the male flowers are spirally arranged
around the rachillae. Burret also presented a partial key to the
sections and species within each. Of 13 species partly keyed
out in the first section, eight are listed as unknown or doubtful;
of four listed in the second section, two are listed as doubtful;
and in the third section the two species are not keyed out and
one is listed as doubtful. In his "Palms of Brazil," Bondar
(1964) listed 14 species of Orbignya, some with brief descriptions,
but without keys.

As previously mentioned in Glassman (1977), Wessels Boer (1965,
1972) treated all species of Orbignya, as well as other closely
related genera in Surinam and Venezuela, as part of the genus
Attalea, sensu lata. Closely related genera to Orbignya (Attalea,
Scheelea, Maximiliana, Parascheelea and Markleya) have been
discussed and differentiated in Glassman (1977).

In preparing this study several facts became evident. In most
cases, type specimens for each species of Orbignya are either
fragmentary or nonexistent, very few additional collections
have been made for each species, and descriptive and illustrative

*This research has been supported by NSF grant BMS 75 09779.

information is usually inadequate.

The following is a description of the genus Orbignya as it is
presently delimited: tall trees mostly with smooth trunks and
inconspicuous leaf scars, or lacking trunks (acaulescent);
leaves usually very long, pinnately compound, leaf base conspicuous, petiole sometimes short, with fibrous margins; pinnae
single for the most part, but clustered in several taxa; plants
monoecious, flowers unisexual, both androgynous and male spathes
woody and deeply sulcate, usually terminating in a fairly long
umbo; androgynous spadices usually with many branches (rachillae),
each branch with few to several female flowers along basal part
forming triads with two male flowers, the terminal portion slender
with male flowers only; female flowers relatively large (2.0 to
4.5 cm long), subtended by two bracts, with 3 subequal or equal
convex imbricate sepals and 3 similar petals, pistil with a
well developed staminodial ring surrounding the ovary, carpels
3-several, fused, stigmas 3-6, style short or absent; male
spadices many branched, male flowers usually arranged on one
side of the rachillae, sometimes spirally arranged; male flowers
with 3 short sepals and 2-5 much longer flattened, curved,
obovate or ovate petals which are often fused and irregularly
notched, stamens 6-24 per flower, included in the petals,
thecas of anthers separate and divergent, irregularly coiled
and inrolled, fruits 1-several seeded, exocarp fibrous, mesocarp
usually pulpy and fibrous, endocarp stony, usually more than twice
as thick as exocarp and mesocarp combined, frequently dotted
with clusters of fibers, persistent perianth and staminodial
ring enlarged in fruit; seeds conforming to size and shape of
locules, endosperm homogeneous.

A total of 30 species has been described or transferred under
the name Orbignya. Of this number, 18 (including six synonyms)
definitely or most probably belong to Orbignya; one species,
O. dubia, is definitely not Orbignya; and the third category
(doubtful or uncertain taxa) encompasses five names. Also
included here are species closely related to Orbignya, but
probably belonging to different genera: Attalea crassispatha,
Markleya dahlgreniana and Parascheelea anchistropetala.

The following key, based on specimens examined plus descriptions
and illustrations, includes 21 taxa (18 species of Orbignya
and the three closely related species mentioned above). One
should be reminded, however, that this is a preliminary study
and that several species are based on incomplete collections
or in some cases only descriptions and illustrations. Only with
further collections can the full range of variability be determined; but in some cases this is not possible because the species

either has become extinct or its original habitat appears to have
been destroyed Subsequent to the species key, each of the four
categories of species mentioned above are arranged alphabetically
with the author and original place of publication. Sometimes,
other pertinent articles are also listed. Complete citations
of most of these plus other articles mentioned in the text are
listed under LITERATURE CITED at the end. Pertinent synonyms
are also listed. The type of each species, when known, is
listed and is then followed by a list of cited specimens examined
by the author. Holotypes, isotypes and lectotypes are specifically
listed as such; however, when its status is uncertain it is merely
called "type." For each specimen, collector's name and collecting
number is followed by a symbol of the herbarium where the collection
is deposited. Abbreviations of herbaria used here are those listed
in "Index Herbariorum" by Holmgren and Keuken (1974).

Key to Species of Orbignya and Related Genera

1. Middle pinnae mostly in clusters of 2-4

    2. Plants acaulescent or with short trunk, middle
        pinnae 15-56 cm long and 2.5-3 cm wide, stamens
        12-24 per flower

        3. Stamens 12-16

            4. Fruit 5 seeded, stamens 12-16 . . . . . <u>O</u>. <u>eichleri</u>

            4. Fruit 2-3 seeded, stamens 12 . . . . . <u>O</u>. <u>humilis</u>

        3. Stamens 16-24

            5. Female flowers 3-3.5 cm long, stamens
                16-18, branched part of androgynous
                spadix 15 cm long, bracts subtending
                female flowers not long acuminate
                . . . . <u>O</u>. <u>campestris</u>

            5. Female flowers 2 5 cm long, stamens
                16-24, branched part of androgynous
                spadix 40 cm long, bracts subtending
                female flowers long acuminate, 3-4 cm
                long . . . . <u>O</u>. <u>longibracteata</u>

    2. Plants with trunk 2-25 m high, middle pinnae
        80-130 cm long and 3.0-8.0 cm wide, stamens
        6-20 per flower

6. Fruits 7-9 cm long, 3-7 seeded

    7. Petals of male flowers lanceolate, broader
    below gradually narrowed above, stamens
    7-10 per flower, fruit 7-8 cm long, 3
    seeded, middle pinnae 7-8 cm wide
    . . . . <u>Markleya dahlgreniana</u>

    7. Petals of male flowers narrowed below,
    abruptly broadened above, stamens 20
    per flower, fruit 8-9 cm long, 3-7
    seeded, middle pinnae 4-5 cm wide
    . . . . <u>O. macrocarpa</u>

6. Fruits 3-6 cm long, 1-3 seeded

    8. Trunk creeping for several m., upright
    part 3-4 m high, fruit 5-6 cm long, middle
    pinnae 5-6 cm wide, stamens 6 or 9-12
    . . . . <u>O. spectabilis</u>

    8. Trunk 20-25 m high, fruit 3-4.5 cm long,
    middle pinnae 3-4 cm wide, stamens 9-11
    . . . . <u>Attalea crassispatha</u>

1. Middle pinnae not clustered, more or less evenly spaced

    9. Plants acaulescent or nearly so

    10. Middle pinnae 4-9 cm wide and 90-140 cm long

        11. Male flowers 10-13 mm long, spirally
        arranged around rachilla, stamens
        16-24 per flower, female flowers 3-4.5
        cm long . . . . <u>O. cuatrecasana</u>

        11. Male flowers 5-10 mm long, arranged on
        one side of rachilla, stamens 6-16 per
        flower, female flowers 1.5-2.5 cm long

            12. Petals of male flowers broader
            below and gradually narrowed above,
            stamens 6-8 per flower, fruits 6-7
            cm long. . . . . <u>Parascheelea anchistropetala</u>

            12. Petals of male flowers usually narrowed
            below and abruptly broadened above,
            stamens 11-16 per flower, fruits 3.5-
            4.5 cm long

13. Male flowers completely encircling
     rachillae of male spadix, stamens
     11-16 per flower, endocarp of
     fruit mostly without fibers
     . . . . <u>O. polysticha</u>

13. Male flowers in 3-5 rows on one
     side of each rachilla of male spadix,
     stamens about 12 per flower, fibers
     in fruit endocarp common
     . . . . <u>O. sagotii</u>

10. Middle pinnae 3-4 cm wide and 40-85 cm long

   14. Male flowers 9-10 mm long, stamens 22
        per flower . . . . <u>O. pixuna</u>

   14. Male flowers 11-14 mm long, stamens
        9-18 per flower

      15. Stamens 9-13 per flower, male
           rachillae 5-6.5 cm long, female flowers
           1.5-2 cm long and 1 cm in diam
           . . . . <u>O. sabulosa</u>

      15. Stamens 15-18 per flower, male
           rachillae 6-12 cm long, female
           flowers 3 cm long and 2 cm in diam
           . . . . <u>O. urbaniana</u>

9. Plants 2-20 m tall

   16. Middle pinnae 2.5-4 cm wide and 60-90 cm long,
        fruit about 7.5 cm long

      17. Stamens mostly 24 per flower, trees about
           20 m tall when full grown
           . . . . <u>O. phalerata</u>

      17. Stamens 18-20 per flower, trees up to
           8 m tall . . . .<u>O. teixeiriana</u>

   16. Middle pinnae 4-7 cm wide and 90-150 cm long,
        fruit 6-12 cm long

      18. Middle pinnae 4-5 cm wide, male flowers
           10-12 mm long, stamens mostly 20 per
           flower

         19. Plants about 10 m tall when full
              grown, fruits 6-9 cm long and 4-4.5
              cm in diam., middle pinnae about

110 cm long; male flowers completely
surround rachilla
. . . . O. guacuyule

19.   Plants 2-5 m tall, fruits about 9 cm
      long and 6.6 cm in diam, middle pinnae
      about 80 cm long; male flowers arranged
      on one side of the rachilla
      . . . . O. macrocarpa

18.   Middle pinnae 5-7 cm wide, male flowers
      13-15 mm long, stamens mostly 24 per flower

      20.   Trees about 20 m tall when mature,
            fruits 9-12 cm long with 3-6 seeds,
            middle pinnae about 150 cm long;
            male flowers arranged in two rows on
            one side of rachilla
            . . . . O. barbosiana

      20.   Trees about 6 m tall when mature, fruits
            7-8 cm long with one seed, middle pinnae
            about 120 cm long; male flowers com-
            pletely surround rachilla
            . . . . O. cohune

ORBIGNYA Mart. ex Endlicher, Gen. Pl. 257. 1837 (Conserved name).
    Orbignya Bertero, Mercurio Chil. 16: 737. 1829 (Euphorbiaceae).
    Type species: Orbignya phalerata Mart.

                        ORBIGNYA

               Alphabetical List of Species

O. barbosiana Burret, Notizbl. 11:690. 1932; H.E. Moore, Prin-
    cipes 7:155. 1963.
    Type: published as a new name for O. speciosa (Mart.)
    Barb. Rodr. mainly because of confusion with O. cohune
    (Mart.) Dahlgr. by Barbosa Rodrigues.
    Attalea speciosa Mart., Hist. Nat. Palm 2:138, t. 96,
    fig 3-6. 1826; Wessels Boer, Indig. Palms Suriname, 164-165.
    1965.
    Orbignya speciosa (Mart.) Barb. Rodr., Pl. Nov. Cult. Jard.
    Bot. Rio de Jan. 1:32, t. 9, fig Bl-9. 1891; Sert. Palm.
    Bras. 1·t. 52-53, 1903; Burret, Notizbl. 10:503-505, t.
    9. 1929.
    Type: Brazil, Maranhão and Pará (no specimens cited)
    O. lydiae Drude, Mart. Fl. Bras. 3:448, t. 102. 1881;
    Lindman, Bih. Sv. Vet. Akad. Handl. 26:fig 8. 1900; Dahlgren,
    pl. 341. 1959. Attalea lydiae (Drude) Barb. Rodr., Sert.
    Palm. Bras. 1:65. 1903.

Lectotype· Brazil, native to Pará, Cult. Rio de Janeiro
(Glaziou 9006-C). c.f. Dahlgren 1959, pl. 341.
O. martiana Barb. Rodr., Palm, Mattogross. 68, t. 22,
t. 23, fig 1-14. 1898.
Type: published as a new name for O. speciosa because
of incomplete descriptions (no flowers) by Martius and
uncertainty of its delimitation by subsequent authors;
in 1903, however, O. martiana was transferred back to
O. speciosa.
O. macropetala Burret, Notizbl. 10:507. 1929.
Holotype: British Guiana, Rupununi (Schomburgk s.n. - B).
O. oleifera Burret, Notizbl. 14:240. 1938; 15:103. 1940.
Holotype: Brazil, Minas Gerais, Pirapora (Burret 19-B).

Specimens examined: Brazil, without locality and collector
(F-614714), (F-614748); Pará, Tapajos, Kuhlmann 2203
(F-611585), Capucho 537 (F); São Luiz, Dahlgren s.n.
(F-615321); Mujuhy dos Campos, near Santarem, Dahlgren s.n.
(F-615318); Ceara, Serra de Baturite, Dahlgren s.n.
(F-613570); Pacoty, Dahlgren s.n. (F-619725); Mato Grosso,
region of Rio Machado, Angustura, Krukoff 1600 (F-620732);
Minas Gerais, Pirapora, Burret 19 (B, holotype of O. oleifera;
Burret 19 & Brade-RB). Surinam, Palaime Kreek, 20 km. W.
of Sipaliwini, Wessels Boer 806 (U); Coeroeni R., in
subhydrophytic forest, Wessels Boer 1588 (U). British
Guiana, Rupununi, Schomburgk s.n. (B, holotype of O. macro-
petala). Cultivated, Brazil, Belem, prop. Alvaro Alfredo,
Dahlgren s.n. (F-615317); British Guiana, Georgetown Bot.
Garden, L.H. Bailey 509 (BH), Dahlgren s.n. (F-610806);
Brazil, Rio de Janeiro, Passeio Publico, Glaziou 9006 (C,
lectotype of O. lydiae; NY, P).
Vernacular names: Babassu, Babaçu, Uaussu, Baguaçu, Guaguaçu.
Distribution: Brazil, Maranhão, Pará, Minas Gerais, Mato
Grosso; Surinam and British Guiana.

Orbignya barbosiana was published as a new name by Burret (1932)
for O. speciosa because the latter species was confused with
O. cohune' (a Central American palm) by Barbosa Rodrigues (p. 32,
t. 9, fig 1-9. 1891; p. 16, t. 5B, 1896) and subsequent authors.
Moore (1963) was one of the first authors to recognize O. barbosiana
as the valid name, whereas Wessels Boer (1965) considered it to
be a superfluous name. There is a fundamental difference between
the two species in question: O. barbosiana has male flowers on
one side of the rachilla, while male flowers surround the rachilla
in O. cohune.

Barbosa Rodrigues (1898) published O. martiana as a new name for
Attalea speciosa Mart. because the latter was based on inadequate

descriptions as well as incomplete collections. Neither Martius
(1826, 1844, 1845, 1853), Wallace (1853), Spruce (1871), Trail
(1876), nor Drude (1881) described or collected flowers from
this species; but in 1898 Barbosa Rodrigues fully described and
illustrated specimens he had personally collected. At the time
the following distributional information was also given: "Brazil;
equatorial and oriental, in silvis Rio Arinos, serra dos Parecis,
Rosario, Rio Cuiyaba, S. Miguel das Areias, Tombador, in Mato
Grosso. Also in woods near Rios Tapajos, Madeira, Purus, near
upper Rio Amazonas; cultivated in Jardim Botanico Rio, no. 1398.
Extends from the Guianas to the forests of Amazonas entering
Mato Grosso (forming large forests) and continuing into Bolivia."
After some deliberation, Barbosa Rodrigues (1903) decided to
transfer O. martiana back to his original combination of O. speciosa.

Barbosa Rodrigues (1898) also considered O. lydiae Drude to be
conspecific with O. martiana because Drude (1881) had described
this species from incomplete collections as well as including
incorrect information on its morphology. Furthermore, Drude
could not adequately compare his material with Attalea speciosa
since it was incompletely known at the time. After Drude published
his article, Barbosa Rodrigues made complete collections (unfortunately none of these specimens has been located) from the
original tree in Passeio Publico, Rio de Janeiro, and after
studying these specimens decided that O. lydiae was synonymous
with the "Baguacu" of Mato Grosso and "Uauassu" of Amazonas.
He also noted that the plant described as "acaulous" had actually
grown into a large tree. In spite of his discussion in 1898,
Barbosa Rodrigues (1903) apparently still recognized O. lydiae
as a distinct species because he transferred it to the genus
Attalea. Lindman (1900) illustrated this species in a palm
forest, with the caption: "Oauassu," the largest and most
beautiful palm in Mato Grosso. Moore (1963) says that this
taxon is incompletely known, however, I am tentatively treating
O. lydiae as a synonym of O. barbosiana because a comparison of
the two species reveals many similarities.

Burret (1938) described O. oleifera as a new species to distinguish
it from O. barbosiana, and further stated this was the Babassu
palm from whose seed oil is extracted. He said he inadvertently
included O. oleifera under his discussion of O. martiana in
1929 when he referred to t. 53, fig. 23-25 (1903) of Barbosa
Rodrigues which is the same as t. 22 (1898) of the same author.
The remaining parts of these plates (t. 53, fig 13-22, 1903;
and t. 23A, 1898) pertain to O. barbosiana. Unfortunately,
Burret did not indicate any significant differences between the
two taxa in either article (1938, 1940); his description of
O. oleifera is rather sketchy (e.g., size of pinnae, male and
female flowers and spadices are lacking, as well as size of the

tree); and type specimens consist only of leaf material. Burret
(1940) also cites Hopp 3013 (B) from Mato Grosso, but I was
unable to locate this specimen. Also there is no information
on the distribution range of this species except for that given
in the cited specimens.

O. campestris Barb. Rodr., Palm. Mattogross. 78. t. 25. 1898,
    t. 50B. 1903.
    Lectotype: Brazil, Mato Grosso, Capão Bonito (t. 25, 1898).
    c.f. Classman 1972, p. 178.
    Vernacular names: Indaya verdadeiro, Indaya redondo.
    Distribution: Brazil, described from the state of Mato Grosso.

Barbosa Rodrigues (1898) lists B.R. 240 under this species,
but no specimens have been located. Therefore, the above
lectotype was designated. Even though authentic specimens have
not been examined, this taxon seems to be distinct based on its
description.

During early September, 1976, I visited Capão Bonito, presumably
the type locality of this species and apparently that of O. longibracteata Barb. Rodr. and O. macrocarpa Barb. Rodr., as well.
This locality is between Sidrolandia and Maracaju within the
boundaries of Fazenda Santa Luzia. It is a heavily wooded area
surrounded on all sides by agricultural land. Although several
kinds of palms grew in the region in the past, none are found
there today (with the exception of scattered specimens of a short
species of Butia). This is another sad example of destruction
of palm habitats by the rapid spread of agriculture in the state
of Mato Grosso.

O. cohune (Mart.) Dahlgren ex Standley, Trop. Woods 30:3. 1932;
    Burret, Notizbl. 11:688. 1932; Standley & Steyermark,
    fig. 46, 1958. Attalea cohune Mart., Palmet. Orbign. 121.
    1844; t. 167. 1845.
    Lectotype: Honduras (Martius t. 167. 1845) c.f. Classman
    1972, p. 23.
    O. Dammeriana Barb. Rodr., Sert. Palm. Bras. 1:62, t. 54.
    1903.
    Lectotype: Brazil, cult. Jard. Bot. Rio (Glaziou 16468-B).

Specimens examined: British Honduras, Punta Gorda, H.W.
Turner s.n. (F); without locality, J.B. Kinloch s.n. (F);
Stann Creek Valley, P.H. Geortle 3234 (B-photo of male spadix).
Honduras, Puerto Sierra, P. Wilson 472 (F); Dept. Atlantida,
Lancetilla Valley, near Tela, wet forest, Standley 53981

(F); vicinity of Lancetilla, forests, T.G. Yuncker 4970 (F).
Guatemala, Dept. Izabal, between Virginia & Lago Izabal,
Steyermark 38771 (F); between Bananera & La Presa, Steyermark
39182, 39210 (F); Dept. Alta Verapaz, woods S.E. of Finca
Yalpemech, Steyermark 45211, 45693 (F). Cultivated, Cuba,
Soledad, Atkins Gardens, Dahlgren 4619 (F), British Guiana,
Georgetown Botanical Gardens, Dahlgren s.n. (F-610577,
610649, 610772, 610697). Brazil, Jard. Bot. Rio, Glaziou
16468 (B, lectotype of O. dammeriana; C. MO, P).
Vernacular Names: Cohune - Honduras. Manaca, Corozo -
Guatemala.
Distribution: British Honduras, El Salvador, Honduras,
Guatemala and? Southern Mexico (Quintana Roo & Campeche).

No specimens were cited by Martius (1844) in his original description nor could any herbarium material be found in Munich;
hence, the selection of t. 167 as the lectotype.

Barbosa Rodrigues (1903) did not cite any specimens for Orbignya
dammeriana, however, Burret (1929) said that Glaziou 16468 (B),
erroneously cited as 16488, probably came from the "type tree"
in Jardim Botanico, Rio de Janeiro. The error of citing Glaziou
16488 was perpetuated by both Dahlgren (1936) and Glassman
(1972). There is no conclusive proof that Glaziou 16468 (B)
actually came from the "type tree," nevertheless I have chosen
it as the lectotype rather than an illustration of the plant.
According to Burret (1929), O. dammeriana was originally included
under O. speciosa (Mart.) Barb. Rodr. when it was transferred
from Attalea to Orbignya by Barbosa Rodrigues (t. 9, fig. 1-9,
1891).

According to Moore (1960), pls. 336-338, listed as O. cohune
by Dahlgren (1959), are actually O. guacuyule. Both species
were previously thought to be synonymous, but were differentiated
by Hernandez Xolocotzi (1949).

O. cuatrecasana Dugand, Caldasia 2:285, fig. p. 286. 1943;
    Cuatrecasas, pl. 2, fig. 2. 1947.
    Holotype: Colombia, Dept. del Valle, Rio Naya (Cuatrecasas
    13980-COL).

    Specimens examined: Colombia, Dept. del Valle, Rio Naya,
    Puerto Merizalde, bosque, Cuatrecasas 13980 (COL, holotype;
    F, isotype); alredores de Puerto Merizalde, I. Barreto &
    L.A. Kairuz s.n. (COL); Rio Calima (Choco region), La
    Trojita, Cuatrecasas 16389 (F); Rio Calima Quebrada de la
    Brea, R.E. Schultes & M. Villareal 7373 (GH); Aqua Dulce,
    Buenaventura, O.F. Cook 81 (US).

Vernacular names· Palma Corozo, Taparo Grande.
Distribution: Endemic to Colombia in forested areas along
Pacific Coast.

This species is the only Orbignya known from Colombia. It is
distinct in being acaulescent, with unclustered pinnae up to
9 cm wide, male flowers spirally arranged around the rachilla,
and stamens 16-24 per flower. Dugand placed it in section
Spirostachys of Burret (1929) characterized by having male flowers
spirally arranged around the rachillae rather than on one side.

O. eichleri Drude, Mart. Fl. Bras. 3:449, t. 103. 1881.
    Lectotype: Brazil, Goias, Sertão d'Amaroleite (Weddell
    2705 - P); c.f. Dahlgren pl. 339. 1959 (excluding leaf).

    Specimens examined: Brazil central (Goias), Sertão d'Amaro-
    leite, Weddell 2705 (P, lectotype - excluding leaf); Maranhão,
    Caxias, Bondar s.n. (F, RB-80812); Maranhão, Ilha dos
    Botes, J. Murça Pires & G.A. Block 1575a (NY).
    Vernacular names: Piassava, Piassaveira, Pindoba.
    Distribution: Native to Brazil in states of Goias and
    Maranhão, and probably Piauhy.

In 1881, Drude cited both Gardner 2755 from Piauhy and Weddell
2705; however, only Weddell 2705 (P) has been located, which
has the following inscription: "Original at Kew." Since Weddell
2705 (P), consisting of a male spadix and whole leaf, is the
only specimen found among those cited by Drude, it has been chosen
as the lectotype. The leaf should be excluded from the type,
however, because it is certainly not an Orbignya, but most
probably is Syagrus flexuosa (Mart.) Becc.

Burret (1929) cited Snethlage 648 (B) from Piauhy under O. eichleri, but after examining this specimen I could not be sure of
its identity because it lacks male flowers and middle pinnae.

Bondar (1954) keyed out three closely related Brazilian species
of Orbignya: O. speciosa, O. teixeirana and O. eichleri. One
of the characteristics he used to distinguish O. speciosa from
the other two taxa was male flowers completely surrounding
rachilla rather than arranged on one side of rachilla. Apparently
Bondar perpetuated the error of confusing O. speciosa (= O. barbosiana) with O. cohune because, in fact, the former species has
male flowers on one side of the rachilla, whereas in O. cohune
they surround the rachilla.

O. guacuyule (Liebm. ex Mart.). Hernandez X, Bol. Soc. Bot.
 _ Mex. 9:17, 1949; Dahlgren, pl. 336-338, 1959. Cocos guacuyule
    Liebm. ex Mart., Hist. Nat. Palm. 3:323. 1853.
    Lectotype: Mexico, Oaxaca, pr. Guatulco (Liebmann 6559-C);
    c.f. Dahlgren, pl. 338. 1959.
    Cocos Cocoyule Mart., Hist. Nat. Palm. 3:324. 1853.
    Lectotype: Mexico, Acapulco (Karwinski s.n.-M).

    Specimens examined: Mexico, Dept. Oaxaca, pr. Guatulco,
    Liebmann 6559 (C, lectotype of Cocos guacuyule); Oaxaca,
    San Benito, 50-60 m tall, B.P. Reno 3462 (US, photo);
    Acapulco, Karwinski s.n. (M, lectotype of C. cocoyule);
    Rio Verde, Pinotepa a Puerto Escondido, deciduous forest,
    T.D. Pennington & J. Sarukhan K. 9488 (NY); Guerrero, near
    El Papayo, H.E. Moore & E. Valiente 6199 (BH); State of
    Nayarit, rich woods outside San Blas, H.E. Moore & V. Cetto
    6405 (BH); Colima, road to Manzanillo, H.E. Moore 8166 (BH).
    Vernacular names: None recorded, but the specific epithet
    guacuyule was probably based on a local native name.
    Distribution: Native to Mexico in the states of Oaxaca,
    Guerrero, Michoacan, Colima, Jalisco and Nayarit.

No specimens were cited in Martius (1853), hence lectotypes
were selected for both C. guacuyule and C. cocoyule.

Many authors considered this taxon to be conspecific with
O. cohune (including Dahlgren, pl. 336-338, 1959), but according
to Hernandez X (1949), they are distinct species with an essentially allopatric distribution. The latter author distinguishes
O. guacuyule from O. cohune mainly by the male flowers having
spatulate, acuminate petals 1.2 cm long and 0.5 cm wide, rather
than oblanceolate, cuspidate petals 1.5 cm long and 0.7-0.9 cm
wide. In both species, however, male flowers completely
surround the rachilla rather than being distributed on one
side of the rachilla, characteristic of most species of Orbignya.

O. humilis Mart., Palmet. Orbign. 129, t. 10-2, t. 32. 1844;
 _ t. 169-1, 1845; t. Z16-3, 1849.
    Type: Bolivia, Prov. Chiquitos, prope Mission S. Anna de
    los Chiquitos, sandy soil (d'Orbigny 22-P, destroyed?).

    Specimens examined: Doubtful, Bolivia, Velasco, Otto Kuntze
    s.n. (NY, US).

Unfortunately, no type material has been located in the herbarium
at Paris. This taxon appears to be closely related to Orbignya
eichleri Drude because descriptions and illustrations of the two

species are similar. Both taxa, however, are incompletely known
(especially information on middle pinnae is lacking); therefore
they cannot be adequately differentiated.

In the above cited specimens (Kuntze s.n.) male and female
flowers seem to match those of illustrations of O. humilis,
but most of the leaf material from (NY) is probably Syagrus
flexuosa rather than Orbignya.

O. longibracteata Barb. Rodr., Palm. Mattogross. 79, t. 26,
    1898, t. 51, 1903, Burret, Notizbl. 15:103. 1940.
    Lectotype: Brazil, Mato Grosso, Capão Bonito, fere Serra
    do Melgaço (t. 51, 1903).

    Specimens examined: Doubtful. Brazil, Mato Grosso, Hopp
    3002 (B) - leaf part and two photos.
    Vernacular names: Indaya mirim, Indaya crespo, Inaja.
    Distribution: Described from Brazil in the state of Mato
    Grosso.

Barbosa Rodrigues (1898) cited B.R. 239 for this taxon, but
no specimens have been located; hence, the selection of t. 51,
1903, as lectotype.

Burret (1940) cited Hopp 3002 with the following information:
140 S. Lat., characteristic palm of the dry forest steppe of Mato
Grosso. I have seen this specimen which is presently represented
only by a leaf part and two photos, one of a living plant growing
out of rocks, and the other photo of a herbarium specimen (probably destroyed) with a leaf part, fruit and androgynous spathe.

Apparently, this species was described as new because of its
long acuminate bracts subtending female flowers; however,
O. macrocarpa also has long bracts, whereas the bracts of
O. campestris are described as "magna minuto." When descriptions
of the three taxa mentioned above are compared, they appear
to be very similar. It is therefore possible that they may be
conspecific, especially since the type locality of all three
is listed as Capão Bonito.

The slight differences between them may be merely due to insufficient information. So far, no authentic specimens have been
seen for any of the species.

O. macrocarpa Barb. Rodr., Palm. Mattogross. 74, t. 23-24B.
   1898; t. 50A. 1903.
   Lectotype: Brazil, Mato Grosso, Capão Bonito prope Serra
   Quebra Cabeça (t. 23-24B, 1898). c.f. Glassman 1972, p. 172.

   Specimens examined: Doubtful. Brazil, Mato Grosso, Fluss-
   gebiet des Amazonas, stemless, W. Hopp 3011-B, destroyed;
   F, photo).
   Vernacular names: Indaya - assu.

Barbosa Rodrigues (1898) cited B.R. 217, but no specimens were
located necessitating the selection of a lectotype from an
illustration (see above).

The specimens cited above (Hopp 3011) consisted of a leaf part
and an androgynous spathe, but it is difficult to determine with
certainty the photograph of this specimen.

As previously mentioned, this taxon may be synonymous with
O. campestris and O. longibracteata because of similar type
locality and similar morphology. Although all three taxa are
recorded from Capão Bonito, Serra do Melgaço is also listed for
O. longibracteata and Serra Quebra Cabeça is mentioned for
O. macrocarpa as well. Since the two additional locales could
not be found on any maps examined, it is not certain if all three
place names recorded from Capão Bonito are in the same general
vicinity of each other or are actually three different, isolated
localities.

O. phalerata Mart., Palmet. Orbign. 126, t. 13-2, 32A. 1844;
   t. 170, 1845; Karsten & Schenck, t. 35-36. 1910.
   Holotype: Bolivia, 12-16° S. lat. in north. part of prov.
   of Chiquitos and in Moxos. Forms immense forests of pure
   stands in the land of the Guarayos covering about 10 square
   miles (d'Orbigny 20-P).

   Specimens examined: Bolivia, Chiquitos, d'Orbigny 20 (P,
   holotype; F, M, isotypes).
   Vernacular name: "Cusi."
   Distribution: Bolivia, in sandy, wet but not flooded soils.

This taxon is the first one described in the genus Orbignya;
hence, it is the type species.

Unfortunately, I have not seen any material referable to this
species except the type collections. The specimen from Paris
consists of an androgynous spadix with sterile (immature?) female

flowers closely matching t. 32 and t. 170 of Martius. The
isotype from Munich contains only male flowers. No leaf
material has been seen by me nor is any illustrated by Martius.

Karsten & Schenck (1910) show two photos of this palm growing
in forested areas. Plate 35 was taken in Velasco. Previously,
I cited a doubtful specimen, O. Kuntze s.n. (NY, US) from Velasco under O. humilis, the only other species of Orbignya
described from Bolivia. According to descriptions and illustrations,
the two species appear to be distinct (the type specimen of
O. humilis could not be found). But since there is a paucity of
herbarium specimens, especially leaf material, it would be difficult
to carefully compare and contrast both taxa.

Burret (1929) referred to O. phalerata, but most of the information
was repeated from Martius's original description.

Martius (1844) mentioned that the seeds yield an excellent oil
for burning and for the hair, and that the leaves make a good
thatch for roofs.

O. pixuna (Barb. Rodr.) Barb. Rodr., Prot. App. 49. 1879; t. 49.
     1903. Attalea pixuna Barb. Rodr., Enum. Palm. Nov. 43. 1875.
     Lectotype: Brazil, Para, calcareous soils of l'Igarape Bom
     Jardim, villa de Itaituba, basin of Rio Tapajos (t. 49,
     1903). c.f. Glassman 1972, p. 26.
     Attalea spectabilis var. polyandra Drude, Mart. Fl. Bras.
     3:440. 1881.
     Type: Brazil, Rio Purus (Wallis s.n. - K, destroyed?).

     Specimens examined: Doubtful. Brazil, Pará, Boa Vista,
     Tapajoz, Capucho 523 (F); Tapajoz, Kuhlmann s.n. (F-611560).
     Vernacular names: Curua-pixuna, Palha preta.
     Distribution: Brazil, in state of Para.

No specimens were cited by Barbosa Rodrigues in any of his articles,
hence the selection of t. 49 as the lectotype.

This species appears to be distinct according to its description
and illustration, however, no authentic specimens have been examined.
The two specimens cited above consist of fruits only and hence
cannot be determined with certainty.

O. polysticha Burret, Notizbl. 11:324. 1932.
     Holotype: Peru, Dept. Loreto, Mishuyacu near Iquitos

(G. <u>Klug</u> <u>205</u> - US).

Specimens examined: Peru: Dept. Loreto, Mishuyacu, near
Iquitos, forest, <u>G</u>. <u>Klug</u> <u>205</u> (US, holotype; F); Loreto,
Santa Rosa, lower Rio Huallaga, <u>E</u>.<u>P</u>. <u>Killip</u> & <u>A</u>.<u>C</u>. <u>Smith</u>
<u>28814</u> (NY, US). Venezuela, Terr. Fed. Amazonas, Rio Orinoco,
San Pedro, sabanita and bosque, <u>G</u>.<u>S</u>. <u>Bunting</u>, <u>L</u>.<u>M</u>. <u>Akkermans</u>
& <u>J</u>. <u>van</u> <u>Rooden</u> <u>3571</u> (U); Brazo Casiquiare, near Solano,
tropical rainforest, on white sandy soil, <u>Wessels</u> <u>Boer</u>
<u>2409</u> (U); near San Carlos de Rio Negro, white sandy soil,
<u>Wessels</u> <u>Boer</u> <u>2273</u> (U).
Vernacular Names: Catirina (Peru), Mavaco (Venezuela).
Distribution: Peru and Venezuela, mostly in Amazon region
in tropical rainforests.

Burret (1932) also cited <u>Killip</u> & <u>Smith</u> <u>28814</u> (see above) as
probably belonging to this species. He also placed <u>O</u>. <u>polysticha</u>
in group <u>Spirostachys</u> where the male flowers completely surrounded
the rachillae of the male spadix. In this respect, <u>O</u>. <u>polysticha</u>
is similar to <u>O</u>. <u>cuatrecasana</u> from Colombia.

Wessels Boer (1972) listed this species as <u>Attalea</u> <u>polysticha</u>
from Venezuela, but the name is invalid because the basionym
(<u>O</u>. <u>polysticha</u>) was not mentioned in the article.

<u>O</u>. <u>sabulosa</u> Barb. Rodr., Vellosia <u>1</u> ed. <u>1</u>:54. 1888; t. 48. 1903;
      Burret, 510. 1929.
      Lectotype: Brazil, Prov. Amazonas, in sandy pastures near
      Rio Tarumauacu, in Rio Negro (t. 48, 1903). c.f. Glassman
      1972, p. 180.

      Specimens examined: <u>Doubtful</u>. Brazil, Amazonas, Manaos,
      Rio Negro, <u>Huebner</u> <u>s.n.</u>, <u>Huebner</u> <u>4a</u>, <u>Huebner</u> <u>100</u>, <u>100a</u>,
      <u>100x</u> (B); Amazonas, basin of Rio Negro - Rio Cuieras,
      savanna forest on sand, <u>Prance</u>, <u>Coelho</u> & <u>Monteiro</u> <u>14830</u> (NY).
      Vernacular names: Curua, Inaya, Pindova
      Distribution: Brazil, Amazon region, in savannas on sandy soil.

No specimens were cited by Barbosa Rodrigues in any of his articles,
hence an illustration (t. 48) was chosen as the lectotype.

Burret (1929) cites <u>Huebner</u> <u>74</u>, <u>74a</u> (4, 4a?), <u>100</u>, <u>100a</u> under
this taxon. I have examined some of these specimens (see above),
but I cannot be certain of their identity because the collections
are incomplete, for the most part. To complicate matters,
Wessels Boer (1965) claims that <u>Huebner</u> <u>4</u> and <u>Huebner</u> <u>100</u> (B) are
actually <u>O</u>. <u>sagotii</u> Trail, a closely related species, because
there was a discrepancy between these specimens and the original

description of O. sabulosa.

Prints of two photographs of O. sabulosa taken by George Huebner
in Manaos in 1935 are deposited in the Field Museum Herbarium.
One photo (460061) shows a whole stand of acaulescent plants
while the other is a closeup of a male spadix and an infructescence
emerging between the leaves at the base of the plant.

O. sagotii Trail ex Im Thurn, Timehari 3:276. 1884. Attalea
    sagotii (Trail ex Im Thurn) Wessels Boer, Indig. Palms
    Suriname 162. 1965.
    Lectotype: French Guiana, Karouany (Sagot 831 - K. Erron-
    eoulsy inscribed on herbarium sheet as 631). c.f. Wessels
    Boer 1965, p. 162.

    Specimens examined: French Guiana, Karouany (Sagot 831-K,
    lectotype; P, isolectotype); Sagot 601 (K, P). Surinam,
    without locality, Wessels Boer 165, 708, 1440, 1493 (U);
    Lindeman 6902 (U); vicinity of Zanderij, wet forest on silt
    loam, Wessels Boer 276 (U); Dist. Brokopondo, high forest,
    Wessels Boer 392 (U); Bakhuis Mts., P.A. Florschutz & P.J.M.
    Maas 2960 (U).
    Vernacular names: Macoupi, Bergi-Maripa, Koeroea.
    Distribution: French Guiana, Surinam and British Guiana.

No specimens were cited by Trail (1884) in his original article;
therefore, a lectotype was chosen by Wessels Boer (1965) in his
book on Surinam palms. Both Sagot 831 (K) and Sagot 601 (K) were
annotated by Trail in 1877 as "O. sagotii n. sp." In addition
to this, Drude (1881) incorrectly cited the above numbers under
Attalea spectabilis var. monosperma.

Wessels Boer (1965) states that this species is apparently close
to O. sabulosa and O. agrestis, both much smaller palms.

O. spectabilis (Mart.) Burret, Notizbl. 10:508. 1929. Attalea
    spectabilis Mart. Hist. Nat. Palm. 2:136. t. 96. fig. 1-2.
    1826; t. Z16. 1849; Wessels Boer, 1965.
    Type: Brazil, Prov. Para, Serra de Baru, near Para and Rio
    Negro (Martius s.n. - M, not seen). c.f. Burret 1929, p. 508.

    Specimens examined: Doubtful. British Guiana, Cult.
    Georgetown Bot. Garden, Dahlgren s.n. (F-610583), Dahlgren
    & Millar s.n. (F-610759), L.H. Bailey 489 (BH). French
    Guiana, Macoupi, Gourdonville, R. Benoist 1707 (P). Surinam,

Rechter Coppename River, on riparian bank, Wessels Boer
1365 (BH,U), near Tafelburg, sandstone rocks, in submesophytic
forest, Wessels Boer 1503 (BH, U).   Brazil, Amazon, Spruce
32 (K); state of Amazonas, Manaus - Itacoatiara Highway,
Reserva Florestal Ducke, forest, G.T. Prance et al 2155 (NY);
Para, Monte Alegre, Krukoff 36 (F-614554).
Vernacular names: Curua piranga, Piuna inquira, Pindoba
das Mattas.
Distribution: Surinam, French Guiana and the Amazon region
of Brazil in wet forests.

No specimens were cited by Martius in any of his articles. According
to Burret (1929), however, he saw a specimen of a rachilla branch
with female flowers collected by Martius and labelled A. spectabilis
in the herbarium at Munich. But Burret said the collection was
actually Orbignya agrestis and noted that it was not determined
by Martius. Unfortunately, neither Wessels Boer (1972) nor I
have been able to find this particular specimen.

Burret also discusses the incomplete and sometimes confusing
description of Martius (1826). He wondered if Martius was
describing two species, especially in the size - "acaulescent
to several feet tall" - and in the number of stamens - "6 as well
as 9-12." It was surprising that Martius did not describe the
stamens in detail (i.e., if the anthers were coiled or straight),
but he may have had sterile flowers and did not realize he was
dealing with a different genus (Orbignya) which he later erected
in 1837. Burret (1929) also mentions that Drude (1881) confused
this taxon with Maximiliana attaleoides.

I am still not certain of the exact delimitation of O. spectabilis.
Martius's description is not clear if the pinnae are clustered
or not. Wessels Boer (1972, & unpublished ms.) describes them
as being in clusters of 2-3, but in several collections examined
they are n t clustered (e.g., Prance et al 2155, Dahlgren & Millar
s.n., L.H.oBailey 489, and R. Benoist 1707). In addition to this,
the number of stamens per male flower is uncertain. Wessels
Boer describes them as 6-9 stamens per flower, but in Prance
et al. 2155, there are 12 stamens in most of the flowers.

O. teixeirana Bondar, Arq. Jard. Bot. Rio de Janeiro 13:58,
      fig. 5, 6-3, 1954.
      Holotype: Brazil, Maranhão, Caxias (Bondar s.n. - RB-80813).

      Specimens examined: Brazil, Maranhão, Caxias, Bondar
      s.n. (RB- 80813, holotype); Bondar s.n. (F-405257).
      Vernacular names: Perinão, Coco de Macacao.

Distribution: Brazil, states of Maranhão and Piauí.

According to Bondar (1954), this species is also found in the
state of Piauí, near Terezina, in margin of Rio Paranaiba. He
also states that it is probably a hybrid between O. barbosiana
(O. speciosa) and O. eichleri, which are both present in the
vicinity of the type locality. More collections of this taxon
should be examined before this can be verified, because O. teixeirana apparently most closely resembles O phalerata from Bolivia,
which is also poorly known.

O. urbaniana Dammer, Engl. Bot. Jahrb. 31, Beibl. 70:23. 1902;
    Dahlgren, pl. 344. 1959.
    Holotype: Brazil, Goias, Serra Dourada, in campis (Glaziou
    22265-C).

    Specimens examined: Brazil, Goias, Glaziou 22265 (C, holo-
    type; F, G, P, isotypes).
    Distribution: Brazil, state of Goias.

Even though Dammer did not designate the herbarium in which
Glaziou 22265 was deposited, the specimen from (C) is inscribed
O. urbaniana U. Dam. n. sp., det. by U. Dammer. The other
specimens from (F, G and P) do not bear such information.
Therefore, the collection from (C) is the holotype.

Burret (1929) cited Glaziou 22265 (B), but this specimen was
probably destroyed, as it could not be found.

Dammer (1902) said that O. urbaniana is close to O. lydiae
(= O. barbosiana), but like many other species of Orbignya, it
is difficult to make comparisons of taxa based on few or incomplete
collections.

Species definitely not Orbignya

Orbignya dubia Mart., Hist. Nat. Palm. 3:304, t. 169-6. 1845
    = Attalea dubia (Mart.) Burret, Notizbl. 10:537. 1929.

Male flowers of this taxon definitely belong to the genus Attalea,
i.e., flattened petals with acute or acuminate tips and straight
rather than coiled anthers.

Species Incertae et Dubiae

Orbignya agrestis (Barb. Rodr.) Burret, Notizbl. 10:511. 1929.
    Attalea agrestis Barb. Rodr., Enum. Palm. Nov. 42. 1875;
    Sert. Palm. Bras. 1:t. 55. 1903.
    Lectotype: Brazil, Amazonas region, sandy soil, Rio Uanincha,
    affluence of Rio Yamunda (t. 55, 1903). c.f. Glassman
    1972, p. 22.

Barbosa Rodrigues (1875) cited Barb. Rodr. 324, but apparently
this specimen has been destroyed; hence the selection of t. 55
as the lectotype.

It is difficult to determine the genus because male flowers
are not mentioned in the descriptions and not illustrated in
t. 55. Burret (1929) transferred this species to Orbignya
because of its resemblance to O. sabulosa Barb. Rodr. He also
cites Huebner 4b (B) from Manaos, which consists of immature
fruits and leaf parts. I have examined this specimen and do not
consider it to be diagnostic.

O. huebneri Burret, Notizbl. 10:501. 1929.
    Holotype: Brazil, Amazonas, Lago Mondurucú, Rio Manacapurú,
    Solimões (Huebner 64-B)

The holotype consists of fruit and leaf material, but no male flowers
were collected or described. Burret suggests a resemblance to
O. speciosa (= O. barbosiana), but says that the two species differ
mainly in the structure of the fruit and in the period of flowering.

Burret (1929) refers to photographs of this taxon, but none were
published in his article. Prints of three photographs of O. huebneri,
taken by George Huebner in Manaos in 1935, are deposited in the
Field Museum Herbarium. One photo (460060) illustrates a
juvenile acaulescent plant with extremely long leaves, whereas
the other two (460058-59) show mature trees about 13 m tall
with unclustered pinnae.

Another collection possibly belonging to this species was
determined by Burret: Mus. Goeldi Garten, Capt. H.A. Johnstone
1038 (B). It consists of mature fruits and naked androgynous
rachillae.

Orbignya huebneri is probably synonymous with O. barbosiana,
but should remain a species dubia until additional material
from the type locality, especially male flowers, can be studied.

O. macrostachya Drude nomen in Scheda ex Barb. Rodr., Sert.
      Palm. Bras. 1 60. 1903, c.f. Burret 1929, p. 513.

This entry is based on a herbarium specimen (Glaziou 16488-BR)
determined by Drude as a new species, and labelled as such,
but a description was never published. Burret (1929) listed
this name as a synonym of O. dammeriana Barb. Rodr. and cited
Glaziou 16488   The latter specimen has been erroneously cited
as the type of O. dammeriana by Dahlgren (1936) and Glassman
(1972), but the type of this species is actually Glaziou
16468-B (see discussion of synonymy under O. cohune).
Glaziou 16488 (BR) consists of two sheets with part of a male
spadix and male flowers, but the specimens are undoubtedly
an undetermined species of Scheelea   Because O. macrostachya has no published description and is actually a species
of Scheelea, it should be designated as nomen nudum et confusum.

O. microcarpa (Mart.) Burret, Notizbl. 10:507. 1929.
      Attalea microcarpa Mart. Palmet. Orbign. 125. 1844;
      t. 168-2, 1845; t. Z16-5. 1849.
      Type: Brazil, Para' (no specimens cited).

Burret (1929) indicated that the spadix illustrated in t. 168
was in the collections at Munich. However, I have not seen
any specimens from that herbarium labelled A. microcarpa.

Burret had no justification for transferring the name to
Orbignya because male flowers and leaves were neither
described nor illustrated. Therefore, its status is uncertain.

O. racemosa (Spruce) Drude, Mart. Fl. Bras. 3:448. 1881;
      Dahlgren, pl. 343. 1959. Attalea racemosa Spruce,
      Journ. Linn. Soc. 11:166. 1871.
      Holotype: Venezuela, between Rio Negro and Guasie
      (Spruce 54-K).

The genus to which this species belongs is uncertain because
male flowers were not described by Spruce (1871) or Drude
(1881). Type specimens from Kew and Paris are without male
flowers, as well.

Wessels Boer (1972) equates Attalea racemosa with A. ferruginea, probably because both have pinnae with ferrugineous

margins and both species come from the Rio Negro region of
Venezuela. At present, there is insufficient evidence to
definitely lump these two species together.

Species closely related to Orbignya, but
which probably belong to distinct genera

Attalea crassispatha (Mart.) Burret, Sv. Vet. Akad. Handl.
    6:23, t. 8-11. 1929b; Bailey, fig. 167-170, 1939.
    Maximiliana crassispatha Mart., Palmet. Orbign. 110.
    1844.
    Lectotype: Haiti (Plumier, Nov. Pl. Amer. Gen. t. 1.
    1703); c.f. Dahlgren, 1936, pp. 209-210.

    Specimens examined: Haiti, Fond des Negres, E. Ekman
    7164 (NY); O. F. Cook s.n. (BH); L. H. Bailey 299 (BH);
    L. Figueiras & P. Louis 2785 (F); between Cavaillon
    and Aux Cayes, H. Loomis & T. Fennell s.n. (US).
    Vernacular names: Carossier, Petit coco.
    Distribution: Endemic to Haiti.

Since no specimens were cited by Martius, Plumier's Plate
was chosen as the lectotype.

Even though this taxon is relatively rare and probably is
confined to one region of Haiti, it is very distinct and
well-known botanically (except for the male spathe and spadix
which apparently has not been described or collected).

Male flowers (from the androgynous spadix) have coiled and
twisted anthers, and fleshy, curved petals, suggesting either
Orbignya or Parascheelea. Wessels Boer (1971) says that this
species, with its 9-11 stamens and twisted anthers, resembles
the Markleya staminate flower type. However, all of the
flowers I examined from androgynous rachillae have only 6
stamens.

Cook (1939) described this taxon under a new genus, Bornoa   .
(which is invalid, because it was published without a Latin
description); and Moore (1963) thought that Cook was perhaps
correct in considering it as a distinct genus (from Attalea
as well as other allied genera).

MARKLEYA Bondar, Arq. Jard. Bot. Rio de Jan. 15:49-55. 1957.
    Type species: Markleya dahlgreniana Bondar.

<u>M. dahlgreniana</u> Bondar, l.c. 50, fig. 1, foto 1-3. 1957.
<u>Attalea dahlgreniana</u> (Bondar) Wessels Boer, Indig.
Palms Suriname 158. 1965.
Holotype: Brazil, Pará, Tracuateua, munic of Bragança
(<u>Bondar s.n.</u> - RB-95829).

Specimens examined: Brazil, Pará, Bragança, <u>Bondar</u>
<u>s.n.</u> (RB-95829, Holotype; F, isotype). Surinam,
Palaime Creek, <u>Wessels Boer</u> <u>805</u> (F, U); near Coeroeni
airstrip, <u>Wessels Boer</u> <u>1587</u> (F, U).
Vernacular name: Perinão.
Distribution: Brazil (state of Pará); and Surinam.

In his original article, Bondar speculates that this taxon
is probably a hybrid between <u>Orbignya speciosa</u> (= <u>O. barbos-</u>
<u>iana</u>) and <u>Maximiliana regia</u> (= <u>M. martiana</u>), because it grows
in conjunction with these two species. Wessels Boer, however,
refutes this idea, because the large uniform populations
he saw in Surinam produced fertile fruits. Moore (1973) says
that <u>Markleya</u> is a possible hybrid and does not list it as a
distinct genus in his article on "Major groups of palms."

Male flowers have twisted and coiled anthers like <u>Orbignya</u>,
but the petals are flat and curved similar to <u>Parascheelea</u>.

PARASCHEELEA Dugand, Caldasia <u>1</u>:10. 1940.
    Type species: <u>Parascheelea anchistropetala</u> Dugand.
    <u>P. anchistropetala</u> Dugand, l.c. 12, fig. 4-5. 1940.
    Holotype: Colombia, Vaupes, Cerro de Circasia (<u>Cuatre-</u>
    <u>casa 7172</u> - COL).
    <u>P. luetzelburgii</u> (Burret) Dugand, Caldasia <u>10</u>:24. 1941.
    <u>Orbignya luetzelburgii</u> Burret, Notizbl. <u>10</u>:1019. 1930.
    Holotype: Brazil, Amazonas, Jutica Varadouro (<u>Luetzel-</u>
    <u>burg 21969</u>-B).

    Specimens examined: Colombia, Vaupes, Circasia, sandy
    savannah quartzite base, <u>Schultes</u> & <u>Cabrera</u> <u>19207</u> (US);
    Vaupes, Cerro de Circasia, <u>Cuatrecasas</u> <u>7172</u> (COL, holo-
    type of <u>P</u>. anchistropetala). Brazil, Amazonas, Jutica
    Varadouro, Urwald, <u>Luetzelburg</u> <u>21969</u> (B, holotype of
    <u>O</u>. luetzelburgii; M, isotype). Venezuela, Terr. Amazonas,
    near Santa Rosa de Amanadona, white sandy soil, <u>Wessels</u>
    <u>Boer</u> <u>2357</u>, <u>2374</u> (F, U).
    Vernacular names: Curua, Yapo (Colombia); Curuaraua
    (Venezuela).
    Distribution: Colombia, Brazil and Venezuela, mostly in
    Amazon region.

Originally described as a distinct genus from Orbignya mainly
because petals of the male flowers are plano-convex and broader
below and gradually narrowed above rather than flattened petals
narrowed below and abruptly broadened above.

According to Wessels Boer (Palms of Venezuela - unpublished
manuscript), Dugand described P. anchistropetala as having
a double branched inflorescence, but the type material shows
a simply branched spadix. Dugand also observed the resemblance
between this species and Orbignya luetzelburgii, but he was
reluctant to unite them because of his misinterpretation of
the inflorescence. Wessels Boer (1972) also made a new combination, Attalea luetzelburgii (misprinted as "wetzelburgii"),
but this name is invalid because the basionym was not listed.
He had intended to list it in a subsequent manuscript with
full taxonomic treatment of Attalea, sensu latu, but unfortunately, this paper never was published.

LITERATURE CITED

Bailey, L.H. 1939. Certain Palms of the Greater Antilles.
    II. Article 19. The Great Carossier. Gentes Herb.
    4:263-265, figs. 167-170.

Barbosa Rodrigues, J. 1875. Enumeratio Palmarum Novarum
    quas Valle Fluminis Amazonum. pp. 1-43. Rio de Janeiro.

    1879. Protesto-Appendice ao Enumeratio Palmarum Novarum.
    pp. 1-48, 2 plates. Rio de Janeiro.

    1888. Palmae Amazonenses Novae. Vellosia 1:33-56.

    1891. Plantas Novas Cultivadas no Jardim Botanico do
    Rio de Janeiro 1:32.

    1896. l.c. 5:16-23.

    1898. Palmae Mattogrossenses Novae vel Minus Cognitae.
    pp. 1-88. Rio de Janeiro.

    1903. Sertum Palmarum Brasilensium, ou Relation des
    Palmiers Nouveaux du Bresil 1:1-140, 91 plates; 2:
    1-114. 83 plates. Bruxelles.

Bondar, G. 1954. Nova Especie de Orbignya, Produtora do Oleo de Babacu. Arq. Jard. Bot. Rio de Janeiro. 13:57-59.

1957. Novo Genero e nova especie de Palmeiras da Tribo Attaleine. l.c. 16·49-55.

1964. Palmeiras do Brasil. pp. 5-159, figs. 1-57. Instituto de Botanica São Paulo.

Burret, M. 1929. Die Palmengattungen Orbignya, Attalea, Scheelea, und Maximiliana. Notizblatt 10:493-543, 651-701.

1929b. Palmae Cubenses et Domingenses a Cl. E.L. Ekman 1914-1928 lectae. Kungl. Svenska Vet. Akad. Handl. ser. 3, 6 (7):3-28.

1930. Palmae Novae Luetzelburgianae. Notizblatt 10: 1013-1026.

1932. Attalea cohune Mart. wirklich eine Orbignya. Notizblatt 11:688-690.

1938. Palmae Brasiliensis. Notizblatt 14:231-260.

1940. Palmae Neogeae XII. Notizblatt 15:99-108.

Cook, O.F. 1939. Bornoa, an endemic palm of Haiti. Nat. Hort. Mag. 18:245-280.

Cuatrecasas, J. 1947. Vistazo a la Vegetacion Natural del Bajo Calima. Rev. Acad. Colomb. Cienc. 7:306-312.

Dahlgren, B.E. 1936. Index of American Palms. Field Mus. Nat. Hist. Bot. 14:1-438.

1959. Index of American Palms. Plates, Field Mus. Nat. Hist. Bot. 14:pl. 1-412.

Dammer, U. 1902. Plantae novae americanae imprimis Glazio-vianae. Engler Bot. Jahrb. 31, Beibl. 70:22-23.

Drude, O. 1881. Palmae in Martius Flora brasiliensis 3: 254-460.

Dugand, A. 1940. Un genero cinco especies nuevas de Palmas. Caldasia 1:10-19.

1941. Notas sobre palmas Colombianas y una del Brasil. Caldasia 1·17-29.

1943. Noticias Botanicas Colombianas II. Especies nuevas y criticas. Caldasia 2:285-293.

Glassman, S.F. 1972. A revision of B.E. Dahlgren's Index of American Palms. 294 pp. J. Cramer. Lehre, Germany.

1977. Preliminary taxonomic studies in the palm genus Attalea. H.B.K. Fieldiana Bot. 38:31-61.

Hernandez Xolocotzi, E. 1949. Estudio Botanico de las Palmas Oleaginosas de Mexico. Bol. Soc. Bot. Mex. 9:13-19.

Holmgren, P. and W. Keuken. 1974. Index Herbariorum. Part I. Herbaria of the world. Sixth ed. 397 pp. Utrecht.

Karsten, G. and H. Schenck. 1910. Vegetationsbilder. 7: t. 35-36. Jena.

Lindman, C.A.M. 1900. Beitrage zur Palmenflora Sudamerikas. Bihang. Kungl. Svenska Vet. Akad. Handl. Band 26 III (no. 5):3-42, t. 1-6.

Martius, C.F.P. von. 1826. Hist. Nat. Palm. 2:91-144.

1837. in Endlicher, Genera Plantarum. Palmae. pp. 243-257. Vienna.

1844. Palmetum Orbignianum in d'Orbigny, Voyage dans l'Amerique meridionale 7(3):1-140. Paris.

1845. Hist. Nat. Palm. 3:261-304.

1849. Hist. Nat. Palm. 3:305-314.

1853. Hist. Nat. Palm. 3:315=350.

Moore, H.E.   1960   B.E. Dahlgren's Index of American Palms,
    Plates (Review). Principes 4:32-33.

    1963. An annotated checklist of cultivated palms.
    Principes 7·119-182.

    1973. The major groups of palms and their distribution.
    Gentes Herb. 11(2)·27-141.

Plumier, C. 1703. Nova Plantarum Americanarum Genera
    pp. 103, t. 1. Paris.

Spruce, R. 1871. Palmae Amazonicae, sive Enumeratio
    Palmarum in Itinere suo per regiones Americae aequat-
    oriales lectarum. Journ. Linn. Soc. 11:65-175.

Standley, P.C. and J.A. Steyermark. 1958. Palmae, in
    Flora of Guatemala I. Fieldiana Bot. 24:196-299.

Trail, J.W.H. 1876. Descriptions of new species and varieties
    of palms collected in the valley of the Amazon in
    north Brasil, in 1874. Journ. Bot. 14 (n.s. vol. 5):
    323-333, 353-359, t. 183.

    1884. in Im Thurn, E.F. Memoranda on the palms of
    British Guiana. Timehari, Journ. Royal Agric. & Comm.
    Soc. Brit. Guiana 3:219-276. Demerara.

Wallace, A.R. 1853. Palm trees of the Amazon and their uses.
    pp. 1-129, pl. 1-48. London.

Wessels Boer, J. 1965. The indigenous palms of Suriname.
    172 pp. Leiden.

    1971. Bactris x moorei, a hybrid in palms. Acta Bot.
    Neerl. 20:167-172.

    1972. Clave Descriptiva de las Palmas de Venezuela. Acta
    Bot. Venez. 6:299-362. (An additional manuscript with
    descriptions and synonymy of the taxa mentioned above
    was prepared by Wessels Boer, but never published.)

NOTES ON NEW AND NOTEWORTHY PLANTS. C

Harold N. Moldenke

LANTANA SCABIOSAEFLORA var. HIRSUTA Moldenke, var. nov.

Haec varietas a forma typica speciei ramis novellis dense patenteque hirsutis recedit.

This variety differs from the typical form of the species in having its younger branches and stems densely hirsute with stiffly wide-spreading ochraceous hairs.

The type of the variety was collected by August Weberbauer (no. 7660) in deciduous bushwood in the mountains east of Hacienda Chicama, province Tumbes, Tumbes, Peru, at 700—800 meters altitude, between February 19 and 24, 1927, and is deposited in the Britton Herbarium at the New York Botanical Garden. The collector describes the plant as a shrub, 1 meter tall, with yellow flowers.

LANTANA SCABIOSAEFLORA var. LIMENSIS (Hayek) Moldenke, stat. nov.

Lantana limensis Hayek in Engl., Bot. Jahrb. 42: 166—167. 1908.

SYNGONANTHUS ELEGANTULUS var. GLAZIOVII Moldenke, Phytologia 36: 35, nom. nud. (1977), var. nov.

Haec varietas a forma typica speciei vaginis dense patenteque hirtellis recedit.

This variety differs from the typical form of the species in having its sheaths densely and conspicuously spreading-hirtellous throughout.

The type of the variety was collected by Auguste François Marie Glaziou (no. 20013) somewhere in Minas Gerais, Brazil, in 1892, and is deposited in the Columbia University herbarium at present on deposit at the New York Botanical Garden.

ADDITIONAL NOTES ON THE GENUS VERBENA. XXIV

Harold N. Moldenke

VERBENA [Dorst.] L.
Additional synonymy: Verbenella Spach, Hist. Nat. Veg. Phan.
9: 237—238. 1840. Glandularia Schau. apud Buek, Gen. Spec. Syn.
Candoll. 3: 199. 1858. Verbera Bert. ex Moldenke, Phytologia 36:
47, in syn. 1977.
Additional & emended bibliography: Apul. Barb., Herb., ed. 1.
1480—1483; Anon., Dialogue des Créatures, dial. 30. 1482; Apul.
Barb,, Herb., ed. 2. 1528; Anon., Bastiment des Receptes, fol. 59
vert. 1544; H. Bock [Tragus], Stirp. Max. Germ. 102, 210, & 211.
1552; Clus., Rar. Stirp. Obser. Hist. 2: 372—373. 1576; Dodoens
[l'Ecluse], Hist. Pl. 96 & 97. 1667; P. Herm., Parad. Batav.
Prodr. [ed. Warton]. 1689; Dill. in Tay, Synop. Meth. Stirp.
Brit., ed. 3, 236. 1724; P. Herm., Mus. Zeyl., ed. 2, 58. 1726;
L., Hort. Cliff., imp. 1, 10—11. 1737; Breyn., Prodr. Fasc. Rar.
Pl., ed. 2, 2: 100 & 104. 1739; Strand in L., Amoen. Acad. 69:
449. 1756; L., Syst. Nat., ed. 10, imp. 1, 2: 851 (1759) and ed.
10, imp. 2, 2: 851. 1760; Chomel, Abrég. Hist. Pl. Usuel., ed. 6,
1: 143 (1761) and ed. 6, 2 (2): 18, 85—87, & 251. 1761; L., Sp.
Pl., ed. 2, 27—29. 1762; Jacq., Obs. Bot. 4: 6—7, pl. 85 & 86.
1771; Ginanni, Istor. Civ. Nat. Pinet. Ravenn. 177. 1774; Chomel,
Abrég. Hist. Pl. Usual, ed. 6 nov., 89, 313, 527, & 637. 1782;
Jacq., Select. Stirp. Amer. Hist. 8. 1788; F. Hernandez, Hist. Pl.
Nuev. Españ., ed. 1, 1: 139 & 439 (1790) and ed. 1, 3: 3 & 486.
1790; Moebch, Suppl. Meth. Pl. 131 & 150—151. 1802; Chomel, Abrég. Hist. Pl. Usuel., ed. 7, 1: 175 & 495 (1803) and ed. 7, 2:
293 & 488. 1803; Stokes, Bot. Mat. Med. 1: 38—39. 1812; A. Rich.,
Bot. Méd. 1: 242—243. 1823; Dierbach, Arzneimit. Hippok. 85 &
270. 1824; A. Rich. [transl. G. Kunze], Med. Bot. 1: 381 (1824)
and 2: 1302. 1826; Sweet, Hort. Brit., ed. 2, 418—419. 1830; G.
Don in Loud., Hort. Brit., ed. 1, 246—247 (1830) and ed. 2, 246—
247. 1832; G. Don in Loud., Hort. Brit. Suppl. 1, imp. 1, 602 &
680. 1832; Loud., Hort. Brit., ed. 2, 552, 553, 575, & 602. 1832;
A. Dietr., Handb. Pharmaceut. Bot. 114 & 412. 1837; D. Dietr.,
Taschenb. Arzneigew. Deutschl. 58 & 262. 1838; Baxt. in Lour.,
Hort. Brit. Suppl. 2: 680. 1839; G. Don. in Loud., Hort. Brit.,
ed. 3, 246—247. 1839; G. Don in Loud., Hort. Brit. Suppl. 2: 704
& 741. 1839; Loud., Hort. Brit., ed. 3, 575 & 602. 1839; Sweet,
Hort. Brit., ed. 3, 768. 1839; Spach, Hist. Nat. Veg. Phan. 9:
227 & 236—241. 1840; Schau. in A. DC., Prodr. 11: 524, 435—557,
614, & 736. 1847; Webb in Hook., Niger Fl. 161. 1849; Baxt. in
Loud., Hort. Brit. Suppl. [3]: 655. 1850; G. Don in Loud., Hort.
Brit. Suppl. 1, imp. 2, 680 & 733. 1850; Anon., Croniqueur du
Périgord 120. 1853; F. Lenormant, Bull. Soc. Bot. France 2: 315—
320. 1855; Schnitzlein, Iconogr. Fam. Nat. 2: 137 Verbenac. [2],
[3], & 137, fig. 4—22 & 30. 1856; Buek, Gen. Spec. Syn. Candoll.

3: 199, 494—496, & 507. 1858; A. Gray, Proc. Am. Acad. 6: 50.
1862; Paine, Ann. Rep. Univ. N. Y. 18: [Pl. Oneida Co.] 109. 1865;
Voss in Vilm., Fl. Pleine Terr., ed. 1, 936—942 (1865) and ed.
2, 2: 973—979. 1866; Symphor Vaudoré, Lettr. Vieux Laboureur 88.
1867; J. Cousin, Secr. Mag. 1868: 7, 37, & 45. 1868; Voss in Vilm.,
Fl. Pleine Terr., ed. 3, 1: 1197—1203. 1870; Buek, Gen. Spec. Syn.
Candoll. 4: 413. 1874; Chenaux, Diable & Ses Cornes 53 & 54. 1876;
Anon., Rev. du Tarn 1877: 39. 1877; Franch., Nouv. Arch. Mus. Hist.
Nat. Paris, ser. 2, 6: 112 [Pl. David. 1: 232]. 1883; Strobl, Oes-
terr. Bot. Zeitschr. 33: 406. 1883; Vilm., Fl. Pleine Terr. Suppl.
3: 195. 1884; Balf. f., Bot. Socotra 232 & 437. 1888; Kuntze, Rev.
Gen. Pl. 2: 502 & 510. 1891; J. Camus, Récept. France in Bull. Soc.
Syndic. Pharmac. Côte-d'Or. 10. 1892; J. Feller, Bull. Folklore 2:
105—109. 1893; Nairne, Flow. Pl. West. India 249. 1894; Voss in
Vilm., Fl. Pleine Terr., ed. 4, 1065—1070. 1894; Engl., Syllab.
Pflanzenfam., ed. 2, 178. 1898; Van Tieghem, Élem. Bot., ed. 3, 2:
372 & 373. 1898; Bidault, Superst. Méd. Morvan 36. 1899; Diels,
Fl. Cent.-China 547. 1902; Engl., Syllab. Pflanzenfam., ed. 3, 187.
1903; Anon., Rev. Tradit. Populaires 1904: 162 (1904) and 1905:
160 & 296. 1905; Druce & Vines, Dill. Herb. 78, 182, & 257. 1907;
Engl., Syllab. Pflanzenfam., ed. 5, 192 (1907) and ed. 6, 198.
1909; Gilg in Engl., Syllab. Pflanzenfam., ed. 7, 314, fig. 413
C—F. 1912; Loes., Verh. Bot. Ver. Brand. 53: 74—75. 1912; Bég.
& Vacc., Ann. di Bot. 12: 118. 1913; Reynier, Bull. Soc. Bot.
France 62 [ser. 4, 15]: 205. 1915; Pamp., Nuov. Giorn. Bot. It.,
ser. 2, 23: 284. 1916; Skottsb., Kgl. Svensk. Vet. Akad. Handl.
56 (5): 291—292. 1916; Gilg in Engl., Syllab. Pflanzenfam., ed.
8, 318, fig. 413 C—F. 1919; Pennell, Bull. Torrey Bot. Club 46:
186. 1919; Knoche, Fl. Balearic., imp. 1, 1: 59. 1921; Fedde in
Just, Bot. Jahresber. 45 (1): 583. 1923; Sydow in Just, Bot. Jahresber. 45 (1): 402. 1923; Gilg in Engl., Syllab. Pflanzenfam.,
ed. 9 & 10, 339, fig. 418 C—F. 1924; Robledo, Bot. Med. 267 &
392. 1924; Cavara, Atti Soc. It. Prodr. Sc. 14: 337. 1925;
Krause in Just, Bot. Jahresber. 44: 1172, 1195, & 1212. 1926;
Pittier, Man. Pl. Usuel. Venez. 395, 398, & 450. 1926; Wangerin
in Just, Bot. Jahresber. 46 (1): 717 & 718. 1926; Fedde in Just,
Bot. Jahresber. 44: 1534. 1927; Wangerin in Just, Bot. Jahresber.
49 (1): 522. 1928; Fedde in Just, Bot. Jahresber. 46 (2): 712.
1929; Freise, Bol. Agric. São Paulo 34: 480 & 494. 1933; Gunther,
Herb. Apul. Barb. [16v], 106, 113, 128, & 129. 1935; E. D. Merr.,
Trans. Am. Phil. Soc., ser. 2, 24 (2): [Comm. Loud.] 15, 331, &
444. 1935; Diels in Engl., Syllab. Pflanzenfam., ed. 11, 339,
fig. 432 C—F. 1936; M. F. Baker, Fla. Wild Fls., ed. 2, imp. 1,
188 & 244. 1938; Rohde, Rose Recipes, imp. 1, 21. 1939; W. Trea-
lease, Pl. Mat. Decorat. Gard. Woody Pl., ed. 5, imp. 1, 144 &
187. 1940; F. Hernandez, Hist. Pl. Nuev. Españ., ed. 2, 653 &
674. 1943; Roi, Atl. Pl. Méd. Chin. [Mus. Heude Not. Bot. Chin.
8:] 96. 1946; A. W. Anderson, How We Got Fls., imp. 1, 90, 168,
& 283. 1951; Conard, Pl. Iowa 44. 1951; Hatta, Kubo, & Watanabe,
List Med. Pl. 14. 1952; Sonohara, Tawada, & Amano [ed. E. H.
Walker], Fl. Okin. 132. 1952; Pételot, Arch. Recherch. Agron. &

Past. Viet. 18: 178, [243], & 244. 1953; Pételot, Pl. Méd. Camb.
Laos & Viet. 2: 178, [243], & 244 (1954) and 4: 21, 39, 70, 170,
184, 193, 208, & 300. 1954; Spencer, Just Weeds, ed. 2, xii & 199-
204, fig. 64 & 65. 1957; R. A. Davidson, State Univ. Iowa Stud.
Nat. Hist. 20 (2): 77. 1959; Hall & Thompson, Cranbrook Inst. Sci.
Bull. 39: 74. 1959; Cooperroder, State Univ. Iowa Stud. Nat. Hist.
20 (5): 70. 1962; Lind & Tallantire, Some Com. Flow. Pl. Uganda,
ed. 1, 145 & 248. 1962; Whitlock & Rankin, New Techn. Dried Fls.
25. 1962; A. W. Anderson, How We Got Fls., imp. 2, 90, 168, & 283.
1966; Banerji, Rec. Bot. Surv. India 19 (2): 75. 1966; Ewan in
Thieret, Southwest. La. Journ. 7: 11, 34, & 42. 1967; P. W. Thomp-
son, Cranbrook Inst. Sci. Bull. 52: 37. 1967; B. C. Harris, Eat
Weeds 39. 1968; Spencer, All About Weeds xii & 199--204, fig. 64 &
65. 1968; W. Trelease, Pl. Mat. Decorat. Gard. Woody Pl., ed. 5,
imp. 2, 144 & 187. 1968; L., Hort. Cliff., imp. 2, 10--11. 1968;
Barker, Univ. Kans. Sci. Bull. 48: 571. 1969; Barriga-Bonilla,
Hernández-Camacho, Jaramillo-T., Jaramillo-Mejía, Mora-Osejo,
Pinto-Escobar, & Ruiz-Carranza, Isla San Andrés 59. 1969; Bolkh.,
Grif, Matvej., & Zakhar., Chrom. Numb. Flow. Pl., imp. 1, 715--
717. 1969; J. Hutchinson, Evol. & Phylog. Flow. Pl. Dicot. 467--
470, fig. 414. 1969; G. W. Thomas, Tax. Pl. Ecolog. Summ. 77 & 78.
1969; Rimpler, Lloydia 33: 491. 1970; Scully, Treas. Am. Ind.
Herbs 145, 159, 193, 207, 215, 249, 254, 270, 283--284, & 289.
1970; Anon., Bioresearch Ind. 7: 1061. 1971; Kashroo, Singh, &
Malik, Bull. Bot. Surv. India 13: 52. 1971; R. E. Harrison, Handb.
Bulbs & Peren. S. Hemisph., ed. 3, 266--267, 281, & 282. 1971;
Hartwell, Lloydia 34: 387 & 437. 1971; Kaul, Bull. Bot. Surv. In-
dia 13: 240. 1971; Lind & Tallantire, Some Com. Flow. Pl. Uganda,
ed. 2, 145 & 248. 1971; Stalter, Castanea 36: 174 & 242. 1971;
Anon., Commonw. Myc. Inst. Index Fungi 3: 824. 1972; Fong, Tro-
jánkova, Trójanek, & Farnsworth, Lloydia 35: 147. 1972; Healy,
Gard. Guide Pl. Names 37 & 225. 1972; Solbrig in Valentine, Tax.
Phytogeogr. & Evol. 91. 1972; Ellison, Kingsbury, & Hyypio, Comm.
Wild Fls. N. Y. [Cornell Ext. Bull. 990:] 19. 1973; Frohne & Jen-
sen, System. Pflanzenr. 203, 261, & 305. 1973; Gillanders, Pater-
son, & Rotherham, Know Your Rock Gard. Pl. 45, 63, & 101. 1973;
Hathaway & Ramsey, Castanea 38: 77. 1973; Hilbig, Wiss. Zeitschr.
Mart.-Luth.-Univ. Halle 22: 56 & 102. 1973; A. & C. Krochmal,
Guide Medic. Pl. U. S. 229--230, 246, 257, & 258, fig. 259. 1973;
Law, Concise Herb. Encycl. 85, 117, 129, 156, 208, 251, 252, 263,
& 266. 1973; Rodhe, Rose Recipes, imp. 2, 21. 1973; Williamson,
Sunset West. Gard. Book, imp. 11, 178, 179, & 437. 1973; Ayensu,
Rep. Endang. & Threat. Spec. 98 & 129. 1974; H. Bennett, Concise
Chem. & Techn. Dict., ed. 3, 256 & 1102. 1974; Bolkh., Grif, Mat-
vej., & Zakhar., Chrom. Numb. Flow. Pl., imp. 2, 715--717. 1974;
E. T. Browne, Castanea 39: 183. 1974; H. B. Carter, Sir Jos.
Banks & Pl. Coll. Kew 340. 1974; El-Gazzar, Egypt. Journ. Bot. 17:
75, 76, & 78. 1974; Ellenberg, Script. Geobot. 9: 80. 1974; Farns-
worth, Pharmacog. Titles 9 (4): x. 1974; R. D. Gibbs, Chemotax.
Flow. Pl. 3: 1752--1755 (1974) and 4: 2295. 1974; S. B. Jones,
Castanea 39: 137. 1974; Knoche, Fl. Balear., imp. 2, Lç 59. 1974;

León & Alain, Fl. Cuba, imp. 2, 2: 279--283, 291--297, fig. 121.
1974; Loewenfeld & Back, Complete Book Herbs & Spices 261--264.
1974; R. W. Long, Fla. Sci. 37: 37. 1974; López-Palacios, Revist.
Fac. Farm. Univ. Los Andes 14: 23. 1974; A. & D. Löve, Cytotax.
Atl. Slov. Fl. 601 & 1241. 1974; Mrs. P. Martin, Am. Horticulturist 53 (5): 33. 1974; Moldenke, Biol. Abstr. 58: 3837. 1974; Moldenke & Neff, Orig. & Struct. Ecosyst. Tech. Rep. 74-18: 40, 51,
57, 101, 116, & 118. 1974; Moser, Natl. Geogr. 146: 510. 1974;
Portères, Journ. Agric. Trop. & Bot. Appl. 21: 6. 1974; Rogerson
& Becker, Bull. Torrey Bot. Club 101: 383. 1974; Rousseau, Géogr.
Florist. Qué. [Trav. Doc. Cent. Étud. Nord 7:] 376--377, 465, 467,
473, 479, 480, 502, 504, 505, 516, 550, 643, 644, & 788, maps
826--829. 1974; Soukup, Biota 10: 231. 1974; Stanley & Linskens,
Pollen 47, 95, & 306. 1974; Stark, Am. Horticulturist 53 (5): 7 &
11. 1974; Sunding, Garcia de Ort. Bot. 2: 20. 1974; L. H. Swift,
Bot. Classif. 312. 1974; Van Saun & Kemp, Bull. Torrey Bot. Club
101: 371. 1974; Welsh, Utah Pl., ed. 3, 232, 354, & 473. 1974;
Whitney in Foley, Herbs for Use & Delight [198] & [207]. 1974;
Täckholm, Stud. Fl. Egypt, ed. 2, 452--454, 817, 830, 876, & 886,
pl. 156, fig. A. 1974; Troncoso, Darwiniana 18: 312. 1974; Anon.,
N. Y. Times D.41, April 6. 1975; Balgooy, Pacif. Pl. Areas 3:
245, 1975; [Bard], Bull. Torrey Bot. Club 102: 430 & 431. 1975;
D. S. & H. B. Correll, Aquat. & Wetland Pl. SW. U. S., imp. 2, 2:
1395--1400 & 1775, fig. 654. 1975; O. & I. Degener & Pekelo,
Hawaii. Pl. Names X.4, X.5, & X.21. 1975; Duncan & Foote, Wildfls.
SE. U. S. 15, 150, [151], & 295. 1975; R. & A. Fitter, Wild Fls.
Brit. & N. Eu. 192, 193, & 336. 1975; Garcia, MacBryde, Molina, &
Herrera-MacBryde, Malez. Preval. Am. Cent. 87, 131, 143, & 161.
1975; Goebel, Act. Bot. Venez. 10: 377 & 379. 1975; Greller, Bull.
Torrey Bot. Club 102: 416. 1975; Hinton & Rzedowski, Anal. Esc.
Nac. Cienc. Biol. 21: 31 & 111. 1975; Hocking, Excerpt. Bot. A.
26: 4--7. 1975; E. H. Jordan, Checklist Organ Pipe Cact. Natl.
Mon. 7. 1975; A. L. Moldenke, Phytologia 31: 415 (1975) and 32:
375. 1975; Moldenke, Biol. Abstr. 59: 6926 (1975) and 60: 68.
1975; López-Palacios, Revist. Fac. Farm. Univ. Los Andes 15: 30,
51, 75--77, 79, & 88--94, [fig. 17]. 1975; Kooiman, Act. Bot.
Neerl. 24: [459] & 463--464. 1975; Mahesh., Journ. Bombay nat.
Hist. Soc. 82: 180. 1975; Moldenke, Phytologia 30: 129--180, 182,
185, 192, 506, & 508--512 (1975), 31: 28, 374--379, 383, 384,
387, 388, 392, 393, 398, & 409--412 (1975), and 32: 52, 229, &
230. 1975; Molina R., Ceiba 19: 95. 1975; Perkins, Estes, &
Thorp, Bull. Torrey Bot. Club 102: 194--198. 1975; J. C. & N. C.
Roberts, Field Guide Baja Calif., ed. 1, 35. 1975; Tovar Serpa,
Biota 10: 286 & 298. 1975; United Communications (Woodmere, N.
Y.), Herbal Visual & Study Chart. 1975; Weberling & Schwantes,
Pflanzensyst., ed. 2 [Ulmer, Uni-Taschenb. 62:] 144. 1975; Whit-
lock & Rankin, Dried Fls. 25. 1975; H. D. Wils., Vasc. Pl. Holmes
Co. Cat. 54. 1975; Anon., Biol. Abstr. 61: AC1.732. 1976; M. F.
Baker, Fla. Wild Fls., ed. 2, imp. 2, 188 & 244. 1976; Burkart,
Journ. Arnold Arb. 57: 224. 1976; L. J. Clark, Wild Fls. Pacif.
Northw. 444--445 & 603. 1976; Duke, Phytologia 34: 27. 1976;

Fleming, Genelle, & Long, Wild Fls. Fla. 15 & 67. 1976; Follman-
Schrag, Excerpt. Bot. A.26: 518. 1976; F. R. Fosberg, Rhodora 78:
113. 1976; Galiano & Cabezudo, Lagascalia 6: 150. 1976; Grimé,
Bot. Black Am. 186, 190—191, 220, & 230. 1976; J. H. Harv.,
Journ. Gard. Hist. Soc. 4 (3): 38. 1976; S. R. Hill, Sida 6: 325.
1976; Hurd & Linsley, Smithson. Contrib. Zool. 220: 10. 1976;
Keys, Chinese Herbs 283--284 & 387. 1976; Lacoursiere, Pontbriand,
& Dumas, Natur. Canad. 103: 174. 1976; Lakela, Long, Fleming, &
Genelle, Pl. Tampa Bay, ed. 3 [Bot. Lab. Univ. S. Fla. Contrib.
73:] 115, 116, 168, 178, & 182. 1976; Long & Lakela, Fl. Trop.
Fla., ed. 2, 733, 740—742, 940, & 961. 1976; López-Palacios, Re-
vist. Fac. Farm. Univ. Los Andes 17: 49—50. 1976; Lousley, Fl.
Surrey 282, map 288. 1976; Moldenke, Phytologia 32: 512 (1976),
33: 374—375, 480, & 512 (1976), and 34: 19—20, 153, 247—252,
254, 256—263, 266—268, 270, 274, 278, 279, 504, & 512. 1976; A.
L. Moldenke, Phytologia 35: 173. 1976; A. R. Moldenke, Phytologia
34: 345, 356, & 361. 1976; Norman, Fla. Scient. 39: 30. 1976;
Park Seed Co., Park Seeds Fls. & Veg. 1976: 63 & 90. 1976; Soukup,
Biota 11: 2, 17—19, & 22. 1976; Van Bruggen, Vasc. Pl. S. Dak.
368—369, 520, 529, 536, & 537. 1976; Vanderpoel, Natl. Wildlife
15 (1): 50. 1976; [Voss], Mich. Bot. 15: 237. 1976; E. H. Walker,
Fl. Okin. & South. Ryuk. 882--884, 1976; Moldenke, Phytologia 35:
512 (1977) and 36: 28—31, 33, 35, 36, 39, 40, 42, 47, 48, 51, &
52. 1977; A. L. Moldenke, Phytologia 35: 173 (1977) and 36: 87 &
88. 1977; F. H. Montgomery, Seeds & Fruits 201, fig. 5 & 6, 202,
fig. 1—3, & 230. 1977; Taylor & MacBryde, Vasc. Pl. Brit. Col.
436 & 751. 1977.

Swift (1974) tells us that "Verbena is an ancient Latin term
for ceremonial foliage rather than a name for a specific class
of plants. In post-classical times the name became applied to
vervain, probably in connection with medicinal uses". Garcia
and his associates (1975) inform us that the vernacular name,
"verbena", in Central America is applied also to _Browallia_ ameri-
cana L. and _Salvia_ occidentalis Sw.

The Fitter work cited in the bibliography above is dated "1974"
on the title-page, but was not actually issued until February 17,
1975.

Don (1830) divides the genus into two sections based on leaf
characters: _Indivisae_ with undivided leaf-blades, and _Trifidae_
with trifid blades.

The following two taxa described by Sloane (1739) are still
puzzling and have not as yet been satisfactorily identified as
far as I know:

"VERBENA _Americana_ procumbens, _Veronicae aquaticae folio_ subro-
tundo, flosculis ad foliorum alas; nobis. _Teucrium Americanum_
procumbens Veronicae aquaticae foliis subrotundis; Hermanni,
Catal. Hort."

"VERBENA _nodiflora_ major _Indica, flore_ niveo; nobis. In Horto
Fageliano, nomine Teucrii & Veronicae, legimus."

Similarly, the following taxon of P. Hermann (1726) is still

unidentified:
"TELKAPALA. Verbena Indica rotundifolia spicis comosis." No
such name as "telkapala" is used in Sri Lanka today according to
local botanists there now.

Roi (1946) lists a Chinese vernacular name for members of this
genus, "ma pien ts'ao" -- probably actually for V. officinalis L.
or xV. hybrida Voss. The Commonwealth Mycological Institute
(1972) lists the fungus, Ascochyta cuneomaculata, as attacking
members of the genus Verbena.

The Endlicher (1838) reference cited in the bibliography of
Verbena is often cited as "1836-1856", but the pages involved
with this genus were actually issued in 1838. The Sibthorpe &
Smith (1809) reference, similarly, is sometimes cited as "1806",
but pages 219--442 of the volume involved here were not actually
published until 1809.

The genus Verbenella Spach, curiously overlooked by the editors of the "Index Kewensis" and by every previous worker on this
group of plants, is based by Spach (1840) on Verbena chamaedryfolia Juss. and would therefore be congeneric with what some present-
day botanists regard as the genus Glandularia (1791).

Gibbs (1974) reports verbenalin (cornin), stachyose, and tannin
present in Verbena, but L-bornesitol (a cyclitol) and raffinose
are absent, while saponins are by some workers reported as probably present, by others as absent or probably absent. Bennett
(1974) defines "Verbena oil" or "verbenalin" as "$C_{17}H_{25}O_{10}$; m.w.
389.2; wh. need.; m.p. 178; s.w.; sl.s.al.; s. acet.; glucoside.
d-verbenone - $C_{10}H_{14}O$; m. w. 150.11; col. oil; sp. gr. 0.997^{20};
m.p. 6.5; b.p. 227; s.w."

From a fancied resemblance, Malva alcea L., a mallow, is often
called the "vervain mallow".

An additional excluded species is
Verbena microcephala López-Palacios, Revist. Fac. Farm. Univ. Los
        Andes 15: 51, sphalm. 1975 = Lantana trifolia f. hirsuta Mol-
        denke

The "verbena" illustrated in color by Moser (1974) is a species
of Abronia in the Nyctaginaceae. The M. E. Jones 359, distributed
as a Verbena sp., is actually Lantana scorta Moldenke, while M. C.
Johnston 2664 and Stuessy 1031 are Priva grandiflora (Ort.) Moldenke and Sohmer 5332 is Salvia occidentalis Sw.

The Degeners and Pekelo (1975) record "hoi" as a Hawaiian name
for the genus Verbena, probably for V. litoralis H.B.K.

Andrew R. Moldenke (1976) has found that members of this genus
(section Glandularia) in California have "strong mechanical or
temporal barriers to inbreeding even though the flowers are genet-
ically self-compatible" and are nearly always very heavily out-
crossed, the pollination being especially effected by butterflies,
but also by Bombyliidae and probably also by short-tongued groups
of insects like many species of Villa, generalist feeding bees in-

cluding many genera in all families and many species of bees
whose males may be common generalists even though the females are
restricted to one genus of plants, including Bombus, Ceratina,
Megachile, Melissodes, and Osmia but not   the 'table-scraping'
sometimes colonial Halictinae which are usually generalist feed-
ers."

VERBENA ABRAMSI Moldenke
    Additional bibliography: Moldenke, Phytologia 28: 344. 1974;
Hocking, Excerpt. Bot. A.26: 5 & 6. 1975.
    Additional citations: CALIFORNIA: Trinity Co.: Moldenke & Mol-
denke 30246 (Ld).

VERBENA ALATA Sweet
    Additional synonymy: Verbena allata Hort. ex Moldenke, Phyto-
logia 34: 278, in syn. 1976.
    Additional bibliography: Sweet, Hort. Brit., ed. 2, 418--419.
1830; Loud., Hort. Brit., ed. 2, 552 & 602 (1832) and ed. 3, 602.
1839; Baxt. in Loud., Hort. Brit. Suppl. [3]: 655. 1850; Buek,
Gen. Spec. Syn. Candoll. 3: 494. 1858; Moldenke, Phytologia 28:
344 (1974) and 34: 278. 1976.
    Ferreira describes this plant as a subshrub and found in it
flower and fruit in July. The corolla color of Ferreira 174 is
said to have been "lilac", while on Hatschbach 17252 it was
"violet".
    The Lindeman & Haas 3010, distributed as V. alata, is actually
V. minutiflora Briq.
    Additional citations: BRAZIL: Minas Gerais: Widgren s.n. [1845]
(Mu--1570). Paraná: L. F. Ferreira 174 (Ld); Hatschbach 17252
(Ld). CULTIVATED: France: Weinkauff s.n. [hort. Paris. 1834]
(Mu--1239). Germany: Herb. Schwaegrichen s.n. [Hort. Lipsiensis]
(Mu--1238); Herb. Zuccarini s.n. [hort. Berol. 1827] (Mu—281).

VERBENA ALBICANS Rojas
    Additional & emended bibliography: Krapovickas, Bol. Soc. Ar-
gent. Bot. 11, Supl. 261. 1970; Moldenke, Phytologia 30: 132
(1975) and 31: 388. 1975.

VERBENA ALBIFLORA Rojas
    Additional & emended bibliography: Krapovickas, Bol. Soc. Ar-
gent. Bot. 11, Supl. 269. 1970; Moldenke, Phytologia 30: 132
(1975) and 31: 388. 1975.

xVERBENA ALLENI Moldenke, Phytologia 34: 153. 1976.
    Bibliography: Moldenke, Phytologia 34: 153, 250, & 279. 1976.
    This plant has been found growing in black calcareous soil
and scattered on roadsides with Cynodon dactylon, Helenium amarum,
Vernonia altissima, and Xanthium strumarium, in and after anthesis
in September. The plants are described as 3—4 feet tall and the

corollas (on <u>Montz 2485</u>) said to have been blue when fresh. Herbarium material has hitherto been misidentified as <u>V. neomexicana</u>
(A. Gray) Small and as <u>V. xutha</u> Lehm. The cited specimens exhibit
many, slender, flexible spikes, some to as much as 45 cm. long,
and apparently with all or almost all of the seeds aborted.

Citations: LOUISIANA: LaSalle Par.: <u>C. A. Brown</u> <u>7409</u> (Lv).
Pointe Coupee Par.: <u>Montz</u> <u>2485</u> (Lv). Saint Helena Par.: <u>C. M. Al-
len</u> <u>1179</u> (Lv--type).

VERBENA AMBROSIFOLIA Rydb.

Additional bibliography: Bolkh., Grif, Matvej., & Zakhar., Chrom.
Numb. Flow. Pl., imp. 1, 716. 1969; G. W. Thomas, Tex. Pl. Ecolog.
Summ. 77. 1969; Bolkh., Grif, Matvej., & Zakhar., Chrom. Numb. Flow.
Pl., imp. 2, 716. 1974; Moldenke, Phytologia 30: 132--133, 139, &
179. 1975; A. L. Moldenke, Phytologia 35: 173. 1977.

Higgins found this plant growing in gravelly to sandy soil in
pinyon-juniper association areas, while Atwood encountered it in
"pinyon-juniper-sage-ponderosa community" and distributed his collection as V. bracteata Lag. & Rodr. Reitzel found it growing in
"clumps 6--7.5 dm. across, 3 dm. tall, at 7700 feet altitude with
<u>Ipomopsis</u>, Rosa, <u>Quercus</u>, <u>Pinus</u> edulis, a few <u>P. ponderosa</u>, <u>Junip-
erus</u>, <u>Mirabilis</u>, <u>Sphaeralcea</u>, and <u>Bromus</u>". Dziekanowski and his
associates report it very scattered in pinyon-juniper woodland
with grassy areas of gramma grass and staghorn cactus.

Other recent collectors have encountered <u>V. ambrosifolia</u> in
rocky washes, "scattered on open sandy-clay flats with <u>Ephedra</u>,
<u>Eriogonum</u>, <u>Lycium</u>, <u>Mentzelia</u>, <u>Psilostrophe</u>, <u>Yucca</u>, etc.", and "in
matorral desértico inerme on alluvial flats in fine calcareous al-
luvium with <u>Prosopis</u> glandulosa, <u>Larrea</u>, and <u>Flourensia cermua</u>".

The corollas on <u>Henrickson 5858</u> are said to have been "light-
purple" when fresh, while those on <u>Correll</u> & <u>Johnston</u> <u>19136</u> were
"magenta" and those on <u>Reitzel</u> <u>27</u> were "pink".

The <u>G. L. Fisher</u> s.n. [Nara Visa, Apr. 21, 1911], distributed as
<u>V. ambrosifolia</u>, actually is <u>V. ciliata</u> var. <u>pubera</u> (Greene) Perry,
while <u>Meebold</u> <u>24224</u> is V. <u>wrightii</u> A. Gray. The <u>Ramirez</u> & <u>Cardenas</u>
<u>13</u>, previously cited as <u>V. ambrosifolia</u> by me in this series of
notes, apparently is a mixture — the University of Texas sheet of
this number definitely is <u>V. ciliata</u> var. <u>longidentata</u> Perry. The
<u>Spellenberg</u> & <u>Spellenberg</u> <u>3062</u> cited by me in 1974 actually is <u>V.
gooddingii</u> var. nepetifolia Tidestr.

Additional citations: TEXAS: Culberson Co.: <u>Correll</u> & <u>Johnston</u>
<u>19136</u> (N). NEW MEXICO: Eddy Co.: <u>Higgins</u> <u>9197</u> (N). Harding Co.:
<u>S. Stephens</u> <u>75643</u> (N). Otero Co.: <u>Reitzel</u> <u>27</u> (N). Rio Arriba Co.:
<u>Atwood</u> <u>6298</u> (N). Torrence Co.: <u>Dziekanowski</u>, <u>Dunn</u>, & <u>Bennett</u> <u>2393</u>
(N). MEXICO: Chihuahua: <u>Henrickson</u> <u>5858</u> (Ld). Nuevo León: <u>Johns-
ton</u>, <u>Wendt</u>, & <u>Chiang</u> C. <u>10212</u> (Ld). Tamaulipas: <u>Kuiper</u> & <u>Kuiper-
Lapré</u> <u>M.17</u> (Ut--328636B).

VERBENA AMBROSIFOLIA f. EGLANDULOSA Perry
    Additional bibliography: Bolkh., Grif, Matvej., & Zakhar.,
Chrom. Numb. Flow. Pl., imp. 1, 716 (1969) and imp. 2, 716. 1974;
Moldenke, Phytologia 30: 132—133 & 139. 1975.
    Howe found this plant in flower and fruit in October.
    The R. Runyon 1782, distributed as this form and so cited by me
in previous publications actually is V. ciliata var. longidentata
Perry.
    Additional citations: ARIZONA: Coconino Co.: D. Howe s.n. [1
October 1968] (Sd—69988).

VERBENA AMOENA Paxt.
    Additional bibliography: Baxt. in Loud., Hort. Brit. Suppl.
[3]: 655. 1850; Moldenke, Phytologia 28: 111 (1974) and 34: 252.
1976.
    Gentry & Arguelles found what appears to be this species grow-
ing along roadsides at 7700 feet altitude, flowering and fruiting in June, and describe the plant as forming large compact
clumps, with secondary rooting at the stem-bases and with showy
purple flowers.
    Additional citations: MEXICO: Chihuahua: Gentry & Arguelles
22955 (Sd—86465).

VERBENA ANDRIEUXII Schau.
    Additional synonymy: Verbena andrieuxii DC. ex Moldenke, Phy-
tologia 34: 278, in syn. 1976.
    Additional bibliography: Buek, Gen. Spec. Syn. Candoll. 3: 494.
1858; Moldenke, Phytologia 23: 214 (1972) and 34: 278. 1976.
    Material of this species has been misidentified and distributed
in some herbaria as V. aubletia Jacq.
    Additional citations: MEXICO: Puebla: Andrieux 138 (Mu—2477—
isotype). CULTIVATED: Germany: Herb. Bot. Staatssamml. Münch. s.
n. (Mu); Herb. Hort. Monac. s.n. (Mu).

VERBENA ARAUCANA R. A. Phil.
    Addtional bibliography: Moldenke, Phytologia 23: 182 (1972),
31: 387 (1975), and 34: 260. 1976.
    Merxmüller found this plant in flower in December, while
Schajovskoy encountered it at 1300 meters altitude.
    Additional citations: CHILE: Malleco: Merxmüller 24966 (Mu).
ARGENTINA: Neuquen: Schajovskoy s.n. [16.I.1967] (Mu).

VERBENA ARISTIGERA S. Moore
    Additional bibliography: Moldenke, Phytologia 30: 133 & 173.
1975.
    Recent collectors describe this plant as decumbent and have
found it in flower in October and November. The corollas are
said to have been "purple" on Krapovickas & al. 26800 and on
Schinini & Cristóbal 9864.
    The Hatschbach 23884, previously cited by me as V. aristigera,

is perhaps better regarded as representing V. tenuisecta Briq.

Additional citations: ARGENTINA: Corrientes: Krapovickas, Cristóbal, Irigoyen, & Schinini 26800 (Ld); Schinini & Cristóbal 9864 (Ld). Entre Ríos: Lorentz 478 (Mu--1565).

VERBENA ATACAMENSIS Reiche

Additional bibliography: Moldenke, Phytologia 28: 344--345 & 441. 1974.

Zöllner encountered this plant growing near the seacoast.

Additional citations: CHILE: Antofagasta: Werdermann 789 (Mu); Zöllner 8310 (Ac). Atacama: Zöllner 9080 (Ld)

VERBENA AURANTIACA Speg.

Additional synonymy: Glandularia aurantiaca Speg. ex Moldenke, Phytologia 34: 274, in syn. 1976.

Additional bibliography: Moldenke, Phytologia 23: 214 (1972) and 34: 274. 1976.

Additional citations: ARGENTINA: Chubut: Kreitbohm 117 (Mu).

xVERBENA BAILEYANA Moldenke

Additional bibliography: Moldenke, Phytologia 28: 345 & 401 (1974) and 34: 270. 1976.

The Herb. Zuccarini specimens cited below are both mixtures with V. officinalis L. and were originally identified and distributed as V. hastata L. and V. scabra Vahl.

Additional citations: CULTIVATED: Germany: Herb. Zuccarini s. n. [Hort. bot. Monac.], in part (Mu, Mu).

VERBENA BAJACALIFORNICA Moldenke

Additional bibliography: Moldenke, Phytologia 23: 214. 1972.

Moran & Reveal refer to this plant as "fairly common" at 300 meters altitude and found it in flower in February. The corollas on their no. 20047 are said to have been "lavender" when fresh.

Additional citations: MEXICO: Baja California: Moran & Reveal 20047 (W--2796936).

VERBENA BALANSAE Briq.

Additional bibliography: Moldenke, Phytologia 28: 345 (1974) and 36: 35. 1977.

Krapovickas and his associates describe this plant as "decumbent", the corollas "violet" in color. Hatschbach refers to it as being xylopodiferous, with lilac (on 37118) or whitish (on 38688) flowers, and found it growing on "campo brejoso de solo vermelho" and "campo limpo", in flower in May.

Additional citations: BRAZIL: Mato Grosso: Hatschbach 38688 (Ld). Paraná: Hatschbach 37118 (Ld). ARGENTINA: Corrientes: Krapovickas, Cristóbal, Schinini, Arbo, Quarín, & González 26144 (Ld).

Additional bibliography: G. W. Thomas, Tex. Pl. Ecolog. Summ. 77. 1969; Moldenke, Phytologia 28: 114. 1974; Hinton & Rzedowski, Anal. Esc. Nac. Cieno. Biol. 21: 111. 1975.

Recent collectors have encountered this plant at 9100 feet altitude in Arizona.

Additional citations: ARIZONA: Cochise Co.: W. W. Jones s.n. [8 August 1967] (Sd—72884). Graham Co.: Williams & Williams 3650 (Z). Santa Cruz Co.: Reeves R.1075 (N).

xVERBENA BLANCHARDI Moldenke

Additional bibliography: Moldenke, Phytologia 28: 114 & 386. 1974.

Additional citations: ILLINOIS: Winnebago Co.: M. S. Bebb s.n. [Fountaindale] (Mu—4249).

VERBENA BONARIENSIS L.

Additional & emended synonymy: Verbena bonariensis altissima, Lavandulae canariensis spica multiplici Dill. in Ray, Synop. Meth. Stirp. Brit., ed. 3, pl. 300, fig. 387. 1724. Verbena bonariensis Willd. ex Druce & Vines, Dill. Herb. 182, in syn. 1907. Verbena bornariensis L. ex Moldenke, Phytologia 36: 47, in syn. 1977.

Additional & emended bibliography: Dill. in Ray, Synop. Meth. Stirp. Brit., ed. 3, pl. 300, fig. 387. 1724; L., Hort. Cliff., imp. 1, 11. 1737; L., Sp. Pl., ed. 2, 28. 1762; G. Don in Loud., Hort. Brit., ed. 1, 246 (1830) and ed. 2, 246. 1832; Loud., Hort. Brit., ed. 2, 552. 1832; G. Don in Loud., Hort. Brit., ed. 3, 246. 1839; Buek, Gen. Spec. Syn. Candoll. 3: 494 & 495. 1858; Kuntze, Rev. Gen. Pl. 2: 510. 1891; Druce & Vines, Dill. Herb. 182. 1907; Krause in Just, Bot. Jahresber. 44: 1195. 1926; Fedde in Just, Bot. Jahresber. 44: 1534. 1927; M. F. Baker, Fla. Wild Fls., ed. 2, imp. 1, 188. 1938; L., Hort. Cliff., imp. 2, 11. 1968; Bolkh., Grif, Matvej., & Zakhar., Chrom. Numb. Flow. Pl., imp. 1, 716. 1969; G. W. Thomas, Tex. Pl. Ecolog. Summ. 77. 1969; R. E. Harrison, Handb. Bulbs & Peren. S. Hemisph., ed. 3, 266. 1971; Fong, Trojánkova, Trojánek, & Farnsworth, Lloydia 35: 147. 1972; Bolkh., Grif, Matvej., & Zakhar., Chrom. Numb. Flow. Pl., imp. 2, 716. 1974; El-Gazzar, Egypt. Journ. Bot. 17: 75 & 78. 1974; R. D. Gibbs, Chemotax. Flow. Pl. 3: 1752—1755 (1974) and 4: 2295. 1974; R. W. Long, Fla. Scient. 37: 37. 1974; D. S. & H. B. Correll, Aquat. & Wetland Pl. SW. U. S., imp. 2, 2: 1396, 1397, & 1775. 1975; Kooiman, Act. Not. Neerl. 24: 463. 1975; Moldenke, Phytologia 30: 134—135, 152, & 167 (1975), 31: 375, 377, 389, & 409 (1975), 33: 374 (1976), and 34: 248, 250, 252, 259, 260, & 278. 1976; M. F. Baker, Fla. Wild Fls., ed. 2, imp. 2, 188. 1976; S. R. Hill, Sida 6: 325. 1976; Long & Lakela, Fl. Trop. Fla., ed. 2, 741—742 & 961. 1976; Soukup, Biota 11: 18. 1976; E. H. Walker, Fl. Okin. & South. Ryuk. 883 & 884. 1976; Moldenke, Phytologia 36: 28, 39, & 47. 1977.

Additional illustrations: Dill. in Ray, Synop. Meth. Stirp. Brit., ed. 3, pl. 300, fig. 387. 1724.

The Dillen pre-Linnean designation for this species is based

on a specimen grown in the Eltham Garden from seed sent from Buenos Aires, Argentina, by Milam in 1726, according to Druce & Vines
(1907), but how this can be possible when Dillen already published
the designation in 1724 is not clear to me. The Eltham specimen
in the Dillen herbarium was at first identified as V. bonariensis
"Willd." by Klinsmann.

Recent collectors describe this plant as a large or "giant"
herb, 1.3—2.5 m. tall, the stems stiff, square, with pronounced
angles and concave sides, hairy, pale- or dark-green, often maroon-
tinged, the leaves "sandpapery" rough, "parchmentaceous-membranous"
or "thickly velvety-herbaceous", glossy dark-green above, with
pale slightly sunken veins, dull light- or gray-green beneath with
prominent veins, the peduncles reddish, the small flowers in
spikes, the calyx green at the base, with purple or purplish tips,
the corolla hairy outside, and the infructescence gray-green.

Goldsmith found V. bonariensis in scattered patches in open
grassland in Rhodesia; Bos refers to it as "abundant" in the Transvaal and "locally dominant" in the Cape Province of South Africa;
Moll calls it a "locally fairly common annual herb" in Natal;
Scheepers says of it "locally frequent at water's edge and environs,
[in] well lit but dense vegetation, widespread as ruderal, especially in moist open spots" in the Transvaal.

Other recent collectors have encountered the species at the
edge of open fields in about 50 percent shade, in rather dry soil
of old fields, and among Baccharis on "spoil banks". Montz reports
it "abundant along railroad 'spoils'" in Louisiana, where Arceneaux
asserts that it is "a very common weed along roads, ditches, etc."
and Hester says it is "plentiful everywhere". Brown comments that
there are "no appendages on anthers" (which is to be expected,
since the species is not a member of the Section Glandularia). Wilbur reports it "common along edge of ditch between road and corn-
field" in North Carolina.

Walker (1976) cites Amano 7373, Hatusima 17571, and Walker 8133
from Okinawa and records the vernacular name, "tachi-ba-bena",
which he explains as "Tachi, erect, ba, leaf, bena, meaning unknown". Arenas encountered the species at the edges of an arroyo
and comments that it "crece en yuyal".

The corollas are described as having been "lilac" in color when
fresh on Ferreira 163, "blue" on Scheepers 93, "violet" on Arenas
949 and Cabrera 12446, "violet-blue" on Bos 1262, "purple" on O. B.
Miller 5184, Moll 745, Sowell s.n., and Wilbur 3947, "rose" on Carauta 686, "purple-mauve" on Goldsmith 71/62, and with "tube purple,
lobes lilac" on Bos 142.

Don (1830) called the species "Buenos Ayres vervain" and dates
its introduction into English gardens as 1732. Thomas (1969) calls
it the "pretty verbena", while Claycomb asserts that in Louisiana
it is known as "blue vervain". Carauta records "pai-joaquim" from
Rio de Janeiro. In Pahang, Malaya, it is said to be a "roadside
weed" in waste places and has been collected in fruit in November.

The Herb. Mus. Bot. Landishuth s.n., cited below, has all its leaves very narrow-oblong and bears very close resemblance to V. inamoena Briq. DeWinter 7176 is a voucher for chemical studies conducted on the species by Dr. P. R. Enslin. The Univ. Calif. Acc. No. 63.740-83, cultivated in California, was grown there from seeds collected in Teneriffe, Canary Islands. Fabris & Marchionni 2392 is a mixture with V. litoralis H.B.K.

Gibbs (1974) reports cyanogenesis and leucoanthocyanin absent from the leaves of this species and syringin absent from the stems.

Material of V. bonariensis has been misidentified and distributed in some herbaria under the designations V. bonariensis Hook. and V. littoralis Kunth, and even as "Labiatae sp.", while T. Rojas 10077 was originally identified as "V. bonariensis L. f. transiens in V. inamoenam Briq.". On the other hand, the Claycomb s.n. [June 13, 1942], Demaree 15103, Keiser 91, D. K. Lowe 31, J. A. Moore 5200, Robinette 239, and R. R. Smith 1732, distributed as V. bonariensis, actually are V. brasiliensis Vell., while Schlieben 7691 is V. litoralis H.B.K., Arceneaux 35a, Beck 453, and Horst JWH.B.281 are V. rigida Spreng., and Bayliss BS. 7344 is V. tenuisecta Briq.

Additional citations: NORTH CAROLINA: Robeson Co.: Moldenke & Moldenke 29992 (Ac), 29994 (Gz, Ld, Tu). Tyrrell Co.: Wilbur 3947 (Mi). SOUTH CAROLINA: Charleston Co.: Curtiss 1963** (Mu—1550). GEORGIA: Spalding Co.: Sowell s.n. [4 August 1967] (Lv). FLORIDA: Polk Co.: Moldenke & Moldenke 29529 (Ac, Gz, Ld, Tu). Putnam Co.: Moldenke & Moldenke 29839 (Ld). LOUISIANA: Cameron Par.: Spindler s.n. [30 Sept. 1973] (Lv). Iberia Par.: Hester 650 (Lv). Jefferson Par.: C. A. Brown 2049 (Lv). Lafayette Par.: Claycomb s.n. [July 6, 1942] (Lv). Livingston Par.: Montz 1824 (Lv). Ouachita Par.: R. D. Thomas & Bot. Class 18637 (Lc). Pointe Coupee Par.: M. Chaney 213 (Lv). Saint Mary Par.: Montz 2289 (Lv). Saint Tammany Par.: Arsène 12534 (Lv). Tangipahoa Par.: Correll & Correll 9259 (Lv). Terrebonne Par.: Arceneaux 35a (Lv); Wurzlow s.n. [May 29, 1914] (Lv). Vermilion Par.: Hester 693 (Lv). TEXAS: Orange Co.: J. A. Churchill s.n. [1 May 1955] (Ln—204155). MEXICO: Federal District: Karwinski s.n. [Chapultepec, Aug. 1827] (Mu—279, Mu—280). State undetermined: Prince Paul of Würtemberg s.n. [1830] (Mu—1549). BRAZIL: Paraná: L. F. Ferreira 163 (Ld). Rio de Janeiro: Carauta 686 [Herb. FEEMA 6811] (Ld). PARAGUAY: T. Rojas 10077 (Mu). URUGUAY: Herter 268 [Herb. Herter 81709] (Mu). CHILE: Arauca: Grau s.n. [31.3. 1968] (Mu). Concepción: Dessauer s.n. [Concepción, Feb. 1870] (Mu—1551). Valdivia: Buchtien s.n. [Valdivia, 7/11/1902] (Mu—3996); Lechler s.n. [Valdivia] (Mu—1568). State undetermined: Leyboldt s.n. [3/1/1860] (Mu—1569). ARGENTINA: Buenos Aires: Cabrera & Fabris 21 (Mu); Herb. Univ. Ludov. Maximil. s.n. (Mu—

277). Corrientes: Arenas 949 (Ld). Entre Ríos: Cabrera 12446
(Mu). San Juan: Fabris & Marchionni 2392, in part (Mu). RHODESIA:
Fries, Norlindh, & Weimarck 4003 (Mu); Goldsmith 71/62 (Mu). SWA-
ZILAND: O. B. Miller 5184 (Mu). SOUTH AFRICA: Cape Province: Bos
142 (Mu); Ecklon 85 (Mu—285); Penther 1795 (Mu—4085). Natal:
Moll 745 (Mu). Transvaal: Bos 1262 (Mu); DeWinter 7176 (Mu); Mee-
bold 12837 (Mu); Scheepers 93 (Mu); Schlieben 7176 (Mu). MASCARENE
ISLANDS: Mauritius: Sieber Fl. Maurit. 86 (Mu—286). INDIA: Khasi
States: Hooker & Thomson s.n. [Mont. Khasia 1-3000 ped.] (Mu—
288). State undetermined: Griffith s.n. [India orientali] (Mu—
287). MALAYA: Pahang: Poore 505 (Kl—505); B. C. Stone 5619 (Kl—
5220), 7236 (Kl—7747). FIJI ISLANDS: Viti Levu: Meebold 16522
(Mu). AUSTRALIA: New South Wales: Lohm s.n. [Sydney, 28.III.10]
(Mu—9138); Meebold 2749 (Mu). GREAT BARRIER REEF: Stradbroke: M.
S. Clemens 44242 (Mi). CULTIVATED: California: Univ. Calif. Acc.
No. 63.740-83 (Mu). France: Weinkauff s.n. [Jard. des Plant. 1834]
(Mu—1243). Germany: Herb. Mus. Bot. Landishuth s.n. (Mu—278);
Herb. Schmiedel s.n. (Mu—272, Mu—273); Herb. Schreber s.n. (Mu—
275, Mu—276); Herb. Schwaegrichen s.n. (Mu—1244); Herb. Zuccarini
s.n. [Hort. bot. Monao. 1832] (Mu—282), s.n. [Hort. bot. Monac.
1887] (Mu—283); Schreber s.n. [hort. bot. Erlang. 1772] (Mu—274).
India: Herb. Hort. Bot. Calcutt. s.n. (Mu—289). Sweden: Collector
undetermined s.n. [H. L. 1812] (Ac); Zetterstedt s.n. [H. L., 10
Sept. 1838] (Ac).

VERBENA BONARIENSIS var. CONGLOMERATA Briq.
    Additional bibliography: Moldenke, Phytologia 28: 346 & 380.
1974.
    Bornmüller encountered this plant at 500 meters altitude. Lin-
deman & Haas describe it as an herb, growing on grassy roadsides,
and the corollas on their no. 3902 are said to have been "deep
purple-blue" when fresh; the corollas were "purple" on Schinini &
Carnevali 10689.
    Additional citations: BRAZIL: Rio Grande do Sul: Bornmüller 255
(Mu—4286); Lindeman & Haas 3902 (Ut—320412); Reineck & Czermak
69 (Mu—3793). ARGENTINA: Corrientes: Ibarrola 2053 (Ut—330563B),
2131 (Ut—330573B). Misiones: Schinini & Carnevali 10689 (Ld).
CULTIVATED: California: Moldenke & Moldenke 30295 (Ac, Gz, Ld, Mu,
Tu, W).

VERBENA BONARIENSIS var. HISPIDA Moldenke, Phytologia 33: 374—
    375. 1976.
    Bibliography: Moldenke, Phytologia 33: 374—375 (1976) and 34:
259. 1976.
    Citations: BRAZIL: Rio Grande do Sul: Bornmüller 647 (Mu—
4302—type, Z—photo of type).

VERBENA BRACTEATA Lag. & Rodr.

Additional synonymy: _Verbena_ _bracteata_ (Michx.) Lag. & Rodr. ex Perkins, Estes, & Thorp, Bull. Torrey Bot. Club 102: 194, sphalm. 1975.

Additional & emended bibliography: G. Don in Loud., Hort. Brit., ed. 1, 247 (1830) and ed. 2, 247. 1832; Loud., Hort. Brit., ed. 2, 552. 1832; G. Don in Loud., Hort. Brit., ed. 3, 247. 1839; Buek, Gen. Spec. Syn. Candoll. 3: 494, 496, & 507. 1858; Kuntze, Rev. Gen. Pl. 2: 510. 1891; Krause in Just, Bot. Jahresber. 44: 1212. 1926; Fedde in Just, Bot. Jahresber. 44: 1534. 1927; Conard, Pl. Iowa 44. 1951; R. A. Davidson, State Univ. Iowa Stud. Nat. Hist. 20 (2): 77. 1959; Hall & Thompson, Cranbrook Inst. Sci. Bull. 39: 74. 1959; Barker, Univ. Kans. Sci. Bull. 48: 571. 1969; Cooperrider, State Univ. Iowa Stud. Nat. Hist. 20 (5): 70. 1962; Bolkh., Grif, Matvej., & Zakhar., Chrom. Numb. Flow. Pl., imp. 1, 716 & 717. 1969; G. W. Thomas, Tex Pl. Ecolog. Summ. 77. 1969; Bolkh., Grif, Matvej., & Zakhar., Chrom. Numb. Flow. Pl., imp. 2, 716 & 717. 1974; Welsh, Utah Pl., ed. 3, 354 & 473, 1974; [Bard], Bull. Torrey Bot. Club 102: 431. 1975; D. S. & H. B. Correll, Aquat. & Wetland Pl. SW. U. S., imp. 2, 2: 1397, 1400, & 1775. 1975; E. H. Jordan, Checklist Organ Pipe Cact. Natl. Mon. 7. 1975; Kooiman, Act. Bot. Neerl. 24: 463. 1975; Moldenke, Phytologia 30: 135—136 (1975) and 31: 376, 377, & 415. 1975; Perkins, Estes, & Thorp, Bull. Torrey Bot. Club 102: 194—198. 1975; H. D. Wils., Vasc. Pl. Holmes Co. Cat. 54. 1975; Anon., Biol. Abstr. 61: AC1.732, 1976; L. J. Clark, Wild Fls. Pacif. Northw. 444—445 & 603. 1976; Grimé, Bot. Black Amer. 191. 1976; Moldenke, Phytologia 34: 248—251, 270, & 278. 1976; Van Bruggen, Vasc. Pl. S. Dak. 369, 529, & 536. 1976; [Voss], Mich. Bot. 15: 237. 1976; Moldenke, Phytologia 36: 29 & 30. 1977; F. H. Montgomery. Seeds & Fruits 201, fig. 5, & 230. 1977; Taylor & MacBryde, Vasc. Pl. Brit. Col. 436 & 751. 1977.

Additional illustrations: F. H. Montgomery, Seeds & Fruits 201, fig. 5. 1977.

Montgomery (1977) describes the seeds of this species as follows: "Nutlets 2.3 x 0.6 mm, oblong 13—14 in l.s., obtriangular 90 in c.s. with the dorsal surface rounded; surface 2—4-ribbed on the lower half and reticulate on the upper half, inner faces papillose".

Wilson (1975) cites Cooperrider 8069 from weedy railroad banks in Holmes County, Ohio. Stevens describes it as "from a woody caudex, branches prostrate" and reports it "common on dry rocky hillsides" in Washington. Higgins encountered it in sandy soil of shin-oak association areas, while Semple & Love found it in dry sandy soil of mesquite shrublands. Other recent collectors have found it growing in bare sandy soil with mesquite dominant, in sidewalk crannies, and "in agrupaciones de malezas (annual herbs mainly) in fine calcareous alluvium of sandy clay loam with Macheranthera parva, Bouteloua barbata, Verbena spp., Marsilea sp., and Wislizenia sp." Moran, in Baja California, found it "prostrate, locally common with Lythrum hyssopifolium in shallow roadside depressions".

Davidson (1959) refers to V. bracteata as a "Weed of roadsides, railways, pastures, barnyards, and other open places; common" in Iowa. In the same state Cooperrider (1962) speaks of it as "Frequent. Roadsides; along railroads; open, waste places" and records it from Clinton, Jackson, and Jones Counties. Hall & Thompson (1959) refer to it as occasional in fields and along open roadsides in Oakland County, Michigan. In Kansas it is said by Barker (1969) to be "Occasional along roadsides, cultivated fields and railway right-of-ways. Found throughout the area." Dziekanowski and his associates found it growing in loess of cultivated areas and very scattered in grassy areas with gramma grass and staghorn cactus in piñyon-juniper woodlands.

The corollas on Semple & Love 297 are said to have been "purple with white center" when fresh, but on Richardson 1580 and W. D. Stevens 1710 they were "blue", while on Moran 23634 it is stated that the corolla-tube was "purplish-pink, the limb white", on Duncan 12696 the corolla was "light-purple" and on Spellenberg & al. 4327 "pale-violet, paler in throat". Taylor & MacBryde (1977) classify the flowers and blue and red, flowering from April to September, and growing as a weed in British Columbia.

Higgins has found V. bracteata in sandy soil in Prosopis-juniper communities and in sandy to clayey soils in the Prosopis-Juniperus-grassland community. Thorne found it growing in association with Gnaphalium purpureum, G. palustre, Navarettia hamata, Amaranthus albus, etc. Moran refers to it as "occasional" near drying pond edges and "towards edge of dry laguna with abundant Sida hederacea" in Baja California. Blakley refers to it as an "annual, common locally, prostrate at moist edge of vernal lake in silty-clay soil", while Demaree found it (the erect form) to be "common" in open rock mountains of Utah. Fosberg encountered it at 20 meters altitude. Churchill encountered it "on eroded sandy slopes in sandy loam soil of upland prairie pasture with Andropogon, Solidago, etc."

Other recent collectors have encountered this species on dry riverbeds and banks, along roadsides, on dry slough bottoms, and in the Shortgrass-Prairie community. Don (1830) calls the plant "bracteose vervain" and dates its introduction into English gardens as 1820 from "Mexico".

Muehlenbach considered his no. 1525 to represent a natural hybrid between V. simplex Lehm. and V. bracteata, but I can see no evidence of such hybridity in the sample of this collection examined by me.

Perkins and his associates (1975) report that in V. bracteata the corollas are rapidly deciduous (caducous) following anthesis, the anthers and stigmas are about 1 mm. apart, the plants are less autogamous than V. halei Small and V. urticifolia L., seven plants studied with 310 potential seeds had a 21.9 percentage of seed-set when bagged, but nine plants with 765 potential seeds had a 66.5 percent seed-set when insect-visited. The flowers were visited by the following insects: Diptera: Exoprosopa sp. (with pollen on head); Hymenoptera: Meusebeckidium obsoletum and Parnopes edwardsii; Lepidoptera: Everes comyntas, Hemiargus isola (with pollen on head), Nathalis iole, Pieris protodice (with pollen on head), Phyciodes phaon (with pollen on head), Pyrgus communis, and Strymon melinus (with pollen on head). The Exoprosopa Hemiargus, Pieris, Pyrgus, and Strymon insects also visited the other Verbena species in the locality, i.e., V. halei, V. stricta, and V. urticifolia. The authors note that V. bracteata "was visited predominantly by butterflies".

It should perhaps be noted here that Edw. Palmer 3411, as least insofar as the Munich specimen is concerned, consists only of extremely young sterile seedlings.

The Atwood 6298, distributed as V. bracteata, is actually V. ambrosifolia Rydb., Drummond s.n. [Saint Louis, 1832] is V. canadensis (L.) Britton, Martens s.n. [Missouri] is xV. deamii Moldenke, Edw. Palmer 342 is V. lasiostachys var. septentrionalis Moldenke, and Meebold 25461 is V. stricta Vent. W. Schumann 1070 is in part V. gracilis Desf. and in part V. canescens H.B.K. Demaree 3571, in the herbarium of Louisiana State University, bears a label inscribed "Dioscorea villosa L." — doubtless a case of mixed labels. The C. A. Brown 20378, Brown & Lenz s.n. [April 7, 1939], M. Chaney 226, N. F. Petersen s.n. [Apr. 10, 1909], and Stotts s. n. [April 12, '41], also distributed as V. bracteata, all are actually V. canadensis (L.) Britton.

Additional citations: GEORGIA: Clarke Co.: Duncan 12696 (Lv). ILLINOIS: Winnebago Co.: M. S. Bebb s.n. [Fountaindale] (Mu). INDIANA: Jefferson Co.: Frazee s.n. [June 6, 1885] (Lc). LaPorte Co.: Moffatt 1685 (Mi). IOWA: Story Co.: Arthur s.n. [Ames, July 17, 1877] (Mu). TENNESSEE: Blount Co.: Hooker s.n. [borders of Ohio river below Louisville] (Mu—293). MICHIGAN: Grand Traverse Co.: J. A. Churchill s.n. [12 August 1955] (Ln—203420). WISCONSIN: Brown Co.: Schuette s.n. [Fort Howard, July 22, 1887] (Mu). SOUTH DAKOTA: Lawrence Co.: N. F. Petersen s.n. [Aug. 17, 1908] (Lv). MISSOURI: Cass Co.: Meebold 24223 (Mu). Saint Louis: Muehlenbach 312 (Mu), 1525 (Mu), 3474 (Ac), ARKANSAS: Carroll Co.: H. H. Rusby 780 1/2 (Mu). Craighead Co.: Demaree 3571 (Lv). County undetermined: F. L. Harvey s.n. [Curtiss 1962] (Mu—1552). LOUISIANA: Lincoln Par.: J. A. Moore 5362 (Lv). UTAH: Utah Co.: Demaree 65401 (Ld). NEVADA: Lyon Co.: Tiehm 1812 (N). Storey Co.: Purpus 5946 (Mu—4287). COLORADO: Archuleta Co.: Weber & Livingston 6259 (Mi). Denver Co.: Cole 5687 [Herb. Kent Sci. Mus. 51385] (Mi). Elbert Co.: S. R. Hill 1277 (N). El Paso Co.: Meebold 12269 (Mu). NEBRASKA: Pierce Co.: N. F. Petersen s.n. [Aug. 11, 1910] (Lv), s.n. [Aug. 12, 1910] (Lv). Polk Co.: S. P. Churchill

6606 (N). IDAHO: Gooding Co.: R. J. Davis 1781 (N). TEXAS:
Bowie Co.: Correll & Correll 12398 (Mi). Martin Co.: Semple & Love
297 (W—2732736). NEW MEXICO: Catron Co.: Pinkava, Lehto, & Reeves P.12492 (N, W—2737086). Curry Co.: Higgins 9050 (N). Dona
Ana Co.: Chiang, Wendt, & Johnston 8621 (Ld). Guadalupe Co.: Higgins 8925 (N). McKinley Co.: Spellenberg, Reitzel, & McKinney
4327 (N). Roosevelt Co.: Higgins 8672 (N). San Miguel Co.: Higgins 8886 (N). Torrence Co.: Dziekanowski, Dunn, & Bennett 2390
(N). Union Co.: Higgins 8810 (N). ARIZONA: Greenlee Co.: Pinkava,
Lehto, & Reeves P.12421 (N). WASHINGTON: Benton Co.: Dziekanow-
ski, Dunn, & Bennett 2518 (N). Klickitat Co.: Doppelbauer & Dop-
pelbauer 663 (Mu). Yakima Co.: W. D. Stevens 1710 (Ln--244892).
OREGON: Malheur Co.: Doppelbauer & Doppelbauer 824 (Mu). CALI-
FORNIA: Los Angeles Co.: F. R. Fosberg S.4105 (Sd—72779); Parish
& Parish 1596 (Mu—1553); Thorne 36516 (Sd—69553). San Luis
Obispo Co.: Edw. Palmer 3411 (Mu). CHANNEL ISLANDS: Santa Cata-
lina: Blakley 5394 (Sd—85037). MEXICO: Baja California: R. V.
Moran 16110 (Sd—71446), 16627 (Sd—72969), 23634 (Ld). Coahui-
la: Richardson 1580 (Au—302675). CULTIVATED: Germany: Herb.
Kummer s.n. [Hort. bot. Monac. 1843] (Mu—1249); Herb. Zuccarini
s.n. [Hort. bot. Monac. 1836] (Mu—295); Schrank s.n. [Hort.
Monac.] (Mu—290). Sweden: Collector undetermined s.n. (Ac).
LOCALITY OF COLLECTION UNDETERMINED: Herb. Reg. Monac. 291 (Mu).

VERBENA BRASILIENSIS Vell.
     Additional synonymy: Verbena litoralis pycnostachya Schau. ex
Moldenke, Fifth Summ. 2: 680, in syn. 1971.
     Additional & emended bibliography: Buek, Gen. Spec. Syn. Can-
doll. 3: 494 & 495. 1858; Bolkh., Grif, Matvej., & Zakhar.,
Chrom. Numb. Flow. Pl., imp. 1, 716. 1969; G. W. Thomas, Tex. Pl.
Ecolog. Summ. 77. 1979; Bolkh., Grif, Matvej., & Zakhar., Chrom.
Numb. Flow. Pl., imp. 2, 716. 1974; D. S. & H. B. Correll, Aquat.
& Wetland Pl. SW. U. S., imp. 2, 2: 1396, 1397, & 1775. 1975;
Moldenke, Phytologia 30: 136 (1975), 31: 375 & 377 (1975), and
34: 248—250, 257, 260, & 270. 1976; Soukup, Biota 11: 18. 1976.
     Ferreira refers to this plant as a shrub, 1.2 m. tall, and
found it growing in "brejo" (sedge meadow). Roivainen found it
with Lupinus arboreus "en la zona humosa-arenisca" in Chile. In
Ecuador it was encountered at 2850 meters altitude by Heinrichs.
Other recent collectors have found it growing in pine forests,
at the edges of ditches, in dry soil in sunny places, in "poorly
drained stiff soil", and in shallow pools on prairies. Monz refers to it as "scarce" and as "scattered but relatively abundant
on roadsides" in Louisiana, Curry found it frequent along banks
and among Colocasia, Kirby reports it abundant on railroad banks,
and Rockett refers to it as common in waste places. Moore calls
it a roadside weed and Claycomb asserts that in Louisiana it is

"a very common roadside weed".

On Ross 66 the label asserts that the leaves and stems were dark-green, while on C. A. Brown 2393 the collector informs us that there were "thick and thin spikes on the same plant". Hebert refers to the species as "abundant in open cornfields". The corollas on Ferreira 126 are said to have been "lilac" in color when fresh, while on Athanasius 64 and Van Wyk 267 they were "blue", on C. Allen 107, C. M. Allen 381, Chandraparya 2, Rockett 125, and Ross 66 they were "purple", on Duque-Jaramillo 2734 they were "light-blue", on Bougere 14 and R. R. Smith 1732 "lavender", and on Krapovickas & al. 26799 "violet".

Additional vernacular names reported for V. brasiliensis are "kudii penkel" and "verbena morada". In addition to months previously reported, the species has been collected in fruit in March. It has been found growing at altitudes of 200 to 3700 meters in Peru. Demaree speaks of it as "common" or "very common" in wet open woods and disturbed areas in Arkansas and in rocky bottoms in Alabama. Werff comments that on Chatham Island, in the Galápagos, its flowers were "smaller and much more intense blue" than those on his no. 2181 (an as yet undetermined species).

The Claycomb s.n. [July 6, 1942] and Montz 1824, distributed as V. brasiliensis, actually are V. bonariensis L., while Hort. Parag. 10054, Romero-Castañeda 10668, and T. Rojas 1889 are actually V. litoralis H.B.K. and Muhammad 259 and Urbatsch 1938 are V. montevidensis Spreng.

Additional citations: NORTH CAROLINA: Robeson Co.: Moldenke & Moldenke 29997 (Ac). SOUTH CAROLINA: Bamberg Co.: Moldenke & Moldenke 29963 (Ac, Ld, Tu). Clarendon Co.: Moldenke & Moldenke 29977 (Tu). Dillon Co.: Moldenke & Moldenke 29989 (Ld). Greenwood Co.: Moldenke & Moldenke 29263 (Ld). Marion Co.: Moldenke & Moldenke 29987 (Kh). Orangeburg Co.: Moldenke & Moldenke 29970 (Ac), 29971 (Gz). GEORGIA: Chatham. Co.: Moldenke & Moldenke 29924 (Ac, Gz, Ld). Dougherty Co.: Moldenke & Moldenke 29361 (Ac). St. Simon's Island: Moldenke & Moldenke 29900 (Ac, Ld, Tu). FLORIDA: Escambia Co.: Meebold 27240 (Mu); R. R. Smith 1732 (Sd—74167); S. M. Tracy 8706 (Ln—70125). Putnam Co.: Moldenke & Moldenke 29845 (Gz). ALABAMA: Montgomery Co.: Demaree 70060 (Ld). MISSISSIPPI: Warren Co.: C. A. Brown 18610 (Lv). ARKANSAS: Cleveland Co.: Demaree 69748a (Ld). Lincoln Co.: Demaree 70225 (Ld). Saint Francis Co.: Demaree 15103 (Lv). LOUISIANA: Acadia Par.: Chandraparya 2 (Lv). Bossier Par.: Robinette 239 (Lv). East Baton Rouge Par.: C. Allen 107 (Lv); C. A. Brown 1008 (Lv); Curry 615 (Lv), 666 (Lv); Hebert 207 (Ld); Ross 66 (Lv). Lafayette Par.: Claycomb s.n. [June 13, 1942] (Lv). Lincoln Par.: Keiser 91 (Lc); D. K. Lowe 31 (Lv); J. A. Moore 5200 (Lv). Orleans Par.: C. A. Brown 238 (Lv), 2393 (Lv); Meebold 27224 (Mu). Ouachita Par.: B. Thompson & Botany

Class 157 (Lc). Plaquemines Par.: C. A. Brown 2309 (Lv). Pointe
Coupee Par.: M. Chaney 111 (Lv). Saint Charles Par.: Montz 56
(Lv), 182 (Lv), 566 (Lv), 637 (Lv, Lv, Lv), 726 (Lv), 790 (Lv).
Saint Helena Par.: C. M. Allen 381 (Lv); Rockett 125 (Lv). Saint
Tammany Par.: Bougere 14 (Lv), 1091 (Lv), 1099 (Lv). Tangipahoa
Par.: Frederick 108 (Lv); Kirby 160 (Lv). Terrebonne Par.: Arce-
neaux s.n. [July 20, '37] (Lv); Bynum, Ingram, & Jaynes s.n.
[Houma, Apr. 23, 1933] (Lv). West Feliciana Par.: Curry, Martin,
& Allen 38 (Lv); Seib 10 (Lv). TEXAS: Jefferson Co.: Stutzenbaker
205 (Mu). CALIFORNIA: Butte Co.: Moldenke & Moldenke 30291 (Ac,
Ld, Mu, W). COLOMBIA: Cundinamarca: Duque-Jaramillo 2707 (N),
2734 (N). ECUADOR: Tunguragua: Heinrichs 65 (Mu). GALAPAGOS IS-
LANDS: Chatham: Werff 218 [1482] (Ld). BRAZIL: Minas Gerais: Ir-
win, Harley, & Onishi 29512 (W—2759076). Paraná: L. F. Ferreira
126 (Ld). Rio Grande do Sul: Reineck & Czermak 63 (Mu). State
undetermined: Martius 1033 (Mu). BOLIVIA: La Paz: M. Bang 136
(Mu—1785). CHILE: Cautin: Roivainen 3054 (Mu). Concepción:
Neger s.n. [1893-96] (Mu—3984). Valdivia: Athanasius 64 (Mu).
Valparaiso: Behn s.n. [Quilpué, 22 Januar 1931] (Mu). ARGENTINA:
Córdoba: Lorentz 131 [Macbride photos 20316] (Mu); Pierotti s.n.
[27/I/1944] (Ut—330559B, Ut—330575B). Corrientes: Krapovickas,
Cristóbal, Irigoyen, & Schinini 26799 (Ld). Mendoza: Ruiz Leal
8393 (Ut—330558B). SOUTH AFRICA: Natal: Meebold 12840 (Mu).
Transvaal: Van Wyk 267 (Ac).

VERBENA CABRERAE Moldenke
    Additional bibliography: Moldenke, Phytologia 28: 120, 195,
197, & 440 (1974) and 31: 388. 1975.
    Additional citations: ARGENTINA: Jujuy: Krapovickas, Schinini,
& Quarín 26689 (Ld).

VERBENA CABRERAE var. ANGUSTILOBATA Moldenke
    Additional bibliography: Moldenke, Phytologia 28: 195, 197, &
440. 1974.

VERBENA CALIFORNICA Moldenke
    Additional bibliography: Moldenke, Phytologia 28: 197. 1974.
    McNeal 925, a topotype, distributed as V. officinalis L., was
collected in moist rocky serpentine soil in streambed crossing
(the actual type locality!), flowering in May. This is apparently the exact spot where my wife and I, as well as our son, observed and collected this plant and where our son originally discovered it.
    Additional citations: CALIFORNIA: Tuolumne Co.: McNeal 925 (N).

VERBENA CALLIANTHA Briq.
    Additional bibliography: Moldenke, Phytologia 30: 136. 1975.
    Troll encountered this species on páramos at 3400 meters alti-

tude. Porto & Oliveira describe it as stoloniferous, the "petals red (5P5/9)".

Additional citations: BRAZIL: Rio Grande do Sul: <u>Porto & Oliveiro ICN.9585</u> (Ut--320460). BOLIVIA: Province undetermined: <u>Troll 281</u> [Altos del Escalon] (Mu).

VERBENA CAMERONENSIS L. I. Davis

Additional & emended bibliography: Bolkh., Grif, Matvej., & Zakhar., Chrom. Numb. Flow. Pl., imp. 1, 716. 1969; G. W. Thomas, Tex. Pl. Ecolog. Summ. 77. 1969; Bolkh., Grif, Matvej., & Zakhar., Chrom. Numb. Flow. Pl., imp. 2, 716. 1974; Moldenke, Phytologia 28: 243. 1974; Hocking, Excerpt. Bot. A.26: 5. 1975.

Martínez-Calderón found this plant growing in "acahual" secondary vegetation, at 6 meters altitude, and the corollas on his <u>no. 1246</u> are said to have been "blue".

Additional citations: MEXICO: Veracruz: <u>Martínez-Calderón 1246</u> (N).

VERBENA CAMPESTRIS Moldenke

Additional bibliography: Moldenke, Phytologia 23: 218 (1972) and 36: 35. 1977.

Kummrow found this plant growing in "orla campo das encosta de pequenos morros", flowering in September, and describes the corollas as white.

Additional citations: BRAZIL: Paraná: <u>Kummrow 1142</u> (Z).

VERBENA CANADENSIS (L.) Britton

Additional synonymy: <u>Verbena lamberti</u> B. M. ex G. Don in Loud., Hort. Brit., ed. 1, 247. 1830. <u>Glandularia aubletia</u> α Spach, Hist. Nat. Veg. Phan. 9: 240. 1840. <u>Glandularia aubletia</u> β Spach, Hist. Nat. Veg. Phan. 9: 240. 1840. <u>Verbena pulchella erinoides</u> Zucc. ex Moldenke, Phytologia 34: 279, in sphalm. 1976.

Additional & emended bibliography: G. Don in Loud., Hort. Brit., ed. 1, 247 (1830) and ed. 2, 247. 1832; G. Don in Loud., Hort. Brit. Suppl. 1: 680. 1832; Loud., Hort. Brit., ed. 2, 552 & 602. 1832; Loud., Hort. Brit., ed. 3, 602. 1839; Baxt. in Loud., Hort. Brit. Supp. 2: 680. 1839; Spach, Hist. Nat. Veg. Phan. 9: 239—240. 1840; Baxt. in Loud., Hort. Brit. Suppl. [3]: 655. 1850; Schnitzlein, Iconogr. Fam. Nat. 2: 137 Verbenac. [3] & 137: fig. 3. 1856; Buek, Gen. Spec. & Syn. Candoll. 3: 199, 494, & 495. 1858; Vilm., Fl. Pleine Terr., ed. 1, 936 (1865), ed. 2, 2: 973—974 (1866), and ed. 3, 1: 1197. 1870; Kuntze, Rev. Gen. Pl. 2: 510. 1891; Voss in Vilm., Fl. Pleine Terr., ed. 4, 1065 & 1070. 1894; Loes., Verh. Bot. Ver. Brand. 53: 75. 1912; Barker, Univ. Kans. Sci. Bull. 48: 571. 1969; Bolkh., Grif, Matvej., & Zakhar., Chrom. Numb. Flow. Pl., imp. 1, 715 & 716. 1969; G. W. Thomas, Tex. Pl. Ecolog. Summ. 77. 1969; Healy, Gard. Guide Pl. Names 225. 1972; Bolkh., Grif, Matvej., & Zakhar., Chrom. Numb. Flow. Pl., imp. 2, 715 & 716. 1974; El-Gazzar, Egyot. Journ. Bot. 17: 75 & 78. 1974; Kooiman, Act. Bot. Neerl. 24: 463. 1975; Moldenke, Phytologia 30: 136—138 (1975) and 31: 376, 377, 409, & 411. 1975; H. D. Wils., Vasc. Pl. Holmes Co. Cat. 54.

1975; Fleming, Genelle, & Long, Wild. Fls. Fla. 15 & 82. 1976;
Moldenke, Phytologia 34: 249, 250,270, 274, & 279. 1976; Vanderpoel, Natl. Wildlife 15 (1): 50. 1976; A. L. Moldenke, Phytologia
36: 88. 1977.

Additional illustrations: Voss in Vilm., Fl. Pleine Terr., ed.
3, 1: 1197 (1870) and ed. 4, 1065. 1894; Vanderpoel, Natl. Wildlife 15 (1): 50 [in color]. 1976.

It should be noted here that Spach's var. ☉ is based on the
original V. aubletia L. and is characterized by him as "a fleurs
pourpres", while his var. 𝛃 is based on V. drummondii (Lindl.)
Baxt. and is characterized as "a fleurs lilas". He regards
Billardiera explanata Moench as a synonym of var. ☉. Of the
species he says "Cette espèce, originaire des provinces méridio-
nales des États-Unis, se cultive fréquement comme plante de par-
terre".

Tans describes the material of V. canadensis which he collected "in recently landscaped roadside gravel in partial shade" as
somewhat sprawling, the corollas rose-purple, the corolla-tube 17
mm. long, its limb 12 mm. wide, and the calyx 12 mm. long. Stephens found the species "in dry rocky limestone soil on prairie
pasture hillside". Waterfall encountered it in openings in oak-
hickory woods on stony hillsides. The corollas on D'Arcy & Beckner 1656 are said to have been "showy blue-purple", while on
LePine 5056 they were "lavender", on Guguch 92 they were "rose",
and on Ware 51 they were "blue".

Healy (1972) lists the horticultural varieties: "Compacta Amethyst", "Miss Susie Double", "Olympia", and "Royal Bouquet". These
are listed as though varieties of V. canadensis, but certainly the
last-mentioned (and perhaps all) is a cultivar of xV. hybrida
Voss. Darlington & Wylie (1956) refer to V. canadensis as native
to "N. & S. Am.", but this is erroneous — it is not known from
South America! Don (1830) calls V. aubletia "Aublet's vervain"
and V. lamberti "Lambert's vervain", both now usually regarded as
synonyms of V. canadensis. He dates the introduction of Aublet's
vervain into English gardens as 1774 from "N. Amer.", but Lambert's vervain he says was introduced from "S. Amer.", date not
known. This latter statement does not seem likely.

In this connection it should be noted here that the Bahama
Islands collection cited below does not indicate on its accompanying label that it originated from cultivated material, but I
assume that it did, since it is most unlikely that V. canadensis
should occur in the wild state on these islands.

The "V. candensis" illustrated by Fleming, Genelle, & Long
(1976) actually is V. tenuisecta Briq.!

Wilson (1975) cites H. D. Wilson 1860 from Holmes County,
Ohio, based on "a single plant".

Recent collectors have encountered V. canadensis on dissected
stream terraces, in cleared thickets on ridges, in mixed woods,
open fields, and wooded areas on black calcareous soil, on "hill-

side rocks", under mixed hardwoods on dry hillsides, in clay soil
of grassy fields with Trifolium, in black or black calcareous
soil of prairies, and in novaculite bottoms. LePine refers to it
as a common perennial, while Guguch calls it an occasional herbaceous perennial. Collectors describe it as erect, with a flat-
topped inflorescence, the stems ascending, 4--5 dm. tall, and the
flowers odorless. Barker (1969) reports it "Common, in rocky upland prairies, on rocky prairie slopes, on rocky roadside embankments and along railroad right-of-ways. Found throughout the
area."

Other recent collectors have found the species in the red soil
of dry pine flatwoods in Texas, while Demaree reports it "common
in dolomite glades" in Arkansas.

The Herb. Staatssamml. Münch. s.n. and Herb. Hort. Monac. s.n.,
originally distributed as V. canadensis, are actually V. andrieuxii
Schau., while J. A. Churchill s.n. [14 May 1954], Herb. Reg. Monac.
s.n. [Missouri], and R. Kral 26422 are V. bipinnatifida Nutt.,
Karwinski s.n. [In imperio mexicano] is V. elegans H.B.K., Purpus
6061 is V. gooddingii Briq., Lindheimer IV.501 is V. pumila Rydb.,
Kirby 116 is V. rigida Spreng., Curtiss 1963 is V. tampensis Nash,
and C. A. Brown 5636 is V. tenuisecta Briq.

Coppor s.n. [June 23, 1959] is anomalous in having the leaves of
V. canadensis and the calyxes of V. ambrosifolia Rydb.

Additional citations: FLORIDA: Alachua Co.: D'Arcy & Beckner
1656 (Sd--85760). ALABAMA: Tuscaloosa Co.: Nevius s.n. (Lv, Lv).
OHIO: County undetermined: Frank s.n. [1837] (Mu--262). TENNESSEE:
Wilson Co.: J. A. Churchill s.n. [24 April 1971] (Ln--236291).
WISCONSIN: Walworth Co.: Tans 1476 (Ts). KANSAS: Franklin Co.:
Coppor s.n. [June 23, 1959] (Lc). Lyon Co.: Sudweeks s.n. [6/13/
60] (Lc). Osage Co.: Roush s.n. [6/16/59] (Lc); Stephens 30597
(Sd--74400). MISSOURI: Barry Co.: Meebold 25570 (Mu). Cooper Co.:
McReynolds 750651 (Lv). Jackson Co.: Meebold 22681 (Mu), 25248
(Mu). St. Louis: Drummond s.n. [Saint Louis, 1832] (Mu--294).
ARKANSAS: Carroll Co.: Demaree 53464 (Ln--236563). Drew Co.: Dem-
aree 16566 (Lv). Hot Springs Co.: Demaree 16920 (Lv). LOUISIANA:
Ascension Par.: Ware 51 (Lv). Bossier Par.: Robinette 67 (Lv).
Caddo Par.: C. A. Brown 20378 (Lv); N. F. Petersen s.n. [Apr. 10,
1909] (Lv). East Baton Rouge Par.: C. A. Brown s.n. [Sept. 30,
1936] (Lv). Natchitoches Par.: C. A. Brown 7147 (Lv). Ouachita
Par.: Garland 200 (Lv); Thomas & Jones 1684 (Kl--11393). Pointe
Coupee Par.: M. Chaney 226 (Lv). Saint John the Baptist Par.:
Guguch 92 (Lv); LePine 5056 (Lv). Saint Tammany Par.: Arsène
11982 (Lv). Winn Par.: Brown & Lenz 7607 (Lv), s.n. [April 7,
1939] (Lv); N. F. Petersen s.n. [4-12-12] (Lv). OKLAHOMA: Creek
Co.: J. A. Churchill s.n. [13 April 1953] (Ln--203426), Mayes
Co.: Waterfall 15292 (Mu). Tulsa Co.: Stutts s.n. [April 12,

'41] (Lv). TEXAS: Harris Co.: <u>J. A. Churchill</u> <u>s.n.</u> [1 May 1955].
(Ln--204153). LOCALITY OF COLLECTION UNDETERMINED: Prince Paul
of Wurtemberg <u>s.n.</u> [America sept. civit. confed. 1831] (Mu--1567).
CULTIVATED: Austria: <u>Herb. Zuccarini</u> <u>s.n.</u> [hort. Hügel, Vindob.
1839] (Mu--269). Bahama Islands: <u>Herb. Schmiedel</u> <u>s.n.</u> [Ins. Ba-
hamensis] (Mu--256). Germany: <u>Herb. Grimm</u> <u>s.n.</u> [1787] (Mu--255);
<u>Herb. Reg. Monac.</u> <u>263</u> (Mu); <u>Herb. Schmiedel</u> <u>s.n.</u> [h. Selton. '87]
(Mu--257); <u>Herb. Schreber</u> <u>s.n.</u> [Hort. Erlang. 1778] (Mu--258), <u>s.</u>
<u>n.</u> [Hort. Erlang. 1785] (Mu--259); <u>Herb. Univ. Ludv. Maximil.</u> <u>s.n.</u>
(Mu--251); <u>Herb. Zuccarini</u> <u>s.n.</u> [Hort. bot. Monac.] (Mu--264, Mu--
265); <u>Schrank</u> <u>s.n.</u> [Hort. Monac.] (Mu--250).

VERBENA CANADENSIS (L.) Britton x V. AMBROSIFOLIA Rydb.
    Additional bibliography: Moldenke, Phytologia 28: 199--200 &
451. 1974.
    <u>Coppor</u> <u>s.n.</u> [June 23, 1959], cited under <u>V. canadensis</u>, has
intermediate characters between that species and <u>V. ambrosifolia</u>
and may thus represent this hybrid.

VERBENA CANADENSIS (L.) Britton x V. ELEGANS H.B.K.
    Additional bibliography: Moldenke, Phytologia 28: 200 & 451.
1974.

VERBENA CANADENSIS (L.) Britton x V. MARITIMA Small
    Additional bibliography: Moldenke, Phytologia 28: 200, 451, &
464. 1974.

VERBENA CANADENSIS (L.) Britton x V. PERUVIANA (L.) Britton
    Additional bibliography: Moldenke, Phytologia 28: 200, 451, &
464. 1974.

VERBENA CANADENSIS (L.) Britton x V. TAMPENSIS Nash
    Additional bibliography: Moldenke, Phytologia 28: 200, 451, &
465. 1974.

VERBENA CANESCENS H.B.K.
    Additional synonymy: <u>Verbena</u> <u>canescens</u> var. <u>canescens</u> [H.B.K.]
apud Thomas, Tex. Pl. Ecolog. Summ. 77. 1969.
    Additional & emended bibliography: Sweet, Hort. Brit., ed. 2,
419. 1830; G. Don in Loud., Hort. Brit., ed. 1, 247 (1830), ed.
2, 247. 1832; Loud., Hort. Brit., ed. 2, 552 & 553. 1832; G. Don
in Loud., Hort. Brit., ed. 3, 247. 1839; Buek, Gen. Spec. Syn.
Candoll. 3: 494. 1858; Loes., Verh. Bot. Ver. Brand. 53: 74. 1912;
Bolkh., Grif, Matvej., & Zakhar., Chrom. Numb. Flow. Pl., imp. 1,
716. 1969; G. W. Thomas, Tex. Pl. Ecolog. Summ. 77. 1969; Bolkh.,
Grif, Matvej., & Zakhar., Chrom. Numb. Flow. Pl., imp. 2, 716.
1974; El-Gazzar, Egypt. Journ. Bot. 17: 75 & 78. 1974; Moldenke,
Phytologia 30: 138, 156, & 165 (1975), 31: 377 & 378 (1975), and
34: 251, 252, & 278. 1976.

Don (1830) calls this the "canescent vervain" and dates its introduction into English gardens from Mexico as 1824, but Loudon (1832) dates its introduction as 1820. Thomas (1969) calls it the "gray verbena".

Recent collectors have encountered this plant in clayish soil of desert scrub, in calcareous loam in chaparral on steep encanyoned conglomerate fans, in calcareous gravelly soil of "matorral desértico microfilo inerme" on limestone hills, in open hilly limestone areas, on rock and gravel in xerophytic canyons, in "matorral desértico inerme y con espinas laterales" on mostly limestone hillsides with some extrusive igneous rock on top, on limestone gravel and rocky limestone slopes, and in "matorral subdesértico inerme y con espinas laterales" on flat areas near the bottom of "bajadas", growing in association with Larrea tridentata, Parthenium incanum, Celtis pallida, Flourensia cernua, Yucca carnerosana, Agave lecheguilla, Vauquelinia corymbosa, Fouquieria splendens, Amelanchier, Acacia, Opuntia, Mortonia, Lycium, Mimosa, and Dasylirion. Henrickson refers to it as "frequent" in some localities and as an "infrequent perennial" in others. Ellis and his associates found it on "oak scrub hills with Lupinus and Populus".

The corollas are said to have been "violet" on Henrickson 6263. Demaree asserts that there were "not many" of these plants where he made his Texan collection. Loesener (1912) cites Seler & Seler 3488 from Coahuila, Mexico.

Material of V. canescens has been misidentified and distributed in some herbaria as "Verbena halii Small". On the other hand, the W. M. Jones 7, distributed as V. canescens, actually is var. roemeriana (Scheele) Perry; W. Schumann 1070 is a mixture with V. gracilis Desf.

Additional citations: TEXAS: Bexar Co.: Meebold 27301 (Mu). Tarrant Co.: Demaree 66202 (Ld). MEXICO: Chihuahua: Ellis, Le Doux, & Watkins 964 (N). Coahuila: Henrickson 11601 (Ld); Johnston, Wendt, & Chiang C. 10276b (Ld); Johnston, Wendt, Chiang C., & Riskind 11740g (Ld); Marsh 1687 (Ld). Jalisco: Schumann 1070, in part (Mu). Nuevo León: Painter, Lucas, & Barkley 14276 (Ld). Oaxaca: Pringle 4784 (Mu—1802). Puebla: Ventura A. 1574 (Sd—78100). San Luis Potosí: Schaffner s.n. [1875-79] (Mu—1557). Tamaulipas: Painter & Barkley 15373 (Ld). Zacatecas: Chiang, Wendt, & Johnston 7901 (Ld), 7920 (Ld); Henrickson 6263 (Ld); Johnston, Chiang, & Wendt 10435 (Ld), 10440 (Ld).

VERBENA CANESCENS var. ROEMERIANA (Scheele) Perry
Additional bibliography: G. W. Thomas, Tex. Pl. Ecolog. Summ. 77. 1969; Moldenke, Phytologia 30: 138, 156, & 165 (1975), 31: 378 (1975), and 34: 251. 1976.

Thomas (1969) calls this the "Roemer verbena".

Additional citations: TEXAS: Brown Co.: J. Reverchon s.n. [Cur-

tiss 1961] (Mu—1556). Caldwell Co.: W. M. Jones 7 (Mu). Cameron
Co.: R. Runyon 629 (Mu). Comal Co.: Lindheimer 1074 (Mu—4086).
Kerr Co.: E. J. Palmer 10002 (Mu).

VERBENA CAROLINA L.
    Additional & emended synonymy: Verbena carolinensis, melissae
folio aspero Dill. in Ray, Synop. Meth. Stirp. Brit., ed. 3, pl.
301, fig. 388. 1724. Verbena polystachya Kunth ex G. Don in
Loud., Hort. Brit., ed. 1, 246. 1830. Verbena carolinensis etc.
Dill. ex Schau. in A. DC., Prodr. 11: 546, in syn. 1847. Verbena
carolinensis &c. Dill. ex Buek, Gen. Spec. Syn. Candoll. 3: 494,
in syn. 1858. Verbena carolineana El-Gazzar, Egypt. Journ. Bot.
17: 75 & 78. 1974. Verbena carolina var. polystachya (H.B.K.)
Schimpff ex Moldenke, Phytologia 34: 278, in syn. 1976.
    Additional & emended bibliography: Dill. in Ray, Synop. Meth.
Stirp. Brit., ed. 3, pl. 301, fig. 388. 1724; G. Don in Loud.,
Hort. Brit., ed. 1, 246 (1830) and ed. 2, 246. 1832; Loud., Hort.
Brit., ed. 2, 552. 1832; G. Don in Loud., Hort. Brit., ed. 3,
246. 1839; Buek, Gen. Spec. Syn. Candoll. 3: 494—496. 1858;
Druce & Vines, Dill. Herb. 182. 1907; Loes., Verh. Bot. Ver.
Brand. 53: 74. 1912; Bolkh., Grif, Matvej., & Zakhar., Chrom.
Numb. Flow. Pl., imp. 1, 716 (1969) and imp. 2, 716. 1974; El-
Gazzar, Egypt. Journ. Bot. 17: 75 & 78. 1974; Garcia, MacBryde,
Molina, & Herrera-MacBryde, Malez. Preval. Cent. Am. 143 & 161.
1975; Hinton & Rzedowski, Anal. Esc. Nac. Cienc. Biol. 21: 111.
1975; López-Palacios, Revist. Fac. Farm. Univ. Los Andes 15: 92.
1975; Moldenke, Phytologia 30: 138—139 (1975) and 31: 378. 1975;
Molina R., Ceiba 19: 95. 1975; Soukup, Biota 11: 18. 1976: Moldenke, Phytologia 34: 270, 278, & 281 (1976) and 36: 47. 1977.
    Additional illustrations: Dill. in Ray, Synop. Meth. Stirp.
Brit., ed. 3, pl. 301, fig. 388. 1724; Garcia, MacBryde, Molina,
& Herrera-MacBryde, Malez. Preval. Cent. Am. 143 (in color).
1975.
    According to Druce & Vines (1907) the Dillen pre-Linnean des-
ignation for this species is represented in his herbarium by a
cultivated specimen from the Eltham Garden, collected in 1726,
and referred to V. caroliniana Willd. by Klinsmann, to V. polystachya H.B.K. by Asa Gray, and finally to V. carolina L.
    Don (1830) calls the species the "Carolina vervain" and gives
1732 as the date of its introduction from "N. Amer." into English gardens — but it was in cultivation there as early as 1726
according to Dillen. The cospecific V. veronicaefolia he calls
the "veronica-leaved vervain" and avers that it was introduced
from Mexico in 1825; V. polystachya he calls the "many-spiked
vervain" and dates its introduction into English gardens from
Mexico as 1820.
    It should be noted here that the specimen from the Bahama Is-
lands, cited below, does not bear any indication on its accompanying label that it came from cultivated material, but I am assuming that it did because V. carolina is not known in the wild

condition on those islands.

Recent collectors have encountered V. carolina in pinelands, along streams in granitic canyons "growing with Cordia, Quercus, Phaseolus, Ipomopsis, grasses, and herbs", and "in pine forests, some parts of which were severely cutover, in basic usually thin sandy soil capped by thick basalt, but with soft bedded volcanic underneath, with caves containing well-preserved Amerind pueblos and relics in the overhang, growing with Pinus spp. and Quercus spp." Henrickson refers to it as an "infrequent annual in disturbed areas". The corollas were "blue" on Contreras 10972 and "violet-blue" on Colaris 1529.

Loesener (1912) cites Seler & Seler 1304 from Mexico. The Karwinski s.n. [in imperio mexicano], distributed as V. carolina, actually is V. ehrenbergiana Schau., while Martínez Calderón 1352 is V. longifolia f. albiflora Moldenke, H. H. Rusby 780 is V. macdougalii Heller, Spencer s.n. [1916.4.25] is V. menthaefolia Benth., Karwinski s.n. is V. recta H.B.K., C. A. Brown 3892, Parish & Parish 1043, and Wurzlow s.n. [June 20, 1912] are V. scabra Vahl, Herb. Schwaegrichen s.n., Herb. Zuccarini s.n. [h. b. Mon. 1819], and Schoepf s.n. [ad New York] are V. simplex Lehm., and Herb. Reg. Monac. 250 is Stachytarpheta angustifolia f. elatior (Schrad.) López-Palacios.

Additional citations: ARIZONA: County undetermined: Stalmach 198 (Au—122071). MEXICO: Chihuahua: McCabe 140 (Ws); Wilson, Johnston, & Johnston 8595 (Ld). Federal District: Barkley, Rowell, & Webster 2199 (Ln—189707). México: Salinas M. 85 (Ws). Oaxaca: Colaris 1529 (Ut—328615B); Pringle 4892 (Mu—1804). Zacatecas: Henrickson 13492 (Ld). GUATEMALA: Baja Verapaz: Contreras 10972 (W—2795345). Guatemala: Kellerman 6540 (Ld). Santa Rosa: Heyde & Lux 3019 (Mu—1808). Sololá: Kellerman 5825 (Ac, Au). CULTIVATED: Bahama Islands: Herb. Schmiedel s.n. [Insul. Bahamens.] (Mu—297). Germany: Schreber s.n. [Hort. Bot. Erlang.] (Mu—298).

VERBENA CAROLINA f. ALBIFLORA Moldenke

Additional bibliography: Buek, Gen. Spec. Syn. Candoll. 3: 495. 1858; Moldenke, Phytologia 28: 203 & 432. 1974.

Additional citations: MEXICO: Veracruz: Gutierrez R. 342 (Ws).

VERBENA CATHARINAE Moldenke

Additional bibliography: Moldenke, Phytologia 28: 203. 1974.

Hatschbach encountered this plant along roadsides and the corollas are said to have been "deep-lilac" on Hatschbach 14968.

Additional citations: BRAZIL: Paraná: Hatschbach 14954 [Herb. Brad. 48008] (Mu), 14968 [Herb. Brad. 48009] (Mu).

VERBENA CHILENSIS Moldenke

Additional bibliography: Moldenke, Phytologia 23: 186 (1972)

and 34: 260. 1976.
     Additional citations: CHILE: Valdivia: Neger s.n. [Villarrica,
1897] (Mu—3983).

VERBENA CILIATA Benth.
     Additional & emended bibliography: Buek, Gen. Spec. Syn. Can-
doll. 3: 494. 1858; Loes., Verh. Bot. Ver. Brand. 53: 75. 1912;
Bolkh., Grif, Matvej., & Zakhar., Chrom. Numb. Flow. Pl., imp. 1,
716. 1969; G. W. Thomas, Tex. Pl. Ecolog. Summ. 77. 1969; Bolkh.,
Grif, Matvej., & Zakhar., Chrom. Numb. Flow. Pl., imp. 2, 716.
1974; Hinton & Rzedowski, Anal. Esc. Nac. Cienc. Biol. 21: 111.
1975; E. H. Jordan, Checklist Organ Pipe Cact. Natl. Mon. 7.
1975; Moldenke, Phytologia 30: 132 & 139 (1975) and 34: 252 & 270.
1976; A. L. Moldenke, Phytologia 35: 173. 1977.
     Tilforth describes this plant as an "annual, branched from base
and above, decumbent to ascending, common but in groups. Other
recent collectors have encountered it "on mezquital and tobosa
flats in alluvial bajadas in fine textured alluvial soil", "in
creosote scrub of Chihuahuan Desert", "in open Chihuahuan Desert
with rocky reddish clay soil", in clay soil of dry lakes, in rocky
clay soil, in small ravines, "in limestone soil of rugged limestone sierra", "in matorral desértico microfilo inerme on broad
alluvial flats, in calcareous adobe alluvium", "in limestone soil
in areas of extensive chaparral and oaks, the lower slopes with
Yucca carnerosana and Dasylirion, the upper slopes with Pinus
ponderosa and Agave macroculmis", "in chaparral with many pines
on the higher slopes, in steep canyon through mountains of igneous rock, mostly intrusive basics, in gravelly, grussy and sandy
soils derived from the igneous rock", "in calcareous gravelly
adobe of matorral subdesértico inerme y con espinas laterales in
flat areas near bottom of bajada", "in gravelly pale alluvial
adobe soil of izotal of Yucca filifera in low flat valley bottoms
between gentle hills", and in open pine-juniper woodland-meadow
areas, growing in association with Prosopis glandulosa, Hilaria
mutica, Agave lecheguilla, Yucca carnerosana, Y. filifera, Larrea
tridentata, Opuntia imbricata, Celtis pallida, Flourensia cernua,
Atriplex, Bahia, Conyza, Salsola, Acacia, Buddleia, Jatropha, Parthenium, Aloysia, Clematis, Sicyos, Quercus, Pinus, Pseudotsuga,
Ceanothus, Cercocarpus, Salvia, Carex, Bidens, Eryngium, pinyon,
juniper, various composites, and numerous annuals. Torke and his
associates found it in "red soil, still heavily cultivated area
with some cacti and associates", at 5750 feet altitude. Stuessy
refers to it as "scarce in Prosopis-Larrea scrub". Davidse encountered it "in Bouteloua grassland with low shrubs and Opuntia".
     Collectors describe the plant as a small decumbent perennial,
6 inches to 1 foot tall. Henrickson refers to it as "frequent"
or "infrequent" and as "a common roadside weed".
     The corollas on Henrickson 13293 and Ventura A. 1646 are said
to have been "purple", while on Tilforth 562 they were "purple

with externally white lobes", on Henrickson 6345c & 11212 they
were "violet", on Henrickson 5950b "light-violet", on Henrickson
8027 "red-violet", on Johnston, chiang, & Wendt 10432 "deep-pink",
on Chiang, Wendt, & Johnston 7987 "dark rose-pink", and on Stuessy
965 "pale-blue". Thomas (1969) calls it the "fringe verbena".
Mrs. Jordan (1975) records the common name, "hairy verbena".
    Loesener (1912) cites Seler & Seler 1120, 1173, 1174, 1379, &
3555 from Guanajuato, Michoacán, Tlaxcala, and Oaxaca, Mexico, and
4056 from Kinney County, Texas.
    The Reeves & Pinkava 11933, distributed as V. ciliata, is actu-
ally V. bipinnatifida Nutt., while J. A. Churchill s.n. [23 April
1953] is V. ciliata var. pubera (Greene) Perry, Edw. Palmer 339,
Reeves & Lehto R.1166, Reeves & Pinkava 11947, and Warren & Turner
68-87 are V. gooddingii Briq., Fugate, McLaughlin, & McManus 652
and Warren & Turner 68-33 are V. gooddingii var. nepetifolia Tidestr., T. Reeves R.1166 is xV. perplexa Moldenke, Lindheimer 1075
is V. pumila Rydb., Croft s.n. [San Diego, 1885] is V. quadrangulata
Heller, Pringle 4180 and Schaffner s.n. [San Luis Potosí, 1875-79]
are V. teucriifolia Mart. & Gal., and Semple & Love 321 and Van
Devender & Van Devender s.n. [28 March 1976] are V. wrightii A.
Gray. Meebold 27286 is a mixture of V. bipinnatifida Nutt. and
Evolvulus sp.
    Additional citations: ARIZONA: Cochise Co.: Tilforth 562 (Mi,
Sd--90751). Pima Co.: J. A. Churchill s.n. [29 March 1972] (Ln--
235661). Final Co.: Thornber s.n. [May 28, 1905] (Ld). MEXICO:
Chihuahua: Henrickson 5723 (Ld), 8027 (Ld), 11212 (Ld); Johnston,
Wendt, & Chiang C. 10540a (Ld), 11307e (Ld); Stuessy 965 (Ws).
Coahuila: Chiang C., Wendt, & Johnston 7738 (Ld); Henrickson
5950b (Ld); Johnston, Wendt, Chiang C., & Riskind 11985m (Ld).
Nuevo León: Chiang C., Wendt, & Johnston 7987 (Ld). Michoacán:
Torke, LeDoux, & Ellis 302 (N). Puebla: Ventura A. 1646 (Sd--
78371). San Luis Potosí: Schaffner s.n. [San Luis Potosí, 1875-
79] (Mu--1561). Veracruz: Kerber 255 (Mi, Mu--1777). Zacatecas:
Chiang C., Wendt, & Johnston 7887 (Ld), 7900 (Ld); Davidse 9944
(Ld); Henrickson 6345c (Ld), 6667 (Ld), 13293 (Ld); Johnston,
Chiang C., & Wendt 10432 (Ld); Johnston, Wendt, & Chiang C.
11566a (Ld). State undetermined: W. Schumann 1071 [Feral] (Mu--
3892).

VERBENA CILIATA var. LONGIDENTATA Perry
    Additional bibliography: G. W. Thomas, Tex Pl. Ecolog. Summ.
78. 1969; Moldenke, Phytologia 28: 204--205 (1974) and 34: 252
& 270. 1976.
    Runyon reports this plant "very abundant in this region [Cam-
eron County, Texas], covers acres of ground and is widespread in
open fields, sandy soil" and describes it as an "erect branching
and spreading herb". The corollas are said by him to have been
"blue-purple" on R. Runyon 1782. a collection which has been mis-

identified and erroneously cited previously by me as V. ambrosifolia f. eglandulosa Perry.

The corollas are described as "lavender-blue" on Henrickson
7630 and "light magenta and purple" on Henrickson 6239. Recent
collectors in Mexico have found the plant growing in rock and
gravel in xerophytic canyons, in low clay roadside ditches, and on
steep limestone-shale southwest-facing slopes, growing in association with Agave, Baccharis, Conyza, Flourensia, Fouquieria, Larrea,
Parthenium, Pectis, Tidestromia, Yucca, cacti, grasses, etc., at
altitudes of 4500—6100 feet. Henrickson refers to it as an "infrequent annual".

Material of this variety has been misidentified and distributed
in some herbaria as V. erinoides Lam. or V. multifida Ruíz & Pav.
The Ramirez & Cardenas 13, cited previously and again below, is a
mixture with V. ambrosifolia Rydb.

Additional & emended citations: TEXAS: Cameron Co.: R. Runyon
1782 (Mu, Rr, Rr). Zavala Co.: Ramirez & Cardenas 13, in part
(Au--245214). MEXICO: Chihuahua: Henrickson 7630 (Ld). Nuevo León: Painter, Lucas, & Barkley 14290 (Ld). Zacatecas: Henrickson
6239 (Ld). CULTIVATED: France: Weinkauff s.n. [Jard. des plant.
1834] (Mu--1254). Germany: Berger s.n. (Mu--310); Herb. Schwaegrichen s.n. [Hort. Lipsiensis] (Mu--1252, Mu--1253); Herb. Zuccarini s.n. (Mu--311, Mu--312).

VERBENA CILIATA var. PUBERA (Greene) Perry
Additional bibliography: G. W. Thomas, Tex. Pl. Ecolog. Summ.
78. 1969; Moldenke, Phytologia 30: 139. 1975.

The Meebold 22491, distributed as V. ciliata var. pubera, is
actually V. wrightii A. Gray.

Additional citations: NEW MEXICO: G. L. Fisher s.n. [Nara Visa,
Apr. 21, 1911] (Mu--4247). ARIZONA: Mohave Co.: J. A. Churchill
s.n. [24 April 1953] (Ln--203425).

VERBENA CLAVATA Ruíz & Pav.
Additional bibliography: Buek, Gen. Spec. Syn. Candoll. 3: 494.
1858; Moldenke, Phytologia 28: 205. 1974; Soukup, Biota 11: 18.
1976.

The Princess Therese of Bavaria 281, distributed as V. clavata,
actually is V. occulta Moldenke.

Additional citations: MOUNTED ILLUSTRATIONS: Ruíz & Pav., Fl.
Peruv. & Chil. 1: pl. 33, fig. b. 1797 (N, Z).

VERBENA CLAVATA f. ALBIFLORA Moldenke
Additional bibliography: Moldenke, Phytologia 23: 191. 1972;
Soukup, Biota 11: 18. 1976.

VERBENA CLAVATA var. CASMENSIS Moldenke
Additional bibliography: Moldenke, Phytologia 23: 191. 1972;

Soukup, Biota 11: 18. 1976.

xVERBENA CLEMENSORUM Moldenke
   Additional bibliography: Moldenke, Phytologia 23: 192 & 435.
1972.

VERBENA CLOVERAE Moldenke
   Additional & emended bibliography: G. W. Thomas, Tex. Pl. Eco-
log. Summ. 78. 1969; Bolkh., Grif, Matvej., & Zakhar., Chrom.
Numb. Flow. Pl., imp. 1, 716 (1969) and imp. 2, 716. 1974; Mol-
denke, Phytologia 30: 139. 1975.
   Thomas (1969) calls this the "Clover verbena".

xVERBENA CORRUPTA Moldenke
   Additional bibliography: Moldenke, Phytologia 30: 139—140
(1975) and 31: 411. 1975.

VERBENA CORYMBOSA Ruíz & Pav.
   Additional bibliography: Buek, Gen. Spec. Syn. Candoll. 3:
494. 1858; Bolkh., Grif, Matvej., & Zakhar., Chrom. Numb. Flow.
Pl., imp. 1, 716 (1969) and imp. 2, 716. 1974; Kooiman, Act. Bot.
Neerl. 24: 463. 1975; Moldenke, Phytologia 30: 140 (1975) and
34: 260. 1976; Soukup, Biota 11: 18. 1976.
   Hollermayer describes this plant as 1 meter or more tall. The
corollas are said to have been "blue" on Hollermayer 62, while
Lindeman says "corola roxa 2P7/6" and encountered the plant in a
"pequeño bañhado".
   Additional citations: BRAZIL: Rio Grande do Sul: Lindeman ICN.
9448 (Ut—320455). Santa Catarina: Lourteig 2164 (N). CHILE:
Bío-Bío: Neger s.n. [S. Juan, 1895-97] (Mu—2980). Concepción:
Merxmüller 24819 (Mu). Valdivia: Buchtien s.n. [Valdivia, 19/2/
1904] (Mu—3995); Hollermayer 62 (Mu). Laxa Island: Poeppig III.
157 (Mu—305). MOUNTED ILLUSTRATIONS: Ruíz & Pav., Fl. Peruv. &
Chil. 1: pl. 33, fig. a. 1797 (N, Z).

VERBENA CRITHMIFOLIA Gill. & Hook.
   Additional & emended synonymy: Verbena critmifolia Gill. &
Hook. ex Moldenke, Lilloa 8: 429, in nota. 1942; Phytologia 9:
46, in syn. 1963. Verbena critmifolia Gill. ex Moldenke, Résumé
Suppl. 3: 37, in syn. 1962.
   Additional & emended bibliography: Baxt. in Loud., Hort. Brit.
Suppl. [3]: 655. 1850; Buek, Gen. Spec. Syn. Candoll. 3: 494 &
507. 1858; Bolkh., Grif, Matvej., & Zakhar., Chrom. Numb. Flow.
Pl., imp. 1, 715 & 716 (1969) and imp. 2, 715 & 716. 1974; Mol-
denke, Phytologia 30: 140. 1975.
   The Lossen 8, distributed as V. crithmifolia, actually is V.
hookeriana (Covas & Schnack) Moldenke.
   Additional citations: ARGENTINA: La Pampa: Krapovickas, Cris-
tóbal, Mroginski, & Fernandez 22321 (N). Neuquen: Ammann 115
(Mu).

VERBENA CUMINGII Moldenke
     Additional bibliography: Moldenke, Phytologia 30: 133 & 140.
1975.

VERBENA CUNEIFOLIA Ruíz & Pav.
     Additional bibliography: Buek, Gen. Spec. Syn. Candoll. 3:
494. 1858; Moldenke, Phytologia 28: 348—349. 1974; Soukup, Biota 11: 18. 1976.
     The corollas on Infantes 5295 are described as having been
"moradas" [purple] and this collector encountered the plant in
flower in February. Ferreyra describes it as a subshrub, 40—
80 cm. tall, and found it growing in puna and pajonal associations, at 3900—4000 meters altitude, flowering and fruiting in
May. López-Palacios describes it as an "hierba rastrera, hojas
de 1—1.5 cm, flores morado claro, pequeñas, espigas delgadas"
and encountered it at 2500 meters altitude.
     Additional citations: ECUADOR: Loja: López-Palacios 4162 (Z).
PERU: Ancash: R. Ferreyra 14393 (W—2740336). La Libertad: In-
fantes 5295 (Mu).

xVERBENA DEAMII Moldenke
     Additional bibliography: Perkins, Estes, & Thorp, Bull. Tor-
rey Bot. Club 102: 194, 195, & 197. 1975; Moldenke, Phytologia
30: 140 (1975) and 34: 250. 1976.
     Perkins and his associates (1975) found a single plant of
this hybrid in a mixed population of verbenas including both
parents, which, when artificially pollinated with V. bracteata
pollen, produced a plant with 133 potential seeds and a 17.3
percentage of actual seed-set — V. bracteata, one parent, had
66.5 percent seed-set and V. stricta, the other parent, had
76.3—87.6 percent seed-set. The only insect observed visiting indiscriminately both parental plants and their hybrid was
Systropus sp. (observed with actual Verbena pollen on the head).
     Additional citations: MISSOURI: County undetermined: Martens
s.n. [Missouri] (Mu—292).

VERBENA DELICATULA Mart.
     Additional bibliography: Buek, Gen. Spec. Syn. Candoll. 3:
494. 1858; Moldenke, Phytologia 23: 106. 1972.

VERBENA DELTICOLA Small
     Additional & emended bibliography: G. W. Thomas, Tex. Pl.
Ecolog. Summ. 78. 1969; Bolkh., Grif, Matvej., & Zakhar.,
Chrom. Numb. Flow. Pl., imp. 1, 716 (1969) and imp. 2, 716.
1974; Moldenke, Phytologia 28: 349 (1974) and 34: 251. 1976.
     Additional citations: TEXAS: Cameron Co.: C. L. Lundell
10758 (N). Nueces Co.: Duncan s.n. [March 16, 1975] (Ac).
MEXICO: Nuevo León: F. A. Barkley 14361 (Ld).

VERBENA DEMISSA Moldenke
   Additional bibliography: Moldenke, Phytologia 23: 222 & 233 (1972), 34: 257 (1976), and 36: 33 & 51. 1977.
   This plant has been found in flower and fruit in March (in addition to the months previously recorded by me). Material has been misidentified and distributed in some herbaria as _V. carolina_ var. _polystachya_ (H.B.K.) Schimpff.
   Additional citations: ECUADOR: Chimborazo: Schimpff 765 (Mu).

VERBENA DEMISSA f. ALBA Moldenke, Phytologia 36: 51. 1977.
   Bibliography: Moldenke, Phytologia 36: 33 & 51. 1977.
   Citations: ECUADOR: Pichincha: López-Palacios 4200 (Z—type).

xVERBENA DERMENI Moldenke
   Additional synonymy: Verbena dermani Mold. ex Soukup, Biota 11: 18. 1976.
   Additional bibliography: Moldenke, Phytologia 30: 140 (1975) and 31: 384. 1975; Soukup, Biota 11: 18. 1976.
   Material of this hybrid has been misidentified and distributed in some herbaria as _V. litoralis_ H.B.K. On the other hand, the Pedersen 9867, distributed as this hybrid, seems to be typical _V. hispida_ Ruíz & Pav.
   Additional citations: PERU: Junín: W. Hoffmann 182 (Mu). BOLIVIA: La Paz: O. Buchtien 185 (Mu).

VERBENA DISSECTA Willd.
   Additional synonymy: Verbena disecta Willd. ex Soukup, Biota 11: 18. 1976. Glandularia disecta (Willd.) Schnack & Covas ex Soukup, Biota 11: 18, in syn. 1976.
   Additional & emended bibliography: Buek, Gen. Spec. Syn. Candoll. 3: 494. 1858; Bolkh., Grif, Matvej., & Zakhar., Chrom. Numb. Flow. Pl., imp. 1, 715 & 716 (1969) and imp. 2, 715 & 716. 1974; Moldenke, Phytologia 28: 349. 1974; Troncoso, Darwiniana 18: 318 & 409. 1974; López-Palacios, Revist. Fac. Farm. Univ. Los Andes 15: 94. 1975; Soukup, Biota 11: 18. 1976; Moldenke, Phytologia 36: 42 & 47. 1977.
   The corollas on Cabrera & al. 17128 are said to have been "lilac" when fresh.
   The Dessauer s.n. and Frömbling s.n. [Chili 1886], distributed as _V. dissecta_, actually are V. berterii (Meisn.) Schau., while Araujo 1256 [Herb. FEEMA 12264] is _V. tenera_ Spreng.
   Additional citations: ARGENTINA: Buenos Aires: Cabrera, Fabris, Torres, & Tur 17128 (Mu). Santiago del Estero: Pierotti "h" [1-IV-944] (Ut—330536B), "h" [6-4-44] (Ut—330564B).

VERBENA DOMINGENSIS Urb.
   Additional bibliography: León & Alain, Fl. Cuba, imp. 2, 2: 281. 1974; Moldenke, Phytologia 30: 140, 159, & 169 (1975) and 34: 19—20 & 254. 1976.

The corollas on <u>Liogier</u> & <u>Liogier</u> 19489 are said to have been "dark-purple" when fresh.

Material of <u>V. domingensis</u> has been distributed in some herbaria as "Labiata sp." and as "Labiatae sp. <u>Subaphylla</u>". On the other hand, the <u>Liogier</u> 16846, distributed as typical <u>V. domingensis</u>, actually is the type collection of f. <u>foliosa</u> Moldenke, while <u>Curtiss</u> 677 and <u>Rugel</u> 212 [856] and perhaps most or all the other Cuban material cited by me in previous installments of these notes are <u>V. officinalis</u> L. [or, perhaps more likely, <u>V. halei</u> Small].

Additional citations: HISPANIOLA: Dominican Republic: <u>Eggers</u> 1828 (Mu--4108), 2175 (Mu—3902); <u>Liogier</u> & <u>Liogier</u> 19489 (N).

VERBENA DOMINGENSIS f. FOLIOSA Moldenke, Phytologia 34: 19--20. 1976.

Bibliography: Moldenke, Phytologia 34: 19--20 & 254. 1976.
Citations: HISPANIOLA: Dominican Republic: <u>Liogier</u> 16846 (N--type).

VERBENA EHRENBERGIANA Schau.

Additional & emended bibliography: Buek, Gen. Spec. Syn. Candoll. 3: 494. 1858; Bolkh., Grif, Matvej., & Zakhar., Chrom. Numb. Flow. Pl., imp. 1, 716 (1969) and imp. 2, 716. 1974; Moldenke, Phytologia 28: 207 (1974) and 34: 251 & 252. 1976.

Mears reports finding this species growing in association with <u>Fuchsia</u>, <u>Reseda</u>, <u>Stillingia</u>, <u>Cheilanthes</u>, and <u>Cornus</u>.

Material of <u>V. ehrenbergiana</u> has been misidentified and distributed in some herbaria as <u>V. officinalis</u> L. On the other hand, the <u>Stalmach</u> 198, previously cited by me as <u>V. ehrenbergiana</u>, seems, on re-examination, to be <u>V. carolina</u> L., thus nullifying the only "record" of <u>V. ehrenbergiana</u> from outside of Mexico.

Additional citations: MEXICO: Hidalgo: <u>Mears</u> 299b (Ln--222097); <u>Moore</u> & <u>Wood</u> 3962 (Mi). Nuevo León: <u>Pringle</u> 1948 (Mu--3893). Tamaulipas: <u>Richardson</u> 188 (Ld). State undetermined: <u>Karwinski</u> s.n. [Taloman, July 1827] (Mu—307, Mu--308, Mu—346), <u>s.n.</u> [in imperio mexicano] (Mu--419).

VERBENA ELEGANS H.B.K.

Additional synonymy: <u>Verbena moranensis</u> H.B.K. apud Buek, Gen. Spec. Syn. Candoll. 3: 495, in syn. 1858.

Additional bibliography: G. Don in Loud., Hort. Brit., ed. 1, 247 (1830), ed. 2, 247 (1832), and ed. 3, 247. 1839; Buek, Gen. Spec. Syn. Candoll. 3: 494 & 495. 1858; G. W. Thomas, Tex. Pl. Ecolog. Summ. 78. 1969; Bolkh., Grif, Matvej., & Zakhar., Chrom. Numb. Flow. Pl., imp. 1, 716 (1969) and imp. 2, 716. 1974; El-Gazzar, Egypt. Journ. Bot. 17: 75 & 78. 1974; Moldenke, Phytologia 28: 349, 451, & 457. 1974; Kooiman, Act. Bot. Neerl. 24: 464. 1975; Moldenke, Phytologia 30: 134 (1975) and 31: 378. 1975.

Recent collectors have encountered this plant on bare hills and

in hillside oak forests, on northeast-facing slopes, in calcareous loam in areas of matorral-chaparral-encinar on slopes of alluvial fans, growing in association with Agave lecheguilla, Parthenium incanum, Opuntia leptocaulis, Hesperaloe, and Cercocarpus, at
550 meters altitude, flowering in May.

Don (1830) calls this species the "elegant vervain" and dates
its introduction into English gardens from Mexico as 1826. Mears
reports finding it growing with Cornus, Drymaria, Dudleya, Lobelia,
Maurandya, Heuchera, Piqueria, and Reseda in Hidalgo. Hendricks
reports it growing "from a stout shallow caudex" and "not common"
at 8000 feet altitude, and on his no. 439 the corollas are said
to have been "lavender, one petal notched and larger".

The Sydow s.n. [Sept. 1900], distributed as typical V. elegans,
actually represents var. asperata Perry.

Additional citations: MEXICO: Coahuila: Johnston, Wendt, & Chiang C. 10161b (Ld); Keil, Meyer, Lewis, & Pinkava 6037 (Ld, Te-
68562). Durango: Hendricks 439 (Ws). Hidalgo: Mears 307c (Mu);
Pringle 6908 (Ln—70156, Mu—3732), 7591 (Mu—4165); Troll 448
(Mu). Sonora: Reeves & Lehto L.18689 (Te—75344). State undetermined: Karwinski s.n. [in imperio mexicano] (Mu—266, Mu—267).

VERBENA ELEGANS H.B.K. x V. PERUVIANA (L.) Britton
    Additional bibliography: Moldenke, Phytologia 28: 208 & 457.
1974.

VERBENA ELEGANS H.B.K. x V. PULCHELLA Sweet
    Additional bibliography: Moldenke, Phytologia 28: 208 & 457.
1974.

VERBENA ELEGANS H.B.K. x V. STELLARIOIDES Cham.
    Additional bibliography: Moldenke, Phytologia 28: 208 & 457.
1974.

VERBENA ELEGANS var. ASPERATA Perry
    Additional bibliography: G. W. Thomas, Tex. Pl. Ecolog. Summ.
78. 1969; Moldenke, Phytologia 30: 134 & 140—141 (1975) and 31:
378. 1975.
    Hendricks encountered this plant in wet mountain meadows.
    Additional citations: MEXICO: Durango: Hendricks 538 (Ws). San
Luis Potosí: Sanderson 262 (Ln—238191). Tamaulipas: Urbatsch
2423 (Ld). CULTIVATED: Germany: Herb. Schönau s.n. (Mu). Sweden:
Sydow s.n. [Sept. 1900] (Ac).

xVERBENA ENGELMANNII Moldenke
    Additional bibliography: Greller, Bull. Torrey Bot. Club 102:
416. 1975; R. A. Davidson, State Univ. Iowa Stud. Nat. Hist. 20
(a): 77. 1959; Moldenke, Phytologia 30: 141 & 174 (1975), 31:
377 (1975), 34: 247 & 248 (1976), and 36: 29. 1977.
    The Wheeler collection cited below exhibits some fasciated in-

florescences. Greller (1975) reports the hybrid from Suffolk County, New York, while Davidson (1959) found it in Louisa and Van Buren Counties, Iowa.

Additional citations: ILLINOIS: Cass Co.: Geyer s.n. [Beardstown, July 1842] (Mu--363--isotype). Winnebago Co.: M. S. Lebb s.n. [Fountaindale] (Mu--4250). MICHIGAN: Ingham Co.: C. F. Wheeler s.n. [Aug. 30, 1890] (Ln--142449).

VERBENA EPHEDROIDES Cham.

Additional bibliography: Buek, Gen. Spec. Syn. Candoll. 3: 494. 1858; Moldenke, Phytologia 30: 141. 1975.

The Lorentz 131, distributed as V. ephedroides, is actually V. brasiliensis Vell.

VERBENA FASCICULATA Benth.

Additional bibliography: Moldenke, Phytologia 23: 230--231 (1972) and 24: 46. 1972; Soukup, Biota 11: 18 & 19. 1976; Moldenke, Phytologia 36: 33. 1977.

Additional citations: PERU: Ica: Ellenberg 4914 (Ac).

VERBENA FERREYRAE Moldenke

Additional bibliography: Moldenke, Phytologia 23: 231. 1972; Soukup, Biota 11: 18. 1976.

VERBENA FILICAULIS Schau.

Additional bibliography: Buek, Gen. Spec. Syn. Candoll. 3: 494 & 495. 1858; Moldenke, Phytologia 30: 141. 1975.

Additional citations: BRAZIL: Minas Gerais: Widgren s.n. [1845] (Mu--1566). Paraná: Dusén 15679 (Mu). São Paulo: F. C. Hoehne 443 (Mu--4317).

VERBENA FLAVA Gill. & Hook.

Additional & emended bibliography: Schau. in A. DC., Prodr. 11: 555--556. 1847; Buek, Gen. Spec. Syn. Candoll. 3: 494. 1858; Bolkh., Grif, Matvej., & Zakhar., Chrom. Numb. Flow. Pl., imp. 1, 715 & 717 (1969) and imp. 2, 715 & 717. 1974; Moldenke, Phytologia 30: 141. 1975.

Additional citations: ARGENTINA: Mendoza: Ruiz Leal 8503 (Ld, Ut--330570B). Neuquen: Ammann 113 (Mu, Mu); Schajovskoy 137 (Mu), 138 (Mu).

VERBENA GALAPAGOSENSIS Moldenke

Additional bibliography: Moldenke, Phytologia 23: 232. 1972; Balgooy, Pacif. Pl. Areas 3: 245. 1975.

López-Palacios describes this plant as "hierba de 70 cm. a 1 m., hojas superiores relativamente estrechas, fl. morado claro" and found it in flower and fruit in February.

Van der Werff's collection, cited below, is placed here tentatively, awaiting his publication on the Galápagos verbenas. He says of it that "The plant was shrubby, with ascending stems to 2 m. long; leaves small, only at [the] base of stems are there dis-

tinctly larger leaves". He found it growing "at crater rim in a .
dry area in no way comparable to the wet fern-sedge zones that are
found on the summits of" Chatham and Indefatigable Islands.
    Additional citations: GALÁPAGOS ISLANDS: Albemarle: <u>López-
Palacios</u> 4298 (Z); <u>Van der Werff</u> 1580 (Z).

VERBENA GENTRYI Moldenke
    Additional bibliography: Moldenke, Phytologia 23: 232 (1972)
and 34: 252. 1976.
    Henrickson has collected what appears to be this species at
7600 feet altitude, where he found it to be frequent in shaded
woods in association with <u>Carex</u>, <u>Bidens</u>, <u>Eryngium</u>, composites,
pines, oaks, etc., in an area of open pine-juniper woodland-meadow,
flowering and fruiting in September.
    Additional citations: MEXICO: Chihuahua: <u>Henrickson</u> 8035 (Ld).

VERBENA GLABRATA H.B.K.
    Additional bibliography: Buek, Gen. Spec. Syn. Candoll. 3: 494.
1858; Balgooy, Pacif. Pl. Areas 3: 245. 1975; López-Palacios, Revist. Fac. Farm. Univ. Los Andes 15: 92. 1975; Moldenke, Phytologia 30: 141. 1975; Soukup, Biota 11: 18. 1976.
    In addition to the months previously reported by me, this spe-
cies has been collected in fruit in March. Recent collectors
have encountered it on valley floors and in volcanic sand and
tufa on basalt rock at 11,600 feet altitude. López-Palacios & Idrobo describe the plant as "Hierba acumbente, de 1—60 cm. Espigas densas. Coronadas de flores lila intenso" or "lila claro"
and as "hierba rastrera", the corollas "morado claro" or "morado
muy claro", found it growing at 2450—3200 meters altitude, and
record the vernacular name, "verbena blanca". The corollas are
said to have been "purple" on Saunders 363.
    Additional citations: COLOMBIA: Cauca: <u>López-Palacios & Idrobo</u>
3724 (Ac, N). Nariño: <u>López-Palacios & Idrobo</u> 3769 (Ld, N). EC-
UADOR: Cotopaxi: <u>Collector undetermined</u> VIII (Mu—1106), <u>XIV</u> (Mu—
1107). Imbabura: <u>López-Palacios</u> 4064 (Ld). Pichincha: <u>Herb.
Univ.</u> Cent. Quito 2341 (Mu), 2342 (Mu), 2344 (Mu); <u>López-Palacios</u>
4206 (Ld), 4227 (Ld). PERU: Lima: <u>S. G. E. Saunders</u> 363 (Ld).

VERBENA GLABRATA var. TENUISPICATA Moldenke
    Additional bibliography: Moldenke, Phytologia 23: 233. 1972;
Balgooy, Pacif. Pl. Areas 3: 245. 1975.
    Van der Werff comments that "This is the <u>Verbena</u> from the wet
fern-sedge zone on Cerro Azul. The stems are crawling-ascending
and the plants no more than 50 cm. tall."
    Additional citations: GALÁPAGOS ISLANDS: Albemarle: <u>Van der
Werff</u> 2286 [1586] (Z).

VERBENA GLANDULIFERA Moldenke
    Additional bibliography: Moldenke, Phytologia 30: 141. 1975.
    The corollas are said to have been "violet" in color on <u>Lossen</u>

<u>72</u> when fresh.
    Additional citations: ARGENTINA: Córdoba: <u>Lossen 72</u> (Mu--4371).

VERBENA GLUTINOSA Kuntze
    Additional bibliography: Moldenke, Phytologia 30: 141. 1975.
    Additional citations: ARGENTINA: Neuquen: <u>Ammann 111</u> (Mu), <u>112</u>
(Mu).

VERBENA GOODDINGII Briq.
    Additional synonymy: Verbena <u>ciliata</u> var. <u>alba</u> Palmer ex Mol-
denke, Phytologia 34: 278, in syn. 1976.
    Additional bibliography: G. W. Thomas, Tex. Pl. Ecolog. Summ.
78. 1969; Bolkh., Grif, Matvej., & Zakhar., Chrom. Numb. Flow. Pl.,
imp. 1, 715 & 717 (1969) and imp. 2, 715 & 717. 1974; R. D. Gibbs,
Chemotax. Flow. Pl. 3: 1753--1755 (1974) and 4: 2295. 1974; E. H.
Jordan, Checklist Organ Pipe Cact. Natl. Mon. 7. 1975; Moldenke,
Phytologia 30: 141--142 & 179 (1975), 31: 415 (1975), and 34: 278.
1976.
    Recent collectors have encountered this species in the juniper
vegetational zone, in sandy soil in the juniper-pinyon-oak association, in a limestone canyon with scattered juniper and mixed
shrubs, in rocky soil, on desert slopes, and by a permanent stream
in grazed oak-Sonoran zone. Moran reports it "common on burns".
Mrs. Jordan (1975) calls it the "Goodding vervain", while Thomas
(1979) calls it the "Goodding verbena".
    Palmer's <u>V. ciliata</u> var. <u>alba</u> seems to be based on <u>Ed. Palmer</u>
<u>339</u> from the Mohave Desert of southern California.
    Gibbs (1974) reports that the Ehrlich test gave positive re-
sults (bright blue-green) in the leaves of this species, but that
cyanogenesis and leucoanthocyanin are absent from the leaves and
syringin is absent from the stems.
    Additional citations: NEVADA: Clark Co.: <u>Purpus 6061</u> (Mu--
4285). ARIZONA: Gila Co.: <u>Higgins 8790</u> (N). Mohave Co.: <u>Atwood</u>
<u>6026</u> (N); <u>J. A. Churchill s.n.</u> [18 April 1953] (Ln--204267); <u>M. E.</u>
<u>Jones s.n.</u> [Hackberry, May 24, 1884] (Ln--70252); <u>Reeves & Pinkava</u>
<u>11947</u> (N, W--2737221). Pima Co.: <u>L. M. Andrews 259</u> (N); <u>Lehto,</u>
<u>Brown, Nash, & Pinkava 10646</u> (W--2736615); <u>Warren & Turner 68-87</u>
(Ld). Pinal Co.: Mrs. <u>R. S. Baker s.n.</u> [Oracle, Spring 1901] (Ln--
120073); <u>Meebold 15588</u> (Mu); <u>Moroz s.n.</u> [Schallert 22814] (Mu).
Santa Cruz Co.: <u>Reeves R.1166</u> (N). CALIFORNIA: San Bernardino
Co.: <u>D. Howe s.n.</u> [24 April 1968] (Sd--70017); <u>Edw. Palmer 339</u> (Mu--
1563). MEXICO: Baja California: <u>R. V. Moran 17739</u> (Sd--75136).

VERBENA GOODDINGII var. NEPETIFOLIA Tidestr.
    Additional bibliography: Moldenke, Phytologia 28: 350 (1974) and
31: 415. 1975.
    Moran found this plant to be "fairly common" at 300 meters alti-
tude in Baja California; also "occasional" and "locally common in
disturbed areas", "occasional in arroyo beds", "occasional in dis-

turbed roadside soil", and on south-facing talus slopes in that
state. Other collectors have found it in short-tree forests and
growing in association with _Washingtonia filifera_.

The corollas are described as having been "lavender" on R. V.
Moran 16405, 20047, 21750, & 22099, "pale-lavender" on R. V. Moran 20638 & 21883, and "pink" on Spellenberg & Spellenberg 3062.

The Spellenberg & Spellenberg 3062, cited below, was previously incorrectly listed by me as V. ambrosifolia Rydb. These two
collectors describe the plants as growing in "clumps with many
stems".

Additional citations: ARIZONA: Cochise Co.: Spellenberg & Spel-
lenberg 3062 (N). Pima Co.: Fugate, McLaughlin, & McManus 652
(Ld); Warren & Turner 68-33 (Sd--78500). Yuma Co.: Phillips,
Phillips, Goldberg, Fugate, & McManus 647 (Ld). MEXICO: Baja Cal-
ifornia: R. V. Moran 16405 (Sd--75481), 20047 (Sd--92459), 20638
(Sd--88939), 21750 (Sd--91255), 21883 (Sd--91260), 22099 (Sd--
19477); Moran, Witham, & Hommersand 16541 (Sd--71562). Sonora:
Arguelles 73 (Ld, Ld).

xVERBENA GOODMANI Moldenke
    Additional synonymy: Verbena stricta x halei Perkins, Estes, &
Thorp, Bull. Torrey Bot. Club 102: 194, in syn. 1975.
    Additional bibliography: Moldenke, Phytologia 23: 237 & 436.
1972; Perkins, Estes, & Thorp, Bull. Torrey Bot. Club 102: 194,
195, & 197. 1975; Moldenke, Phytologia 34: 279. 1976.
    Perkins and his associates (1975) found one plant of this hy-
brid among a mixed population of verbenas including both parents,
which, when artificially cross-pollinated with V. halei pollen
produced 149 potential seeds or a 26.8 percent seed-set, whereas
12 insect-visited V. halei plants produced 468 seeds (a 68 percent
seed-set) and insect-visited V. stricta plants produced a 76.3--
87.6 percent seed-set. They tell us that xV. goodmani and xV. illicita Moldenke are the two most common naturally occurring hybrids in a mixed population of the parental species; 10 plants of
the former and 20 of the latter were found in the area tested.

VERBENA GRACILESCENS (Cham.) Herter
    Additional & emended bibliography: Bolkh., Grif, Matvej., &
Zakhar., Chrom. Numb. Flow. Pl., imp. 1, 717 (1969) and imp. 2,
717. 1974; López-Palacios, Revist. Fac. Farm. Univ. Los Andes 15:
92. 1975; Moldenke, Phytologia 30: 142 & 159. 1975.
    Recent collectors have found this plant in fruit in November.
The corollas are said to have been "violet" in color when fresh
on Krapovickas & al. 26687 and "white" on Schinini & Cristóbal
9865. Lindeman and his associates encountered it in a "beira do
río inundada periodicamente, solo argilo duro" and describe the
corolla color as "azul 10PB6/8".

    Additional citations: BRAZIL: Rio Grande do Sul: Lindeman, Ir-

gang, & Valls ICN.8423 (Ut--320458). PARAGUAY: Fiebrig 4432 (Mu).
URUGUAY: Herter 1058 [Herb. Herter 82656] (Mu). ARGENTINA: Córdoba: Lorentz 113 (Mu--1572). Corrientes: Cabrera 11784 (Mu); Schinini & Cristóbal 9865 (Ld). Jujuy: Krapovickas, Schinini, & Quarín 26687 (Ld). Salta: Schreiter 11466 (Ld). Santiago del Estero:
Luillo 4 (Ut--330562B).

VERBENA GRACILIS Desf.
    Additional bibliography: Buek, Gen. Spec. Syn. Candoll. 3: 494
& 495. 1858; Hinton & Rzedowski, Anal. Esc. Nac. Cienc. Biol. 21:
111. 1975; A. L. Moldenke, Phytologia 31: 415. 1975; Moldenke, Phytologia 30: 142 (1975) and 34: 270. 1976.
    Henrickson encountered this plant in clay soil of pasture land,
growing with Agave, Buddleia, Ephedra, Mimosa, Opuntia, Yucca,
etc., while Hendricks found it in "weedy yards" and in "sandy bar-
ren areas dominated by mesquite and Acacia".
    The W. Schumann 1070 collection, cited below, is a mixture with
V. canescens H.B.K., while Hendricks 332 is a mixture with a com-
posite. The Cabrera 11784 and Herter 1058 [Herb. Herter 82656],
distributed as V. gracilis, actually are V. gracilescens (Cham.)
Herter, while J. A. Churchill s.n. [7 April 1972] is V. neomexicana
(A. Gray) Small.
    Additional citations: MEXICO: Durango: Hendricks 332, in part
(Ws), 450 (Ws). Federal District: Pringle 6539 (Mu--1829). Jalis-
co: W. Schumann 1070, in part (Mu--3891). San Luis Potosí: Hen-
rickson B.6381 (Ld); Schaffner s.n. [1875-79] (Mu--1559). CULTI-
VATED: Germany: Herb. Kummer s.n. [Hort. bot. Monac.] (Mu--1245);
Herb. Zuccarini s.n. [Hort. bot. Monac. 1835] (Mu--296).

VERBENA GRISEA Robinson & Greenm.
    Additional bibliography: Moldenke, Phytologia 23: 239. 1972;
Balgooy, Pacif. Pl. Areas 3: 245. 1975.

VERBENA GYNOBASIS Wedd.
    Additional bibliography: Moldenke, Phytologia 23: 240. 1972;
Soukup, Biota 11: 18. 1976.
    The Troll collection, cited below, in the Berlin herbarium
bears a label indicating that it was collected in Bolivia, while in
the Munich herbarium it is inscribed "Chile". It is not certain
which locality is correct, but Bolivia is the more likely.
    Additional citations: BOLIVIA: Santa Cruz: Troll 3312 (Mu).

VERBENA GYNOBASIS var. STRIGOSA Wedd.
    Additional bibliography: Moldenke, Phytologia 23: 240. 1972;
Soukup, Biota 11: 18. 1976.

                        [to be continued]

BOOK REVIEWS

Alma L. Moldenke

"WINTER FLOWERS IN GREENHOUSE AND SUN-HEATED PIT" Revised Edition
by Kathryn S. Taylor & Edith W. Gregg, xix & 281 pp., illus.,
Charles Scribner's Sons, New York, N. Y. 10017. 1976.
$4.95 paperbound.

Back in 1941 this book first appeared without "Greenhouse" in
the title and consideration. By 1969 this highly successful
book's supply needed replenishment. It was revised with greenhouse consideration added and a few other changes made. Now this
is the welcome paperbound edition with all its tested, detailed,
interestingly written guidance which is even more pertinent today
because of our increased interest in and need for conservation of
energy and growing things for food or fun after summer days have
waned. The illustrations are convincing and encouraging.
Lantana sellowiana is often so-called in the horticultural
trade, but its correct name is Lantana montevidensis.

"THE COMPLETE BOOK OF TERRARIUM GARDENING" by Jack Kramer, ix &
146 pp., illus., Charles Scribner's Sons, New York, N. Y.
10017. 1976. $5.95 paperbound.

From plastic trapezoids, leaded glass pyramids, Belljar covers
on metal platters, old bottles, aquarium bowls suspended or in-
verted, from size ranges small enough for miniature mosses to
large enough for Nerium oleander, from explanations of how and
why terraria thrive when the same exposed plants well potted on
the window-sill do not, from careful construction and remedial
procedures, and from carefully listed and described plants for
various types of living collections just about any information
needed by the enthusiastic novice or the experienced horticulturist can be culled. But this does not justify the word "complete"
in the title of this, or, indeed, of almost any other, book!
On page 140 two scientific names are misspelled.

"NÓMINA DESCRIPTIVA DE LAS GRAMINEAS BOLIVIANAS HASTA HOY CONOCI-
DAS" by Adolfo M. Jiménez f.s.c., 294 pp., illus., Herbario
La Salle, Facultad de Ciencias Agricolas 'Martin Cardenas',
Universidad Mayor de San Simon, Cochabamba, Bolivia. 1976.
Paperbound.

This worthwhile publication is one of the many efforts in di-
verse fields to commemorate the sesquicentennial of the Republic
of Bolivia. The author is eminently qualified to prepare this

well keyed clearly descriptive manual, having spent his professional life in teaching and in field and herbarium studies. Species in 119 grass genera are presented. The introduction and
text can be read quite facilely. The printing is far freer of
error than that in many South American journals and scientific
publications even though the names of Eric Asplund and Boris
Krukoff are misspelled.

The paper cover, not the title-page quoted above, records
this work as "Flora Boliviana - GRAMINEAS (Gramineae)", and
thus offers problems to cataloguers.

"THIS HUNGRY WORLD" by Ray Vickers, ix & 270 pp., Charles Scrib-
     ner's Sons, New York, N. Y. 10017. 1975. $9.95.

The author supervises the London Bureau of "The Wall Street
Journal" covering Europe, the Middle East and Africa, attended as
a well-informed journalist the World Food Conference in Rome in
1974, and observed firsthand the drought-famine conditions in the
Sahel , among a hundred other countries visited. These experi-
ences and the knowledge that the United States is the greatest
food producer in the world and that people are being born, especially in the less developed countries far, far in excess of the
food supply, provide the careful analyses of this serious ines-
capable situation. Politics, inefficiency and fraud in myriad
forms, places and levels have hindered several constructive efforts.

"America's actions to meet the world's food crisis should be
focussed upon:
     pushing food production
     offering technical assistance to the Third World, cooperat-
          ing with agroindustry and the United Nations
     advancing family planning
     promoting rational environmental controls
     increasing nutritional, weather, and other research
     building food reserves and improving crop information data
     liberalizing agricultural trade"
This is an important book.

"CONIFERS FOR YOUR GARDEN" by Adrian Bloom, 147 pp., illus.,
     Charles Scribner's Sons Inc., New York, N. Y. 10017. 1975.
     $8.95.

After introductory remarks on their nature, horticultural use
and care, over 200 beautifully color-illustrated garden prospects
from Abies balsamea 'Hudsonia' to Tsuga heterophylla are described
as to possible height and growing conditions.

The author is a member of a family of famous English horticul-
turists. He has specialized in dwarf conifers such as the first
one that was found in the White Mountains of New Hampshire, U.S.A.

PHYTOLOGIA

Designed to expedite botanical publication

CONTENTS

Published by Harold N Moldenke and Alma L Moldenke

303 Parkside Road
Plainfield, New Jersey 07060
U S A.

Price of this number, $3, per volume, $9 75 in advance or after close of the volume, 75 cents extra to all foreign addresses, 512 pages constitute a volume

Contribución a la Flora Liquenológica de Venezuela

Manuel López Figueiras
Departamento de Botánica, Facultad de Farmacia
Universidad de los Andes, Mérida
Venezuela

Durante los años 1975-1976 he realizado una serie de excursiones a los páramos de los Andes Venezolanos para recolectar líquenes como parte de un proyecto que el Departamento de Botánica de la Facultad de Farmacia de la Universidad de los Andes, Mérida, Venezuela, adelanta en relación con la flora andina del país.

En ese período de tiempo colecté alrededor de 4700 números de macrolíquenes, de los cuales un juego casi completo de duplicados se enviaron a la Smithsonian Institution. Aprovechando el disfrute de mi año sabático que la Universidad de los Andes me concedió me trasladé a Washington y comencé inmediatamente el estudio de los mismos bajo la dirección y cooperación del Dr. Mason E. Hale.

Resultado de estos estudios preliminares es la siguiente lista de géneros y especies que no fueron citados previamente para la Flora de Venezuela en la literatura liquenológica hasta la fecha y que por lo tanto son nuevos para nuestra flora.

Antracothecium Hampe	Microphiale Zahlbr.
Blastenia Mass.	Normandina Nyl.
Buellia De Not.	Ocellularia Mey.
Cetraria Ach.	Parmeliopsis Nyl.
Chaenotheca (Th. Fr.) Th. Fr.	Phaeophyscia Moberg
Collema Web.	Phyllopsora Müll. Arg.
Corella Wain.	Protoblastenia Steiner
Erioderma Fée	Psora Hoffm.
Glossodium Nyl.	Solorina Ach.
Herpothallon Tobler	Tylophoron Nyl.
Lopadium Körb.	Umbilicaria Hoffm.

Buellia modesta (Krmphbr.) Müll. Arg.
Cetraria rassadinae Karnefelt, inédita
Chiodecton sphaerale Ach.
Cladina lopezii Ahti, inédita
Cladina substellata Vainio
Collema laeve H. f. & Tayl., fma

Corella zahlbruckneri Schffn.
Glossodium aversum Nyl.
Herpothallon sanguineum (Sw.) Tobler
Heterodermia dendritica (Pers.) Poelt.
Heterodermia galactophylla (Tuck.) W. Culb.
Heterodermia japonica (Sato) Swinsc. & Krog, fma
Heterodermia lepidota Swinsc. & Krog
Heterodermia magellanica (Zahlbr.) Swinsc. & Krog
Heterodermia microphylla (Durok.) Swinsc. & Krog
Heterodermia obscurata (Nyl.) Trev.
Heterodermia podocarpa (Bel.) Awas.
Heterodermia speciosa (Wulf.) Trev.
Heterodermia vulgaris (Vain.) Follm. & Redon
Lopadium leucoxanthum (Spreng.) Zahlbr.
Normandina pulchella Nyl.
Ocellularia glaucula (Nyl.) Zahlbr.
Ocellularia pachystoma (Nyl.)
Parmelina versiformis (Kremp.) Hale
Parmeliopsis aleurites (Ach.) Nyl., fma
Phyllopsora corallina Müll. Arg.
Phaeophyscia endococcinea (Koerb.) Moberg
Psora rufonigra Schneid.
Relicina subabstrusa (Gyelnik) Hale
Tylophoron protudens Nyl.
Umbilicaria hyperborea (Ach.) Hoffm.

Evidentemente esta lista tiene caracter parcial debido a que muchas colecciones o se están estudiando o serán estudiadas en el porvenir preferentemente por algun especialista.

BIBLIOGRAFIA CONSULTADA

Hale, Mason E. Jr.
     1975. A Revision of the Lichen Genus Hypotrachyna in Tropical
            America. Smithsonian Contributions to Botany, 25:1-73.

     1976. Synopsis of the New Lichen Genus Eveniastrum Hale
            (Parmeliaceae). Mycotaxon, III (3):345-353.

     1976. A Monograph of the Lichen Genus Relicina (Parmeliaceae).
            Smithsonian Contributions to Botany, 26:1-32.

     1976. A Monograph of the Lichen Genus Pseudoparmelia Lynge
            (Parmeliaceae). Smithsonian Contributions to Botany,
            31:1-62.

1976.   A Monograph of the Lichen Genus Bulbothrix Hale
        (Parmeliaceae). Smithsonian Contribution to Botany,
        32:1-29.

1976.   A Monograph of the Lichen Genus Parmelina Hale
        (Parmeliaceae). Smithsonian Contributions to Botany,
        33:1-60.

1977.   New Species in the Lichen Genus Parmotrema Mass.
        Mycotaxon, V (2).423-448.

Poelt, J.
1974.   Zur Kenntnis der Flechtenfamilie Candelariaceae.
        Phyton, 16 (1-4):189-210.

Ramirez Reyes, C.
1974.   Nota adicional acerca del Catálogo de los Líquenes de
        Venezuela. Bryologist, 77:248-249.

Vareschi, V.
1973.   Catálogo de los Líquenes de Venezuela. Acta Botanica
        Venezuelica, 8 (1-4):177-245.

AGRADECIMIENTOS

Se agradece al Departamento de Botánica de la Smithsonian
Institution y al Dr. Mason E. Hale las facilidades recibidas para
este estudio y al C. D. C. H. de la Universidad de los Andes,
Mérida, Venezuela, el soporte económico para las exploraciones
que precedieron al mismo.

NOTES ON NEW AND NOTEWORTHY PLANTS. CI

Harold N. Moldenke

CITHAREXYLUM FRUTICOSUM f. SUBSERRATUM (Sw,) Moldenke, stat. nov.
Citharexylum subserratum Sw., Prodr. Veg. Ind. Occ. 91. 1788.

CITHAREXYLUM FRUTICOSUM f. SUBVILLOSUM (Moldenke) Moldenke, stat.
nov.
Citharexylum fruticosum var. subvillosum Moldenke, Feddes Repert. Spec. Nov. 37: 223. 1934.

LANTANA HATSCHBACHII Moldenke, sp. nov.
Frutex 1 m. altus, ramis ramulisque gracilibus inermibus tetragonis fusco-pubescentibus; foliis decussato-oppositis parvis; petiolis gracillimis 5—8 mm. longis irregulariter pilosulis sparse resinosis; laminis foliorum firme chartaceis late ellipticis vel subrotundis 1.5 cm. latis longisque, apice rotundatis basin breviter cuneatis margine perspicue regulariterque serratis, supra rugosis nigrescentibus, subtus densiuscule pilosulis resinosis reticulo venularum prominente; inflorescentiis terminalibus capitatis parvis.
Shrub, about 1 m. tall, much branched; branches and branchlets slender, tetragonal, unarmed, rather densely pubescent with brownish-fuscous hairs, the internodes apparently short; leaves decussate-opposite, small; petioles very slender, 5—8 mm. long, irregularly pilosulous and resinous; leaf-blades firmly chartaceous, broadly elliptic or subrotund, about 1.5 cm. long and wide, apically rounded, basally shortly cuneate, marginally regularly and conspicuously serrate, rugose and nigrescent above, rather densely pilosulous and resinous beneath with fuscous-brown hairs, brunnescent in drying; inflorescence terminal, capitate, small, rather few-flowered; peduncles very slender, 2—3 cm. long, rather densely pilosulous with wide-spreading hairs; heads hemispheric, about 1.5 cm. wide in anthesis; bracts lanceolate, 7—9 mm. long, apically attenuate or acute, about 1 mm. wide at the base, rather densely pilosulous and resinous on the back, the margins more or less ciliolate; corolla rose-colored, hypocrateriform, the tube very slender, 8—9 mm. long, the limb about 5 mm. wide in anthesis.
The type of this species was collected by Gert Hatschbach (no. 39651) — in whose honor it is named — at Morrão, in the municipality of Morro do Chapeu, Bahia, Brazil, on January 15, 1977, and is deposited in my personal herbarium.

VERBENA TENUISECTA f. ALBA (Benary) Moldenke, comb. nov.
Verbena erinoides alba Benary ex Wittmack, Gartenfl. 49: 585. 1900.

VITEX LEUCOXYLON f. ZEYLANICA (Moldenke) Moldenke, stat. nov.
Vitex leucoxylon var. zeylanica Moldenke, Phytologia 21: 419. 1971.

GARYSMITHIA BIFURCATA
A NEW GENUS AND SPECIES OF LESKEACEAE
(MUSCI) FROM ALASKA AND COLORADO

W.C. Steere[1]

During my identification of the bryophytes collected by Dr. Gary Smith in Arctic Alaska during the summer field season of 1966, I encountered a sterile pleurocarpous moss that was totally unfamiliar to me. Even the family to which it belonged was not readily apparent. Soon thereafter, I received an unknown moss for identification from Dr. F.J. Hermann, collected in Mt. McKinley National Park in 1967, which proved to be the same thing. Among a group of my own Arctic Alaskan collections that I had segregated out as serious puzzles for future study, I found another specimen of this unknown moss, collected in the same general area as Smith's specimen. More recently, Dr. Hermann sent me a specimen of this same moss from Colorado, which he had collected in 1976, but did not immediately recognize. Realizing that this moss had a wider geographical distribution than had seemed likely at first, I sent a sample to Barbara M. Murray, at the University of Alaska, with the request that she send me anything that matched it, and received, almost by return mail, two specimens which she had collected in 1976 in the eastern Brooks Range. If this moss has the same pattern of geographical distribution as, for example, Oreas martiana, it should also occur in the Canadian Rockies. However, when I sent a specimen to Dr. Dale Vitt, at the University of Alberta, for comparison with his unknowns, he replied that it was a species which he had never before seen.

GARYSMITHIA BIFURCATA Steere, gen. et sp. nov. Leskeacearum. Plantae caespitosae, paulum ramosae. Caules subturgidi vel julacei. Folia ovato-deltoidea vel cordata, acuta vel acuminata, imbricata ac appressa, homomalla vel subsecunda, praecipue ad apicem caulis substratum versus curvata, distincte decurrentia. Costa e basi valida indivisa deinde plerumque bifurcata, longitudinem folii 1/2 vel (raro) 2/3 attingens. Folii cellulae omnes verruculosae (subtiliter papillosae), medianae brevi- vel elongato-rhombicae, marginales abbreviatae, centrales elongatae, et eae utriusque anguli basalis quadratae valde delineatae secus marginem per folii longitudinem 1/4 extensae. Gametangia capsulaeque desunt.

---

[1]New York Botanical Garden, Bronx, New York 10458.

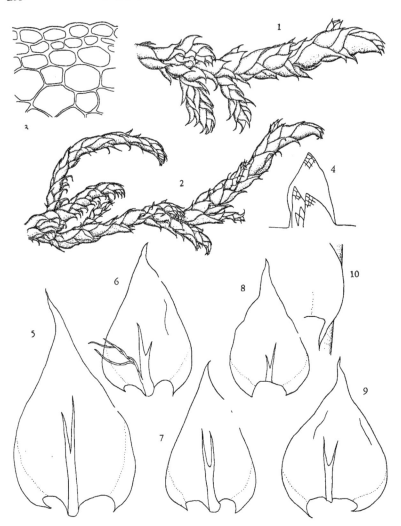

Plants small to medium in size, caespitose, dull yellowish to pale brown, not at all green and not shining, branching sparse and irregular, never pinnate. Leafy stems interwoven, 1-2cm long, 0.4-0.6mm in diameter, usually appearing julaceous, especially when dry, because of the imbricate-appressed leaves. Leaves convex, ovate-deltoid to cordate, acute to acuminate, somewhat homomallous to subsecund, turned toward the substratum, more conspicuously so at the stem apex, 1-1.5(-2)mm long, 0.5-0.7mm wide, distinctly decurrent at basal corners as a narrow wing 1-2 cells wide and 5-7 cells long. Costa variable from leaf to leaf on the same plant, reaching 1/2 (rarely 2/3) the length of the leaf, usually stout at base, bifurcating above into two nearly parallel to divergent, equal or nearly equal branches, which disappear at their apex into the cells of the lamina, occasionally branching from the base of the leaf, only rarely undivided, the stout basal part of the costa occasionally producing from one to many conspicuous yellowish-brown rhizoids along its length. Leaf margins plane, entire or minutely serrulate from the outward projection of cell corners, especially in the upper half and at the apex. Leaf apex short- or long-acuminate, infrequently consisting of a filiform series of 1-4 elongated single cells, the leaf tip incurved, recurved or curved to one side, depending on the orientation of the leaf on the stem with respect to the substratum. Leaf cells thick-walled, short- to elongated-rhomboidal, (20-)26(-34) microns long, (7-)10(-13) microns wide, shorter at the margins, more elongated toward, and at the center, the basal angles of the leaf filled with a large, conspicuous, and well demarcated area of crowded, rectangular cells that runs up the basal leaf margin to approximately 1/4 of the leaf length, at least some of the basal cells transversely elongated, 10-16 microns wide and 7-11 microns high, gradually becoming isodiametric, usually quadrate, eventually merging with the short-rhombic to rectangular cells above. Cells over the basal part of the costa on both sides of leaf much longer, narrower, and thicker-walled than the cells of the lamina; cells over the branches of the costa identical to the cells of the lamina in size and shape but usually more strongly colored, all leaf cells finely papillose, most con-

---

FIG. 1-10. Garysmithia bifurcata. Fig. 1-2, habit drawings of stems, showing homomallous leaves, X9. Fig. 1 in moist condition, Fig. 2 in dry condition. Fig. 3, detail of cross-section of stem, showing smaller, thicker outer cells, X390. Fig. 4, pseudoparaphyllia of stem, X99. Fig. 5-9, leaves, showing variation of costa, size and shape, X39. Fig. 6 shows rhizoids on basal stouter part of costa. Fig. 10, decurrent basal corner of leaf, X99.

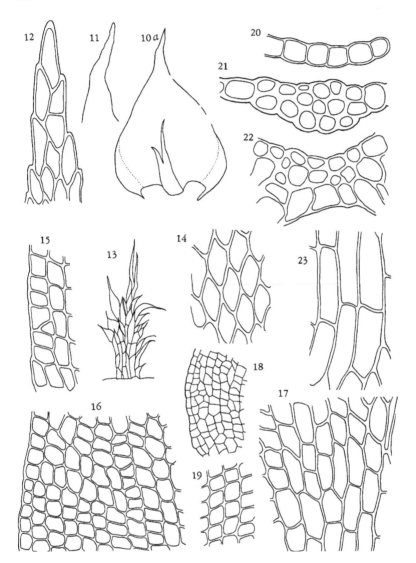

spicuously so near the costa, with numerous papillae per cell. Sexual organs and sporophyte not found on any of the specimens.

On non-calcareous rock faces or in rock crevices.

TYPE ALASKA Ogotoruk Creek, Cape Thompson, Brooks Range, rock ledge, with Orthotrichum pylaisii Brid. (2 July 1966, G. L. Smith A304) NY (HOLOTYPE), ALA, ALTA, HIRO, MICH, NICH.

OTHER SPECIMENS EXAMINED ALASKA Ogotoruk Creek, near Cape Thompson, W end of Brooks Range (Point Hope Quadrangle), 68° 06'N, 165°45'W, on thin soil over rock on mountainside (21 July 1965, W. C. Steere 650721-12) NY, Yukon River-Prudhoe Bay Haul Road just E of Galbraith Lake (Philip Smith Mountains Quadrangle), 68° 30'N, 149°25'W, on conglomerate outcrop, 1220m alt (20 July 1976, Barbara M. Murray 76-290B, 76-306) ALA, NY, Mt. McKinley National Park, Just W of Polychrome Pass, on face of granite outcrop (Marmot Rock), 3800ft alt. (31 July 1967, F. J. Hermann 21533) NY. COLORADO Hinsdale County, Cebolla Creek, vertical face of granite bluff, Cebolla Campground, 15mi E of Lake City, 9300ft alt. (20 July 1976, F. J. Hermann 27230) NY.

In the absence of sporophytes, it is difficult to assign this plant to any known genus, or for that matter, to any particular family of mosses. I have placed it in the Leskeaceae largely because of its minutely papillose cells. It has been assigned there more as a matter of convenience than through conviction, since the Julaceous stems and

---

FIG. 10A-23. Garysmithia bifurcata. Fig. 10A, entire individual leaf, X39. Fig. 11, leaf apex, X99. Fig. 12, apex of leaf, showing cellular detail, X390. Fig. 13, young rhizoids arising from dorsal side of costa, X99. Fig. 14, cellular detail of upper central part of leaf, X390. Fig. 15, cellular detail of upper part of leaf margin, X390. Fig. 16, cellular detail of upper part of specialized alar group, X375. Fig. 17, cellular detail of leaf base near costa, X375. Fig. 18, cellular detail of lower part of specialized alar group, X197. Fig. 19, enlarged drawing of same tissue as Fig. 18, X390. Fig. 20, cross-section of leaf at upper margin, X390. Fig. 21, cross-section of leaf at upper center, through one branch of costa, X390. Fig. 22, cross-section of leaf near base, through lower and stouter part of single costa, X390. Fig. 23, epidermal cells of branch, X390.

Original pencil drawings for all illustrations were made by Dr. Zennoske Iwatsuki with a camera lucida, and the inking, stippling, and composing of the plates was done by Miss N. Ando, both of the Hattori Botanical Laboratory, Nichinan, Japan.

the cellular areolation, especially the dense alar groups of quadrate cells, suggest the Leucodontaceae.

A specimen of Garysmithia was sent to Dr. H. Ando, of Hiroshima University, who has monographed the genus Homomallium, since the homomallous leaves suggest the possibility of relationship with that genus. His response is so interesting that I take the liberty of quoting from it (pers. comm ,. 13 April 1977)· 'In microscopical observation, your specimen is quite different and is certainly a new moss which I have never seen. I do not think it is a Homomallium. As to the family position I cannot give a decisive conclusion. It is connected with the Leucodontaceae through Pterogonium, on the other hand, with the Leskeaceae through Pseudoleskeella tectorum. At any rate I don't think that it belongs in the Hypnaceae.'

All of the Alaskan specimens were collected without being recognized in the field as something of special interest, so that it would be difficult indeed to find the exact localities again. However, Dr. Hermann believes that he can relocate the Colorado station, and it is possible that sporophytes will eventually be found in populations that are kept under observation over a period of several years.

SPHAGNUM RECURVUM[1]

G. L Smith[2]

During a brief visit to Paris (PC) in the spring of 1974, I examined the Polytrichaceae and Sphagna contained in a bundle of specimens labelled "Muscinées de l'Amérique du Nord de l'herbier de L.C. Richard. Types du Flora boreali-americana." This portfolio of specimens has since been studied critically by Dr. Geneva Sayre, and authenticated as Richard's North American bryophyte herbarium, arranged by F. A. Camus (1852-1922), its last private owner (cf Sayre, 1976, for details) Richard's original labels are intact. The herbarium contains several specimens of North American mosses from Palisot de Beauvois, including one which is evidently an isotype of Sphagnum recurvum Beauv., published in his Prodrome (1805) An annotation on the specimen by Camus says that the specimen was given by Palisot to Richard shortly after the publication of the Flora boreali-americana of Michaux in 1803. The label on the specimen reads "S. acutifolium Hedw./ Carolina m.-" The type of S. recurvum was collected in South Carolina (Carolina meridionalis) by Louis A G Bosc, as stated in the protologue. Bosc was French vice-consul in Carolina from 1798 to 1800 (Burdet, 1972). Unfortunately, the label does not bear the name Sphagnum recurvum or the name of the collector. A branch, several stem leaves, and a fragment of the stem cortex were removed from the specimen for careful study at a later date. Permanent slides prepared from this material have since been returned to PC.

The name Sphagnum recurvum has been in common use for over a century, but it has been used in different senses, and it would be particularly useful to have a specimen that could serve as the type. After due consideration of the nomenclatural consequences, which are discussed below, I have designated this specimen as the lectotype of Sphagnum recurvum. As treated by Andrews (1913), S. recurvum includes four closely related taxa which are recognized by most other

[1]Supported in part by National Science Foundation Grant GB37662.
[2]New York Botanical Garden, Bronx, New York 10458.

171

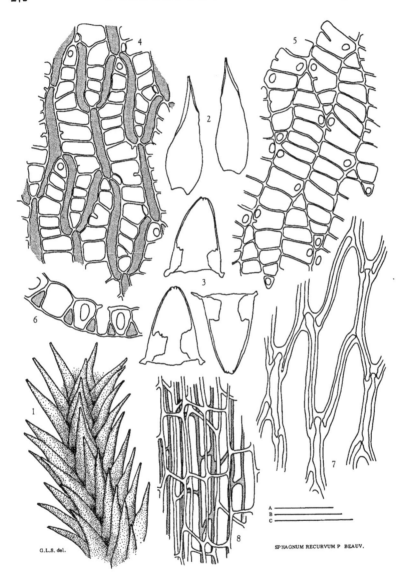

G.L.S. del.

SPHAGNUM RECURVUM P BEAUV.

sphagnologists as distinct species   Sphagnum fallax (H. Klinggr.) H.
Klinggr. [=S. mucronatum (Russow) Zickendr., S. apiculatum H.
Lindb.], Sphagnum angustifolium (Russow) C. Jensen [=S. parvifolium
(Sendtn.) Warnst.], Sphagnum flexuosum Dozy & Molk. [=S. amblyphyllum (Russow) Zickendr.], and Sphagnum pulchricoma C. Mull.
The nomenclature is that of Isoviita (1966). Andrews recognized S.
angustifolium at the varietal level as S. recurvum var. tenue   All
except S. pulchricoma are widely distributed boreal taxa. Those who
have treated all four as species have usually associated the name S
recurvum with the "apiculatum" form (S. fallax, e. g. Warnstorf,
1911). Crum (1973, p 32) treats the first three as S. recurvum var.
recurvum, var. tenue, and var. amblyphyllum, respectively, but includes S. pulchricoma in his concept of the var. amblyphyllum (pers.
comm.).

In his nomenclatural revision of the European Sphagna, Isoviita
(1966, p. 242) suggests that the name Sphagnum recurvum probably
applies to the exclusively American S. pulchricoma, and not to any
European species. Andrus (1974) agrees that S. pulchricoma is the
only recurvum-segregate likely to be collected in South Carolina.
According to Andrus, S. pulchricoma is a species of the Atlantic and
Gulf coastal plain, extending from New Jersey south to Florida and
Louisiana, although recorded from as far north as Nova Scotia. The
type of S. pulchricoma came from Brazil (Müller, 1848).

As the accompanying illustrations show (Figs. 1-8), the lectotype
of Sphagnum recurvum belongs to the taxon currently known as S.
pulchricoma C. Müll., which is characterized by 1) chlorophyll cells
of the branch leaves well-included on the concave surface, 2) a fairly
well-differentiated, 2-3-layered stem cortex, and 3) rather narrow,
distinctly 5-ranked branch leaves. An example was distributed by
Andrus and Vitt as Sphagnotheca Boreali-americana 21.

At least some of the South American specimens of Sphagnum
pulchricoma at NY seem to be indistinguishable from the type of S.
recurvum, including an authentic specimen from Brazil. This collection, from Itajahi (Pabst, s.n.), is cited as S. pulchricoma by

---

Sphagnum recurvum Beauv. 1, Portion of strong branch, with
distinctly 5-ranked leaves, 2, Branch leaves, 3, Stem leaves, 4,
Outer (convex) surface of branch leaf, 5, Inner (concave) surface of
branch leaf, with chlorophyll-cells entirely included. 6, Cross-section
of branch leaf, 7, Median cells of stem leaf, 8, Stem cortex, surface
view. (Figs. 1-8 from the lectotype, PC). Fig. 1 A=1mm, Figs.
2, 3· B=1mm, Figs. 4-7 C=0.05mm, Fig. 8 C=0.1mm.

Müller in the supplement to his Synopsis Muscorum (1851). The type specimen of S. pulchricoma has not been examined.

The stem leaves of Sphagnum recurvum sens. strict. are similar to those of S. flexuosum. Andrus (1974) considers S. pulchricoma to be a good species, distinguished from S. flexuosum by its geographical distribution and by the characters listed above. Judging from my own experience with this handsome plant in the Pine Barrens of New Jersey and eastern Long Island, New York, I am convinced that our southern Atlantic and Gulf coastal plain S. recurvum (S. pulchricoma) is a distinct taxon, whatever rank one wishes to give it.

The typification of the name Sphagnum recurvum leaves the former "var. recurvum" (S. fallax) without a name at the varietal level. The basionym, S. cuspidatum var. fallax H. Klinggr., of 1872, cannot be used because of the existence of an S. recurvum var. fallax Warnst., of 1884, a synonym of S. obtusum (Warnstorf, 1911); von Klinggraef's var. fallax was not transferred to S. recurvum until 1939. Isoviita (1966) indicates that he has seen "authentic material" of S cuspidatum var. brevifolium Lindb. ex Braithw., of 1878, and that it is S. fallax. The date of Braithwaite's Sphagnaceae is generally given as 1880, but Dr. W. C. Steere owns a copy of an earlier printing of this work, which is dated 1878 on the title page. A glance at the Index Muscorum shows that there are many varietal epithets to choose from which might be S. fallax, but that var. brevifolium, which dates from 1878, is older by several years than any of these. I have not seen any of the specimens cited in the protologue, Braithwaite's Sphagnaceae Brittanicae Exsiccatae 53 is missing from the set at NY, which is otherwise complete. Warnstorf (1911, p. 215), having seen the "original" of this variety, makes it a form of S. balticum, but Isoviita was presumably dealing with material from Lindberg's own herbarium, and this should be a more reliable indication of the correct use of the name. The stem leaf of var. brevifolium illustrated by Braithwaite (1878, Pl. 27, figs. 5, 5a), does not look like S. balticum.

Sphagnum recurvum var. amblyphyllum (Russow) Warnst., which is used by Crum (1973) for S. flexuosum, dates from 1890 as a varietal epithet. Isoviita lists no varieties as possible synonyms for this species. Of all the possible varietal epithets listed in the Index Muscorum, the oldest which can be applied to S. flexuosum, to the best of my knowledge, is S. recurvum var. majus (Ångstr. ex Warnst.) Warnst. of 1883, originally published by Warnstorf in 1881 as S. var - iabile var. intermedium f. majus Ångstr. "non Russow." I have examined Gravet's Sphagnotheca Belgica, 26 and 27 (FH!), which are the only specimens mentioned in the protologue of f. majus, and they are both S. flexuosum.

The following are what seem to be the correct names for the segregates of Sphagnum recurvum sens. lat. as species, as subspecies,
and as varieties. The situation at the varietal level is unsettled, and
only those names discussed above are included in the synonymy. A
detailed consideration of this knotty problem is beyond the scope of
this paper. At least, the name of S. angustifolium at the varietal
level seems to be reasonably secure the var. tenue H. Klinggr. has
no rivals, as far as I know. The nomenclature at the subspecific
level presents no such difficulties and has the added appeal of famil-
iarity, since the epithets mucronatum, angustifolium and amblyphyllum have been, until recently, in general use for these taxa

Sphagnum recurvum Beauv., Prodr. Aethéog. 88. 1805.
       LECTOTYPE. "S. acutifolium Hedw./ Carolina m -"Herb.
       Richard (PC!).
       Sphagnum pentastichon Brid., Musc. Recent Suppl. 1 16.
       1806.
       Sphagnum pulchricoma C. Müll., Syn. 1 102. 1848.

Sphagnum fallax (H. Klinggr.) H. Klinggr., Topogr Fl. Westpr. 128.
     1880.
       Sphagnum recurvum subsp. mucronatum Russow, Sitz.-ber.
       Nat.-Ges. Dorpat 9 109. 1889.
       Sphagnum cuspidatum var. brevifolium Lindb. ex Braithw.,
       Sphag. 84. 1878.
       Sphagnum recurvum var. brevifolium (Lindb. ex Braithw.)
       Warnst., Flora 67 608. 1884.

Sphagnum angustifolium (C. Jensen ex Russow) C. Jensen, Bih. Sv.
     Vet.-Akad. Handl. III. 16 48. 1891.
       Sphagnum recurvum subsp. angustifolium C. Jensen ex Russow
       Sitz.-ber. Nat.-Ges. Dorpat 9 112. 1889.
       Sphagnum recurvum var. tenue H. Klinggr., Schr. Phys.-ök.
       Ges. Konigsb. 13 5. 1872.

Sphagnum flexuosum Dozy & Molk., Prodr. Fl. Batav. 2(1) 76. 1851.
       Sphagnum recurvum subsp. amblyphyllum Russow, Sitz.-ber.
       Nat.-Ges. Dorpat 9 112. 1889.
       Sphagnum variabile var. intermedium f. majus Ångstr. ex
       Warnst., Eur. Torfm. 65. 1881.
       Sphagnum variabile var. majus (Ångstr. ex Warnst.) Warnst.,
       Flora 65 550. 1882.
       Sphagnum recurvum var. majus (Ångstr. ex Warnst.) Warnst ,
       Flora 66 374. 1883.

The apex of the stem leaves of Sphagnum recurvum sens. strict. varies from narrow and almost entire to broad and lacerate, as a result of the progressive resorption of the walls of the hyaline cells. This variation can often be observed along the length of a single stem. The loss of the inner and outer cell walls allows the chlorophyll-cell mesh to spread, resulting in a broadly lacerate leaf apex.

LITERATURE CITED

Andrews, A. L. 1913. Sphagnaceae. N. Am. Fl. 15· 1-31.

Andrus, R. E. 1974. The Sphagna of New York State. 1-421. Ph. D. thesis, State Univ. N. Y. Syracuse. Univ. Microfilms 74-24023.

Burdet, H. M. 1972. Cartulae ad botanicorum graphicem. Candollea 27 307-340.

Crum, H. A. 1973. Mosses of the Great Lakes forest. Contr. Univ. Mich. Herb. 10· 1-404.

Isoviita, P. 1966. Studies on Sphagnum L. I. Nomenclatural revision of the European taxa. Ann. Bot. Fenn. 3 199-264.

Müller, C. 1848, 1851. Synopsis muscorum frondosorum omnium hucusque cognitorum. 1 1-812, 2· 1-772.

Palisot de Beauvois, A. M. F. J. 1805. Prodrome des cinquième et sixième familles de l'Aethéogamie. 1-114. Paris.

Sayre, G. 1976. The type herbarium of the Flora Boreali-americana. Revue Bryol. Lichénol. 42· 677-681.

Warnstorf, C. 1911. Sphagnales-Sphagnaceae (Sphagnologia universalis). In Engler, Pflanzenreich 51· 1-546.

NOTES ON THE GENUS AND SPECIES LIMITS OF

PSEUDOGYNOXYS (GREENM.) CABRERA

(SENECIONEAE, ASTERACEAE).

Harold Robinson and Jose Cuatrecasas
Department of Botany
Smithsonian Institution, Washington, DC. 20560.

The neotropical genus Pseudogynoxys was raised to
generic rank by Cabrera (1950) and has been generally
accepted by students of the Senecioneae since that
time (Cuatrecasas, 1955; Afzelius, 1966; and Nordenstam, 1977). The genus has been sharply defined on
the basis of the scandent habit, the alternate leaves,
the membranaceous to subchartaceous leaf blades, the
radiate heads, and the styles of the disk flowers with
pointed hirsute appendages. The flowers are notably
orange-colored becoming reddish with age, and some
specimens have been noted as fragrant. At least two
species have been cultivated and the distribution of
P. chenopodioides might be partially the result of
human intervention. Unfortunately, taxonomic treatments of the genus have consisted largely of transfers
and synonymizations of names and descriptions of new
entities without any complete survey of the diversity
and limitations of all the known species. Perhaps
partially for this reason a recent treatment for the
Flora of Guatemala (Williams, 1976) has reduced the
genus to synonymy under Senecio and has placed all the
Central American species in synonymy under Senecio
chenopodioides HBK. The present effort attempts to
correct the primary inaccuracies and deficiencies of
past studies.
    The placement of Pseudogynoxys in Senecio cannot
be considered truly traditional nor natural. From the
time of Cassini (1827) to the time of Greenman (1902)
the species were generally described under the
neotropical genus Gynoxys because of the pointed tips
of the styles. The placement was unsatisfactory since
Gynoxys is a genus of shrubby plants with usually
opposite coriaceous leaves. It was Greenman (1902)
who transferred the group to Senecio and established
the subgenus Pseudogynoxys. Greenman's effort did
not include South American material and it included a
number of confusing species names that were never
validated.

177

Pseudogynoxys is without close relatives among
the American genera of the Senecioneae and there is no
obvious integration of characters. Actual relationship seems to be to the genus Gynura Cassini of the
Eastern Hemisphere. A few species of Gynura possess
habits similar to Pseudogynoxys and the style branches
are pointed. The Old World genus is distinguished by
the lack of rays in the head and by the exact shape of
the style appendages which are much longer with only
short hairs.

In reviewing the species of Pseudogynoxys taxonom-
ically valid distinctions have been seen in the
straightness of the stem, the pubescence, the leaf
shape and venation, the position of the inflorescence,
the stoutness of the pedicels, the structure of the
calyculus, and the tips of the involucral bracts.
Some differences in floral structure also occur, but
the only ones noted in this study are the nearly
glabrous ray styles and the distinctive pappus of
P. cabrerae, and the anther collars of P. scabra.

A number of names have been reviewed for placement
in the genus including approximately 21 that have been
transferred into the genus as valid species. Only 13
species are recognized here. An additional three names
described in the genus Gynoxys by Turczaninow (1851)
share some described features of Pseudogynoxys. The
three species were based on Jameson collections from
Ecuador and isotypes have been located in the U. S.
National Herbarium. The three prove to be members of
the genus Senecio section Aetheolaena (Cass.) Hoffm.
and are disposed as follows: Gynoxys prenanthifolia
Turcz., Bull. Soc. Nat. Mosc. 24 (pt. 1): 207. 1851.
Type: In Andibus Quitensibus Jameson 636; and
G. auriculata Turcz. ibid. (pt. 2): 86. 1851. Type:
In alpe Pichincha alt. 14000 ped. Jameson s.n. both
prove to be Senecio patens (HBK.) DC; G. heterophylla
Turcz. ibid. (pt. 2): 85. 1851. Type: In Andibus
Quitensibus. Jameson 894-896 proves to be Senecio
pindilicensis Hieron. In the latter case the Turczaninow name is older but the combination Senecio hetero-
phyllus is preoccuppied. The identity of G. prenanthifolia has already been noted by Weddell (1855-1857,
p.92). The styles of the disk flowers of section
Aetheolaena do not actually have pointed tips but have
an apical tuft of hairs.

Pseudogynoxys (Greenm.) Cabrera, Brittonia 7: 54.
     1950.
     Senecio subg. Pseudogynoxys Greenm., Bot. Jahrb.
     32: 23. 1902.

Plants suffrutescent, scandent. Stems coarsely striated, green or pale brownish Leaves alternate; petioles slender, sometimes with stipuliform bases; lamina membranaceous to subchartaceous, ovate to oblong ovate, base broadly acute to cordate, margins minutely serrulate to coarsely dentate, apex acute to short-acuminate. Inflorescence terminal or axillary with one to many heads. Head campanulate to hemispherical; calyculus of ca. 10-30 distinct bracts; involucre uniseriate, with short-acute to long-attenuate tips, extreme tips densely pubescent; receptacle glabrous. Corollas glabrous, deep orange, becoming reddish or purplish with age, rays ca. 6-15; styles with 2 stigmatic lines, style tips acute, usually sparsely pubescent; disk corollas with long basal tube, throat narrowly funnelform, lobes narrowly oblong-lanceolate, median resin duct evident; anther collars with larger or thinner-walled cells below; anther thecae tapering or slightly cordate at base, thecial cells elongate with single minute nodular thickenings at upper and lower ends, few to many rows nearest connective also with minute thickenings along vertical walls; anther appendage lanceolate with narrow tip; style appendages short- to long-acute with numerous hairs often more prominent around base and at tip. Achenes cylindrical with ca. 10 ribs, hirtellous, surface slightly to strongly papillose with projecting small cells; carpopodium short, incurved at lower margin, not sharply demarcated above, with many rows of small cells; pappus of 3-5 series of capillary bristles; bristles sometimes flattened, scabrous, distally 20-25μ wide. Pollen mostly 30-35μ in diameter.

Type species: Gynoxys cordifolia Cass.

The species of Pseudogynoxys can be distinguished by the following key. Further clarification can be obtained in the appended local keys for species of Guatemala and Ecuador.

Key to the species of Pseudogynoxys

1. With stipuliform expansions at bases of the petioles 2

2. Leaves subrhombic-lanceolate and sharply dentate (Peru) . . . . P. filicalyculata

2. Leaves rounded or slightly cordate at the base, margins denticulate (Colombia) . . P. bogotensis

1. Without stipuliform expansions at bases of the
    petioles 3

3. Heads with ten or less involucral bracts (Guat.,
    Mex.) . . . . P. fragans

3. Heads with more than ten involucral bracts . . 4

4. Leaves essentially glabrous below on veins and
    on undersurface, stems and involucres nearly to
    completely glabrous (Mexico to Colombia, West
    Indies) . . . . P. chenopodioides

4. Leaves sparsely minutely appressed puberulous to
    tomentose on lower surface, stems or involucres
    glabrous to hirtellous 5

  5. Heads mostly single or in groups of 2-3 on
      stout pedicels 5-20 cm long 6

    6. Bracts of calyculus mostly 1.5-2.0 mm wide
        (Ecuador) . . . . P. sodiroi

    6. Bracts of calyculus less than 1 mm wide . . 7

      7. Leaves trinervate from at or near base of
          lamina, usually cordate (Argentina, Brasil,
          Paraguay) . . . . P. cabrerae

      7. Leaves subpinnately veined, bases usually
          broadly acute to truncate (Mexico to
          Venezuela) . . . . P. cummingii

  5. Heads clustered, pedicels mostly less than 5 cm
      long, slender 8

    8. Undersurface of leaf thinly tomentose . . . 9

      9. Inflorescence terminal on leafy stems; leaf
          base mostly broadly acute to rounded, margins
          sharply serrate (Ecuador, Peru)
                                        P. sonchoides

      9. Inflorescences on short axillary branches;
          leaf base cordate, margins broadly and
          shallowly dentate (Peru) . . . P. cordifolia

    8. Undersurface of leaf puberulous to scabrellous
                                                    10

10. Calyculus densely pubescent (Mexico, Cent.
Amer.) . . . . P. haenkei.

10. Calyculus sparsely puberulous, hairs mostly
on margins of bracts (S.Amer.) 11

   11. Stems slender and obviously deflected at
nodes; leaves ovate; involucral bracts acute
(Ecuador) . . . . P. engleri

   11. Stems mostly stout and straight; leaves
often oblong-ovate; involucral bracts
attenuate 12

     12. Leaves with erect hairs on surfaces
(Ecuador, Peru) . . . . P. scabra

     12. Leaves with only minute appressed hairs
(Peru) . . . . P. poeppigii

Key to species in Mexico and Central America

1. Heads with ten or less involucral bracts
P. fragans

1. Heads with more than ten involucral bracts . . . 2

  2. Plants mostly glabrous; involucral bracts glabrous,
calyculus sparsely pubescent; leaves usually with
remote sharp teeth . . . . . . P. chenopodioides

  2. Plants distinctly pubescent on stems, leaves or
involucre, calyculus densely pubescent; leaves
usually with crowded or blunt serrations . . . 3

   3. Heads single or in groups of 2-4 on stout
pedicels mostly over 5 cm long; involucral bracts
attenuate and often reddish at tip
P. cummingii

   3. Heads numerous in clusters, pedicels mostly less
than 3 cm long, slender; involucral bracts acute
or only slightly attenuate . . . . . P. haenkei

Key to Ecuadorian species

1. Heads single or in groups of 2-4, often 2 cm wide;
pedicels usually over 5 cm long, stout; bracts of

calyculus 1.5-2.0 mm wide . . . . P. sodiroi

1. Heads in clusters, usually 1 cm or less broad;
    pedicels usually less than 5 cm long, slender;
    bracts of calyculus less than 1 mm wide 2

    2. Inflorescences terminal on leafy stems or branches;
        leaves with 3-4 pairs of prominent secondary veins
        congested near base . . . . P. sonchoides

    2. Inflorescences mostly on short axillary branches;
        leaves with only 1 or 2 pairs of prominent second-
        ary veins near base 3

      3. Stems slender and obviously deflected at nodes;
          leaves ovate, usually gradually narrowed to a
          sharply acute tip; involucral bracts acute
                                              P. engleri

      3. Stems stout and straight; leaves oblong-ovate,
          becoming short-acuminate; involucral bracts
          attenuate . . . . P. scabra

The recognized species of Pseudogynoxys and their
synonyms are as follows.

Pseudogynoxys bogotensis (Spreng.) Cuatr., Brittonia
      8: 156. 1955.
    Senecio macrophyllus HBK., Nov. Gen. & Sp. 4: 140.
        1818, ed folio. Not. S. macrophyllus Bieb.
    Senecio bogotensis Spreng., Syst. 3: 556. 1826.
    Senecio moritzianus Klatt, Leopoldina 24: 127. 1888.
      The synonymy follows that of Cuatrecasas (1955).
The species is represented in the U.S. National Herbarium by seven specimens. COLOMBIA: Cundinamarca:
Fusagasugá. André s.n.; 4 kms NW of Sasaima along
highway to Villeta, banks of Río Dulce. Barclay,
Juajibioy & Gama 3677; Entre Sasaima y Villeta. Dugand
& Jaramillo 3933; La Vega. Pérez Arbeláez & Cuatrecasas
5341; Santander: Río Suratá valley, between El Jabon-
cillo and Suratá. Killip & Smith 16429; Between El
Roble and Tona. Killip & Smith 19419; Tolima: Ibague
to Rio Coello, New Quindo trail. Hazen 9644.

Pseudogynoxys cabrerae H.Robinson & J.Cuatrecasas, sp.
      nov.
      Plantae suffrutescentes scandentes laxe ramosae.
Caules angulato-striati sparse vel dense puberuli.
Folia alternata, petiolis 1-3 cm longis base non

auriculatis; laminae herbaceae vel membranaceae ovatae
vel late ovatae ca. 6.0-11.5 cm longae et ca. 3.0-7.5
cm latae base plerumque cordatae vel subcordatae fere
ad basem trinervatae margine argute serratae vel grosse
dentatae apice breviter acuminatae supra sparse puberulae subtus densius puberulae, nervis secundariis
principalibus plerumque remotis et parallelis, basilar-
ibus plerumque mox ramosis. Inflorescentiae in ramis
foliatis terminales 1-4 capitatae, pedicellis plerumque
5-15 cm longis raro brevioribus crassis distincte
minute puberulis. Capitula 15-20 mm alta et 25-30 mm
lata; bracteae calyculi 20-30 lineares ca. 7-10 mm
longae et 1 mm latae plerumque dense puberulae;
bracteae involucri 25-30 uniseriatae lineari-lanceolatae 10-13 mm longae 1.0-1.5 mm latae base gibbosae non
angulatae apice longe attenuatae saepe rubro-tinctae
extus plerumque puberulae; receptacula plana, inter-
stitiis non vel breviter lobuliferis. Flores radii
12-14; corollae aurantiacae deinde rubrae; tubis ca. 7
mm longis glabris, limbis oblongis vel vix obovatis
ca. 15 mm longis et ca. 6.5 mm latis glabris; appendices stylorum glabrae vel subglabrae. Flores disci
ca 100; corollae aurantiacae deinde rubrae 11-14 mm
longae anguste infundibulares glabrae, tubis 7-11 mm
longis, faucis 1.5-1.8 mm longis, lobis lineari-
lanceolatis 2.0-2.5 mm longis sub medio equilatis ca.
0.4 mm latis. Achaenia ca. 3 mm longa et ca. 0.8 mm
lata plerumque 10-costata ubique minute puberula base
truncata; setae pappi 80-110 longiores ca. 12 mm
longae 4-5-seriatae complanatae vel percomplanatae
base subintegrae superne pertenues et scabrellae.
     TYPE: ARGENTINA: Corrientes: Dep. Empedrado,
Estancia "Las Tres Marias", Dry woodland on the bank
of the Rio Parana, soil hard clay, shrub or subshrub,
growing to a height of one to a couple of metres,
supported by other shrubs. 6/11 1952. Pedersen 1888
(Holotype, US). PARATYPES: ARGENTINA: Chaco:
Jörgensen 2019 (US); Corrientes: Dep. Mburucuyá,
Estancia "Santa Maria", 30/8 1962. Pedersen 6506 (US);
Jujuy: Dep. Ledesma, Yuto, Vinalito. 7-VII-1937.
Cabrera 4049 (US); Quinda pr. Laguna de La Brea. 1/6
1901. Fries 37 (US); Salta: Dep. Oran, Rio Pescado.
16/IX/1938. Cabrera 4584 (US); Dep. Oran, Yaguani.
3-XII-1941. Maldonado 759 (US); Dep. Oran, Embarcación.
Dic. 20, 1926. Venturi 5108 (US); BRASIL: São Paulo:
Campinas. 20 Nov. 1938. Carvalho & Mendes 2942 (US);
Campinas. Campos Novaes 139 (US); Mogi-Guassi, Fazenda
Campininha. 3/II/1955. Kuhlmann 3516 (US); Along river
at Usina, 9 km west of Santa Cruz do Rio Pardo. 10-12-
1936. Archer 4195 (US); PARAGUAY: 10 km north of

Porto Gibaja, banks of Arr. M-boi-ci, banks of Rio
Parana. Aug. 21, 1952. Beetle 2166 (US); Villarrica.
Jorgensen 7501 (US); Central Paraguay, In regione lacus
Ypacaray. July 1913. Hassler 11845 (US); Pilcomayo
River. 1888-1890. Morong 848 (US); Morong 842 (US);
N. Paraguay. IX 1892. Kuntze (US).

Both Baker (1884) and Cabrera (1950) evidently
recognized the species as distinct from the Central
American P. cummingii, but all the names used are
based directly on the Central American type or are
derived nomenclaturally from it. Grisebach (1879)
first treated material of P. cabrerae as Senecio
benthami Griseb., but the latter was a nom. nov. based
on Gynoxys cummingii Benth. and represented a broad
concept including both species. Baker (1884) attempted
to use the Grisebach name in a more restricted sense
which, however, excluded the typical element. Cabrera
(1950) unfortunately chose to refer to his Pseudogynoxys
benthami as a nom. nov. rather than as a new species and
therefore it also is tied nomenclaturally to the name
used by Baker rather than to the description.

The new species resembles P. cummingii and in spite
of some divergence of forms the shape of the leaves on
superficial examination does seem to overlap. The
totally discontinuous distribution and the extremely
cordate and dentate leaf-form frequent in Argentinian
material does strongly indicate a separate species is
involved. Pseudogynoxys cummingii does occur in South
America, but only in the northern parts of Colombia
and Venezuela. The most northern P. cabrerae specimens
are from southern Brasil and Paraguay. Specimens of
similar habit from areas between prove to be the
distinctive P. bogotensis and P. sodiroi. More critical
examination of P. cabrerae shows consistent differences
in two significant characters, the essentially tri-
nervate venation of the leaves, and the 80-110 setae of
the pappus in 4-5 series. The pappus setae are more
flattened than in other species of the genus and often
are bent or broken in the distal portions. In P.
cummingii the leaves have subpinnate venation, and the
pappus setae are in ca. 3 series as is conven-
tional for the tribe. The individual setae are less
flattened though occasional isolated setae may be
extremely broad and flat. Even in immature material
where the bases of the setae cannot be seen properly,
the less flattened condition can be noticed in compar-
isions. The mature specimens of P. cummingii often
seem to have lost most or all of the pappus while the
P. cabrerae pappus seems more persistent. The individual setae do not appear less fragile but the extra

series apparently provide more resistance.

The puberulence of P. cummingii is coarser than in P. cabrerae especially on the undersurface of the leaves, but this is only obvious in direct comparison of material. The style appendages of the ray flowers of P. cabrerae are notable for the lack or near lack of hairs. A few short hairs often form a very short apical tuft. All other species of Pseudogynoxys have more hairs on the style appendages of the rays. One specimen of P. cabrerae (Hassler 11845) has more hairs on the styles of the rays which probably represents a partial failure of the normal differentiation between the style of the ray and disk flowers.

Pseudogynoxys chenopodioides (HBK.) Cabrera, Brittonia
        7: 5 . 1950.
    Senecio chenopodioides HBK., Nov. Gen. & Sp. 4: 140.
        1818, ed folio.
    Gynoxys berlandieri DC., Prodr. 6: 326. 1837.
    Gynoxys cordifolia Neaei DC., Prodr. 6: 326. 1837.
    Senecio confusus Britten, J. Bot. 36: 260. 1898.
    Pseudogynoxys berlandieri (DC.) Cabrera, Brittonia
        7: 56. 1950.
    The species is common in Mexico, Central America and the West Indies and has been cultivated in such places as Hawaii. There are only two specimens in the U.S. National Herbarium from South America. COLOMBIA: Valle: Cali. Planta cultivada en jardines. Patiño 599; Magdalena: Ciénaga. Romero Castañeda 1907.

Pseudogynoxys cordifolia (Cass.) Cabrera, Brittonia
        7: 54. 1950.
    Gynoxys cordifolia Cass., Dict. Sci. Nat. 48: 456.
        1827. Type in Herb. Jussieu No. 8939 under unpublished name Senecio scandens Juss.
    Senecio volubilis Hook., Bot. Misc. 2: 226. 1831.
    Senecio jussieui Klatt, Ann. K.K. Naturhist.
        Hoffmuseums 9: 367. 1894.
    Pseudogynoxys volubilis (Hook.) Cabrera, Brittonia
        7: 56. 1950.
    The synonymy follows that of Cabrera (1959). The species is endemic to the coastal ranges of Peru and ranges from Departments of Lambayeque and Cajamarca in the north to Lima in the south.

Pseudogynoxys cummingii (Benth.) H.Robinson & J.Cuatr-
        ecasas, comb. nov.
    Gynoxys cummingii Benth. ex Oerst., Kjoeb. Vidensk.
        Meddel. Dansk. Naturhist. Foren. 1852: 106.
        1852.

Senecio benthami Griseb., Goett. Abhand. 24: 206.
    1879.
Senecio calocephalus Hemsl., Biol. Cent. Amer., Bot.
    2: 237. 18881. Not. S. calocephalus Poepp. &
    Endl.
Senecio hoffmannii Klatt, Leopoldina 25: 106. 1889.
Pseudogynoxys benthami Cabrera, Brittonia 7: 56.
    1950.
Pseudogynoxys hoffmannii (Klatt) Cuatr., Brittonia
    8: 156. 1955.
    The species is widely distributed in Central
America from Panama northward to southern Mexico.
Only two specimens from South America are in the U.S.
National Herbarium. COLOMBIA: Magdalena: In coffee
grove above Manaure, alt. about 600 m. Haught 3980;
VENEZUELA: Zulia: Perija, alt. 1175 m. Gines 1384.

Pseudogynoxys engleri (Hieron.) H.Robinson & J.Cuatr-
    ecasas, comb. nov.
    Senecio engleri Hieron., Bot. Jahrb. 28: 644. 1901.
        Syn. cited, S. jussieui Klatt sensu Hieron.,
        Bot. Jahrb. 19: 69. 1894.
    Senecio almagroi Cuatr., An. Univ. Madrid 4 (fasc.
        2): 238. 1935.
    The species occurs at lower elevations in west-
central Ecuador. Specimens in the U.S.National Herbarium are as follows. ECUADOR: Canar: Along the road
to Canar, ca. 17 kms ESE of El Triunfo, elev. ca. 300
ft. King 6995; Guayas: Manglaralto, elev. 0-50 m, low
semi-arid hills back from beach. Dodson & Thien 1656;
Along stream 12 km north of Pedro Carbo. Alt. prob.
about 150 m. Haught 3058; Terecita, Stevens, 47, 130;
Los Rios: Hacienda Clementina on Rio Pita, marsh.
Asplund 5425.

Pseudogynoxys filicaliculata (Cuatr.) Cuatr., Britton-
    ia, 8: 156. 1955.
    Senecio filicaliculatus Cuatr., Collect. Bot. 3:
        29. 1953.
    The species is apparently still known only from
the type from Peru (Weberbauer 7721).

Pseudogynoxys fragans (Hook.) H.Robinson & J.Cuatre-
    casas, comb. nov.
    Gynoxys fragans Hook., Bot. Mag. 76, t. 1511. 1850.
    Senecio skinneri Hemsl., Biol. Cent. Amer., Bot.
        2: 247. 1881.
    The type grown at Kew was apparently originally
collected by Skinner in Guatemala. The original
description is accompanied by a detailed illustration.

No specimens fitting the description have been seen,
but a photograph distributed by the Field Museum of
plants in the Berlin Herbarium under the name Senecio
convolvuloides Greenm. shows an Ehrenberg collection
from Mexico that is apparantly P fragans.

Pseudogynoxys haenkei (DC.) Cabrera, Brittonia 7: 54.
      1950.
   Gynoxys haenkei DC., Prodr. 6: 326. 1837.
   Gynoxys oerstedii Benth. ex Oerst., Kjoeb. Vidensk.
      Meddel. Dansk. Naturhist. Foren. 1852: 107.
      1852.
   Senecio kermesinus Hemsl., Biol. Cent. Amer., Bot.
      2: 242. 1881.
   Senecio chinotegensis Klatt, Leopoldina 24: 125.
      1888.
   Senecio rothschuhianus Greenm., Bot. Jahrb. 60: 370.
      1926.
   Pseudogynoxys oerstedii (Benth.) Cuatr., Brittonia
      8: 156. 1955.
   In addition to the synonyms given, the following
unvalidated names of Greenman on the basis of annotated
specimens seem to apply to this species: Senecio
bernoullianus, S. convolvuloides, and S. trixioides
all nomen, Bot. Jahrb. 32: 22. 1902.

Pseudogynoxys poeppigii (DC.) H.Robinson & J.Cuatre-
      casas, comb nov.
   Gynoxys poeppigii DC., Prodr. 6: 326. 1837.
   Senecio sprucei Klatt, Leopoldina 24: 128. 1888.
   Pseudogynoxys sprucei (Klatt) Cabrera, Brittonia
      7: 56. 1950.
   All specimens seen have been from the Department
of San Martin in Peru. The species is obviously
closely related to P. scabra, but differs primarily
by the minute appressed hairs that give the leaves
a glabrous appearance. The species also seems to be
restricted to comparatively low elevations on the
eastern slopes of the Andes, while P. scabra is from
low elevations to the west of the Andes.

Pseudogynoxys scabra (Benth.) Cuatr., Brittonia 8: 156.
      1955.
   Gynoxys scabra Benth., Voy. Sulphur 121. 1836.
   Senecio eggersii Hieron., Bot. Jahrb. 28: 645.
      1901.
   Pseudogynoxys eggersii (Hieron.) Cabrera, Brittonia
      7: 56. 1950.
   Senecio neovolubilis Cuatr., Repert. Sp. Nov. 55:
      141. 1953.

Pseudogynoxys neovolubilis (Cuatr.) Cuatr., Brittonia
    8: 156. 1955.
Pseudogynoxys asplundii K.Afzelius, Bot. Notis. 119:
    233. 1966.
Pseudogynoxys chongonensis K.Afzelius, Bot. Notis.
    119: 237. 1966.
The species is characterized by the straight main
stems and the densely paniculate axillary inflorescences, a habit well illustrated in the photographs of the
type specimens of Afzelius (1966). The habit is particularly distinct from that of the sympatric P. engleri
with its deflected stem and less dense corymbose
inflorescences. The type of Gynoxys scabra has not
been seen and Bentham does not describe the stem, but
other described features indicate identity with
Senecio eggersii Hieron. Most anther collars that have
been examined show less differentiation of the basal
cells than in other species of the genus. Enlarged
cells are evident in the type specimen of S. neovolubilis, however. Specimens in the U.S. National Herbarium are as follows. ECUADOR: Junction of Guayas,
Cañar, Chimborazo & Bolivar: Foothills of the western
cordillera near the village of Bucay; 1000-1250 ft.
elev. Camp E-3964; Guayas: Road from Guayaquil to
Cuevedo; km 78; elev. 100 m. Dodson & Thien 1273;
Guayaquil, alt. 0-50 m. Hitchcock 19966, 20126;
Between Guayaquil and Salinas; near sea level. Mexia
6767; Prope Guayaquil. Mille 218; Guayaquil, along
road to Aguas Piedras. Rowlee & Mixter 1108; 8 km
north of Guayaquil; dry loam of hillside, alt. 5 m.
Stork, Eyerdam & Beetle 8968, 8969; Guayaquil and
vicinity, elev. 0-20 m. Valverde 342; Loja: Sabiango.
Townsend 885 (Holotype of Senecio neovolubilis);
Manabi: Roadsides near Santa Ana, alt. 100 m. Haught
3504. PERU: Piura: Canchaque upper limits of town
(several km above town, rd to Huancabamba, alt. 1350
m. Hutchison & Wright 6660.

Pseudogynoxys sodiroi (Hieron.) Cuatr., Ciencia 23:
    150. 1964.
    Senecio sodiroi Hieron., Bot. Jahrb. 29: 73. 1900.
    Senecio viridifluminis Cuatr., Repert Sp. Nov. 55:
        152. 1953.
    Pseudogynoxys viridifluminis (Cuatr.) Cuatr.,
        Brittonia 8: 157. 1955.
    Pseudogynoxys guarumalensis K.Afzelius, Bot. Notis.
        119: 237. 1966.
    Pseudogynoxys pastazensis K. Afzelius, Bot. Notis.
        119: 239. 1966.
    The synonymy is emended from Cuatrecasas (1964).

The specimens seen by the authors are as follows.
ECUADOR: Rio Verde, Pachano 235 (US, holotype of S.
viridifluminis); Cotopaxi: Cordillera Occidental;
Cordillera de Angamarca y Zumbagua, above Pilaló.
H.Barclay & Juajibioy 8071 (US); Pastaza: Road 3 km
E of Rio Blanco, elev. 1700 m. Dodson & Thien 2006
(US); Pastaza, 1200 m alt. Rimbach 273 (F); Pichincha:
About 84 km east of Quevedo, en route to Quito, alt.
2275 m. Maguire & Maguire 44259 (US); Tungurahua:
Valley of Rio Pastaza, Hacienda Rio Verde Grande, alt.
1500 m. Asplund 7838 (US, isotype of P. pastazensis);
Region near hot water pool at Baños, 1750 m alt.
Penland & Summers 34 (F).

Pseudogynoxys sonchoides (HBK.) Cuatr., Brittonia 8:
    157. 1955
    Senecio sonchoides HBK., Nov. Gen. & Sp. 4: 139.
        1818, ed folio.
    Gynoxys sinclairi Benth., Voy. Sulphur 120. 1836.
    Senecio jamesoni Spruce ex Klatt, Leopoldina 24:
        127. 1888.
    Senecio sinclairi (Benth.) Hieron., Bot. Jahrb. 19:
        68. 1894.
    Pseudogynoxys chiribogensis K.Afzelius, Bot. Notis.
        119: 235. 1966.
    The synonymy is emended from that of Cuatrecasas
(1955). Specimens in the U.S.National Herbarium are
as follows. ECUADOR: Bolivar: Balzapamba, alt. 800 m.
Haught 3311; Chimborazo: Cañon of the Rio Chanchan
near Huigra; 4000-4500 ft. elev. Camp E-3049; Huigra,
alt. 1200 m. Hitchcock 20348; Along the road to
Riobamba, ca 11 kms NE of Bucay, elev. ca. 1600 ft.
King 6957; Vicinity of Huigra, mostly on the Hacienda
de Licay, Rose & Rose 22188; Südwestlich Huigra, 1200
m. Schimpff 463; Cotopaxi: Road between Pilaloa and
Macuchi, alt. about 2400 m. Haught 2960; El Oro: Along
Rio Amarillo, upstream from Portovelo, alt. 640-760 m.
Steyermark 54076; Guayas: Near Bucay, alt. about 300 m.
Haught 2890; Loja: Between Loja and San Lucas, alt.
2100-2600 m. Hitchcock 21492; Sabiango, elev. 3000 ft.
Townsend A.96; Manabi: Road from Chone to Pichincha;
km 82, elev. 450 m. Dodson & Thien 1772; Pichincha:
Road from Quito to Santo Domingo de los Colorados; km
95, elev. 1000 m. Dodson & Thien 1216; s.l. Jameson
835   PERU: Lambayeque: Km 28 E of Olmos on Maranon
highway, vicinity of restaurant "El Salvador", alt.
1150 m. Hutchison & Wright 3424; Piura: Prov. Paita,
Talara, Haught 68; Santa Rosa (Abajao de Canchaque),
alt. 800 m. Sagástegui, Cabanillas & Dios 8288.

Literature Cited

Afzelius, K. 1966. Some species of Pseudogynoxys
    from Ecuador. Botaniska Notiser 119 (2): 233-242.

Baker, J. G. 1884. IV. Helianthoideae, Helenioideae,
    Anthemideae, Senecionideae, Cynaroideae, Ligulatae,
    Mutisiaceae. in Martius, Flora Brasiliensis 6 (3):
    136-412, pl. 45-108.

Cabrera, A. 1950. Notes on the Brazilian Senecioneae.
    Brittonia 7: 53-74.

_____. 1959. Notas sobre tipos de Compuestas
    Sudamericanas en herbarios Europeos. I. Bol. Soc.
    Argent. Bot. 7: 233-246.

Cassini, H. 1827. Senecionées. Dict. Sci. Nat. 48:
    446-466.

Cuatrecasas, J. 1955. A new genus and other novelties
    in Compositae. Brittonia 8: 151-163.

_____. 1964. Miscelánea sobre Flora Neotropica, I.
    Ciencia, Méx. 23 (4): 137-151.

Greenman, J. 1902. Monographie der nord- und central-
    amerikanischen Arten der Gattung Senecio. Bot.
    Jahrb. 32: 1-33.

Grisebach, A. H. R. 1879. Symbolae ad floram
    argentinam. Abhand. Kön. Ges. Wiss. Goett. Phys.
    Cl. 24 (1): 1-345.

Nordenstam, B. 1977. Systematics of the Liabeae and
    Senecioneae. in The Biology and Chemistry of the
    Compositae. in press.

Turczaninow, N. 1851. Synanthereae Quaedam hucusque
    indescriptae. Bulletin de la Société impériale
    de Naturalistes de Moscou. 24 (pt.1): 166-214;
    (pt.2): 59-95.

Weddell, H. A. 1855-1857. Ord. I. Compositae.
    Chloris andina 1: 1-232, pl. 1-42.

Williams, L. O. 1976. Tribe VIII. Senecioneae. in
    Flora of Guatemala. Fieldiana: Botany 24 (pt.
    12): 392-393, 395-423.

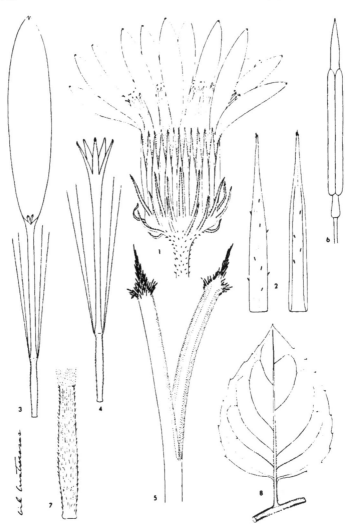

Figs. 1-8. Pseudogynoxys cummingii (Benth.) H.Robinson & J.Cuatrecasas, from Haught 3980, Colombia. 1. Head, X 2. 2. Involucral bracts, X $4\frac{1}{2}$. 3. Ray flower, X $4\frac{1}{2}$. 4. Disk flower, X $4\frac{1}{2}$. 5. Style branches of disk flower, X 27. 6. Anther, X 18. 7. Achene, X 9. 8. Leaf, X 4/9.

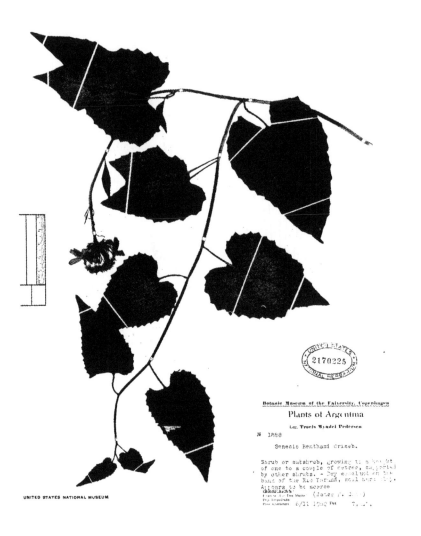

Botanic Museum of the University, Copenhagen

Plants of Argentina

Leg. Troels Myndel Pedersen

№ 1898

Senecio Benthami Griseb.

Shrub or subshrub, growing to a height of one to a couple of metres, supported by other shrubs. — Dry woodland on the bank of the Río Paraná, soil here dry. Appears to be scarce.

Pseudogynoxys cabrerae H.Robinson & J.Cuatrecasas, Holotype, United States National Herbarium. Photo by Victor E. Krantz, Staff Photographer, National Museum of Natural History.

STUDIES IN THE EUPATORIEAE (ASTERACEAE). CLXIII.

ADDITIONS TO THE GENUS FLEISCHMANNIOPSIS.

R. M. King and H. Robinson
Smithsonian Institution, Washington, D.C. 20560

The genus Fleischmanniopsis was originally established
for three species of Critonioid Eupatorieae (King & Robinson, 1971). Continuing research both in the field and in
the herbarium have yielded the following two new species
and new combination presented below. The type species
of Fleischmanniopsis, F. leucocephala (Benth.) K & R,
is notable for the white color of the involucral bracts,
a condition that is evident very early in the developing inflorescence. There has been a tendency to dis-
miss all such material as one species, but specimens
from southern Guatemala and El Salvador prove to be
distinct. A single specimen with white involucres from
Chiapas proves to be a third distinct species.
Immaturity of specimens is a problem in the genus.
The white involucres show from very early stages in
development and many specimens are collected in the
erroneous assumption they are mature. This factor
obscures some of the species limits since the smaller
head size and early deciduous primary leaves in F.
anomalochaeta are not evident in comparisons of immature
specimens. Williams' concern with immaturity in F.
mendax seems unwarranted, however, since an isotype(GH)
is sufficiently developed and shows larger heads than
any seen in F. leucocephala. Williams was perhaps mis-
lead by the apparently characteristic greenish rather
than whitish involucre in F. mendax.
The concept of the genus is expanded here to contain 5 species. The increased diversity of the species
level reenforces the generic distinctions and some new
characters can be added. Fleischmanniopsis is regarded
as Critonioid and because of the broadened and thickened clavate tips of the style branches a relationship
might be suggested to Critoniadelphus. There is some
resemblence in habit between that genus and F. nubigenoides (B.L.Robinson) K. & R., a resemblance that led
to the species name of the latter. Differences, however, discourage any thought of close relationship.
In Fleischmanniopsis, the inner involucral bracts are
usually persistent, the pollen is only 18-20μ in diameter, the anther collars are annulated, the corolla

lobes are nearly glabrous, the pappus is in a single
uncongested row of slender setae, the carpopodium is
pale and tapering with a sharp upper rim, the anthers
are partly to completely pinkish as seen through the
thin corollas, the anther appendage is less than half
as long   as wide, the ribs of the achene are narrow
and not noticeably corticated, the leaves are trinervate-
ly rather than pinnately veined, and the corollas have
veins ending below the lobes. Fleischmanniopsis was
named after the resemblance of the achene and the shape
of the corolla lobes to those of Fleischmannia. The
latter genus does not seem particularly closely related,
however, differing by the corolla lobes papillose on
both surfaces with papillae on the upper ends of the
cells, the veins of the corollas extending into the
corolla lobes and being greatly thickened in the lower
throat, the corolla having a distinct short basal tube,
the anthers not being reddish, the style branches lacking nodular tips but being densely covered with elongate
papillae, the head usually containing 20 or more flowers
with 10 in only one species, and the anther appendage
being longer than wide.

        The five species of Fleischmanniopsis may be
distinguished by the following key.

1. Inflorescence corymbose or pyramidally paniculate;
      leaves trinervate from well above base, second-
      ary veins parallel with basal margin ---------- 2

    2. Involucre brownish at maturity; leaves membran-
          ous, dark when dry, acuminatium less than 1/5
          of leaf length; pappus ca. 3 mm long; corolla
          glabrous inside   F. nubigenoides

    2. Involucre whitish; leaves herbaceous, green when
          dry; acumination ¼ of leaf length; pappus ca. 2
          mm long; corolla with hairs inside at the bases
          of the filaments--------------- F. langmaniae

1. Inflorescence thyrsoid-paniculate, elongate cylind-
      rical; leaves trinervate from at or near the base,
      secondary veins diverging from basal margins--- 3

    3. Heads ca. 7 mm long, involucre greenish at mat-
          urity; leaf lamina rounded at base, trinervate
          from base, tip abruptly acuminate--- F. mendax

3. Heads 4-6 mm long, involucre white at maturity;
    leaf lamina acute at base, trinervate from above
    base, tip gradually acuminate--------------- 4

   4. Heads ca. 4 mm long; tips of pappus setae
       contorted and irregularly barbellate; achene
       scabrous throughout; primary leaves usually
       lacking at anthesis------ F. anomalochaeta

   4. Heads mostly 5-6 mm long; tips of pappus setae
       straight and antrorsely scabrid; achene
       usually glabrous below; primary leaves per-
       sistent through anthesis - F. leucocephala

Fleischmanniopsis anomalochaeta R. M. King & H. Robin-
    son, sp. nov. Plantae fruticosae 1.0-1.5 m altae
multo ramosae. Caules glabri in nodis pauce minute
puberuli teretes vel subtiliter hexagonales flavo-
fulvescentes. Folia opposita, petiolis 0.5-2.5 cm
longis; folia primaria ca. 10-11 cm longa et ca. 4 cm
lata; saepe per anthesin decidua; folia ramosa plerumque 2-5 cm longa et ca. 1 cm lata; laminae ovatae vel
late lanceolatae base acutae vel late cuneatae fere ad
basem valde trinervatae margine utrinque argute 5-13-
serratae apice sensim anguste acuminatae supra et subtus sparse puberulae. Inflorescentiae elongatae
thyrsoideo-paniculatae ampliatae, ramis et ramulis
puberulis vel dense puberulis. Capitula in ramulis
corymbosis congesta ca. 4 mm alta et 1.5-1.8 mm lata.
Squamae involucri ca. 3-seriatae ca. 15 albae ovatae
vel oblongae 1-3 mm longae 0.5-0.9 mm latae margine
minuto fimbriatae extus glabrae vel subglabrae exter-
iores apice minute apiculatae interiores obtusae vel
rotundatae. Flores 7-9; corollae albae 2.0-2.2 mm
longae anguste infundibulares, tubis angustatis indistinctis, lobis ca. 0.3 mm longis et 0.25 mm latis;
filamenta in parte superiore ca. 0.3 mm longa parietibus cellularum dense annulatis; thecae antherarum
ca. 0.5 mm longae, appendices truncatae ca. 0.05 mm
longae et 0.15 mm latae. Achaenia 1.1-1.5 mm longa
ubique sparse setifera; setae pappi ca. 18-20 non contiguae 2.0-2.5 mm longae apice contortae saepe tenues
patentiter vel retrorse spiculiferae. Grana pollinis
18-19µ diam. minute papillosa.

   TYPE: GUATEMALA: Sacatepequez: along the dirt road
to Antigua, ca. 13 kms generally N of Escuintla. Elevation ca. 3,200 ft. 23 January 1977. R. M. King 7179

(Holotype US).   Paratypes: GUATEMALA: Sacatepequez
along the dirt road to Antigua, ca. 11 kms generally
N of Escuintla. Elevation ca. 2,500 ft. 23 January
1977, King 7177 (US). Dry secondary forest, lower
slopes Volcan de Fuego, 3 km southwest of Alotenango,
1,200-1,300 m January 15,,1974 , Williams & Williams
43469 (US). Dry thickets between hills Agua and Fuego
Volcanoes, road to Alotenango, elevation 1,200 m
Molina et al 16666 (US). Escuintla: along the road to
Escuintla, ca. 16 kms generally SW of Amatitlan. Elevation ca. 2,750 ft, King 7169 (US). Amatitlan: Palin,
alt. 3,560 ft. John Donnell Smith 2843 (US). EL SALVADOR
Ahuachapan: Sierra de Apaneca, in the region of Finca
Colima, Jan. 17-19, 1922, Standley 20160 (US). Without
precise locality, all Sisto Alberto Padilla 250, 280,
281 (All US).

    The new species has the general aspect of the
common Fleischmanniopsis leucocephala but is most distinct in the contorted , tenuous and uniquely barbellate
tips of the pappus setae. In the typical form the
barbs of the setae are proliferated into papillae and
bifid tips that appear almost haustorial. The character
is weakly developed in only one of the collections seen.
The concept is reenforced by the smaller heads and the
more setiferous achenes. The heads are 4 mm long while
mature heads of F. leucocephala are 5-6 mm long. The
achenes are prominently scabrous to the base while
those of F. leucocephala are almost glabrous. A few
specimens of the latter species from central Mexico
have achene pubescence as in F. anomalochaeta but
apparently represent a parallel variation.
    The new species can be recognized in the field by
subtile differences in aspect. The smaller heads con-
tribute to the appearance but might be confused with
immature specimens of F. leucocephala. Specimens of
F. anomalochaeta also are usually distinctive in the
numerous lateral shoots with small leaves. Larger
leaves occur on the main stems but are apparently
usually lost before anthesis. In contrast, F. leucocephala has persistent primary leaves and branchlets
bearing smaller leaves are usually not prominent.

Fleischmanniopsis langmaniae R. M. King & H. Robinson,
    sp. nov. Plantae fruticosae ca. 1 m altae laxe
ramosae. Caules glabri vel subglabri teretes vel subtiliter hexagonales fulvescentes. Folia opposita,
petiolis 0.5-1.5 cm longis; laminae ovatae plerumque

4.5-10.0 cm longae et 1.5-3.5 cm latae base acutae
margine 5-9 argute serratae apice sensim anguste acuminatae supra et subtus sparse puberulae distincte supra
basem trinervatae, nervis secundariis marginis basil-
aribus parallelis.   Inflorescentiae pyramidaliter
paniculatae ampliatae, ramis et ramulis puberulis vel
dense puberulis.   Capitula in ramulis subcorymbosis
subcongesta ca. 5 mm alta et 3 mm lata.   Squamae
involucri ca. 18 albae 1-3-purpureo-lineatae orbiculatae
vel oblongae 0.7-3.7 mm longae et 1.0-1.4 mm latae
margine minute fimbriatae extus glabrae apice rotundatae.   Flores 7-9; corollae albae in nervis purpureis
ca. 2.5 mm longae anguste tubulares superne infundib-
laris, tubis indistinctis, faucis base intus sparse
puberulis, lobis ca. 0.35 mm longis et 0.3 mm latis
extus 0-2-setiferis; filamenta in parte superiore ca.
0.25 mm longa base indistincta, parietibus cellularum
dense annulatis; thecae antherarum ca. 0.55 mm longae,
appendices truncatae vel retusae ca. 0.05 mm longae et
0.15 mm latae.   Achaenia 1.5-1.7 mm longa glabra vel
apice pauce spiculifera; setae pappi 25-35 contiguae
ca. 2 mm longae ubique aequicrassae et aequiscabrae.
Granna pollinis ca. 17µ   diam. minute papillosa.

TYPE:MEXICO: Chiapas: between San Fernando and
Plan de Ayala, 4/17/49, Ida K. Langman 3914 (Holotype
US).

Fleischmanniopsis langmanae is most closely related to F. nubigenoides in the more broadly paniculate
inflorescence and in the venation of the leaves.   The
trinervation of the leaves is farther from the base of
the lamina and parallel to the basal margine as in F.
nubigenoides.   The two species also seem to share somewhat broader basal tubes on the corollas and pappus
setae of even width and equally distributed scabrosity.
The new species is more like F. leucocephala by the
whitish involucres and differs from F. nubigenoides
also by the shorter florets and shorter pappus.   The
new species is unique in the genus by the hairs inside
the corollla at the bases of the filaments.

The following variety should be added to the
genus.

Fleischmanniopsis leucocephala (Benth.) R.M.King & H.
     Robinson var. anodonta (B.L.Robinson) R.M.King &
     H.Robinson, comb. nov. Eupatorium leucocephala
     Benth. var. anodontum B.L.Robinson, Proc. Amer.
     Acad. 51:534. 1916. Mexico.

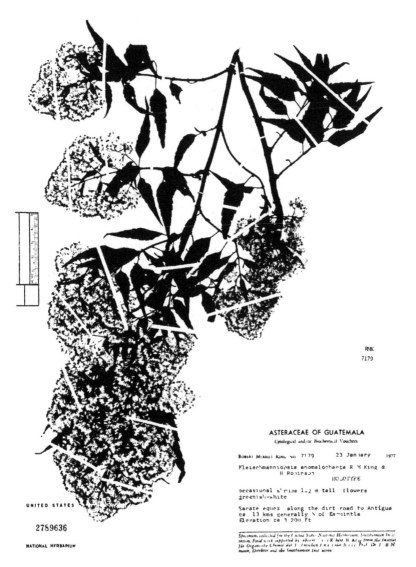

RMK
7179

ASTERACEAE OF GUATEMALA
Cytological and/or Biochemical Vouchers

Robert Merrill King no 7179   23 January   1977

Fleischmanniopsis anomalochaeta R M King &
H Robinson
HOLOTYPE

occasional shrub 1-2 m tall flowers
greenish-white

Sacate aguez along the dirt road to Antigua
ca 13 kms generally N of Escuintla
Elevation ca 3 200 ft

UNITED STATES

2789636

NATIONAL HERBARIUM

Fleischmanniopsis anomalochaeta R. M. King & H. Robinson, Holotype, United States National Herbarium.   Photos
by Victor E. Krantz, Staff Photographer, National Museum
of Natural History.

Fleischmanniopsis langmaniae R. M. King & H. Robinson, Holotype, United States National Herbarium.

Enlargements of heads of Fleischmanniopsis. Top.
F. anomalochaeta; bottom. F. langmaniae.

Literature Cited

King, R. M. and H. Robinson 1971. Studies in the
     Eupatorieae (Asteraceae). XLV. A new genus,
     Fleischmanniopsis. Phytologia 21: 402-404.

STUDIES IN THE HELIANTHEAE (ASTERACEAE). VIII.

NOTES ON GENUS AND SPECIES LIMITS

IN THE GENUS VIGUIERA.

Harold   Robinson
Department of Botany
Smithsonian Institution, Washington, D.C. 20560.

The genus Viguiera contains the greatest diversity
of species of any genus in the subtribe Helianthinae,
and both the generic and species limits have been sub-
ject to question. The genus has been studied on a
classical basis by Blake (1918) and comparatively few
additiona have been made since that time. Of particular interest are the more herbaceous species of the
genus which form the typical element and which could be
included in a more narrowly circumscribed genus concept. Some immediate relatives have been placed in
segregate genera and the policy is followed here that
was initiated by D'Arcy (1975) in the Flora of Panama,
the herbaceous species with 1-3 seriate involucre and
shortly but distinctly appendaged styles of the disk
flowers are placed in Viguiera regardless of the
presence or absence of differentiated awns and squamellae on the achenes. In addition to the new species
described below, a number of details of floral structure have been noted for other species. The group
seems to have many characters in its "floral anatomy"
that are useful at the species level, though such
characters are comparatively rare in other Heliantheae.
    Viguiera dentata (Cav.) Spreng. includes the type
species of the genus, V. helianthoides HBK., and is
distributed widely in the West Indies, Mexico and
Central America. The involucral bracts tend to be
distinctive by the base being nearly filled by a pair
of prominent costae and by the distinct linear tip.
The species proves far more distinctive and unique in
the genus by the presence of hairs on the filaments of
the anthers which usually form a dense pubescnce
visible in the opened corollas under the dissecting
microscope. The disk corollas also have the basal
tubes nearly glabrous, the throat densely scabrous
below, and the throat less than twice as long as the
lobes, a combination of features differing from most
members of the genus.

Viguiera molinae H.Robinson, sp. nov.
      Plantae herbaceae 1½-2 m altae laxe ramosae.
caules brunnescentes vel rubro-tincti teretes et minute
striati sparse antrorse strigosi. Folia alternata,
basilaris opposita?, petiolis angustis plerumque 5-10
mm longis dense strigosis; laminae ovatae vel lanceolatae plerumque 4-8 cm longae et 1-4 cm latae base
breviter acutae vel vix acuminatae margine 6-17 argute
serratae sensim in foliis superioribus minoribus sub-
integra et integra apice longe anguste acuminatae supra
sparse appresse strigosae et dense minute scabridae
subtus densius strigosae fere ad basem valde trinervatae. Inflorescentiae laxe subcymosae, pedicellis
tenuibus 1.5-3.0 cm longis dense strigosis. Capitula
ca. 8 mm alta et 5-6 mm lata. Squamae involucri ca.
8-10 plerumque uniseriatae erectae herbaceae lanceolatae 4-5 mm longae et 1.0-1.2 mm latae apice anguste
acutae extus dense strigosae. Paleae chartaceae
oblongae ca. 5 mm longae et 2 mm latae superne irregulariter breviter serratae apice breviter acutae vix
erectae apiculatae extus variabiliter puberulae in
nervis mediis strigosae viridivittatae. Flores radii
0-1 in capitulo steriles; corollae flavae, tubis tenuibus ca. 1.2 mm longis superne parce puberulis, limbis
late ellipticis ca. 2.5 mm longis et 1.8 mm latis
subtus in nervis strigosis. Flores disci ca. 20;
corollae flavae 3.5-4.3 mm longae, tubis 1.0-1.3 mm
longis parce spiculiferis base pauce glanduliferis,
glandulis non capitatis, faucis cylindricis 1.8-2.0 mm
longis inferne dense et superne sparse spiculiferis
intus sparse antrorse papillatis, lobis triangularibus
ca. 0.6 mm longis et ca. 0.4 mm latis extus dense
strigosis setis perverrucosis intus dense uniformiter
breviter papillatis; filamenta glabra in partibus
superioribus ca. 0.35 mm longa e thecis plerumque
exerta; thecae 1.0-1.2 mm longae; appendices antherarum
flavae ovatae ca. 0.4 mm longae et 0.17 mm latae base
in fasciculo glanduliferae; rami stylorum breviter
distincte appendiculati. Achaenia 3.0-3.3 mm longa et
ca. 1.2 mm lata distincte compressa minute albo-macul-
ata dense subsordide sericea; aristae pappi 2 ca. 3 mm
longae inferne scariose alatae, squamellae pappi ca. 6
aliquantum latae 0.7-1.0 mm longae base breviter
connatae distaliter valde laceratae. Grana pollinis
ca. 23-25μ diam. dense hispida.
      Type: NICARAGUA: Dept. Estelí: La Guava, Estelí
River 22 km north of Esteli, alt. 650 m. Nov. 23-26,
1973. L.O.Williams & A.Molina R. 42374 (Holotype, US).
The specimen was received as Aldama dentata var. dentata
and duplicates should be sought in other herbaria under
that name.

Viguiera molinae is in the group Blake (1918)
referred to as sect. Diplostichis and is obviously
closely related to V. gracillima Brandegee of Oaxaca
and Chiapas.   The new species has the same canescent-
strigose lanceolate involucral bracts and combined
strigose and scabrid upper surfaces of the leaves that
occur in V. gracillima, and in an initial sorting was
thought to be only a range extension of that species.
The more robust plant with the extensive many-headed
inflorescence and the narrower long-acuminate leaves
was at great variance, however.   Careful examination
proved the species was thoroughly distinct in the
rayless or nearly rayless condition of the heads.
Numerous heads have been examined leaving no doubt that
the heads are basically rayless.   The anther collars
of V. molinae are unusually long, also, and they
extend far below the basal points of the anther thecae.
Most members of the genus have anther collars scarcely
longer than the bases of the thecae.

Other Central American species placed in the same
section by Blake include V. tenuis A.Gray, V. strigosa
Klatt and V. sylvatica Klatt.   The first of these has
involucral bracts similar to V. molinae, but it differs
by the mostly opposite subsessile leaves with non-
acuminate tips, upper surfaces of the leaves with
denser ascending strigose hairs and no spicules, the
corolla having a very short basal tube, the corolla
lobes being only as long as wide, and the lobe inner
surface having much more elongate papillae.   Both V.
strigosa and V. sylvatica of Costa Rica differ by the
glandular dots on the undersurfaces of the leaves and
by the broader thicker involucral bracts without
canescent-strigose pubescence.   The two species have
been confused in identifications and some notes suggest
the two are synonyms.   An isotype of V. strigosa and
one other specimen (Standley & Valerio 44883) have been
examined and prove to be notably distinct in the
shorter lobes of the disk corolla that are smooth
rather than papillose on the inner surface, the anthers
are mostly or completely included in the mature
corollas, and the anther collars are elongate, extending well below the bases of the thecae.   Specimens
seen of V. sylvatica all show very deeply cut disk
corolla lobes with sparse papillae inside and long
dense papillae forming a mass of hair inside the
throat below the lobes, the anthers are exserted at
maturity and the anther collars are short so as to
scarcely extend below the bases of the thecae.

Haplocalymma microcephala (Greenm.) Blake is very
closely related to the species of section Diplostichis
having lanceolate canescent-strigose involucral bracts.
The species has leaf pubescence precisely like that of
V. gracillima and V. molinae. The only significant
differences seem to be the coarsely dentate leaves and
the smaller more densely clustered heads. The evenly
spaced uniseriate 5 involucral bracts do not seem to
warrant a generic distinction and the species should be
known as V. microcephala Greenm. Contrary to Blake's
characterization the pappus tends to have differentiated
though short awns.

Viguiera woronowii (Blake) H.Robinson, comb. nov.
    Haplocalymma woronowii Blake, Proc. Biol. Soc.
Wash. 43: 163. 1930. The value of Blake's concept of
Haplocalymma is essentially disproven by the nature of
this second species described by him. While the species is in the general relationship, it would certainly
not seem to be the closest relative of V. microcephala
Greenm. The species is most distinctive in the high-
conical receptacle which resembles Jaegeria. It is a
similarity of some concern since the recognition of
sterile rays in one species of Jaegeria, J. sterilis
McVaugh. The place of V. woronowii is proven, however,
by the presence of a single continuous stigmatic surface
on each style branch as in all Helianthiinae, there are
two stigmatic lines in Jaegeria. Also, V. woronowii
has a small style appendage as in typical Viguiera and
the distinctly papillose appressed hairs common in
Viguiera and many other Heliantheae, both characters
differing from Jaegeria.

    In the Flora of Panama, D'Arcy (1975) adopts a
broad concept of Viguiera which includes Wedelia
cordata Hook.& Arn. and which thus essentially synony-
mizes the genus Hymenostephium. A lectotype is also
chosen by D'Arcy for the genus Gymnolomia (see below
under Viguiera rudbeckioides). The broad concept was
not explained by D'Arcy, but it is thoroughly justified.
Blake (1918) in his monograph of Viguiera held Hymenostephium distinct for those members of the relationship
having a pappus with squamellae but withous differentiated awns. The distinction was very subtle since
Blake retained in Viguiera species such as the S.E. U.S.
endemic V. porteri (A.Gray) Blake and V. quitensis
which have no pappus. The disk achenes of the latter
had setae on their surfaces indicating the lack of a
pappus was not the calvous form found erratically in
individual specimens of pappose species throughout the

family Asteraceae. Blake provided no distinction
between the two genera *Viguiera* and *Hymenostephium* in
cases where calvous achenes lacking both pappus and
lateral hairs occur. The latter condition is particularly common in *Hymenostephium* and not rare in *Viguiera*.
There are also cases such as V. *lepidostephana* Cuatr.
where the awns are scarcely different from the squamellae. The distinction between the genera must therefore be regarded as unworkable. There is also ample
evidence that the distinction between *Viguiera* and
Hymenostephium creates an unnatural division between
closely related species.

Viguiera *cordata* (Hook.& Arn.) D'Arcy, distributed
from Mexico to Panama, shows most of the basic features
found in all species that have been placed in *Hymenostephium*, the cylindrical throats of the disk corollas
twice or more as long as the lobes, the reddish tint of
the lobes of older disk flowers, the short-triangular
shape of the lobes with dense papillosity on the inner
surface, the anther collars not or scarcely extending
below the bases of the thecae, and the anther appendages being yellow. A few of the more restricted traits
are the abruptly broadened base of the throat of the
disk corollas, and a basal tube nearly 1 mm long that
is essentially as scabrid as the base of the throat.
There is variation in the species in the erect versus
appressed pubescence used to distinguish H. *guatemalense* (B.L.Robins.& Greenm.) Blake, and in the shape of
the paleae. Short-tipped paleae are predominent in
Guatemala and long-tips are found in most Costa Rican
specimens, but both types are found throughout the
range of the species. The supposed difference of
Hymenostephium microcephalum (Less.) Blake (including
H. *mexicanum* Benth., the type of the genus; not V.
microcephalum Greenm.), the cylindrical rather than
campanulate shape of the heads, would seem mostly to
distinguish immature specimens from mature material.
The primary test of the species, however, seems to be
the achene which has a high proportion of calvous
forms and identical pappose forms throughout the range
in specimens otherwise identifiable as H. *cordatum*,
H. *microcephalum* and H. *guatemalense*. The pappose
forms all show achenes with long-sericeous setosity on
the sides and a distinctive stringy form of squamellae.
Of all the material under the name *Hymenostephium* seen
from Mexico and Central America, only the following
seem to be distinct from *Viguiera cordata*.

Viguiera hintonii H.Robinson, sp. nov.
     Plantae herbaceae vel suffrutescentes 1-2 m altae

laxe ramosae, ramis erectis valde patentibus. Caules
tenues brunnescentes sparse minute puberuli vel glabres-
centes. Folia plerumque opposita, petiolis tenuibus
brevibus 3-13 mm longis sparse strigosis; laminae
oblongo-lanceolatae vel anguste ovatae plerumque 3-8 cm
longae et 1.2-2.7 cm latae base rotundatae margine 3-10
plerumque remote serratae apice anguste acuminatae supra
et subtus sparse appresse scabrae fere ad basem trinervatae. Inflorescentiae laxe ramosae cymosae pauce
capitatae, pedicellis tenuibus 1-10 mm longis sparse vel
dense strigosis. Capitula 4.5-5.0 mm alta et 2.0-2.5 mm
lata. Squamae involucri ca. 6 plerumque uniseriatae
erectae herbaceae ovate lanceolatae 2-3 mm longae et ca.
0.9 mm latae apice acuminatae extus subglabrae vel dense
strigosae. Paleae chartaceae oblongo-ovatae 3.0-3.5 mm
longae et ca. 1.5 mm latae apice breviter acuminatae
margine remote dentatae in costis resinosis numerosis
aureo-striatae glabrae vel minute puberulae. Flores
radii 5 in capitulo steriles; corollae flavae, tubis
0.3-0.5 mm longis puberulis, limbis late ellipticis 3-6
mm longis et 2.0-2.3 mm latis subtus in nervis minute
strigosis. Flores disci 5-10; corollae flavae vel
superne rubro-tinctae ca. 2.5 mm longae, tubis ca. 0.5
mm longis extus dense scabris, faucis ca. 1.5 mm longis
base leniter demarcatis inferne scabris superne glabris,
lobis triangularibus ca. 0.5 mm longis et 0.45 mm latis
extus dense minute strigosis intus dense uniformiter
breviter papillosis; filamenta glabra in partibus
superioribus ca. 0.2 mm longa e basis thecarum non vel
vix exerta; thecae ca. 1.2 mm longae; appendices antherarum flavae ovatae ca. 0.2 mm longae et 0.17 mm latae
base in fasciculo glanduliferae; rami stylorum breviter
distincte appendiculati. Achaenia 2.0-2.5 mm longa et
ca. 0.8-1.0 mm lata distincte compressa ubique nigra et
dense breviter setifera, setis erecto-patentibus; pappus
nullus vel subnullus. Grana pollinis 23-25 µ diam.

     Type: MEXICO: Michoacan: Steep hills about 25 km
south of Arteaga, road to Playa Azul; forest of Quercus
macrophylla; elev. 600-650 m. abundant, shrub 1-2 m
high; flowers yellow. 27 Feb. 1965. McVaugh 22637
(Holotype, US). Paratypes: MEXICO: Guerrero: Vallecitos,
Montes de Oca. Oak woods. Flower yellow. 7-17-37.
Hinton et al. 10611 (US); Chilacayote 1675, Mina. Shady
mixed forest. Flower yellow. 4-20-39. Hinton et al.
14182 (US).

     The flowering heads of V. hintonii are the smallest
of any seen in the genus and this combined with the
general aspect has caused specimens to be placed under
the name Hymenostephium microcephalum (Less.) Blake.
The achenes are totally distinct in the short setae on

the lateral surfaces and the essential lack of pappus.
Both the pappus and the long sericeous lateral setae
are distinctive in V. cordata. The new species has
achenes technically more like V. kingii McVaugh of
Nayarit in Mexico, but the latter is a much more robust
species with very elongate pedicels. One specimen of
the new species (Hinton 10611) has been annotated
apparently by Blake as Haplocalymma n.sp. with reference to H. woronowii. The involucral bracts are not
strictly in a single series but usually have a single
extra bract as in a few other members of Viguiera sect.
Diplostichis. The shape of the leaves is somewhat
distinctive by being rather oblong-lanceolate with
often remote teeth.

     The South American species placed in Hymenosteph-
ium and the related species of Viguiera are as follows.

Viguiera anomala Blake of Colombia was not seen in this
     study but is supposedly distinct among those
members of the group having awns by the sordid or rufous
pubescence of the inflorescence and by the glabrous
lateral surfaces of the achenes.

Viguiera cabrerae H.Robinson, sp. nov.
     Plantae herbaceae annuae? ca. 0.5 m altae laxe
ramosae. Caules tenues brunnescentes vel rubri sparse
strigosi. Folia plerumque alternata basilaria opposita,
petiolis tenuibus 5-17 mm longis sparse strigosis;
laminae ovatae plerumque 2-6 cm longae et 1-4 cm latae
base obtusae vel acutae margine utrinque ca. 6-7 serratae apice breviter acutae supra et subtus appresse strigosae fere ad basem trinervatae. Inflorescentiae laxe
ramosae pauce capitatae, pedicellis tenuibus 2.0-5.5 cm
longis dense strigosis. Capitula 7-9 mm alta et 5-7 mm
lata. Squamae involucri 8-9 plerumque uniseriatae
erectae herbaceae lineari-lanceolatae 5-6 mm longae et
1.0-1.3 mm latae apice attenuatae extus dense strigosae.
Paleae chartaceae oblongae ca. 5 mm longae et 1.5 mm
latae apice obtusae vel breviter acutae margine breviter
dense spiculiferae extus virides vel viridivittatae
inferne puberulae superne subglabrae. Flores radii 8-9
in capitulo steriles; corollae flavae, tubis ca. 1 mm
longis minute sparse puberulis, limbis late ellipticis
ca. 8 mm longis et 4 mm latis subtus in nervis strigosis. Flores disci ca. 15; corollae flavae ca. 4 mm
longae superne et in lobis in ductis resinosis pluribus
striatis, tubis 0.5-1.0 mm longis dense spiculiferis,
faucis cylindricis 2.0-2.5 mm longis inferne dense et
superne sparse spiculiferis intus sparse antrorse pap-

illatis, lobis triangularibus ca. 1 mm longis et 0.8 mm
latis extus dense strigosis setis verrucosis intus
dense uniformiter breviter papillatis; filamenta glabra
in partibus superioribus ca. 0.2 mm longa e basis
thecarum vix exerta; thecae ca. 1.5 mm longae; appendices antherarum flavae ovatae ca. 0.35 mm longae et ca.
0.25 mm latae non vel pauce glanduliferae; rami stylorum breviter distincte appendiculati. Achaenia ca. 3
mm longa et ca. 1.3 mm lata distincte compressa ubique
nigra et dense appresse puberula; pappus nullus. Grana
pollinis 23-25μ diam.

Type: ARGENTINA: Salta: Dep. Candelaria, Rio del
Potrero, en la orilla del rio, alt. 1420 m. Flor
amarillo. Abril 8, 1925. S. Venturi 3675 (Holotype,
US). Paratypes: ARGENTINA: Salta: Dep. Capital, Cerro
San Bernardo. Capitulos amarillos. 27 V 1933. A.L.
Cabrera 3017 (US); Cerro San Bernardo. 31 V 1933, A.L.
Cabrera 3082 (US).

The species is the southernmost of Viguiera sect.
Diplostichis. The Cabrera specimens had been annotated
in the U.S. National Herbarium as Hymenostephium, but
they possess the combination of setiferous achenes and
no pappus that Blake was inclined to retain in Viguiera.
The combination of achene characters might relate the
species to V. quitensis (Benth.) Blake, but the Argentinian plants are much more delicate, being perhaps
annuals, and the heads are much fewer on long slender
pedicels. The leaves also differ from most members of
the relationship by the fewer coarser teeth on the
margins and by the acute rather than acuminate tips.

Viguiera goebelii (Klatt) H.Robinson, comb. nov.
Gymnolomia goebelii Klatt in Goebel, Pflanzenbiol.
Schilderung. 2: 49, 1891. No distinction is evident
between this and Hymenostephium meridense Blake which
was also described from the paramos near Merida in
Venezuela. The species is notable for the narrowly
ovate leaves with extremely dense usually appressed
pubescence that nearly covers the surfaces, and by the
2-3 headed branches of the inflorescence.

Viguiera lepidostephana Cuatr. is obviously a member of
the Hymenostephium-Diplostichis series in South
America. The species is distinctive by the large
squamose pappus segments with the awns being only
slightly larger than the squamellae. The species was
described from the Depart. Tumbes, Peru (Ellenberg
1423). Ferreyra 12259 represents an additional collection from the same area collected at nearly the same
time.

Viguiera leptodonta Blake is among the species described having a pappus with distinct awns, but in the
calvous-achened condition the species would be indistinguishable from material called Hymenostephium guatemal-
ensis in Venezuela. The latter specimens seem best
placed in the present species in view of the less
expanded throats of the disk corollas which indicate
they are not the same as the Central American material.

Viguiera mucronata Blake occurs in Venezuela and adjacent northern Colombia. The species has an awned
pappus and is similar to the Central American V. tenuis-
V. molinae-V. microcephala series discussed above
having very densely strigose involucral bracts. The
species is distinct among the strongly awned species of
South America by the lax herbaceous habit with long
pedicels and mostly alternate upper leaves.

Viguiera quitensis (Benth.) Blake is notable for the
pubescent achenes with no pappus. The species has
been known primarily from Ecuador. New records from
Colombia (Metcalf 30029) and Venezuela (Pittier 12662;
Steyermark & Dunsterville 98837) had previously been
identified as Viguiera mucronata, Hymenostephium cordatum and H. guatemalense. Viguiera quitensis tends to
be more robust with larger heads and longer rays than
in the related species.

Viguiera rudbeckioides (HBK.) H.Robinson, comb. nov.
Gymnolomia rudbeckioides HBK., Nov. Gen. & Sp. 4:
172, t. 574. 1818. ed folio. A specimen (Townsend A90)
from Sabiango, Ecuador has been seen and seems to match
all described and illustrated features of the type
specimen collected in immediately adjacent Peru (Piura:
Ayavaca). The species seems distinctive in the slender
habit, the very short basal tube of the disk corolla,
and in the long corolla lobes nearly twice as long as
wide. The pappus is a fringe of minute squamellae. In
selecting the species as lectotype of Gymnolomia,
D'Arcy (1975) was unaware of the transfer of the species
to Hymenostephium by Blake (1924, p. 630). In the same
Blake paper (p. 620) the other three original species
of Gymnolomia were also transferred, G. triplinervia to
Aspilia triplinervia (HBK.) Blake and both G. tenella
and G. hondensis HBK. to Aspilia tenella (HBK.) Blake.
In view of the consistent references by Blake (1918)
and D'Arcy (1975) to the Aspilia nature of Gymnolomia
it would I believe require rejection of G. rudbeckioides and selection of G. tenella HBK as the lectotype of
Gymnolomia.

Viguiera serrata (Rusby) H.Robinson, comb. nov.
     Montanoa serrata Rusby, Desc. New Sp. S. Amer. Pl.
151. 1920. The South American material under the name
Hymenostephium cordatum does not seem the same as that
from Central America. The corolla is not as abruptly
expanded at the base of the throat, the hairs of the
achene are shorter and the pappus is different. The
squamellae in V. serrata are broad with lacerations
distally. The squamellae of V. cordata are divided
nearly to the base into slender segments that are often
partly hidden by the denser covering of setae on the
sides of the achene. In addition to the type (Colombia:
Magdalena: Santa Marta, H.H.Smith 516) specimens with
identical form of pappus have been seen from central
Colombia (Cundinamarca: Guasca, Bro. Ariste-Joseph A538)
and Venezuela (Trujillo: Vicinity of Escuque, Pittier
13133). Some specimens seen with calvous achenes may
also represent this species.

Viguiera viridis (Steyermark) H.Robinson, comb. nov.
     Hymenostephium viride Steyerm., Fieldiana: Bot.
28 (3): 641. 1953. Syn. H. angustifolium Benth., not
Viguiera angustifolia (Hook.& Arn.) Blake. The
Venezuelan species is distinctive in the very short
blunt involucral bracts which cover only the bases of
the mature paleae. The pubescence of the leaves is
very sparse compared to V. goebelii.

                    Literature Cited

Blake, S. F. 1918. A revision of the genus Viguiera.
     Contr. Gray Herb. n.s. 54: 1-205.

_____. 1924. New American Asteraceae. Contr. U.S.
     National Herbarium 22 (8): 587-661, i-xi, pl. 54-
     63.

D'Arcy, W. G. 1975. 73. Viguiera, in Flora of Panama.
     Annals Missouri Bot. Gard. 62: 1156-1161.

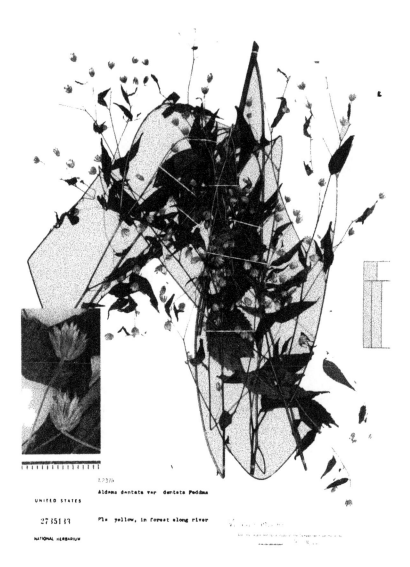

Viguiera molinae H.Robinson, Holotype, United
States National Herbarium. Photo by Victor E. Krantz,
Staff Photographer, National Museum of Natural History.

PLANTS OF MICHOACAN

Hymnostephium microcephalum (Less.) Blake

Abundant

Shrub 1-2 m high, flowers yellow

ROGERS McVAUGH No. 22637   27 FEBRUARY 1960

Viguiera hintonii H.Robinson, Holotype, United
States National Herbarium.

Viguiera hintonii, enlargement of heads.

Viguiera cabrerae H.Robinson, Holotyoe, United States National Herbarium.

Viguiera _cabrerae_, enlargement of heads.

ADDITIONAL NOTES ON THE GENUS VERBENA. XXV

Harold N. Moldenke

VERBENA HALEI Small

Additional synonymy: Verbena halii Small ex Moldenke, Phytologia 36: 47, in syn. 1977.

Additional & emended bibliography: Loes., Verh. Bot. Ver. Brand. 53: 74. 1912; G. W. Thomas, Tex. Pl. Ecolog. Summ. 78. 1969; Bolkh., Grif, Matvej., & Zakhar., Chrom. Numb. Flow. Pl., imp. 1, 717 (1969) and imp. 2, 717. 1974; [Bard], Bull. Torrey Bot. Club 102: 431. 1975; E. H. Jordan, Checklist Organ Pipe Cact. Natl. Mon. 7. 1975; Moldenke, Phytologia 30: 142—143 & 159 (1975) and 31: 375, 377, & 378. 1975; Perkins, Estes, & Thorp, Bull. Torrey Bot. Club 102: 194—198. 1975; Anon., Biol. Abstr. 61: AC1.732. 1976; Hurd & Lindl., Smithson. Contrib. Zool. 220: 10. 1976; Ziegler & Sohmer, Contrib. Herb. Univ. Wisc. LaCrosse 13: 16. 1976; Moldenke, Phytologia 34: 20, 250, 251, 270, & 279 (1976) and 36: 47, 128, 134, 135, 143, 152, & 157. 1977; A. L. Moldenke, Phytologia 36: 87. 1977.

Lewis and his associates encountered this plant along a streamside in Coahuila. Demaree reports it as "common" on low ridges, at 670 feet altitude, in Texas. The Ellisons assert that it was "abundant in local population in full sun in sandy soil with V. brasiliensis, no hybridization evident", the corolla "bluish-lavender".

Other recent collectors refer to this plant as an upright perennial herb, 2 feet tall, and have encountered it in old rice fields, in sandy soil along roadsides, in open dry gravelly soils on railroad embankments, in "nearly level stiff or buckshot soils", on the river side of levees, in sandy dry open areas, and in "nearly level to gently sloping front lands", at 375 meters altitude. Brown refers to it as common on road shoulders in marshes; Allen found it abundant in open areas with V. tenuisecta and Cynodon; Montz reports it infrequent on levee spillways with Ambrosia; and Bougere found it "not abundant, in small compact patches on roadsides". Ajour found it "abundant in very dry soil in shade with a lot of sedges and grasses".

The corollas are said to have been "deep-blue" on Ajour 11, "lavender" on Bougere 1087, Correll & Johnston 22127, and Luke s.n., "blue" on Allen 675 & 990, and Correll & Correll 12421, "purple" on Curry, Martin, & Allen 437, and "purple to lavender" on Killmer 35. Wendt and his associates found it growing "in saline and probably gypsiferous fine alluvial soil in matorral desértico inerme y con espinos laterales with fine mineralized alluvial soil in bajada", growing with Prosopis glandulosa, Koeberlinia spinosa, Condalia, Lycium, and Selinocarpus. Eger reports it common "in full sun in sandy soil of pine woods with Oxalis, Aster, and Lobelia". Higgins found it among "mixed grassland

216

shrubs with mesquite and oak predominating".

Brown & McFarlin note that there are "no appendages on the anthers". Mrs. Jordan (1975) calls the species "Hale's vervain".
Perkins and his associates (1975) inform us that V. halei is
highly autogamous (like V. urticifolia), the number, frequency,
and pollen-carrying ability of insect vectors favor crosses of V.
stricta Vent. with V. urticifolia L. and V. halei. They found
that 12 plants of V. halei which were insect-visited and with 462
potential seeds had a 68 percent seed-set, while 15 bagged plants
with only 391 potential seeds had a 54.5 percent seed-set. They
found that the anthers and stigmas in V. halei are less than 1 mm.
apart and the corolla-tube is straight, the plants are short (25—
70 cm. tall) with an inflorescence of 10—58 branches (mean 26.4),
each with only 2 flowers about 3.4 mm. apart at anthesis, and observed the following insects visiting the flowers: Diptera: Allograpta sp., Baocha sp. (with Verbena pollen on head), Systropus
sp. (with pollen on head), and Villa sp.; Hymenoptera: Calliopsis
andreniformis, Ceratina sp. (with pollen on head), Augochlorella
striata, and Megachile sp. (with pollen on head); and Lepidoptera:
Hemiargus isola, Phycoides phaon (with pollen on head), and Strymon melinus.

Loesener (1912) cites Seler 3460 as "V. officinalis L. forma?"
from Atascosa County, Texas. The C. M. Allen 1179, distributed
as V. halei, is the type collection of xV. alleni Moldenke, while
Ellis, LeDoux, & Watkins 964 is V. canescens H.B.K.

Additional citations: SOUTH CAROLINA: Aiken Co.: Ellison & El-
lison 1010 (Sd—77488). GEORGIA: Sumter Co.: Moldenke & Molden-
ke 29341 (I-A). FLORIDA: Levy Co.: Moldenke & Moldenke 29444 (Ld).
LOUISIANA: Acadia Par.: D. E. Ellis 58 (Lv). Allen Par.: Eger 141
(Ld). Bossier Par.: Robinette 147 (Lv), 203 (Lv). Calcasieu Par.:
J. A. Churchill s.n. [2 May 1955] (Ln—204089); Wurzlow s.n. [Sept.
15, 1917] (Lv). Cameron Par.: C. A. Brown 9271 (Lv). East Baton
Rouge Par.: C. A. Brown 1156 (Lv), s.n. [April 20, 1935] (Lv);
Brown & McFarlin 2118 (Mi); N. F. Petersen s.n. [May 7, 1909] (Lv,
Lv). Jefferson Davis Par.: D. Dickinson s.n. [June 8, 1918] (Lv).
Lafayette Par.: Claycomb s.n. [April 15, 1943] (Lv). Lincoln
Par.: Thomas & Gremillion 2487 (Kl—10265). Ouachita Par.: Pick-
ett & Bot. Class 50 (Lc). Plaquemines Par.: V. Keller s.n. [Aug.
24, 1917] (Lv, Lv). Pointe Coupee Par.: M. Chaney 399 (Lv).
Saint Charles Par.: Monts 3089 (Lv). Saint Helena Par.: C. M. Al-
len 675 (Lv), 990 (Lv). Saint Mary Par.: Bynum, Ingram, & Jaynes
s.n. [Apr. 18, 1933] (Lv). Saint Tammany Par.: Arsène 12242 (Lv);
Bougere 1087 (Lv), 1096 (Lv). Tangipahoa Par.: Correll & Correll
9254 (Lv). Terrebonne Par.: Wurzlow s.n. [May 1, 1912] (Lv). Vermilion Par.: C. A. Brown 18279 (Lv), 21409 (Lv); Killmer 35 (Lv).
Vernon Par.: Turba s.n. [April 22, 1935] (Lv). West Feliciana
Par.: Curry, Martin, & Allen 437 (Lv). TEXAS: Austin Co.: Wurzlow

s.n. [1905] (Lv). Bowie Co.: Correll & Correll 12421 (Mi). Bra-
zos Co.: Ajour 11 (N). Cameron Co.: R. Runyon 4857 (Mu). Comal
Co.: Charette 814 (Mu); Lindheimer 1076 (Mu—4089). Galveston
Co.: Lindheimer s.n. [Galveston, May 1843] (Mu—354). Harris Co.:
Luke s.n. [2 April 1972] (Lv). Orange Co.: J. A. Churchill s.n.
[1 May 1955] (Ln—204154). San Patricio Co.: R. Runyon 4720 (Mu).
Tarrant Co.: Demaree 68284 (Ld). Wichita Co.: L. C. Higgins 10022
(N). Young Co.: Correll & Johnston 22127 (N). MEXICO: Coahuila:
Lewis, Lehto, Keil, Meyer, LeBounty, & Pinkava 5876 (Te—68563);
Wendt, Chiang C., & Johnston 10147 (Ld). CULTIVATED: Missouri:
Prince Paul of Wurttemberg s.n. [Hort. Mergentheim] (Mu—1573).

VERBENA HALEI f. PARVIFLORA Moldenke, Phytologia 34: 20. 1976.
   Bibliography: Moldenke, Phytologia 34: 20 & 251. 1976.
   Citations: TEXAS: Galveston Island: Johnston, Johnston, Saus-
trup, Darr, & Darr 12436a (Ac—isotype, Z—type).

VERBENA HALEI f. ROSEIFLORA (Benke) Moldenke
   Additional bibliography: Moldenke, Phytologia 28: 213 & 362.
1974.

VERBENA HASSLERANA Briq.
   Additional bibliography: Moldenke, Phytologia 30: 143. 1975.
   Pedersen comments that this species is found in moist meadows,
on the banks of small streams, and elsewhere in Corrientes. He
found it in flower and fruit in January, and the corollas on
Pedersen 988 are described as having been "lilac" in color when
fresh.
   Additional citations: ARGENTINA: Corrientes: Pedersen 988 (N).

VERBENA HASSLERANA var. GLANDULOSA Moldenke
   Additional bibliography: Moldenke, Phytologia 28: 351 & 464.
1974.

VERBENA HASTATA L.
   Additional synonymy: Verbena pinnatifida Ph. ex G. Don in
Loud., Hort. Brit., ed. 1, 247. 1830. Verbena laciniosa Schwae-
gr. ex Moldenke, Phytologia 34: 279, in syn. 1976. Verbena pan-
iculata L. ex Moldenke, Phytologia 34: 279, in syn. 1976. Ver-
bena hastata var.paniculata Lam. ex Moldenke, Phytologia 34: 279,
in syn. 1976.
   Additional & emended bibliography: G. Don in Loud., Hort.
Brit., ed. 1, 246 & 247 (1830) and ed. 2, 246 & 247. 1832; Loud.,
Hort. Brit., ed. 2, 552. 1832; G. Don in Loud., Hort. Brit., ed.
3, 246 & 247. 1839; Buek, Gen. Spec. Syn. Candol. 3: 495. 1858;
Paine, Ann. Rep. Univ. N. Y. 18: [Pl. Oneida Co.] 109. 1865;
Kuntze, Rev. Gen. Pl. 2: 510. 1891; Conard, Pl. Iowa 44. 1951; R.
A. Davidson, State Univ. Iowa Stud. Nat. Hist. 20 (2): 77. 1959;
Hall & Thompson, Cranbrook Inst. Sci. Bull. 39: 74. 1959; Cooper-

rider, State Univ. Iowa Stud. Nat. Hist. 20 (5): 70. 1962; P. W. ·
Thompson, Cranbrook Inst. Sci. Bull. 52: 37. 1967; Barker, Univ.
Kans. Sci. Bull. 48: 571. 1969; G. W. Thomas, Tex. Pl. Ecolog.
Summ. 78. 1969; Bolkh., Grif, Matvej., & Zakhar., Chrom. Numb.
Flow. Pl., imp. 1, 717. 1969; Rimpler, Lloydia 33: 491. 1970;
Scully, Treas. Am. Ind. Herbs 283. 1970; Anon., Bioresearch Ind.
7: 1061. 1971; Ellison, Kingsbury, & Hyypio, Comm. Wild Fls. N.
Y. [Cornell Ext. Bull. 990:] 19. 1973; Hathaway & Ramsey, Castanea
38: 77. 1973; A. & C. Krochmal, Guide Medic. Pl. U. S. 229—230,
246, 257, & 258, fig. 259. 1973; Bolkh., Grif, Matvej., & Zakhar.,
Chrom. Numb. Flow. Pl., imp. 2, 717. 1974; El-Gazzar, Egypt.
Journ. Bot. 17: 75 & 78. 1974; R. D. Gibbs, Chemotax. Flow. Pl. 3:
1753—1755 (1974) and 4: 2295. 1974; Mrs. P. Martin, Am. Horticulturist 53 (5): 33. 1974; Rousseau, Géogr. Florist. Qué. [Trav.
Doc. Cent. Étud. Nord 7:] 376, 473, 502, 550, 643, & 788, map 826.
1974; Stark, Am. Horticulturist 53 (5): 11. 1974; Welsh, Utah Pl.,
ed. 3, 354 & 473. 1974; Whitney in Foley, Herbs Use & Delight
[198]. 1974; D. S. & H. B. Correll, Aquat. & Wetland Pl. SW. U. S.,
imp. 2, 2: 1396, [1398], 1399, & 1775, fig. 654 g—k. 1975; Kooiman, Act. Bot. Neerl. 24: 464. 1975; Moldenke, Phytologia 30: 143—
148 & 174 (1975) and 31: 374, 376—378, & 409. 1975; A. L. Moldenke, Phytologia 31: 415. 1975; United Communications (Woodmere, N.
Y.), Herbal Visual & Study Chart n.p. 1975; H. D. Wils., Vasc. Pl.
Holmes Co. Cat. 54. 1975; Grimé, Bot. Black Amer. 191. 1976; La-
coursière, Pontbriand, & Dumas, Naturl. Canad. 103: 174. 1976;
Moldenke, Phytologia 34: 247—251, 270, & 279. 1976; Van Bruggen,
Vasc. Pl. S. Dak. 369 & 536. 1976; [Voss], Mich. Bot. 15: 237.
1976; Moldenke, Phytologia 36: 28, 29, 47, & 126. 1977; F. H.
Montgomery, Seeds & Fruits 201, fig. 6, & 230. 1977; Taylor & Mac
Bryde, Vasc. Pl. Brit. Col. 436 & 751. 1977.

Additional illustrations: Ellison, Kingsbury, & Hyypio, Comm.
Wild Fls. N. Y. [Cornell Ext. Bull. 990:] 19. 1973; A. & C. Krochmal, Guide Medic. Pl. U. S. 230, fig. 259. 1973; D. S. & H. B.
Correll, Aquat. & Wetland Pl. SW. U. S., imp. 2, 2: [1398], fig.
654 g—k. 1975; F. H. Montgomery, Seeds & Fruits 201, fig. 6.
1977.

Montgomery (1977) describes the seeds of this species as "Nutlets 2.0 x 0.6 x 0.5 mm, similar to the previous species [V.
bracteata], margins ridged, dorsal surface longitudinally 3—5-
ribbed on the lower half and reticulate near the apex, inner
faces papillose".

Wilson (1975) reports V. hastata frequent on streambanks and
in low pastures and marshes in Holmes County, Ohio. Other recent
collectors have found it on the shingle shores of lakes, in muck
soil along drainage ditches, on prairies, by lakes with peat bogs
along the shores, in meadows and ditches, on floodplains and open
pond banks, in marshy land, and on gravel bars and rocky ground
along creeks.

The corolla is said to have been "deep-purple" on W. D. Stevens
1633 and this collection also exhibits perfectly hastate leaves.
On Correll & Correll 30034 the corolla was "lavender" and on Blake

11178 it was "violet-purple".

Hathaway & Ramsey (1973) record V. hastata from Pittsylvania County, Virginia. Churchill & Sutherland encountered it "in marshy banks of small pond depressions in cultivated fields with Salix, Cornus, Scirpus, and Asclepias in Otoe County, Nebraska. Thompson (1967) records it from Leelenau County, Michigan, noting that it grows there "along roadsides and in meadows. Common"; Hall & Thompson (1959) found it in Oakland County in the same state. Cooperrider (1962) reports it "frequent.....Marshy places; stream banks; wet ditches" in Clinton and Jones Counties, Iowa. Davidson (1959) says that it is common "Usually in wet to moist open places, occasionally in upland woods and openings....Plants of dry soil, not recognizable as hybrids, differ considerably from those found in moist situations, those from the dry places being smaller with less incised leaves and more sparingly branched inflorescences." This is a very valid comment and I have frequently observed these differences, too, among others, in the field. I think that the dryland plants showing these characters deserve a form name.

Barker (1969) refers to V. hastata as "Occasional, along prairie drainage areas, in low prairie canyons, along margins of impoundments and streams. Occurs throughout the area [Kansas]." Stark (1974) asserts that it is usually found in "rich soils with high humus levels" and a pH level of 6.0, while Martin (1974) recommends it for "most open garden in sun with pH of 5—6.5". Taylor & MacBryde (1977) classify the corolla-color as "blue and violet" and gives its normal blooming period as June to September in British Columbia, where it is certainly var. scabra to which they are referring.

Don (1830) calls V. paniculata the "panicled vervain" and says that it was introduced into English gardens from "N. Amer." in 1800; he calls V. hastata, which he keeps as a separate species, the "halberd-leaved vervain" and gives its date of introduction from "Canada" as 1710, while he says that what he calls V. pinnatifida, the "pinnatifid vervain", came from "N. Amer." in 1810. Curiously, he places V. paniculata in his section Indivisae and the other two in section Trifidae.

Verbena laciniosa is based on an unnumbered collection from the Schwaegrichen herbarium now deposited in the herbarium of the Staatssammlung at Munich. The plant was originally cultivated in the Leipzig botanical garden and exhibits remarkably laciniate-lobed leaves. Probably it is deserving of form rank. It should also be noted here that very plainly hastate leaves are seen on the Herb. Schreber s.n. [Hort. Erl. 1770] & s.n. [Insul. Bahamensis] specimens cited below, clearly showing why Linnaeus applied the epithet "hastata" to this species and also showing Farwell's reason for separating the commonly found form of the species as var. paniculata. Personally, I feel that form rank would be more justified.

Rimpler (1970) reports the isolation of a new iridoid, hasta-toside, from V. hastata. Gibbs (1974) reports that cyanogenesis

is absent in the shoots of this species, leucoanthocyanin is ab-.
sent from the leaves, and syringin is absent from the stems, while
the Ehrlich test gives negative results in the leaves and the
Juglone test is negative in the stems and leaves but produces a
blue fluorescence.

Scully (1970) reports that the Amerinds of North America used
"vervain" in the treatment of colds and quite commonly against
dropsy, with or without milkweed and decocted cottonwood leaves.
Against jaundice they used it as a decoction alone or with any of
the following where available: cinquefoil, parsley, oregon-grape,
nettle-root, or columbine. Mixed with fat, it was used in treating swelling or hardening of the testicles, especially if accompanied by pain. As an infusion for quieting the nerves, a sponge
bath and tea of verbena leaves is still used today among some
tribes; for scurvy an infusion alone or mixed with wormwood or
birch leaves is used. It is drunk as a tea in cases of smallpox
to mitigate the suffering; as a tea also for sore throat and in-
testinal worms, especially in children. But Scully points out
that in almost all cases vervain was not the first choice of
possible medications. It is probable that several species of the
genus are involved here.

Material of V. hastata has been misidentified and distributed
in various herbaria as "V. hasta L." On the other hand, the Bresinsky s.n. [Lafayette, 12.7.1967], Correll & Correll 39844, Gilkey
s.n. [July 31, 1945], Higgins 9712, MacDougal 566, A. R. Moldenke
1297, E. L. Reed 4034, Tharp 4504, Woodcox 55, and York & Rodgers
363, distributed as typical V. hastata, actually represent var.
scabra Moldenke, Herb. Zuccarini s.n. [Hort. bot. Monac.] is a
mixture of xV. baileyana Moldenke and V. officinalis L.; Chandra-
panya 2 and Kirby 160 are V. brasiliensis Vell.; C. A. Brown s.n.
[Sept. 30, 1936] is V. canadensis (L.) Britton; Engelmann s.n.
[Banks of the Mississippi, July 1842] and Herb. Staatsherb. Münch.
s.n. [Hort. Bot. Monac. 28.7.53] are xV. engelmannii Moldenke;
Iltis, Bell, Melchert, Patman, & Witt 12361 is xV. perriana Moldenke; Hillebrand 1863 is V. robusta Greene; Meebold 19175 is V.
simplex Lehm.; Herb. Kummer s.n. [Mississippi] is V. urticifolia
L.; and Schroer 71 is V. xntha Lehm.

Additional citations: MAINE: York Co.: F. T. Hubbard s.n.
[VIII/15/1901] (Ld). VERMONT: Grand Isle Co.: Moldenke & Molden-
ke 31103 (Ac, Ld). Lamoille Co.: Moldenke & Moldenke 31100 (Gs)·
MASSACHUSETTS: Hampshire Co.: Ahles 77872 (Mu). Norfolk Co.:
Blake 11178 (Ld). NEW YORK: Monroe Co.: H. Ernst 1102 (Mu). Os-
wego Co.: Moldenke & Moldenke 31132 (Ld). Schuyler Co.: J. A.
Churchill s.n. [23 August 1937] (Ln—213612); Moldenke & Moldenke
31135 (Tu). Yates Co.: Moldenke & Moldenke 31133 (Lv). NEW JER-
SEY: Morris Co.: Moldenke & Moldenke 25633 (Ld, Ld). County un-
determined: Hillebrand s.n. (Mn). PENNSYLVANIA: Berks Co.: Herb.
Zuccarini s.n. [Reading] (Mu—321). Bucks Co.: Mayer s.n. [Qua-

kertown, Aug. '76] (Mu). Lycoming Co.: Moldenke & Moldenke 31149
(Mu, Ut). Monroe Co.: Swinerton s.n. [Pocono Mts., Aug. 1896]
(Mu). Northampton Co.: Herb. Schreber s.n. [Bethlehem] (Mu—316),
s.n. [Nazareth] (Mu—317). Tioga Co.: Moldenke & Moldenke 31139
(W). Union Co.: Moldenke & Moldenke 31150 (Lv, Tu, W, Ws). Coun-
ty undetermined: Schweinitz s.n. (Mu—1257). ILLINOIS: Cass Co.:
Geyer s.n. [Beardstown, July 1842] (Mu—324, Mu—362, Mu—1676).
Cook Co.: Solereder s.n. [Chicago, Aug. 1893] (Mu—4140). OHIO:
Auglaize Co.: Purpus 242 (Mu). MICHIGAN: Alger Co.: J. A. Church-
ill s.n. [9 July 1964] (Ln—204262). Branch Co.: W. D. Stevens
1633 (Ln—237063). Ingham Co.: R. D. Bradbury 32 (Ln—161013).
Macomb Co.: J. A. Churchill s.n. [24 July 1954] (Ln—203431).
Otsego Co.: Bresinsky s.n. [Hardwood Lake] (Mu). Wayne Co.: Far-
well 8461 1/2 (Mu); G. Stewart s.n. [Aug. 1898] (Ln—142428, Ln—
142430). WISCONSIN: LaCrosse Co.: Demaske 2220 (Ld). MINNESOTA:
Hennepin Co.: Sandberg 152 [36] (Mu). KANSAS: Lyon Co.: J. L.
Watson 7 (Lc). MISSOURI: Clark Co.: E. J. Palmer 43737 (Ld).
Saint Louis: Eggert 7558 (Mu); Muhlenbach 1035 (Mu). County un-
determined: Martens s.n. (Mu—322). NEBRASKA: Otoe Co.: Church-
ill & Sutherland 3945 (N). OKLAHOMA: Muskogee Co.: Wallis 7728
(Au—170667). Sequoyah Co.: Wallis 5557 (Au—169341). TEXAS:
Hemphill Co.: Correll & Correll 30034 (Ld). CULTIVATED: Bahama
Islands: Herb. Schreber s.n. [Insul. Bahamensis] (Mu—315). Eng-
land: Herb. Grimm s.n. [H. Kew.] (Mu—314). France: Thouin s.n.
[Hort. Paris.] (Mu—355, Mu—356). Germany: Herb. Schreber s.n.
[Hort. Erl. 1770] (Mu—319); Herb. Schwaegrichen s.n. [Hort. Lip-
siensis] (Mu—1365); Herb. Zuccarini s.n. [h. b. E.] (Mu—359),
s.n. [Hort. bot. Monac.] (Mu—357, Mu—358). LOCALITY OF COLLEC-
TION UNDETERMINED: Herb. Grimm s.n. (Mu—1255); Herb. Mus. Bot.
Landishmth. s.n. (Mu—320); Herb. Reg. Monac. s.n. (Mu—313);
Herb. Schreber 17 (Mu—318); Herb. Schwaegrichen s.n. (Mu—1256);
Hooker s.n. [United States] (Mu—320).

VERBENA HASTATA f. ALBIFLORA Moldenke
  Additional bibliography: Moldenke, Phytologia 28: 217 (1974)
and 30: 176. 1975.

VERBENA HASTATA f. CAERULEA Moldenke
  Additional synonymy: Verbena americana, urticae foliis angustioribus, flore coeruleo P. Herm., Hort. Acad. Lugd.-Bat.
Cat. 699. 1687. Verbena americana urticae foliis angustioribus,
spica multiplici flore caeruleo P. Herm., Fl. Lugd.-Bat. 54—55.
1690. Verbena urticae fol. angustiore flore coerul. Herb. ex
Rivin., Introd. Gen. Rem Herb. Ord. Fl. Irreg. Monop. [24], pl.
[57]. 1690. Verbena altissima americana spica multiplici,
urticae foliis angustis, floribus caeruleis P. Herm., Parad. Bat.,

ed. 1, 242. 1698. Verbena americana urticae foliis angustioribus, flore caeruleo Moris., Pl. Hist. Univ. Oxon. 3: "408" [=418]. 1699. Verbena altissima americana, spica multiplici, urticae foliis angustis, floribus coeruleis Herm. apud Ray, Hist. Plant. 3: Suppl. 286. 1704. Verbena altissima americana spica multiplici, urticae foliis angustis, floribus coeruleis P. Herm., Parad. Bat., ed. 2, 242. 1705. Verbena; americana; altissima; urticae foliis angustioribus; spicis brevioribus; floribus caeruleis Herm. apud Boerh., Ind. Alt. Plant. Hort. Acad. Lugd., ed. 2, 1: 186. 1720. Verbena americana altissima, urticae foliis angustioribus, spicis brevibus. floribus caeruleis Boerh. apud L., Hort. Cliff., imp. 1, 11, in syn. 1737. Verbena americana altissima, urticae foliis angustioribus, spicis brevioribus, floris caeruleis Boerh. apud A. van Toyen, Fl. Leyd. Prod. 327, in syn. 1740. Verbena americana altissima, spica multiplici, urticae foliis angustis, floribus caeruleis Ray apud L., Hort. Ups. 8, in syn. 1748. Verbena americana altissima, spici multiplici, urticaefoliis angustis, floribus caeruleis Herm. ex L., Sp. Pl., ed. 1, imp. 1, 1: 20, in syn. 1753. Verbena americana altissima, spica multiplici, urticae foliis angustis, floribus coeruleis Ray apud J. F. Gronov., Fl. Virg., ed. 2, 4, in syn. 1762. Verbena americana, altissima, spicâ multiplici, urticaefoliis angustis, floribus caeruleis Herm. apud Poir. in Lam., Encycl. Méth. Bot. 8: 546, in syn. 1808. Verbena americana altiss., spica multipl., urticae fol. angustis, fl. coeruleis Herm. apud H. E. Richter, Cod. Bot. Linn. 35, in syn. 1835. Verbena amer. urticae fel. angustiorib., fl. caeruleo Moris apud H. E. Richter, Cod. Bot. Linn. 35, in syn. 1835. Verbena americana, spica multiplici, foliis urticae angustissimis, floribus caeruleis Herm. ex Moldenke, Résumé Suppl. 4: 14, in syn. 1962.

Additional bibliography: L., Hort. Cliff., imp. 1, 11 (1737) and imp. 2, 11. 1968; Moldenke, Phytologia 28: 217, 426, & 427 (1974) and 31: 409. 1975.

VERBENA HASTATA f. ROSEA Cheney
Additional bibliography: Moldenke, Phytologia 28: 352, 451, 464, & 465. 1974.

VERBENA HASTATA var. SCABRA Moldenke
Additional bibliography: D. S. & H. B. Correll, Aquat. & Wetland Pl. SW. U. S., imp. 2, 2: 1396, 1399, & 1775. 1975; A. L. Moldenke, Phytologia 31: 415. 1975; Moldenke, Phytologia 30: 146—148 (1975), 31: 374 & 376—378 (1975), and 34: 248—251. 1976.
Recent collectors have encountered this plant in moist soil, moist sandy loam, and moist loam at seeping springs in open grassland, at the edge of ponds, in Populus-Prosopis-Tamarix communities, and in the drier areas of bogs (as the typical form),

but also in "badly overgrazed pastures" (as the field form) — in
fact, it has been described by some as a "weed in horse pastures",
a typical field form habitat.

In addition to the months previously reported, it has been
found in flower in October and at 4600 feet altitude (in Utah).
Crutchfield reports it attaining a height of 6 feet (like the
typical form and unlike the usual "field form". Material has sometimes been misidentified and distributed in herbaria as V. stricta
Vent. and as "V. hasta L."

The corollas are said to have been "purple" on Crutchfield 3525
and "deep-purple" on Crutchfield 3551 when fresh, as in the typical form.

The Spellenberg & Spellenberg 2082, distributed as V. hastata
var. scabra, actually is V. macdougalii Heller.

Additional citations: INDIANA: Tippecanoe Co.: Bresinsky s.n.
[Lafayette, 12.7.1967] (Mu). KANSAS: Duckinson Co.: A. R. Molden-
ke 1297 (Ld). UTAH: Utah Co.: Woodcox 55 (Au—122283). NEBRASKA:
Pierce Co.: N. F. Petersen s.n. [Aug. 10, 1910] (Lv, Lv). OKLAHO-
MA: Ottawa Co.: Correll & Correll 39844 (Ld). TEXAS: Hemphill Co.:
Crutchfield 3525 (Ld); E. L. Reed 4034 (Au—122282); Tharp 4504
(Au—122281). Hutchinson Co.: Crutchfield 3551 (Ld). Oldham Co.:
York & Rodgers 363 (Au—201798). Potter Co.: Higgins 9712 (N).
ARIZONA: Coconino Co.: MacDougal 566 (Au—122291). WASHINGTON:
Yakima Co.: Moldenke & Moldenke 2123 (Ld). OREGON: Multnomah
Co.: Gilkey s.n. [July 31, 1945] (Au—122290).

VERBENA HATSCHBACHI Moldenke

Additional bibliography: Moldenke, Phytologia 28: 352. 1974.
The corollas are said to have been "violet" in color when
fresh on Hatschbach 8558.
Additional citations: BRAZIL: Paraná: Hatschbach 8558 [Herb.
Brad. 15182] (Mu).

VERBENA HAYEKII Moldenke

Additional bibliography: Moldenke, Phytologia 28: 218 & 252.
1974; Soukup, Biota 11: 18. 1976.
Richardson refers to this as a scattered prostrate plant in the
rocky soil of roadsides, and the corollas on Richardson 2066 are
said to have been "blue".
Additional citations: PERU: Junín: Richardson 2066 (Ld).

VERBENA HERTERI Moldenke

Additional bibliography: Moldenke, Phytologia 28: 352. 1974.
Additional citations: URUGUAY: Herter 979 [Herb. Herter 82378]
(Mu—isotype).

VERBENA HIRTA Spreng.

Additional bibliography: Buek, Gen. Spec. Syn. Candoll. 3: 495.
1858; Moldenke, Phytologia 28: 352 (1974), 33: 480 (1976) and 34:
259. 1976.

Recent collectors have encountered this plant in secondary forests on laterite soil, on campos, in open scrub, in ruderal grasslands, and among ruderal vegetation on hills, flowering and fruiting in May, October, and November. Araujo refers to it as a heliophilous herb "crescendo em pasto a beira do precipício". The Tryons describe it as "woody at base, 0.5 m. tall". The corollas are said to have been "lilac" on Dziewa 3, Ferreira 97, and Kummrow 646 & 1049, "blue-purple" on Tryon & Tryon 6713, "blue-purple (5P5/8)" on Lindeman & Haas 5137, "purple" on Lindeman & Haas 300 "purple (2 1/2 P6/6" on Lindeman & Haas 4008, "purple (2 1/2 P6/8)" on Lindeman & Haas 18, "purple (2 1/2 P7/6" on Lindeman & Haas 2444, "purple, tube slightly paler (2 1/2 P5/8-6/4)" on Lindeman & Haas 2460, and "red" on Araujo 1262.

On Lindeman & Haas 2460 the entire plant is pronouncedly cinnamon-colored, but this is probably an effect produced by a layer of dust from its roadside habitat.

The Dusén s.n. [11.12.903] and Smith & Klein 13885, distributed as and previously cited by me as typical V. hirta, are better regarded as representing var. dusenii Moldenke, of which the former is the type collection.

Additional citations: BRAZIL: Paraná: Dziewa 3 (Ld); L. F. Ferreira 97 (Ld); Kummrow 646 (Ac), 1049 (Ld), 1123 (Ld); Lindeman & Haas 18 (Ws), 300 (Ac), 2444 (Ut—320417), 2460 (Ut—320416), 4008 (Ld). Rio de Janeiro: Angeli 107 [Herb. FEEMA 345] (Ld); A. Castellanos s.n. [3.XII.1964; Herb. FEEMA 7165] (Ld); Dusén s.n. [Oct. 1903] (Mu—4251); Lindeman & Haas 5137 (Ut—320428); Tryon & Tryon 6713 (N). Rio Grande do Sul: Araujo 1262 [Herb. FEEMA 12280] (Pf). Santa Catarina: A. Castellanos 24675 [Herb. Cent. Pesq. Florest. 3417] (Fe). State undetermined: J. E. Pohl s.n. (Mu—571).

VERBENA HIRTA var. DUSENII Moldenke, Phytologia 33: 480. 1976.
    Bibliography: Moldenke, Phytologia 33: 480 (1976) and 34: 259. 1976.
    Collectors have found this plant growing on campos, on campos near the borders of planaltos, and in low woods, at 900—1000 m. altitude, flowering in November and December. Hitherto it has been confused with, and material has been distributed as, typical V. hirta Spreng.
    Citations: BRAZIL: Paraná: Dusén s.n. [11.12.903] (N—type); Lindeman & Haas 3251d (Ut—320415). Santa Catarina: Smith & Klein 13885 (Ac, N).

VERBENA HIRTA var. GRACILIS Dusén
    Additional bibliography: Moldenke, Phytologia 28: 352. 1974.
    In addition to the months previously reported, this plant has been found in fruit in October. The corollas are said to have been "dark-lilac" on Hatschbach 35191.
    The Reitz & Klein 17616 previously cited by me and distributed

as this variety actually is V. strigosa Cham.
     Additional citations: BRAZIL: Paraná: Hatschbach 35191 (Ld).

VERBENA HISPIDA Ruíz & Pav.
     Additional & emended bibliography: G. Don in Loud., Hort. Brit.,
ed. 1, 247 (1830) and ed. 2, 247. 1832; G. Don in Loud., Hort.
Brit. Suppl. 1: 680. 1832; Loud., Hort. Brit., ed. 2, 552. 1832;
Baxt. in Loud., Hort. Brit. Suppl. 2: 680. 1839; G. Don in Loud.,
Hort. Brit., ed. 3, 247. 1839; Baxt. in Loud., Hort. Brit. Suppl.
[3]: 655. 1850; Buek, Gen. Spec. Syn. Candoll. 3: 494 & 495. 1858;
Bolkh., Grif, Matvej., & Zakhar., Chrom. Numb. Flow. Pl., imp. 1,
717 (1969) and imp. 2, 717. 1974; R. D. Gibbs, Chemotax. Flow. Pl.
3: 1753—1755 (1974) and 4: 2295. 1974; Kooiman, Act. Bot. Neerl.
24: 464. 1975; Moldenke, Phytologia 30: 148. 1975; Soukup, Biota
11: 18. 1976; Moldenke, Phytologia 36: 33 & 151. 1977.
     Don (1830) calls this the "hispid vervain" and says that it was
introduced into English gardens from Peru in 1816. The corollas
are described as having been "violet" in color when fresh on Krapovickas, Schinini, & Quarín 26557. López-Palacios describes the
plant as "hierba rastrera de hójas sésiles y muy pilosas, espigas
cilíndricas y relativamente engrosadas".
     Gibbs (1974) reports that cyanogenesis and leucoanthocyanin
are absent from the leaves of this species and syringin is absent
from the stems, while the Juglone test gives negative results in
the leaves and bark.
     The Pedersen 9867, cited below, collected in clay soil along
roadsides in the Chaco, is said by Troncoso perhaps to be xV. dermeni Moldenke, but I fail to discern the hybrid characters in the
specimen cited. The Pavon collection cited below may be part of
the type collection. The Herb. Kummer s.n. [Hort. bot. Monac.] is
a mixture with V. rigida Spreng.
     The Buchtien s.n. [Valdivia, 7/11/1902], distributed as V. his-
pida, actually is V. bonariensis L., while W. Forster s.n. [8.I.
1954] is V. parvula Hayek.
     Additional citations: ECUADOR: Loja: López-Palacios 4163 (Ld).
PERU: Province undetermined: Pavon s.n. (Mu—1257). BOLIVIA: La
Paz: O. Buchtien 8426 (Mu). CHILE: Valpariso: Zöllner 8100 (Gz).
ARGENTINA: Chaco: Pedersen 9867 (N). Jujuy: Cabrera, Ancibor, Ré,
Tello, & Torres 15080 (Mu); Krapovickas, Schinini, & Quarín 26557
(Ld). Mendoza: Semper s.n. [12-18/III/944] (Ut—330561B). CULTI-
VATED: Germany: Herb. Kummer s.n. [Hort. bot. Monac.] (Mu—1276);
Herb. Zuccarini s.n. [Hort. bot. Monac. 1836] (Mu—326), s.n.
[Hort. bot. Monac. 1837] (Mu—284), s.n. [Hort. bot. Monac.] (Mu—
325).

VERBENA HOOKERIANA (Covas & Schnack) Moldenke
     Additional & emended bibliography: Buek, Gen. Spec. Syn. Can-
doll. 3: 494. 1858; Bolkh., Grif, Matvej., & Zakhar., Chrom. Numb.
Flow. Pl., imp. 1, 715 & 717 (1969) and imp. 2, 715 & 717. 1974;

Moldenke, Phytologia 30: 148 (1975) and 36: 149. 1977.
    The corollas on Lossen 8 are said to have been "violet" in color when fresh.
    Additional citations: ARGENTINA: Córdoba: Lossen 8 (Mu—4370). Río Negro: O'Donell 1553 (Ut—330530B, Ut—33071B).

VERBENA HUMIFUSA Cham.
    Additional bibliography: Buek, Gen. Spec. Syn. Candoll. 3: 495. 1858; Moldenke, Phytologia 28: 353. 1974.
    Reineck & Czermak 21 is a mixture with V. marrubioides Cham.
    Additional & emended citations: BRAZIL: Rio Grande do Sul: Reineck & Czermak 21, in part [Herb. Osten 4160] (Mu, N, N—photo, Po—63874, Po—63876, S, Ug).

xVERBENA HYBRIDA Voss in Vilm., Fl. Pleine Terr., ed. 1, 936. 1865 [not V. hybrida Bicknell, 1941].
    Additional & emended synonymy: Verbena hybrida Vossler apud López-Palacios, Revist. Fac. Farm. Univ. Los Andes 15: 89. 1975; Moldenke, Phytologia 30: 149, in syn. 1975. Glandularia hybrida (Vossl.) López-Palacios, Revist. Fac. Farm. Univ. Los Andes 15: 89. 1975. Verbena genii Hort. ex Moldenke, Phytologia 34: 279, in syn. 1976.
    Additional & emended bibliography: Vilm., Fl. Pleine Terr., ed. 1, 939—942 (1865), ed. 2, 976—979 (1866), and ed. 3, 1: 1200—1203. 1870; Vilm., Fl. Pleine Terr. Suppl. 195. 1884; Cooke, Fl. Presid. Bombay, ed. 1, 3: 437. 1906; Knoche, Fl. Balear., imp. 1, 1: 59. 1921; Wangerin in Just, Bot. Jahresber. 46 (1): 717. 1926; A. W. Anderson, How We Got Fls., imp. 1, 90 & 283. 1951; Conard, Fl. Iowa 44. 1951; Cooke, Fl. Presid. Bombay, ed. 2, imp. 1, 517—518. 1958; A. W. Anderson, How We Got Fls., imp. 2, 90 & 283. 1966; Cooke, Fl. Presid. Bombay, ed. 2, imp. 2, 2: 517—518. 1967; Bolkh., Grif, Matvej., & Zakhar., Chrom. Numb. Flow. Pl., imp. 1, 717. 1969; G. W. Thomas, Tex. Pl. Ecolog. Summ. 78. 1969; R. E. Harrison, Handb. Bulbs & Peren. S. Hemisph., ed. 3, 266—267. 1971; Healy, Gard. Guide Pl. Names 225. 1972; Williamson, Sunset West. Gard. Book, imp. 11, 437. 1973; Bolkh., Grif, Matvej., & Zakhar., Chrom. Numb. Flow. Pl., imp. 2, 717. 1974; Knoche, Fl. Balear., imp. 2, 1: 59. 1974; Hocking, Excerpt. Bot. A.26: 5. 1975; Kooiman, Act. Bot. Neerl. 24: 464. 1975; López-Palacios, Revist. Fac. Farm. Univ. Los Andes 15: 89. 1975; Moldenke, Phytologia 30: 148—151 & 163 (1975), 31: 398 & 410—412 (1975), and 34: 263, 270, & 279. 1976; Park Seed Co., Park Seeds Fls. & Veg. 1976: 63 & 90. 1976; Soukup, Biota 11: 18. 1976; Moldenke, Phytologia 36: 40 & 140. 1977.
    Additional illustrations: Voss in Vilm., Fl. Pleine Terr., ed 3, 1: 1200 & 1201. 1870; R. E. Harrison, Handb. Bulbs & Peren. S. Hemisph., ed. 3, 267. 1971; Park Seed Co., Park Seeds Fls. & Veg. 1976: 90 (in color). 1976.
    Misra (1970) makes the remarkable statement that xV. hybrida is a "Weed in shade" in Bihar, India, but Mukherjee also says of

it: "annual herb of waste places" in West Bengal. It would be interesting to know what cultivars are involved here and if they are breeding true or reverting to one of the several ancestral species. Blakeslee (1926) discusses the observed unlike reactions of different human individuals to the fragrance in Verbena flowers.

López-Palacios (1975), who found the red-flowered cultivar in cultivation at 1600 m. altitude in Ecuador and who credits the name to Vossler instead of to Voss, comments that "Esta hermosa planta jardánica está extendida por todo el territorio nacional [Venezuela]. Schauer, al hablar de la V. peruviana, de la que proceden las razas de flores rojas de la V. hybrida, dice: 'Planta floribus magnis splendide scarlatines maxime spectabilis, hortorum europeorum nunc tamen eximium decus.....eleganti colorum et foliorum varietate excelentes," Apud DC. Prodromus 9: 537, sub V. chamaedrifolia. Lo que puede decirse no sólo de los jardines europeos, sino de todas las partes del mundo." Actually, of course, V. peruviana is NOT widely cultivated in gardens any more and has been replaced by the very common xV. hybrida, among whose multitudinous color forms I have never yet seen anything to match the brilliant splendor of the true wild V. peruviana!

Stewart (1972) cites the accepted binomial for this species as "V. hybrida Hort. ex Vilm. Fl. Pl. Terre Suppl. 195. 1865", but the only Supplement to this work existing either in the New York or Washington libraries is a supplement to edition 3 of 1884.

Knoche (1921) reports that in the Balearic Islands this plant is called "carmelita" and is there cultivated. Duque-Jaramillo found it in flower and fruit in March at 2620 meters altitude in Colombian gardens. Thomas (1969) calls it the "hybrid verbena". Williamson (1973) lists the very modern horticultural varieties "Amethyst" and "Miss Susie".

The Ardoin 21, distributed as xV. hybrida, actually is V. rigida Spreng.

Additional citations: COLOMBIA: Cauca: López-Palacios & Idrobo 3832 (Ac). INDIA: West Bengal: Mukherjee s.n. [12.9.74] (Ld). CULTIVATED: Colombia: Duque-Jaramillo 2990 (N); López-Palacios 3616 (Ld, N). Czechoslovakia: Presl s.n. (Mu—4372). Ecuador: López-Palacios 4177 (Ld). Germany: Olin s.n. [June 1893] (Ac).

xVERBENA ILLICITA Moldenke

Additional bibliography: Perkins, Estes, & Thorp, Bull. Torrey Bot. Club 102: 194, 195, & 197. 1975; Moldenke, Phytologia 30: 151 (1975), 34: 250 (1976), and 36: 157. 1977.

Perkins and his associates (1975) report finding this hybrid to be the most abundant of four natural hybrids occurring in a single area in Oklahoma, there having been about 20 individuals of it as compared to 10 of xV. goodmani Moldenke, 1 of xV. deamii Moldenke, and 1 of xV. perriana Moldenke. A single artificially cross-pollinated plant (using V. urticifolia L. pollen) with 168 potential seeds had a 12.5 percent seed-set, while the parental species, V. urticifolia, had 47.3—66.5 percent seed-set and V. stricta Vent. (when insect-pollinated) had 76.3—87.6 percent.

Additional citations: ILLINOIS: Cass Co.: Geyer s.n. [Beardstown, July 1842] (Mu—411). MISSOURI: Saint Louis: Engelmann s. n. [St. Louis] (Mu—412).

VERBENA INAMOENA Briq.

Additional bibliography: Moldenke, Phytologia 30: 151 (1975), 34: 279 (1976), and 36: 131. 1977.

T. Rojas 10077, cited herein under V. bonariensis L., has been annotated by an unknown hand as "Verbena bonariensis L. f. transiens in V. inamomam Briq." The Herb. Mus. Bot. Landishuth s.n., also cited by me as V. bonariensis, has all its leaves very narrow-oblong in shape and may represent V. inamoena instead.

VERBENA INCISA Hook.

Additional synonymy: Verbena arenaria Hügel ex Moldenke, Phytologia 34: 278, in syn. 1976 [not V. arenaria Moldenke, 1961]. Verbena arenariana Kummer ex Moldenke, Phytologia 34: 278, in syn. 1976.

Additional bibliography: G. Don in Loud., Hort. Brit. Suppl. 1: 680. 1832; Loud., Hort. Brit., ed. 2, 552. 1832; Baxt. in Loud., Hort. Brit. Suppl. 2: 680. 1839; G. Don in Loud., Hort. Brit. Suppl. 2: 704. 1839; Baxt. in Loud., Hort. Brit. Suppl. [3]: 655. 1850; Vilm., Fl. Pleine Terre, ed. 1, 939 (1865), ed. 2, 2: 976 & 977 (1866), ed. 3, 1: 1200 (1870), and ed. 4, 1067. 1894; Moldenke, Phytologia 30: 150—152, 163, & 172 (1975), 31: 392 & 412 (1975), and 34: 270 & 278. 1976.

The corollas on Cristóbal 1210, Fabris 4688, and Krapovickas & al. 25759 & 27068 are described as having been "red", while on Herzog 1217 they were "cinnabar-red" and on Pflanz 951 "scarlet-red".

The cheironymous V. arenaria and V. arenariana listed in the synonymy above are based on specimens in the Munich herbarium collected, respectively, by Hügel in the Vienna and by Kummer in the Munich botanical garden. Don (1839) calls this species "Lady Arran's verbena".

Material of this species has been misidentified and distributed in some herbaria as V. phlogiflora Cham.

Additional citations: BRAZIL: Rio de Janeiro: Cabral s.n. [28. I. 1963; Herb. FEEMA 5192] (Fe). Rio Grande do Sul: Bornmüller 143 (Mu—4290). BOLIVIA: Santa Cruz: Herzog 1217 (Mu). Tarija: Pflanz 951 (Mu). PARAGUAY: T. Rojas 3406 [Hort. Parag. 11793] (Mu). URUGUAY: Herb. Herter 84884 (Mu); Herter 1057 [Herb. Herter 82941] (Mu). ARGENTINA: Catamarca: Rodriguez Vaquero 349 (Ut-330568B). Corrientes: Cristóbal 1210 (Ld); Krapovickas, Cristóbal, Irigoyen, & Schinini 27068 (Ld); Krapovickas, Cristóbal, Schinini, Arbo, Quarín, & González 25759 (Ld). Jujuy: Fabris 4688

(Mu). Misiones: Bertoni s.n. [Herb. Inst. M. Lillo 98412] (Ld);
Montes 14662 (N). CULTIVATED: Austria: Hügel s.n. [hort. Hügel
Vindob. 1839] (Mu—303). Germany: Herb. Kummer s.n. [Hort. bot.
Monac. 1840.IX.13] (Mu—1260, Mu—1261), s.n. [Hort. bot. Monac.]
(Mu—1259). Sweden: Zetterstedt s.n. [H. L. 10 Oct. 1839] (Ac).

VERBENA INTEGRIFOLIA Sessé & Moc.
    Additional & emended bibliography: Bolkh., Grif, Matvej., &
Zakhar., Chrom. Numb. Flow. Pl., imp. 1, 717 (1969) and imp. 2,
717. 1974; Moldenke, Phytologia 28: 246. 1974.

xVERBENA INTERCEDENS Briq.
    Additional bibliography: Moldenke, Phytologia 28: 246 & 440.
1974.

VERBENA INTERMEDIA Gill. & Hook.
    Additional & emended bibliography: Buek, Gen. Spec. Syn. Can-
doll. 3: 494—496. 1858; Bolkh., Grif, Matvej., & Zakhar., Chrom.
Numb. Flow. Pl., imp. 1, 717 (1969) and imp. 2, 717. 1974; Moldenke, Phytologia 30: 152 (1975) and 31: 387 & 409. 1975.
    Pedersen encountered this plant on "black earth" grasslands in
Corrientes and Job found it on dunes in Buenos Aires. The corollas on Herter 1155 are said to have been "lilac-blue" when fresh.
    Additional citations: URUGUAY: Herter 1155 [Herb. Herter
83295] (Mu). ARGENTINA: Buenos Aires: Job 1597 (Ut—330533B).
Corrientes: Pedersen 4660 (N).

VERBENA JORDANENSIS Moldenke
    Additional bibliography: Moldenke, Phytologia 28: 354 (1974)
and 30: 192. 1975.

VERBENA KUHLMANNII Moldenke, Phytologia 31: 29. 1975.
    Bibliography: Moldenke, Phytologia 31: 29 & 387. 1975; Anon.,
Biol. Abstr. 61: AC1.732. 1976.
    Citations: BRAZIL: São Paulo: M. Kuhlmann 3717 [Herb. Inst.
Bot. S. Paulo 79535] (W—2748267—type, Z—photo of type, Z—
photo of type).

VERBENA LACINIATA (L.) Briq.
    Additional bibliography: Sweet, Hort. Brit., ed. 2, 419. 1830;
G. Don in Loud., Hort. Brit., ed. 1, 247 (1830) and ed. 2, 247.
1832; G. Don in Loud., Hort. Brit. Suppl. 1: 680. 1832; Loud.,
Hort. Brit., ed. 2, 552. 1832; Baxt. in Loud., Hort. Brit. Suppl.
2: 680. 1839; G. Don in Loud., Hort. Brit., ed. 3, 247. 1839;
Baxt. in Loud., Hort. Brit. Suppl. [3]: 655. 1850; Buek, Gen.
Spec. Syn. Candoll. 3: 494 & 495. 1858; Vilm., Fl. Pleine Terr.,
ed. 1, 937 (1865), ed. 2, 2: 974 (1866), ed. 3, 1: 1197—1198
(1870), and ed. 4, 1066 & 1070. 1894; A. W. Anderson, How We Got
Fls., imp. 1, 168 & 283 (1951) and imp. 2, 168 & 283. 1966;
Bolkh., Grif, Matvej., & Zakhar., Chrom. Numb. Flow. Pl., imp. 1,

715—717 (1969) and imp. 2, 715—717. 1974; Rinton & Rzedowski,
Anal. Esc. Nac. Cienc. Biol. 21: 111. 1975; Kooiman, Act. Bot.
Neerl. 24: 464. 1975; López-Palacios, Revist. Fac. Farm. Univ. Los
Andes 15: 94. 1975; Moldenke, Phytologia 30: 152—153 & 172 (1975),
31: 383, 410, & 411 (1975), and 34: 259 & 260. 1976; Soukup, Biota
11: 19. 1976; E. H. Walker, Fl. Okin. & South. Ryuk. 884. 1976;
Moldenke, Phytologia 36: 128, 139, & 148. 1977.

Additional illustrations: Voss in Vilm., Fl. Pleine Terr., ed.
4, 1066. 1894.

Fosberg encountered this plant in grassy places grazed by goats
on the top of a low hill and the corollas on no. 27645 are said to
have been "lavender" when fresh, while on Grandjot s.n. [XI.32]
they were "rose-violet" in color.

Don (1830) lists V. multifida Ruíz & Pav. as a synonym of V.
erinoides Willd. and implies that it, like V. erinoides, was introduced into English gardens from Peru in 1818. Tawada (1967)
reports "V. erinoides Lamarck" as cultivated in Okinawa, but probably this is an error in identification for V. tenuisecta Briq.
It should be noted here that Walker (1976) gives "1968" as the
correct publication date for Tawada's work.

The Lorentz 478, distributed as V. laciniata, actually is V.
aristigera S. Moore, while Merxmüller 24804 is V. berterii
(Meisn.) Schau., Collector undetermined s.n. [H. L. 1840], is V.
bipinnatifida Nutt., Lossen 72 is V. glandulifera Moldenke, Herter
1805 and Herb. Herter 96556 are V. pulchella Sweet, Martius s.n.
[ad S. Joaõ d'El Rey, Febr.] is V. regnelliana Moldenke, Herter
181 and Herb. Herter 79174 are V. selloi Spreng., and Brixle s.n.,
Herb. Hort. Monac. s.n., and Herb. Merxmüller 14336, as well as
Kupper s.n. [cult. h. b. M.] are V. tenuisecta Briq.

Additional citations: ECUADOR: Cañar: Herb. Univ. Cent. Quito
2350 (Mu). Chimborazo: F. R. Fosberg 27645 (N). BOLIVIA: La Paz:
O. E. White s.n. [2-18-1963] (W—2774548). CHILE: Concepcion:
Neger s.n. [1893—96] (Mu—3981). Malleco: Baeza s.n. [19.XII.
1913] (Mu—4330). Santiago: Grandjot s.n. [XI.32] (Mu, Mu). Valparaíso: Behn s.n. [Quilpué, 2.X.1932] (Mu), s.n. [Valparaíso, 1
Okt. 1922] (Mu); O. Buchtien s.n. [8.IX.1895] (Mu—1837); Kausel
s.n. [Limache, 16.I.27] (Mu). Province undetermined: Dusén s.n.
[Chili australis 1896—97] (Mu—2982); Reuca s.n. [1889] (Mu—
4322). MOUNTED ILLUSTRATIONS: Ruíz & Pav., Fl. Peruv. & Chil. 1:
pl. 33, fig. 2. 1797 (N, Z).

VERBENA LACINIATA var. CONTRACTA (Lindl.) Moldenke
Additional bibliography: G. Don in Loud., Hort. Brit., ed. 1,
247 (1830), ed. 2, 247 (1832), and ed. 3, 247. 1839; Baxt. in
Loud., Hort. Brit. Suppl. [3]: 655. 1850; Buek, Gen. Spec. Syn.
Candoll. 3: 494 & 495. 1858; Moldenke, Phytologia 30: 153 (1975)
and 31: 410 & 411. 1975.

Don (1830) calls V. erinoides the "Erinus-like vervain" and

claims that it was introduced into English gardens from Peru in 1818 probably in this contracted form.

VERBENA LASIOSTACHYS Link

Additional bibliography: Sweet, Hort. Brit., ed. 2, 418 & 419. 1830; G. Don in Loud., Hort. Brit., ed. 1, 246 & 247 (1830) and ed. 2, 246 & 247. 1832; Loud., Hort. Brit., ed. 2, 552. 1832; G. Don in Loud., Hort. Brit., ed. 3, 246 & 247. 1839; Buek, Gen. Spec. Syn. Candoll. 3: 495. 1858; Kooiman, Act. Bot. Neerl. 24: 464. 1975; Moldenke, Phytologia 30: 153 (1975), 34: 251 (1976), and 36: 135. 1977.

Don (1830) places V. lasiostachys in his Section Indivisae (with undivided leaves), calls it the "hairy-spiked vervain", and says that it was introduced into English gardens from California in 1826, while V. prostrata is placed in his Section Trifidae, calls it the "prostrate vervain", and gives the date of its introduction from "N. Amer." as 1794.

The Herb. Schwaegrichen s.n. [1837], Herb. Zuccarini s.n. [Hort. Bot. Monac. 1835, 1836], Raven 2951, and Thorne & Tilforth 39918, distributed as typical V. lasiostachys, actually seem to represent var. septentrionalis Moldenke, while Meebold 20234 is V. robusta Greene.

Additional citations: CALIFORNIA: Humboldt Co.: Moldenke & Moldenke 30232 (Ac, Gz, Kh, Ld, Ln, Mu, Tu, Ut, W). Santa Barbara Co.: Meebold 22111 (Mu). Santa Cruz Co.: M. E. Jones 2215 (Mu—1575).

VERBENA LASIOSTACHYS var. SEPTENTRIONALIS Moldenke

Additional bibliography: Moldenke, Phytologia 30: 153 (1975) and 36: 135. 1977.

Recent collectors have encountered this plant in dry places and "a few feet above high-tide line on coastal cliffs", at altitudes of 2—30 meters, and describe it as a widely branching herb, 6 dm. tall. The corollas are said to have been "blue" on Witham 508, "purplish" on Thorne & Tilforth 39918, and "purple" on Beauchamp 2523.

Material of this variety has been misidentified and distributed in some herbaria as V. bracteata Lag. & Rodr.

Additional citations: OREGON: Josephine Co.: Baker & Ruhle 434 (N). CALIFORNIA: Alameda Co.: Meebold 19930 (Mu); Michener & Bioletti s.n. [Oakland, June 1891] (Mi). Butte Co.: Moldenke & Moldenke 30339 (Gz, Mu, Tu, Ut). Los Angeles Co.: Gallup s.n. [8/13/1949] (Sd—72252); Meebold 20050 (Mu); Thorne & Tilforth 39918 (K1—16018). Nevada Co.: M. E. Jones 2598 (Ln—70251, Mu—1576). San Diego Co.: Beauchamp 2523 (Sd—85664); M. F. Spencer 1037 (Mu—4319); Witham 1454 (Sd—80427). San Luis Obispo Co.: Edw. Palmer 342 (Mu—1555); Raven 2951 (Ac); Witham 508 (Sd—75714). Shasta Co.: Moldenke & Moldenke 30260 (Ac, Ld, W). CULTIVATED: Germany:

Herb. Schwaegrichen s.n. [1837] (Mu—1267); Herb. Zuccarini s.n.
[Hort. Bot. Monac. 1835] (Mu—367), s.n. [Hort. Bot. Monac. 1836]
(Mu—368). s.n. [Hort. Bot. Monac.] (Mu—369).

VERBENA LILACINA Greene
     Additional bibliography: Moldenke, Phytologia 30: 153. 1975.
     Recent collectors describe this species as a dense bush or
subshrub, 2—5 feet tall, growing in the steep north slopes of
canyons, along the sides of large granite boulders, at altitudes
of 5—50 meters, flowering and fruiting in March, April, and
July. Moran reports it as "common in arroyos". The corollas
are said to have been "lavender" in color on Moran 17123, 17127,
& 17185.
     Additional citations: MEXICO: Baja California: Bostic s.n. [2
July 1969] (Sd—70839); R. V. Moran 17123 (Sd—76988), 17127 (Sd—
76989), 17185 (Sd—76987).

VERBENA LINDBERGI Moldenke
     Additional bibliography: Moldenke, Phytologia 23: 288 (1972)
and 31: 387. 1975.
     Merxmüller encountered this species at 2350 meters altitude,
flowering in December.
     Additional citations: BRAZIL: Rio de Janeiro: Merxmüller
25555 (Mu).

VERBENA LIPOZYGIOIDES Walp.
     Additional bibliography: Buek, Gen. Spec. Syn. Candoll. 3: 495.
1858; Moldenke, Phytologia 28: 354. 1974.

VERBENA LITORALIS H.B.K.
     Additional synonymy: Verbena atriota Pabst ex Moldenke, Phyto-
logia 34: 278, in syn. 1976.
     Additional & emended bibliography: G. Don in Loud., Hort. Brit.
Suppl. 1: 680. 1832; Baxt. in Loud., Hort. Brit. Suppl. 2: 680
(1839) and [3]: 655. 1850; Buek, Gen. Spec. Syn. Candoll. 3: 494 &
495. 1858; Robledo, Bot. Med. 392. 1924; Barriga-Bonilla, Hernández-Camacho, Jaramillo-T., Jaramillo-Mejía, Mora-Osejo, Pinto-
Escobar, & Ruiz-Carranza, Isla San Andrés 59. 1969; G. W. Thomas,
Tex. Pl. Ecolog. Summ. 78. 1969; Bolkh., Grif, Matvej., & Zakhar.,
Chromb. Numb. Flow. Pl., imp. 1, 717. 1969; Hartwell, Lloydia 34:
387. 1971; Bolkh., Grif, Matvej., & Zakhar., Chrom. Numb. Flow.
Pl., imp. 2, 717. 1974; Gibbs, Chemotax. Flow. Pl. 3: 1753 & 1754.
1974; Balgooy, Pacif. Pl. Areas 3: 245. 1975; O. & I. Degener &
Pekelo, Hawaii. Pl. Names x.4, x.21, & x.22. 1975; Hinton & Rze-
dowski, Ann. Esc. Nac. Cienc. Biol. 21: 31 & 111. 1975; Kooiman,
Act. Bot. Neerl. 24: 464. 1975; López-Palacios, Revist. Fac. Farm.
Univ. Los Andes 15: 51 & 90—93, fig. [17]. 1975; Molina R., Ceiba
19: 95. 1975; Tovar Serpa, Biota 10: 286 & 298. 1975; Moldenke,
Phytologia 30: 136 & 153—154 (1975), 31: 378, 379, 383, & 392
(1975), and 34: 256, 260, 267, 270, & 278. 1976; López-Falacios,
Revist. Fac. Farm. Univ. Los Andes 17: 50. 1976; Soukup, Biota 11:

19. 1976; E. H. Walker, Fl. Okin. & South. Ryuk. 883 & 884. 1976;
Moldenke, Phytologia 36: 31, 33, 47, 51, 52, 122, 131, 136, 137,
& 151. 1977.

Additional illustrations: López-Palacios, Revist. Fac. Farm.
Univ. Los Andes 15: fig. [17]. 1975.

Molina R. refers to this species as a "weed common in sugarcane plantations" in Nicaragua. Walker (1976) calls it "A weed
of roadsides and waste places" on Okinawa, while Schlieben reports
it "very abundant" on bush savannas in the Transvaal. On Saipan
island Stone found it to be a "common weed with Asclepias curas-
savica, Conyza, Cardiospermum, etc." Herbst speaks of it as "uncommon" on Hauai island. Werff found it in the moist zone on
Chatham island in the Galápagos and comments that there its flowers were "not as intensely colored as those of 2182 [V. brasiliensis Vell.]". Bianco describes it as 0.8--1.3 meters tall and
"medicinal" in Venezuela; Taylor found it on steep roadside banks
in Costa Rica.

The inflorescence tips on Molina R. 27244 are much congested
because they are insect-galled; the corollas are said to have been
"lilac" in color when fresh, as they were also on his no. 11508,
on Hatschbach & Kummrow 35764, Romero-Castañeda 10668, and Schlieben 7691. They are said to have been "violet" on Pabst 7372 and
Schinini & al. 10259, "blue" on Behn s.n. [14 Decbr. 1930], "blue-
violet" on Plowman & Davis 4889, "bluish-violet" on Stone 5233,
"pretty purple" on Clemens 42170, and "pale-blue" on Herbst 2296.
On López-Palacios 4220 & 4332 they are described as "moradas", on
4196 "morado lila", on 4100 "lila", and on 4040 "azul morado hasta morado muy claro".

Hatschbach & Kummrow report V. litoralis being used medicinally
in Brazil, while in Peru, according to Plowman & Davis, the dried
ground-up leaves are placed on wounds.

It is not certain to what species Gibbs (1974) is referring
when he uses the name "Verbena arborea", but such a binomial has
been used for V. litoralis in the horticultural trade.

López-Palacios (1975) comments that "En París, en el Herbario
HB, existe el N. 658 [of Bonpland], determinado como V. caracasana, que bien puede ser el tipo o, cuando menos, el esótipo" [of
var. caracasana]. He continues: "Yo he examinado el material de
Willdenow y no encuentro diferencia alguna con la V. litoralis
HBK. El especimen 11134 Willd., Verbena lanceolata, corresponde
al 638 de Humboldt (P), rotulado quizás por él mismo, o por Bonpland, V. caracasana, pero en mi concepto, no creo que alconce ni
siquiera a una diferencia de variedad de la V. litoralis. También
las poblaciones son aguales, y no se observa diferencia entre las
de Caracas y las de Mérida, p. e. De la V. glabrata no hay con-
stancia segura de que exista en el territorio venezolano. El
ejemplar de Mocquerys, Duaea 893 ? No o fecha?), determinado por
Doña N. Troncoso como V. glabrata es, simplemente, V. litoralis.

Muchos las consideran como coespecíficas.....Para mí es difícil
establecer la diferencia entre las dos especies. [actually V. glabrata H.B.K. as seen so abundantly in Ecuador is very easily distinguished in general aspect from V. litoralis!]. De existir en
Venezuela, sería probablemente en Táchira y Zulia, hasta donde pue-
den extenderse las poblaciones colombianas del Norte de Santander,
en donde la V. glabrata ha sido registrada."

    López-Palacios 3638 is a close match for the type illustration
of V. litoralis, but his 3948 is the very widespread loosely fruit-
ed form, while his 3623 is the very dense-flowered and -fruited
form now passing as var. caracasana. He reports that the plants
growing in his own garden and represented by his no. 3974 were 80—
100 cm. tall. In a letter to me, dated January 16, 1976, he says:
"Por correo le envés 5 ejemplares de Verbena litoralis colectados
de una misma planta que yo tengo cultivada. Este taxon es poli-
mórfico y variable no sólo en la población sino en la misma planta,
como Ud. podrá observar; en hojas y en espigas la variación es
grande. Yo ví en Herb. Willdenow el tipo de la var. caracasana,
pero en mi concepto todo ello es simplemente V. litoralis. Naturalmente esto es una cosa subjetiva y se deba a mi formación, influ-
ida por las escuelas europeas (Holandesa y Alemana) que tienen un
criterio muy amplio de la especie. Naturalmente el dibujo de HBK
sólo muestra un estadio de crecimiento y por tanto no puede regis-
trar todas las variaciones de la especie." He suggests that his
4161 may be a form of V. brasiliensis Vell., and this is, indeed,
very possible. It is also most probable that these two species
hybridize when growing in close proximity, as they often do. His
no. 4302, from the Galápagos islands, is described as "hierba de
hojas medianas, espigas congestas luego alargadas".

    Walker (1976) reports for V. litoralis the vernacular name,
"hime-kuma-tsusura" [=small delicate V. officinalis] on Okinawa.
Witham collected the species at 1000 m. altitude in Hawaii, describing it as a "tough perennial" with blue flowers. Tovar Serpa
(1975) records the vernacular name, "wirwena", for it in Peru. The
Degeners & Pekelo (1975) list the names, "ha'uōwī", "'oi", and
"ōwī", in Hawaii; Thomas (1969) calls it the "coast verbena".

    Gibbs (1974) reports that in what he calls "Verbena arborea" cy-
anogenesis is absent from the leaves and syringin is absent from
the stems. It is not definitely known if he refers here to V. litoralis or to Petrea arborea, for both of which taxa the name "Verbena arborea" has in the past been used in literature. Hartwell
(1971) reports that in Mexico V. litoralis is called "verbena del
campo" and that the twigs are used to make a decoction drunk in
the treatment of internal tumors. Krapovickas and his associates
report its use in Salta, Argentina, "para golpe, pasa sangre".
Krapovickas, Schinini, & González 28440 represents the dense-
spiked form of the species.

    The Widgren s.n. [1845], distributed as V. litoralis, actually

is V. alata Sweet, while Lechler s.n. [Valdivia], Leyboldt s.n.,
and Monts 2289 are V. bonariensis L., Fabris & Marchionni 2392
is a mixture of V. litoralis and V. bonariensis, Behn s.n. [Quil-
pué, 22 January 1931], Bougere 14, 1091, & 1099, C. A. Brown 1008,
2309, 2381, & 18610, Bynum, Ingram, & Jaynes s.n. [Houma, Apr. 23,
1933], M. Chaney 111, Claycomb s.n. [June 13, 1942], Heinrichs
65, Kirby 160, D. K. Lowe 31, Meebold 27224 & 27240, Monts 637, J.
A. Moore 5200, Neger s.n. [1893—96], Robinette 239, Rockett 125,
Roivainen 3054, and Stutzenbaker 205 are V. brasiliensis Vell.,
Heyde & Lux 3019 is V. carolina L., O. Buchtien 185 is xV. dermeni
Moldenke, Claycomb s.n. [April 15, 1943] is V. halei Small, C. A.
Brown 18767, Herb. Herter 81713, Herter 269, and Thibodeaux 236,
260, 284, 297, 321, & 417 are V. montevidensis Spreng., J. Taylor
17625 is V. parvula Hayek, and Schimpff 132 is V. sedula Moldenke.
    Additional citations: LOUISIANA: Terrebonne Par.: Wurzlow s.n.
[May 5, 1914] (Lv, Lv). MEXICO: Oaxaca: Pringle 4877 (Mu—1803).
Veracruz: Kerber 311 (Mi, Mu—1791). GUATEMALA: Guatemala: L. M.
Andrews 507 (N). Santa Rosa: Heyde & Lux 4370 (Mu—1790). HON-
DURAS: Cortes: Molina R. 11508 (W—2735773). NICARAGUA: Estelí:
Molina R. 27244 (N, W—2735237). COSTA RICA: San José: J. Taylor
17446 (N). COLOMBIA: Arauca: López-Palacios 3948 (Ld, N). Cun-
dinamarca: López-Palacios 3623 (Ld, N), 3638 (Ld, N); López-
Palacios & Jaramillo Mejía 3674 (Ld, N). Magdalena: Romero—Cas-
tañeda 10668 (N). VENEZUELA: Aragua: Vogl 938 (Mu, Mu, Mu).
Mérida: Oberwinkler & Oberwinkler 12196 (Mu). Miranda: Pittier
442 (Mu). ECUADOR: Carchi: López-Palacios 4040 (Ld). Chimborazo:
Collector undetermined s.n. [September 1858] (Mu—1105). El Oro:
López-Palacios 4100 (Ld). Guayas: Eggers 14372 (Mu—3882). Imba-
bura: López-Palacios 4072 (Ld). Loja: López-Palacios 4161 (Ld).
Pichincha: López-Palacios 4196 (Ld), 4220 (Ld), 4332 (Ld). GAL-
ÁPAGOS ISLANDS: Chatham: Schimpff 142 (Mu); Werff 2183 [1483]
(Ld), 2186 [1486] (Ld). Narborough: López-Palacios 4302 (Ld).
PERU: Cuzco: Plowman & Davis 4889 (Oa). BRAZIL: Minas Gerais:
Irwin, Harley, & Onishi 28721 (W—2759077). Rio de Janeiro:
Pabst 7372 (Mu). BOLIVIA: La Paz: M. Bang 204 (Mu—1788). PARA-
GUAY: Pedersen 8625 (N); T. Rojas 1889 (Mu). CHILE: Valparaíso:
Behn s.n. [14 Decbr. 1930] (Mu); O. Buchtien s.n. [18.X.1895]
(Mu—1838). Province undetermined: Dessauer s.n. [Chile, VI-IX-
87] (Mu); Frömbling s.n. [Chili, 1886] (Mu—1789). ARGENTINA:
Jujuy: Schinini, Quarín, Arbo, & Pire 10259 (Ld). Salta: Krapo-
vickas, Schinini, & González 28440 (Ld). San Juan: Fabris &
Marchionni 2392, in part (Mu). Santiago del Estero: Lullo 4
(Ut—220576B); Pierotti "h" [6-II-1944] (Ut—330535B). SOUTH
AFRICA: Transvaal: Meebold 12839 (Mu); Scheepers 334 (Mu); Schlie-

ben 7691 (Mu). MARIANAS ISLANDS: Saipan: B. C. Stone 5233 (Kl)..
AUSTRALIA: Queensland: M. S. Clemens 42170 (Mi); Meebold 7818 (Mu).
HAWAIIAN ISLANDS: Hawaii: Witham 1713 (Sd—83745). Kauai: Herbst
2296 (N). Oahu: Meebold 8304 (Mu); Schmer s.n. [9/13/69] (Lc).
CULTIVATED: Brazil: Hatschbach & Kummrow 35764 (Ld). Germany:
Herb. Kummer s.n. [Hort. bot. Monac. 1839] (Mu—1248). Venezuela:
Bianco 110 (N); López-Palacios 3974 (Ac, Gz, Mu, Tu).

VERBENA LITORALIS var. ALBIFLORA Moldenke
   Synonymy: Verbena littoralis var. albiflora Moldenke ex Hinton
& Rzedowski, Anal. Esc. Nac. Cienc. Biol. 21: 111. 1975.
   Additional bibliography: Moldenke, Phytologia 28: 252, 432, &
438. 1974; Hinton & Rzedowski, Anal. Esc. Nac. Cienc. Biol. 21:
31 & 111. 1975; Soukup, Biota 11: 19. 1976; Moldenke, Phytologia
36: 47. 1977.

VERBENA LITORALIS var. CARACASANA (H.B.K.) Moldenke
   After extensive field and herbarium studies, including an examination of the type collection, López-Palacios has shown that this
taxon cannot be distinguished from typical V. litoralis H.B.K.
All my previous notes in this series under this heading should
therefore be transferred to typical V. litoralis.

VERBENA LITORALIS f. MAGNIFOLIA Moldenke, Phytologia 36: 51—52.
   1977.
   Bibliography: Moldenke, Phytologia 36: 33 & 51—52. 1977.
   Citations: ECUADOR: Napo: López-Palacios 4188 (Z—type).

VERBENA LOBATA Vell.
   Additional bibliography: Buek, Gen. Spec. Syn. Candoll. 3: 494
& 495. 1858; Moldenke, Phytologia 30: 154. 1975.
   The Hatschbach HH.14883 and Herb. Brad. 48010, distributed as
V. lobata, actually represent var. sessilis M°ldenke.
   Additional citations: BRAZIL: Minas Gerais: Dusén 242 (Mu—3998).
Rio de Janeiro: A. Castellanos 25682 [Herb. FEEMA 4520] (Ld). Rio
Grande do Sul: Bornmüller 602 (Mu—4295).

VERBENA LOBATA var. HIRSUTA Moldenke
   Additional bibliography: Moldenke, Phytologia 28: 253. 1974.
   The corollas are said to have been "lilac" in color when fresh
on Hatschbach 35664.
   Additional citations: BRAZIL: Paraná: Hatschbach 35664 (Ld).

VERBENA LOBATA var. SESSILIS Moldenke
   Additional bibliography: Moldenke, Phytologia 28: 355. 1974.
   The corollas on Hatschbach HH.14883 are said to have been
"dark-lilac" in color when fresh, and this collector encountered
the plant in "brejo" (sedge meadow), flowering in October. It was
distributed in some herbaria as typical V. lobata Vell.
   Additional citations: BRAZIL: Paraná: Hatschbach HH.14883 [Herb.

Brad. 48010] (Mu).

VERBENA LONGIFOLIA Mart. & Gal.
    Additional synonymy: Verbena longifolia H.B.K. ex Moldenke,
Phytologia 36: 47, in syn. 1977.
    Additional bibliography: Buek, Gen. Spec. Syn. Candoll. 3: 495.
1858; Moldenke, Phytologia 23: 296 (1972), 34: 252 (1976), and 36:
47 & 145. 1977.
    The corollas are said to have been "white to pale-pink" on
Ernst 2355a.
    Additional citations: MEXICO: Oaxaca: Ernst 2355a (Mi).

VERBENA LONGIFOLIA f. ALBIFLORA Moldenke
    Additional bibliography: Moldenke, Phytologia 28: 253 (1974),
34: 252 (1976), and 36: 145. 1977.
    Martínez Calderón refers to this plant as an annual herb which
he encountered at 5 m. altitude in "suelo arcilloso-arenoso in
acahual" and which he misidentified and distributed as V. carolina
L. The Ernst 2355a, cited under typical V. longifolia (above), is
said to have come from a population with white to light-pink flowers so it may, in part, at least, also represent this form.
    Additional citations: MEXICO: Veracruz: Martínez Calderón 1352
(N).

VERBENA LUCANENSIS Moldenke
    Additional bibliography: Moldenke, Phytologia 23: 297. 1972;
Soukup, Biota 11: 19. 1976.

VERBENA MACDOUGALII Heller
    Additional & emended bibliography: Bolkh., Grif, Matvej., &
Zakhar., Chrom. Numb. Flow. Pl., imp. 1, 717. 1969; G. W. Thomas,
Tex. Pl. Ecolog. Summ. 78. 1969; Fong, Trojánkova, Trojánek, &
Farnsworth, Lloydia 35: 147. 1972; Bolkh., Grif, Matvej., & Zak-
har., Chrom. Numb. Flow. Pl., imp. 2, 717. 1974; R. D. Gibbs,
Chemotax. Flow. Pl. 3: 1753 & 1754 (1974) and 4: 2295. 1974; D. S.
& H. B. Correll, Aquat. & Wetland Pl. SW. U. S., imp. 2, 2: 1397,
1399—1400, & 1775. 1975; Kooiman, Act. Bot. Neerl. 24: 464. 1975;
A. L. Moldenke, Phytologia 31: 415. 1975; Moldenke, Phytologia 30:
154 (1975) and 36: 145. 1977.
    The Spellenbergs encountered this plant "along dirt road with
grasses, in Douglas fir, Pinus, Quercus, and Holodiscus" association. Pinkava and his associates found it along roadsides in ponderosa pine forests — a habitat apparently identical to that in
which my wife, my son, and I saw it in great abundance and which
seems to be its favorite habitat. Higgins reports finding it in
sandy soil of "short-grass prairie community", in sandy soil of
"spruce-fir-pine community", and in "coarse sandy to gravelly soil
in mountain brush and mixed evergreen community and aspen scattered in patches". Dziekanowski and his associates observed it in
"very rocky yellow pine forests".
    The corollas are said to have been "violet, fading to blue" on

Spellenberg & Spellenberg 2082. Thomas (1969) calls the species
the "Macdougal verbena".

Gibbs (1974) reports cyanogenesis absent in the leaves of V.
macdougalii and the Ehrlich test negative, but syringin is doubtfully present in the stems.

Material has been misidentified and distributed in some herbaria as V. hastata var. scabra Moldenke.

Additional citations: COLORADO: Archuleta Co.: C. F. Baker s.n.
[Arboles, 7-10-99] (Mu—3912). NEW MEXICO: Otero Co.: Spellenberg
& Spellenberg 2082 (N). Lincoln Co.: Higgins 8604 (N). San Miguel Co.: Higgins 8881 (N). Taos Co.: Higgins, Higgins, & Rook
10040 (N); Waterfall 12250 (Mi). Yavapai Co.: H. H. Rusby 780, in
part (Mu). ARIZONA: Apache Co.: Lehto, McGill, Nash, & Pinkava
11506 (W—2734642); Pinkava, Lehto, & Reeves P.12352 (N). Coconino Co.: Dziekanowski, Dunn, & Bennett 2395 (N).

VERBENA MACDOUGALII f. ALBIFLORA Moldenke

Additional bibliography: Moldenke, Phytologia 28: 253, 254, &
431. 1974.

VERBENA MALMII Moldenke

Additional bibliography: Moldenke, Phytologia 23: 298. 1972;
Troncoso, Darwiniana 18: 311 & 412. 1974.

Hatschbach describes this species as growing from a xylopodi-
um. He found it on dry campos, flowering in December, and the
corollas on his no. 35553 are said to have been "lilac" in color
when fresh.

Additional citations: BRAZIL: Paraná: Hatschbach 35553 (Ld).

VERBENA MARITIMA Small

Additional synonymy: Verbena maritima Sm. ex Norman, Fla. Scientist 39: 30. 1976.

Additional bibliography: M. F. Baker, Fla. Wild Fls., ed. 2,
imp. 1, 188. 1938; Ayensu, Rep. Endang. & Threat. Pl. Spec. 98 &
129. 1974; Moldenke, Phytologia 28: 254, 451. & 464 (1974) and 34:
248 & 279. 1976; M. F. Baker, Fla. Wild Fls., ed. 2, imp. 2, 188.
1976; Long & Lakela, Fl. Trop. Fla., ed. 2, 741 & 961. 1976; Norman, Fla. Scientist 39: 30. 1976; Moldenke, Phytologia 36: 142.
1977.

Ayensu (1974) has officially listed this as an endangered or
threatened species. With the rapidity of the commercialization
of the Florida beaches, the survival of this species, limited to
that specialized habitat, is certainly in great doubt.

Churchill has encountered V. maritima in pine flatwoods on
oolitic limestone and on the lee side of dunes, as well as in
dune hollows, in flower and fruit in March and June. Norman (1976)
calls it the "seaside verbena" and justifiably refers to it as
already "rare".

Additional citations: FLORIDA: Brevard Co.: Curtiss 1963* (Mu—
1545). Broward Co.: Meebold 27688 (Mu). Dade Co.: J. A. Church-

ill s.n. [12 March 1956] (Ln—204149). Martin Co.: J. A. Churchill s.n. [18 June 1968] (Ln—225090). Long Key: J. K. Small 8123
(Mu).

VERBENA MARRUBIOIDES Cham.
    Additional bibliography: Buek, Gen. Spec. Syn. Candoll. 3: 495.
1858; Moldenke, Phytologia 30: 154. 1975; Soukup, Biota 11: 19.
1976.
    Recent collectors have encountered this plant along roadsides
on campos and in "campo com pequeño banhado". The corollas are
said to have been "red (5P6/8)" on Lindeman, Irgang, & Valls ICN.
8805 and "blue-purple (10PB5/8)" on Lindeman & Haas 2459.
    Reineck & Czermak 21 is a mixture of V. marrubioides and V.
humifusa Cham.
    Additional citations: BRAZIL: Paraná: Lindeman & Haas 2459
(Ld). Rio Grande do Sul: Lindeman, Irgang, & Valls ICN.8805
(Ut—320456); Reineck & Czermak 21, in part (Mu).

VERBENA MEDICINALIS Rojas
    Additional & emended bibliography: Krapovickas, Bol. Soc. Ar-
gent. Bot. 11, Supl. 269. 1970; Moldenke, Phytologia 30: 154—155
(1975) and 31: 388. 1975.

VERBENA MEGAPOTAMICA Spreng.
    Additional synonymy: Verbena phlogiflora β macilenta Cham. ex
Buek, Gen. Spec. Syn. Candoll. 3: 495. 1858.
    Additional & emended bibliography: Loud., Hort. Brit., ed. 2,
552. 1832; Buek, Gen. Spec. Syn. Candoll. 3: 495. 1858; Bolkh.,
Grif, Matvej., & Zakhar., Chrom. Numb. Flow. Pl., imp. 1, 715
(1969) and imp. 2, 715. 1974; Moldenke, Phytologia 30: 155 & 178
(1975) and 36: 47. 1977.
    The label on Krapovickas & Cristóbal 28956 bears the statement
"en campos pantanosos, erecta, flores amarillas" — if the corolla
color given here is correct this represents a remarkable undescribed color-form, but it seems more probable to me that it represents
a mistake in memory or transcription.
    The Hort. Parag. 11793 and T. Rojas 3406, distributed as V.
megapotamica, actually are V. incisa Hook., while Duarte 6309, Herb.
Brad. 16885 & 22512, Pabst 6093, and E. Pereira 6266 are V. phlogiflora Cham.
    Additional citations: ARGENTINA: Corrientes: Krapovickas & Cris-
tóbal 28956 (Ld).

VERBENA MEGAPOTAMICA Spreng. x V. PULCHELLA Sweet
    Additional bibliography: Moldenke, Phytologia 28: 255, 451, &
464. 1974.

VERBENA MENDOCINA R. A. Phil.
    Additional & emended bibliography: Bolkh., Grif, Matvej., & Zak-

har., Chrom. Numb. Flow. Pl., imp. 1, 715 & 717 (1969) and imp.
2, 715 & 717. 1974; Moldenke, Phytologia 30: 155. 1975.

VERBENA MENTHAEFOLIA Benth.
    Additional bibliography: Buek, Gen. Spec. Syn. Candoll. 3:
495 & 496. 1858; G. W. Thomas, Tex. Pl. Ecolog. Summ. 78. 1969;
Moldenke, Phytologia 28: 355 & 362 (1974) and 30: 159. 1975;
Hinton & Rzedowski, Anal. Esc. Nac. Cienc. Biol. 21: 31 & 111.
1975; Moldenke, Phytologia 36: 145. 1977.
    Recent collectors have encountered this plant on rocky hills
with thin gravelly soil and oak-pine grassland cover. Mears
found it growing in association with Cassia, Mimosa, Juniperus,
Solanum, Cuphea, Quercus, Indigofera, and Phoradendron. In Baja
California it is reported by Moran as "occasional", "abundant in
roadside depressions", "common in roadside ditches", and "locally common in several places on dry open southeast slopes", at 10
to 200 meters altitude, describing it as a "decumbent bush" or
"prostrate". The corollas are described as having been "blue"
on Moran 16643 & 18675, "blue-violet" on Moran 16098, 18563, &
21824, "purple" on Witham 783, and "light-blue, with white center" on Moran 22459. Thomas (1969) calls it the "mintleaf verbena", a singularly inappropriate name since its leaves do not
resemble those of any typical mint with which I am familiar.
The Baja California material ascribed to this species needs to
be more carefully compared to V. comonduensis Moldenke, a closely related taxon.
    Additional citations: CALIFORNIA: San Diego Co.: R. V. Moran
16098 (Sd—71707); Spencer s.n. [4.25.1916] (Mu—4318); Witham
783 (Sd—79855). MEXICO: Baja California: R. V. Moran 16643
(Sd—73069), 18563 (Sd—80229), 18675 (Sd—80255), 21824 (Sd—
91272), 22459 (Sd—91088). Federal District: Barkley & Rowell
7464 (Ln—166003). Hidalgo: Mears 259d (Ln—222126), 326d (Ln—
222197). México: Pringle 8534 (Mu—3989). Zacatecas: Taylor &
Taylor 6230 (W—2734032).

VERBENA MICROPHYLLA H.B.K.
    Additional bibliography: Buek, Gen. Spec. Syn. Candoll. 3: 495.
1858; Moldenke, Phytologia 30: 155. 1975; Soukup, Biota 11: 19.
1976.
    Schultes has placed an interesting note on the M. Wagner s.n.
[Sept. 1858] sheet at Munich: "Verbena erinoidi Lam. proxima sed
ipse vix videtur. An V. multifida R. & Pavon quae a Schauero
Verbenas erinoidi subjungitur".
    Legname & Vervoorst refer to V. microphylla as a "prostrate
hemicryptophyte" and encountered it in "terreno arenoso-arcilloso"
— the corollas on their no. 101 were "pale-lilac" when fresh.
    Additional citations: ECUADOR: Chimborazo: Schimpff 720 (Mu);
M. Wagner s.n. [Sept. 1858] (Mu—1104). Cotopaxi: M. Wagner s.n.
[October 1858] (Mu—1262). Province undetermined: M. Wagner s.n.

[Tacunga, Octob.--Nov. 1858] (Mu--1263). PERU: Cuzco: W. Hoffmann 307 (Mu). BOLIVIA: La Paz: O. Buchtien 1102 (Mu); K. Graf
453 (N). Oruro: Troll 2919 (Mu). Potosí: Fiebrig 2613 (Mu--
4088). Province undetermined: K. Graf 599 [Taurichambi] (N).
ARGENTINA: Catamarca: Legname & Vervoorst 101 (N). Jujuy: Cabrera, Ancibor, Ré, Tello, & Torres 15474 (Mu); Ellenberg 4259
(Ld), 4261 (Ac). Province undetermined: Princess Therese of Bavaria 282 (Mu).

VERBENA MINUTIFLORA Briq.
    Additional bibliography: Moldenke, Phytologia 28: 356 & 383
(1974) and 36: 36 & 123. 1977.
    In addition to the months previously reported by me, this
plant has been collected in fruit in March and October. It has
been described by Lindeman & Haas as an almost leafless shrub,
1.5 m. tall, and the corollas on their no. 3010 are said to have
been "purple" when fresh. On Lindeman ICN.9446 the collector
notes "arbusto de 1.70 m., 1.5 cm. diam., corola roxa 10PB7/6",
and it was encountered by him in a "pequeño banhado quase seco".
    Davidse and his associates report that in Santa Catarina it
is used in the treatment of stomach and digestive ailments.
    Material of V. minutiflora has been misidentified and distrib-
uted in some herbaria as V. alata Sweet.
    Additional citations: BRAZIL: Paraná: Hatschbach 37374 (Ld);
Lindeman & Haas 3010 (Ws). Rio Grande do Sul: Lindeman ICN.9446
(Ut--320459). Santa Catarina: Davidse, Ramamoorthy, & Vital
11089 (Ld). ARGENTINA: Toledo Island: Ibarrola 739 (Ut--330572B).

xVERBENA MOECHINA Moldenke
    Additional synonymy: Verbena moenchina Moldenke ex R. A. Dav-
idson, State Univ. Iowa Stud. Nat. Hist. 20 (2): 77, sphalm.
1959.
    Additional bibliography: R. A. Davidson, State Univ. Iowa
Stud. Nat. Hist. 20 (2): 77. 1959; Cooperrider, State Univ. Iowa
Stud. Nat. Hist. 20 (5): 70. 1962; Moldenke, Phytologia 28: 356,
386, 387, 429, & 465 (194), 34: 249 (1976), and 36: 29 & 47.
1977.
    Tans encountered this hybrid along roadsides and in an aban-
doned quarry on limestone gravel in association with native
prairie plants such as Asclepias verticillata, Andropogon gerardi, Eragrostis spectabilis, Kuhnia eupatorioides, Ratibida pinnata, Solidago nemoralis, and Verbena stricta, with Verbascum
and Ambrosia invading. Davidson (1959) records the hybrid from
Louisa and Muscatine Counties, Iowa, where he found it to be "infrequent" in dry sandy soil.
    The Herb. Zuccarini s.n. [Hort. bot. Monac.] collection, cited
below, is a mixture with V. stricta Vent.
    Additional citations: WISCONSIN: Rock Co.: Tans 1431 (Ts, Ts,

Ts). MISSOURI: Reynolds Co.: Meehold 25420 (Mu). CULTIVATED:
Germany: Herb. Zuccarini s.n. [Hort. bot. Monac.] (Mu—375, Mu—
376).

VERBENA MONACENSIS Moldenke
Additional bibliography: Moldenke, Phytologia 28: 356, 394, &
451 (1974), 30: 133 (1975), and 34: 270. 1976; López-Palacios,
Revist. Fac. Farm. Univ. Los Andes 17: 50. 1976; Moldenke, Phyto-
logia 36: 40. 1977.
López-Palacios refers to this plant as an "hierba rastrera de
unos 30 cms. Flores rosadas" and found it being cultivated at
1650 meters altitude, flowering in August.
Additional citations: CULTIVATED: Colombia: López-Palacios 3618
(Ld, N), 3862 (Tu); López-Palacios & Idrobo 3833 (Ac).

VERBENA MONTEVIDENSIS Spreng.
Additional & emended bibliography: Buek, Gen. Spec. Syn. Can-
doll. 3: 495. 1858; Bolkh., Grif, Matvej., & Zakhar., Chrom. Numb.
Flow. Pl., imp. 1, 717 (1969) and imp. 2, 717. 1974; Moldenke,
Phytologia 30: 155 (1975), 31: 377 (1975), and 36: 137. 1977.
The corollas on Ferreira 98, Hatschbach 35653, and Kummrow 764
are said to have been "lilac" in color when fresh, on Schinini &
Carnevali 10471 they were "purple", and on Fiebrig 4635, Herb.
Brad. 22518, Pabst 6146, and E. Fereira 6319 they were "violet".
Ferreira found the plant growing in "orla de brejo". Others
have encountered it on high or marshy campos, in rough grassland,
along roadsides, on headlands of ricefields and sugarcane fields,
in bottomland soil, and in "brejo" (sedge meadows). Muhammad
refers to it as "an erect perennial herb, infrequent in open
fields maintained by fire", while Urbatsch found it "in roadside
and railway right-of-way, aquatic marsh, and dryland habitats".
The corollas on Schinini & Cristóbal 9707 were "violet" in
color when fresh, those on Quarín, Schinini, & González 2460 were
"purple", those on Krapovickas, Cristóbal, & Schinini 26513 were
"white-lilac", and those on Pedersen 9816 were "white".
Lindeman and his associates encountered this plant on a "campo
estilo pomar com árvores baixas de espinilo Acacia caven" and say
on the label of their no. 8468 "corola 5RP8/4, calice em botao
5RP5/4". The vernacular name, "quina", is reported by Kummrow.
Pedersen found the plant in flower and fruit in April.
Additional citations: LOUISIANA: Calcasieu Par.: Thibodeaux
260 (Lv). Cameron Par.: Thibodeaux 236 (Lv). East Baton Rouge
Par.: C. A. Brown 18767 (Lv). Jefferson Davis Par.: Thibodeaux
417 (Lv). Lafayette Par.: Thibodeaux 297 (Lv), 321 (Lv). Saint
Landry Par.: Thibodeaux 428 (Lv). Tangipahoa Par.: Muhammad 259
(Lv); Urbatsch 1938 (Lv). Vermilion Par.: Thibodeaux 284 (Lv).
BRAZIL: Paraná: Dusén 10856 (Mu); L. F. Ferreira 98 (Ld); Hatsch-
bach 35653 (Ld); Kummrow 764 (Tu). Rio Grande do Sul: Lindeman,
Irgang, & Valls ICN.8468 (Ut—320457). Santa Catarina: Pabst 6146

[E. Pereira 6319; Herb. Brad. 22518] (Mu). PARAGUAY: Fiebrig
4635 (Mu—4144). URUGUAY: Herter 269 [Herb. Herter 81713] (Mu).
ARGENTINA: Corrientes: Krapovickas, Cristóbal, & Schinini 26513
(Ld); Pedersen 9816 (N); Quarín, Schinini, & González 2460 (Ld);
Schinini & al. 11864 (Ld); Schinini & Carnevali 10471 (Ld); Sohi-
nini & Cristóbal 9707 (Ld). Misiones: Montes 14719 (N), 27576
(N).

VERBENA MONTICOLA Moldenke
   Additional bibliography: Hocking, Excerpt. Bot. A.26: 6. 1975;
Moldenke, Phytologia 30: 155 (1975) and 31: 384. 1975.

VERBENA MULTICAULIS Raf.
   This taxon is probably the same as V. simplex var. eggerti Moldenke. All previously published notes in this series under this
heading should be transferred to that variety.

VERBENA NANA Moldenke
   Additional bibliography: Moldenke, Phytologia 30: 155. 1975.
   The corollas on Fiebrig 4371 are said to have been "violet-
rose" when fresh.
   Additional citations: PARAGUAY: Fiebrig 4371 (Mu).

VERBENA NEOMEXICANA (A. Gray) Small
   Additional synonymy: Verbena neomexicana var. neomexicana [A.
Gray] apud Thomas, Tex. Pl. Ecolog. Summ. 78. 1969.
   Additional & emended bibliography: G. W. Thomas, Tex. Pl. Ecolog. Summ. 78. 1969; Bolkh., Grif, Matvej., & Zakhar., Chrom. Numb.
Flow. Pl., imp. 1, 717 (1969) and imp. 2, 717. 1974; E. H. Jordan,
Checklist Organ Pipe Cact. Natl. Mon. 7. 1975; Moldenke, Phytolo-
gia 30: 138, 155-156, & 180 (1975), 31: 378 (1975), 34: 252 & 279
(1976), and 36: 124 & 158. 1977.
   Semple refers to this plant as "rare, in small clumps in dry
stream bed". Mrs. Jordan (1975) calls it the "New Mexican vervain"
and Thomas (1969) names it the "New Mexico verbena". Urbatsch and
his associates aver that it is "uncommon in sandy soil" in Baja
California.
   Material of this species has been misidentified and distributed
in some herbaria as V. gracilis Desf. or V. plicata Greene. On the
other hand, the Spellenberg & Spellenberg 3984, distributed as
typical V. neomexicana, seems better placed as var. hirtella Perry,
while Reeves R.1131 and Wentworth 1061 are var. xylopoda Perry, C.
A. Brown 7409 and Monts 2485 are xV. alleni Moldenke, Taylor & Taylor 6230 is V. menthaefolia Benth., Hess & Stickney 3406 and Higgins 9228 are V. perennis Wooton, S. Walker 76H37 is V. pinetorum
Moldenke, and Meebold 26696 is V. racemosa Eggert.
   Additional citations: TEXAS: Brewster Co.: Semple 357 (W—
2732729). ARIZONA: Pima Co.: J. A. Churchill s.n. [7 April 1972]
(Ln—235702). Santa Cruz Co.: Reeves R.1198 (N). MEXICO: Baja

California: <u>Urbatsch</u>, <u>Clark</u>, & <u>Betkouski</u> 1436 (Ld). Coahuila:
<u>Barkley</u>, <u>Webster</u>, & <u>Rowell</u> 7189 (Ln—189725).

VERBENA NEOMEXICANA var. HIRTELLA Perry
   Additional bibliography: G. W. Thomas, Tex. Pl. Ecolog. Summ.
78, 1969; Moldenke, Phytologia 30: 138 & 156 (1975) and 34: 252.
1976.
   Arnold found this plant growing on "steep gravelly hills and
creek beds" in Coahuila. In Chihuahua the Spellenbergs encountered the plant "on grassy knolls with mostly ocotillo, creosotebush,
<u>Yucca</u> <u>torreyi</u>, sotol, and <u>Hedyotis</u> <u>rubra</u>". In Baja California
Moran found it "occasional in disturbed roadside soil" and "occas-
ional on sandy flats", at altitudes of 575—1600 meters. Correll
& Pollins encountered it "on gravel knolls along roadsides" in Texas.
   The corollas are said to have been "deep-blue" on <u>Moran</u> 21749,
"blue" on <u>Correll</u> & <u>Rollins</u> 23652 and <u>Moran</u> 20727, "blue-violet"
on <u>Moran</u> 20748 and <u>Spellenberg</u> & <u>Spellenberg</u> 3984, and "lavender"
on <u>Moran</u> 16893; they were also "blue" on <u>Henrickson</u> 5944
   Other recent collectors have found <u>V. neomexicana</u> var. <u>hirtella</u>
growing in "calcareous gravel in chaparral on very steep slopes of
limestone sierra", "in calcareous gravelly soil in matorral desértico microfilo on limestone slopes and limestone-conglomerate fan",
"in rocky calcareous soil in crasirosulifolio espinosos izotal on
steep slopes of metamorphosed shaly limestone", "in dark, sandy,
grassy, gravelly loam on gentle slopes of extrusive igneous rock",
"in sandy alluvium in matorral desértico inerme on gravel fans", in
"calcareous gravelly soil in matorral con espinas laterales in canyons through limestone", "on rocky northeast-facing slopes", in
"rocky soil of matorral on steep slopes of igneous rocks with chaparral and encinares (oak woods) higher up", in "rocky reddish clay
soil of deserts", and in "limestone outcroppings in open Chihuahuan
Desert", in association with <u>Yucca</u> <u>carnerosana</u>, <u>Agave</u> <u>parrasana</u>, <u>A.
lecheguilla</u>, <u>Condalia</u> <u>warnockii</u>, <u>Berberis</u> <u>trifoliolata</u>, <u>Mammillaria
melacantha</u>, <u>Opuntia</u> <u>rufida</u>, <u>O. lindheimeri</u>, <u>Acacia</u> <u>rigidula</u>, <u>Vigui-
era</u> <u>stenoloba</u>, <u>Sagaretia</u> <u>wrightii</u>, <u>Dasylirion</u>, <u>Quercus</u>, <u>Lindleya</u>,
<u>Krameria</u>, <u>Mimosa</u>, <u>Muhlenbergia</u>, <u>Pinus</u>, <u>Larrea</u>, <u>Jatropha</u>, <u>Parthenium</u>,
<u>Leucophyllum</u>, <u>Ptelea</u>, <u>Garrya</u>, <u>Juglans</u>, <u>Flourensia</u>, <u>Buddleia</u>, <u>Fouqui-
eria</u>, <u>Nama</u>, grasses, and numerous animals. Henrickson found it "in-
frequent along highways".
   Additional citations: TEXAS: Presidio Co.: <u>Correll</u> & <u>Rollins</u>
23652 (N). MEXICO: Baja California: R. V. <u>Moran</u> 16893 (Sd—76990),
20727 (Sd—88938), 20748 (Sd—88937), 21749 (Sd—91271). Chihuahua: <u>A. A. Heller</u> s.n. [April 6, 1897] (Ln—93653); <u>Henrickson</u> 7617
(Ld); <u>Johnston</u>, <u>Wendt</u>, & <u>Chiang</u> <u>C.</u> 10773b (Ld); <u>Spellenberg</u> & <u>Spel-
lenberg</u> 3984 (N); <u>Wilson</u>, <u>Wilson</u>, <u>Johnston</u>, & <u>Johnston</u> 8510 (Ld).
Coahuila: <u>E. T. Arnold</u> 32 (Te—68564); <u>Henrickson</u> 5944 (Ld), 6132
(Ld); <u>Johnston</u>, <u>Wendt</u>, & <u>Chiang</u> <u>C.</u> 10284d (Ld), 10500c (Ld), 11687

(Ld). Nuevo León: Johnston, Wendt, & Chiang C. 10235b (Ld). Zacatecas: Johnston, Wendt, & Chiang C. 10489 (Ld).

VERBENA NEOMEXICANA var. XYLOPODA Perry
Additional bibliography: G. W. Thomas, Tex. Pl. Ecolog. Summ.
78. 1969; Moldenke, Phytologia 30: 156 & 180. 1975.
Recent collectors have encountered this variety on north-facing
slopes with rock outcrops, calcareous stones and soils with scattered oaks, on rocky outcrops with Fouquieria, and in Cercocarpus
breviflorus scrub on limestone bedrock.
Other collectors have found it growing in rocky limestone soil,
in "sandy, grussy, gravelly loam in small creek canyons through
extensive igneous rock", in "rocky soil in pastizal, badly degraded
and in places invaded by shrubs", on "hill of igneous extrusives
with partly volcanic ash red of red color", and in "grussy, gravelly, thin soil on rather steep hills of extrusive igneous rocks",
growing in association with Parthenium incanum, Lippia graveolens,
Jatropha dioica, Acacia neovernicosa, Bouteloua gracilis, Larrea,
Quercus, Pinus, and Pseudotsuga.
In Baja California Moran reports this variety "locally common
on open upper south slopes", "scarce on ridges", "occasional in
granitic soil on rocky hillsides", and "a small colony on dry
rocky slope", at altitudes of 1025—1240 meters. The corollas
were "blue" on Moran 20983 & 22170, "blue-violet" on Moran 18264,
"light-blue, paler in the center" on Moran 17658, and "purple" on
Powell, Turner, & Sikes 2479.
The Wentworth collection cited below is a voucher for ecologic
studies.
Additional citations: ARIZONA: Cochise Co.: Wentworth 1061 (N).
Pinal Co.: Lehto, Hensel, & Pinkava 11033 (W—2736741). Santa
Cruz Cc.: Reeves R.1131 (N). MEXICO: Baja California: R. V. Moran
17658 (Sd—75054), 18264 (Sd—77109), 20983 (Sd—83866), 22170
(Sd—91462). Chihuahua: Chiang C., Wendt, & Johnston 8311a (Ld);
Johnston, Wendt, & Chiang C. 10524 (Ld); Powell, Turner, & Sikes
2479 (Ld); Wilson, Wilson, Johnston, & Johnston 8483 (Ld). Tamau-
lipas: Kuiper & Kuiper-Lapré M.15 (Ut—328637B).

VERBENA NIGRICANS Rojas
Additional & emended bibliography: Krapovickas, Bol. Soc. Argent. Bot. 11, Supl. 269. 1970; Moldenke, Phytologia 30: 156 (1975)
and 31: 388. 1975.

VERBENA NIVEA Moldenke
Additional synonymy: Glandularia nivea Mold. ex Moldenke, Phytologia 34: 274, in syn. 1976.
Additional bibliography: Moldenke, Phytologia 28: 357 (1974)
and 34: 274. 1976.
In addition to the months previously reported by me, this plant
has been collected in fruit in February.

The Legname & Vervoorst 101, distributed as V. nivea, actually
is V. microphylla H.B.K.

VERBENA OCCULTA Moldenke
    Additional bibliography: Moldenke, Phytologia 23: 376—377
(1972) and 34: 258. 1976; Soukup, Biota 11: 19. 1976; Moldenke,
Phytologia 36: 148. 1977.
    Material of this species has been misidentified and distribu-
ted in some herbaria as V. calcicola Walp. and V. clavata Ruíz &
Pav.
    Additional citations: PERU: Arequipa: Princess Therese of Ba-
varia 281 (Mu).

VERBENA OCCULTA f. ALBA Moldenke
    Additional bibliography: Moldenke, Phytologia 23: 377. 1972;
Soukup, Biota 11: 19. 1976.

VERBENA OCCULTA f. AURANTIACA Moldenke
    Additional bibliography: Moldenke, Phytologia 23: 377. 1972;
Soukup, Biota 11: 19. 1976.

VERBENA OFFICINALIS L.
    Additional synonymy: Verbena vulgaris folio variegato Breyn.,
Prodr. Fasc. Rar. Pl., ed. 2, 2: 104. 1739.
    Additional & emended bibliography: Apul. Barb., Herb., ed. 1.
1480-1483; Anon., Dialogue des Créatures, 30th dial. 1482; Apul.
Barb., Herb., ed. 2. 1528; Anon., Bastiment des Receptes fol.
59 vert. 1544; H. Bock [Tragus], Stirp. Max. Germ. 102. 1552;
Dill. in Ray, Synop. Meth. Stirp. Brit., ed. 3, 236. 1724; L.,
Hort. Cliff., imp. 1, 11. 1737; Breyn., Prodr. Fasc. Rar. Pl.,
ed. 2, 2: 104. 1739; Strand in L., Amoen. Acad. 69: 449. 1756;
Chomel, Abrég. Hist. Pl. Usuel., ed. 6, 2 (2): 85—87 & 251.
1761; Ginanni, Istor. Civ. Nat. Pinet. Ravenn. 177. 1774; Chomel,
Abrég. Hist. Pl. Usuel., ed. 6 nov., 313 & 637. 1782; F. Hernandez, Hist. Pl. Nuev. Españ., ed. 1, 1: 139 & 439 (1790) and ed. 1,
3: 3 & 486. 1790; R. A. Salisb., Prodr. 71. 1796; Chomel, Abrég.
Hist. Pl. Usuel., ed. 7, 1: 495 (1803) and ed. 7, 2: 488. 1803;
Stokes, Bot. Mat. Med. 40—41. 1812; A. Rich., Bot. Méd. 1: 242—
243. 1823; Dierbach, Arzneimit. Hippok. 85 & 270. 1824; A. Rich.
[transl. G. Kunze], Med. Bot. 1: 381 (1824) and 2: 1302. 1826; G.
Don in Loud., Hort. Brit., ed. 1, 247 (1830) and ed. 2, 247.
1832; Loud., Hort. Brit., ed. 2, 552. 1832; A. Dietr., Handb.
Pharmaceut. Bot. 114 & 412. 1837; D. Dietr., Taschenb. Arzneigew.
Deutschl. 58 & 262. 1838; G. Don in Loud., Hort. Brit., ed. 3,
247. 1839; Spach, Hist. Nat. Veg. Phan. 9: 237. 1840; Webb in
Hook., Niger Fl. 161. 1849; Anon., Chroniqueur du Périgord 120.
1853; F. Lenormant, Bull. Sic. Bot. France 2: 315—320. 1855;
Schnitzlein, Iconofr. Fam. Nat. 2: 137 Verbenac. [2] & 137, fig.
4—22 & 30. 1856; Buek, Gen. Spec. Syn. Candoll. 3: 495 & 496.
1858; Symphor Vaudoré, Lettr. Vieux Laboureur 88. 1867; J. Cousin, Secr. Mag. 1868: 7, 37, & 45. 1868; Chenaux, Le Diable & Ses

Cornes 53 & 54. 1876; Anon., Rev. du Tarn 1877: 39. 1877; Franch.,
Nouv. Arch. Mus. Hist. Nat. Paris, ser. 2, 6: 112 [Pl. David. 1:
232]. 1883; Strobl, Oesterr. Bot. Zeitschr. 33: 406. 1883; Kuntze,
Rev. Gen. Pl. 2: 510. 1891; J. Camus, Récept. Franç. in Bull. Soc.
Syndic. Pharmac. Côte-d'Or. 10. 1892; J. Feller, Bull. Folklore 2:
105—109. 1893; Nairne, Flow. Pl. West. India 249. 1894; Van Tieghem, Élém. Bot., ed. 3, 2: 373. 1898; Bidault, Superst. Méd. Morvan 36. 1899; Diels, Fl. Cent.-China 547. 1902; Anon., Rev. Tradit.
Populaires 1904: 162 (1904) anf 1905: 160 & 296. 1905; Druce &
Vines, Dill. Herb. 78. 1907; Rolland, Fl. Populaire 8: 38—43.
1910; Gilg in Engl., Syllab. Pflanzenfam., ed. 7, 314, fig. 413C.
1912; Loes., Verh. Bot. Ver. Brand. 53: 74. 1912; Gilg in Engl.,
Syllab. Pflanzenfam., ed. 8, 318, fig. 413C (1919) and ed. 9 & 10,
339, fig. 418C. 1924; Robledo, Bot. Med. 267. 1924; Krause in Just,
Bot. Jahresber. 44: 1172. 1926; Fedde in Just, Bot. Jahresber. 44:
1534. 1927; Freise, Bol. Agric. São Paulo 34: 480 & 494. 1933;
Gunther, Herb. Apul. Barb. [16v], [35v], 106, 128, 129, & 133.
1935; E. D. Merr., Trans. Am. Phil. Soc., ser. 2, 24 (2): [Comm.
Lour.] 331 & 444. 1935; Diels in Engl., Syllab. Pflanzenfam., ed.
11, 339, fig. 432C. 1936; F. Hernandez, Hist. Pl. Nuev. Españ., ed.
2, 653 & 674. 1943; Roi, Atl. Pl. Méd. Chin. [Mus. Heude Bot. Bot.
Chin. 8:] 96. 1946; Hatta, Kubo, & Watanabe, List Med. Pl. 14.
1952; Sonohara, Tawada, & Amano [ed. E. H. Walker], Fl. Okin. 132.
1952; Pételot, Arch. Recherch. Agron. & Past. Viet. 18: [253].
1953; Pételot, Pl. Méd. Camb. Laos & Viet. 2: 243 (1954) and 4: 21,
39, 70, 170, 184, 193, 208, & 300. 1954; L., Hort. Cliff., imp. 2,
11. 1968; J. Hutchinson, Evol. & Phyleg. Flow. Pl. Dicot. 470, fig.
414. 1969; Rimpler, Lloydia 33: 491. 1970; Saxena, Bull. Bot. Surv.
India 12: 56. 1970; Scully, Treas. Am. Ind. Herbs 283. 1970; Anon.,
Bioresearch Ind. 7: 1061. 1971; Kachroo, Singh, & Malik, Bull. Bot.
Surv. India 13: 52. 1971; Kaul, Bull. Bot. Surv. India 13: 240.
1971; Gilmour, Thom. Johnson 31, 50, 78, 106, 107, & 122. 1972;
Healy, Gard. Guide Pl. Names 37 & 225. 1972; Frohne & Jensen, System. Pflanzenr. 203, 261, & 305. 1973; Hilbig, Wiss. Zeitschr.
Mart. Luth. Univ. Halle 22: 56 & 102. 1973; Law, Concise Herb. Encycl. 85 & 263. 1973; El-Gazzar, Egypt. Journ. Bot. 17: 75 & 78.
1974; Ellenberg, Script. Geobot. 9: 80. 1974; Farnsworth, Pharmacog. Titles 9 (4): x. 1974; R. D. Gibbs, Chemotax. Flow. Pl. 3:
1752—1755 (1974) and 4: 2295. 1974; León & Alain, Fl. Cuba, imp.
2, 2: 281. 1974; Loewenfeld & Back, Complete Book Herbs & Spices
261—264. 1974; A. & D. Löve, Cytotax. Atl. Slov. Fl. 601 & 1241.
1974; Portéres, Journ. Agric. Trop. & Bot. Appl. 21: 6. 1974;
Stanley & Linskens, Pollen 47, 95, & 306. 1974; Sunding, Garcia de
Ort. Bot. 2: 20. 1974; Täckholm, Stud. Fl. Egypt, ed. 2, 454. 1974;
Whitney in Foley, Herbs Use & Delight [207]. 1974; R. & A. Fitter,
Wild Fls. Brit. & N. Eu. 192, 193, & 336. 1975; Kooiman, Act. Bot.
Neerl. 24: 464. 1975; López-Palacios, Revist. Fac. Farm. Univ. Los
Andes 15: 88, 90, & 93. 1975; Weberling & Schwantes, Pflanzensyst.,
ed. 2 [Ulmer, Uni-Taschenb. 62:] 144. 1975; Moldenke, Phytologia
30: 156—161 (1975), 31: 410 & 412 (1975), and 34: 249, 254, 260—
262, 266, & 279. 1976; Anon., Biol. Abstr. 61: AC1.732. 1976; Gal-

iano & Cabezudo, Lagascalia 6: 150. 1976; Keys, Chinese Herbs 283—284 & 387. 1976; Lakela, Long, Fleming, & Genelle, Pl. Tampa Bay, ed. 3 [Bot. Lab. Univ. S. Fla. Contrib. 73:] 116 & 182. 1976; Lousley, Fl. Surrey 282, map 288. 1976; Soukup, Biota 11: 19. 1976; E. H. Walker, Fl. Okin. & South. Ryuk. 883—884. 1976; Moldenke, Phytologia 36: 40, 126, 138, & 152. 1977; A. L. Moldenke, Phytologia 36: 87. 1977.

Additional illustrations: H. Bock [Tragus], Stirp. Max. Germ. 102. 1552; Schnitzlein, Iconogr. Fam. Nat. 2: 137 Verbenac. fig. 4—22 [partly in color] & 30. 1856; Van Tieghem, Élém. Bot., ed. 3, 2: 373. 1898; Gilg in Engl., Syllab. Pflanzenfam., ed. 7, 314, fig. 413 C (1912), ed. 8, 318, fig. 413 C (1919), and ed. 9 & 10, 339, fig. 418 C. 1924; Gunther, Herb. Apul. Barb. [16v]. 1935; Diels in Engl., Syllab. Pflanzenfam., ed. 11, 339, fig. 432 C. 1936; Roi, Atl. Pl. Méd. Chin. [Mus. Heude Not. Bot. Chin. 8:] 96. 1946; J. Hutchinson, Evol. & Phylog. Flow. Pl. Dicot. 470, fig. 414. 1969; Loewenfeld & Back, Complete Book Herbs & Spices [262]. 1974; R. & A. Fitter, Wild Fls. Brit. & N. Eu. 193, fig. 7 (in color). 1975; Keys, Chinese Herbs 283. 1976.

Recent collectors have encountered this plant "along water channels", in irrigated wheat fields, and "on granite substrate".

Lakela and her associates (1976) aver that in the Tampa Bay [Florida] area V. officinalis inhabits "trails, wooded lots, [and] burns", flowering from sprint to fall, but it is most probable that the reference here is to V. halei Small.

Mrs. Clemens refers to V. officinalis as a "common weed" in Queensland. Hendricks 590, cited below, was "purchased in market place" in Durango, Mexico, and is questionably referred here — it may, instead, actually represent the top portion of a native Mexican species.

Hutchinson (1969) regards the Verbenaceae (of which Verbena officinalis is the type species of the type genus) as the culmination of the "fundamentally woody phylum, Lignosae" and therefore "at the end of the author's system" of classification of flowering plants (dicotyledons).

Sunding (1974) records V. officinalis from Santiago Island in the Cape Verde Islands, citing Sunding 2849, while Webb (1849) cites Hooker f. 120 from Santo Jacobi island. Saxena (1970) reports it as "Rare in open places" in India, citing Indorkar 11146. Kaul (1971) refers to it as an "Annual herb. Flowers pinkish white" and found i to be "rare" in Kashmir, India, flowering and fruiting there from June to August, citing Kaul 4624S. Strand (1756) records it from Palestine. Sonohara and his associates (1952) refer to it as "A perennial herb, common on plains; used for drugs" in Okinawa and re cords the common name, "kumatsuzura". Walker (1976) cites Hatusima 24199 from Yonaguni island. Lousley (1976) reports that in Surrey (England) it is "locally frequent" in "Chalk downs, quarries, waste places and roadsides, in dry places on chalk or gravel soils. Most common on the chalk."

The illustration given by Loewenfeld & Back (1974), purporting to depict V. officinalis, is horribly poor and most unrepresentative of that taxon. The illustration given by Schnitzlein (1856), purporting to be of V. supina L., seems to represent V. officinalis instead.

Friese (1933) comments that "A familia das Verbenaceas conta no Estado do Espirito Santo [Brazil] uma representante não descripta ainda, pertencente ao genero Verbena, bem affim á especie V. officinalis L., em fórma de subarbusto escandente, parcamente armado de espinhos; folhas oppostas e denteadas; inflorescencia em fórma de espiga com flóres amarellas ou brancas; fruto drupaceo". This he follows with a description of the medicinal uses for the leaves, but it is most uncertain to what plant he is here referring: possibly a Lantana, but most certainly not Verbena officinalis!

Petélot (1953) affirms that V. officinalis "Répandue dans toutes les régions tempérées et même tropicales" [in Indochina]. "C'est une plante amère légèrement tonique. Au Centre-Vietnam, la plante est considérée comme amère et aromatique et Loureiro... signals qu'elle est employée en décoction contre l'hydropisie et en cataplasmes sur les tumeurs du scrotum. D'après le R. P. Robert, elle passe pour régulariser les menstrues et pour guérir la 'boule hystérique'. Pour cela, on en prépare une purée que l'on fait cuire et qui se prend avec de l'alcohol de riz." The Chinese name for the plant there is "ma pien ts'ao", the Vietnamese name is "cò roi ngựa", and in Mexico the Mayan name is "chichiantic".

Linnaeus (1737) says of this species "Crescit juxta areas & vias inque locis ruderatis per Belgium, Angliam, Galliam, &c." Breyne (1739) says of it "VERBENA vulgaris folio variegato; nobis In Horto Honestissimae Sapientissimaeque Matronae, Domine de Flines, collegimus. Verbenae Notae: 1) Flores tubulosi, in extremo vix galeati & labiati, in caulium nec non ramorum summo, vel etiam in virgulis longis e foliorum sede exsuntibus spicatim ut plurimum dispositi; 2) Calix foliolis constructus; 3) Semina quaterna oblonga."

In addition to the several hundred vernacular names recorded by me in previous installments of this series of notes, Rolland (1910) lists the following: "aelius", "auricula vervicina", "ayàn nouthay", "barbàntano", "barbénéga", "barbáno", "barbèra", "barlenn", "benerea", "beneria", "berbeana", "bérbés", "bèrbèn", "berbenaca", "bèrbéno", "bèrbiéno", "bèrmày'no", "bèrmèno", "biscopwurtil", "bona herba Veneris", "bonion", "bordèno", "botanica", "bouono barbéno", "brébouane", "centrum galli", "cincinalis", "clumbeina", "columbaire", "columbaria", "columbaris", "columbina recta", "columbyne", "créy'jéta", "créy'sèta", "crijéta", "crista gallinacia", "crous", "dametra", "darbèno", "demedria", "demetina", "demetria", "diosatim", "diosatin", "eisebrich", "emagallis", "erba colombina", "erba de san-Gioan", "erba milzea", "erba minsaea", "erbo crousado", "erbo crusàdo", "èrbo dé la mèrbèlho", "èrbo dé la rato", "erbo dé lo bèrbèno".

[to be continued]

HELICONIA IN NICARAGUA

Robert R. Smith
Hartwick College, Oneonta, N.Y.

The flora of Nicaragua is presently being studied by Mr. Frank
C. Seymour of The Herbarium, University of Florida. Mr. Seymour
and companions have made six expeditions to Nicaragua between 1968
and 1975. Several collections of Heliconia were made on these
expeditions. With the encouragement of Mr. Seymour, the author,
utilizing the Seymour collection and information accumulated in a
thesis on the genus Heliconia of Middle America, decided to con-
tribute to the flora of Nicaragua by writing the "Heliconia in
Nicaragua".

The format of this paper is similar to "Cassia in Nicaragua"
by Mr. Seymour. It consists of an introduction, generic description,
artificial key to the species of Heliconia in Nicaragua, an annotated
systematic list. Since there are a new species and two new varieties
whose ranges include Nicaragua, a section on new names and combinations
follows the annotated systematic list.

The genus Heliconia is included in the family Musaceae. It
is characterized in the following manner: perennial, herbaceous,
erect plants of various heights (1-5 m., occasionally taller);
musoid or cannoid habit. Leaves frequently large, distichous or
appear as such. Inflorescence terminal, erect or pendulous, consisting of few to many, usually bright colored, boat-shaped branch-
bracts, either distichous or spirally arranged. Branch-bracts
enclose clusters of flowers. Each flower subtended by a floral-
bract. Flowers perfect. Perianth consists of two whorls; calyx
with connate abaxial sepals, free adaxial sepals; corolla with
connate petals except for free margins opposite adaxial sepal.
Stamens 6, 5 functional and one staminode. Anthers linear. Stigma
lobed, clavate or subclavate. Style filiform, straight or genic-
ulate. Overy inferior; fruit 3-loculed, berry-like (fleshy schizocarp). Seed stony.

ARTIFICIAL KEY TO THE SPECIES

A. Branch-bracts overlapping forming close-knit spike, rachis
   covered.. B
   B. Inflorescence pendent......................... H. mariae
   B. Inflorescence erect............................ H. imbricata

A. Branch-bracts separated (distant on rachis) with rachis showing,
or partially overlapping with thick branch-bracts 12-18 cm. long
.... C
C. Mature inflorescence pendent............................. D
   D. Branch-bracts spiralled.................. H. collinsiana
   D. Branch-bracts distichous................. H. longa
C. Mature inflorescence erect................................. E
E. Inflorescence sessile or subsessile, branch-bracts thick,
deep boat-shaped, may be partially overlapping with rachis
diameter 1 cm. or more....................... H. wagneriana
E. Inflorescence usually peduncled, branch-bracts not thick,
moderate to shallow boat-shaped with rachis diameter less
than 1 cm. thick... F
   F. Plants with cannoid habit; leaves sessile or nearly so;
   leaf-blade not more than 55 cm. long.................. G
      G. Branch-bracts orange, flowers white or cream-white
      .... H
         H. Flowers shorter than or as long as branch-
         bracts; perianth less than 5 cm. long........
         .... H. aurantiaca
         H. Flowers longer than branch-bracts; perianth
         more than 5 cm. long............. H. longiflora
      G. Branch-bracts red, flowers red or yellow......... I
         I. Flowers red, perianth puberulent to hirsute...
         .... H. hirsuta var. rubiflora
         I. Flowers yellow, often with green apices, perianth
         glabrous........................ H. vaginalis
   F. Plants with musoid habit; leaves petioled; if cannoid
   habit, lower leaf-blades more than 55 cm. long........
   .... J
      J. Branch-bracts spiralled........................... K
         K. Branch-bracts yellow with red apices or
         yellow-orange; rachis glabrous... H. latispatha
         K. Branch-bracts red; rachis tomentose..........
         .... H. tortuosa
      J. Branch-bracts distichous......................... L
         L. Branch-bracts 12-22, close together, strongly
         curved upward.................... H. librata
         L. Branch-bracts 4-11, widely spaced, extend out
         horizontally or reflexed..................... M
            M. Branch-bracts deep red to orange scarlet,
            perianth 4.0-4.5 cm. long, slightly pubescent;
            petioles short............... H. osaensis
            M. Branch-bracts yellow to orange, perianth
            2.5-3 cm. long, villous; petioles long....
            .... H. schiedeana var. spissa

ANNOTATED SYSTEMATIC LIST

1. H. aurantiaca Ghiesbr. in Lemaire's L'Illustr. Hortic. Pl. 332.
    1862. Synonyms: Bihai aurantiaca (Ghiesbr.) Griggs, Bull. Torr.
    Bot. Club 31: 445. 1904. H. brevispatha Hook. in Curtis's Bot.
    Mag. t. 5416. 1864. H. choconiana S. Wats. in Proc. Amer. Acad.
    23: 284. 1888. Bihai choconiana (S. Wats.) Griggs, Bull. Torr.
    Bot. Club 31: 445. 1904. H. crassa Griggs, Bull. Torr. Bot.
    Club 30: 646. 1903. Bihai crassa (Griggs) Griggs, Bull. Torr.
    Bot. Club 31: 445. 1904. Range: s. Mexico to Costa Rica.
    NICARAGUA: DEPT. ZELAYA: Comarca del Cabo, San Mateo (circa
    de Rio Wawa), A. Molina 15065 (F); El Recreo, L. E. Long 54 (F).

2. H. collinsiana Griggs, Bull. Torr. Bot. Club 30: 648. 1903.
    Synonym: Bihai collinsiana (Griggs) Griggs, Bull. Torr. Bot.
    Club 31: 445. 1904. Range: s. Mexico to Costa Rica.
    NICARAGUA: DEPT. MANAGUA: between El Curcero and house of
    Finca Santa Julia, P. C. Standley 8380 (F).

3. H. hirsuta L. f. var. rubiflora R. R. Smith var. nov.
    See section on treatment of new names and combinations.

4. H. imbricata (O. Ktze.) Baker, Ann. Bot. 7: 191. 1893. Synonym:
    Bihai imbricata O. Ktze., Rev. Gen. Pl. 2: 684. 1891. Range:
    Nicaragua to Panama.
    NICARAGUA: DEPT. ZELAYA: Comarca del Cabo, Miguel Bikon,
    S. B. Robbins 5863a (MO, SEY).

5. H. latispatha Benth., Voy. Sulph. 170-171. 1844. Synonyms:
    Bihai latispatha (Benth.) Griggs, Bull. Torr. Bot. Club 31:
    445. 1904. H. meridensis Kl. in Linnaea 20: 463. 1847.
    Bihai meridensis (Kl.) O. Ktze., Rev. Gen. Pl. 2: 685. 1891.
    Range: s. Mexico to n. S.A.
    NICARAGUA: DEPT. CHINANDEGA: Chinandega, C. F. Baker 2016 (GH,
    MO, UC, US); Ameya, W. R. Maxon, A. D. Harvey, and A. T. Valentine
    7188 (US). DEPT. GRANADA: Mombacho Volcano, L. O. Williams and
    A. Molina 20027 (F). DEPT. MANAGUA: Managua, Garnier 862 (MICH,
    US); Los Nubes, s. of Managua, W. R. Maxon, A. D. Harvey and
    A. T. Valentine 7500 (US); Tipitapa, F. C. Seymour and J. T.
    Atwood 2827 (BM, ENAG, F, GH, MICH, MO, NY, SEY, SMU, UC, WDP).
    DEPT. RIVAS: Penas Blancas, J. T. Atwood 1809 (BM, ENAG, F, GH,
    NO, NY, SEY, SMU, UC, WDP). DEPT. ZELAYA: El Recreo,
    R. B..Hamblett 328 (GH, SEY); Rama, S. A. Marshall and D. A.
    Neill 6460 (SEY); Comarca del Cabo, San Mateo, A. Molina 15096
    (F); Corn Island, F. C. Seymour and J. T. Atwood 4266 (BM, ENAG,
    GH, SMU).

6. H. librata Griggs, Bull. Torr. Bot. Club 30: 649. 1903. Synonym:
    Bihai librata (Griggs) Griggs, Bull. Torr. Bot. Club 31. 1904.

Range: s. Mexico to Nicaragua.
NICARAGUA: DEPT. ZELAYA: Comarca del Cabo, Slima Sia, A. Molina
1470 (F); Comarca del Cabo, Miguel Bikon, S. B. Robbins 5863 (SEY)

7. H. longa (Griggs) Winkl. in Eng. and Prantl. Nat. Pflanzenf. 2
    Aufl. 15A: 536. 1930. Synonym: Bihai longa Griggs, Bull. Torr.
    Bot. Club 31: 446. 1904. Range: Nicaragua to Panama.
    NICARAGUA: DEPT. ZELAYA: 5 mi. w. of Bonanza, J. T. Atwood and
    D. A. Neill 6995 (MO, SEY); El Recreo, P. C. Standley 19407 (F).

8. H. longiflora R. R. Smith sp. nov.
    See section on treatment of new names and combinations.

9. H. mariae Hook. f., Jour. Linn. Soc. Bot. 7: 68-69. 1864.
    Synonyms: Bihai mariae (Hook. f.) O. Ktze., Rev. Gen. Pl. 2:
    684. 1891. H. elegans Peters. in Mart. Fl. Bras. 33: 12. 1890.
    Bihai elegans (Peters.)). Ktze., Rev. Gen. Pl. 2: 684. 1891.
    H. punicea (Griggs) L. B. Smith, Contr. Gray Herb. 124: 6. 1939.
    Bihai punicea Griggs, Bull. Torr. Bot. Club 42: 321. 1915.
    Range: Guatemala and Belize to S. A.
    NICARAGUA: To date there are no recorded specimens of this
    species for Nicaragua. The range of the species indicates it
    should be present.

10. H. osaensis Cuf., Archivio Bot. 9: 189. 1933. Range: Nicaragua
    to Panama.
    NICARAGUA: DEPT. ZELAYA: San Antonio de Susun, region of
    Braggman's Bluff, F. C. Englesing 138 (F, US); El Recreo, L. E.
    Long 45 (F); Area Del Ocotal, Rio Grande, A. Molina 2332 (F, GH);
    El Recreo, P. C. Standley 19363 (F).

11. H. schiedeana Kl. var. spissa (Griggs) R. R. Smith var. nov.
    See section on treatment of new names and combinations.

12. H. tortuosa Griggs, Bull. Torr. Bot. Club 30: 650. 1903.
    Synonym: Bihai tortuosa (Griggs) Griggs, Bull. Torr. Bot. Club
    31: 445. 1904. Range: Guatemala to Costa Rica.
    NICARAGUA: DEPT. JINOTEGA: Las Mercedes, e. of Jinotega, P. C.
    Standley 10724 (F). DEPT. MADRIZ: Volcan Somoto, s. of Somoto,
    L. O. Williams and A. Molina 20274 (F). DEPT. MATAGALPA: along
    road to La Fundadora, L. O. Williams, A. Molina and T. P.
    Williams 24827 (F).

13. H. vaginalis Benth., Bot. Voy. Sulph. 171. 1844. Range: s.
    Mexico to n. S. A.
    NICARAGUA: DEPT. MATAGALPA: about 6-10 km. n. e. of Matagalpa,
    L. O. Williams, A. Molina and T. P. Williams 23823 (F).

DEPT. ZELAYA: Braggman's Bluff, F. C. Englesing 254 (F, US);
El Recreo, L. E. Long 65 (F); Montana Esquipulas, P. J. Shank
and A. Molina 4728 (F); Comarca del Cabo, Miguel Bikon, B. W.
Taylor 4554 (F).

14. H. wagneriana O. G. Peters. in Mart. Fl. Bras. 3 pt. 3: 12.
1890. Synonyms: Bihai wagneriana (Peters.) O. Ktze., Rev.
Gen. Pl. 2: 685. 1891. H. elongata Griggs. Bull. Torr. Bot.
Club 30: 653. 1903. Bihai elongata (Griggs) Griggs, Bull.
Torr. Bot. Club 31: 445. 1904.
NICARAGUA: DEPT. ZELAYA: Comarca del Cabo, San Mateo, cerca
de Rio Wawa, A. Molina 15077 (F).

TREATMENT OF NEW NAMES AND COMBINATIONS

The following new names and combinations are those belonging
to Central American species of Heliconia. The new taxa discussed
here are H. longiflora, H. hirsuta var. rubiflora and H. schiedeana
var. spissa. Their ranges include Nicaragua. A brief treatment of
the species H. schiedeana and H. hirsuta, are included with the
discussion of each variety.

Heliconia longiflora R. R. Smith, sp. nov.

Planta cannoidis, 2-5 m. alta. Lamina lanceolato-elliptica,
17-29 cm. longa, 5-7 cm. lata; apice acuminata. basi obtusa. utrique
viridis. Inflorescentia erecta, 10-15 cm. longa, glabra, breviter
pedunculata, rhachi leviter flexuoso. Bractae 4-9, aurantiacae,
glabrae, lanceolatae, non profunde cymbiformes, 4.5-9 cm. longae.
Bractae internae florales 4 cm. longae, deciduae, lanceolatae,
membranaceae, glabrae. Flores in bractearum axillis 3-8, albi ad
cremei, 1-1.5 cm. longe pedicellati; perianthium 5.5-6.0 cm longum,
glabrum. Fructus caeruleus, ca. 1 cm. diametro.

Slender plants 2-5 m. tall, cannoid habit. Leaf-blade lanceo-
late-elliptic, 17-29 cm. long, 5-7 cm. wide, long acuminate apex,
obtuse base; upper and lower surfaces green; petiole nearly absent
or leaf-blade sessile. Inflorescence erect, 10-15 cm. long,
glabrous; peduncle short, if present 0.3 cm. diameter; rachis
slightly flexuose, 0.2 cm. diameter. Branch-bracts 4-9, orange,
glabrous, lanceolate, shallow boat-shaped; lower branch-bracts 9 cm.
long; upper branch-bracts 4.5-7.0 cm. long; all branch-bracts 0.6 cm.
side width; internode between branch-bracts ca. 1 cm. Floral-bracts
few, deciduous, lanceolate, membranous, 4 cm. long, ca. 0.5 cm. wide,
glabrous. Flowers 3-8, white to cream, glabrous, 7 cm. long; perianth
5.5-6.0 cm. long; pedicel 1.0-1.5 cm. long. Fruit blue, subglobose to
3-sided, 0.8-1.0 cm. diameter. Seeds not seen.

Type: PANAMA. PROV. COLON: vicinity of Camp Pina, alt. 25 m., 11 July 1946, P. H. Allen 3590 (Holotype: US; Isotypes F, NY, UC) (Fig. 1).

Distribution: Edge of moist forest and along river margins, low altitudes. It appears to be restricted to Central America in the countries of Nicaragua, Costa Rica and Panama.

This species was previously included with H. aurantiaca, since it also possesses the orange branch-bracts. The flowers in H. aurantiaca are cream to yellow colored. In H. longiflora the flowers are white, and then change to cream color when they mature. The flowers of H. longiflora are nearly 2 cm. longer than those of H. aurantiaca. As noted in the key, H. aurantiaca possesses flowers which are normally shorter than or as long as the bracts. The flowers of H. longiflora are normally longer than the bracts.

Representative Specimens:

COSTA RICA: PROV. ALAJUELA: lowland rain forest between Los Chiles and Venecia, Llanura de San Carlos, alt. 100 m., 20 Feb. 1966, A. Molina R., L. O. Williams, W. Burger, B. Wallenta 17574 (F). PROV. PUNTARENAS: edge of forest between Golfo Dulce and Rio Terraba, alt. 30 m., Dec. 1947, A. F. Skutch 5397 (US); in forest, basin of El General, alt. 675-900 m., March 1940, A. F. Skutch 4761 (GH, NY).

NICARAGUA: DEPT. ZELAYA: open bush, vicinity of El Recreo, 12 Aug. 1947, L. E. Long 59 (F).

PANAMA: CANAL ZONE: n.w. part of Canal Zone, area w. of Limon Bay, Gatun Locks and Gatun Lake, 8 Nov. 1955, I. M. Johnston 1625 (GH); forests around Puerto Obaldia, San Blas coast, alt. 0-50 m., Aug. 1911, H. Pittier 4285 (GH). PROV. COCLE: La Mesa, 31 Aug. 1941, P. H. Allen 2692 (GH, MO). PROV. COLON: vicinity of Camp Pina, alt. 25 m., 11 July 1946, P. H. Allen 3590 (F). PROV. DARIEN: Rio Chico across from Boca de Tesca along the top of a ridge, 18 July 1962, J. A. Duke 5209 (MO); ascent of Cerro Pirre from Rio Pirre s. of El Real, 600-750 m., 11 Aug. 1962, J. A. Duke 5304 (MO); along Pam Am Highway between Pucro and Rio Punusa, 3 Aug. 1962, J. A. Duke 5303 (MO) Cana-Cuasi Trail (Camp 2) Chepigana district, alt. 2000 ft., 9 March 1940, M. E. Terry and R. A. Terry 1423 (F). PROV. PANAMA: ca. 7 mi. n. Cerro Azu on road to Cerro Jefe, elev. ca. 2600 ft., 13 Nov. 1965, K. E. Blum, R. K. Godfrey, and E. Tyson 1843 (FSU); on trail to Cerro Campana, 23 Aug. 1967, J. H. Kirkbride, Jr. and Sister Hayden 297 (MO).

Heliconia hirsuta L. f., Suppl. Syst. Veg. 158. 1781. Synonyms: Bihai hirsuta (L. f.) O. Ktze., Rev. Gen. Pl. 2:684. 1891. H. straminea (Griggs) Standley, Jour. Wash. Acad. Sci. 17:162. 1927.

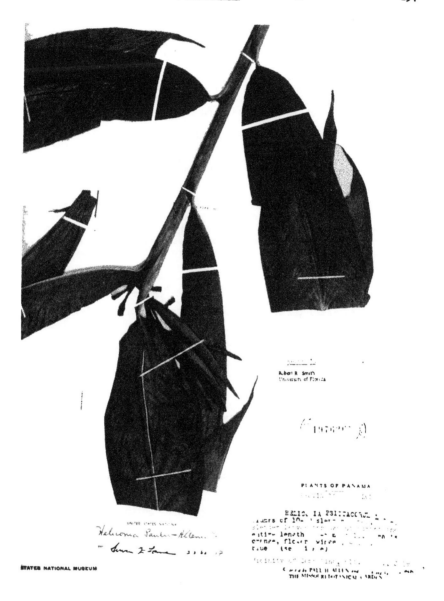

Fig. 1. Holotype of <u>H</u>. <u>longiflora</u> (US).

Bihai stramınea Griggs, Bull. Torr. Bot. Club 42:327. 1915.

    Slender plant, 1-3 m. tall; cannoid habit. Inflorescence
erect, nearly glabrous to pubescent, 6-9 cm. long. Branch-bracts
4-9, dark, brick-red or straw-yellow wıth greenish tinge, shallow
boat-shaped. Floral-bracts deciduous, ca. 1.5-1.8 cm. long.
Flowers 4-12 per branch-bract, yellow with green apices or completely red, ca. 3.5-4.5 cm. long; perianth 2.2-3.0 cm. long;
pedicel 0.4-1.8 cm. long, slightly pubescent to hirsute.

    Distrıbution: Moist thickets and woods, along banks of rivers
and edges of forest swamps. Heliconia hirsuta is common in the
northern part of South America, and extends up into Panama. The
varıety is found farther north in the Central American countries
of Belıze and Nicaragua.

    Heliconia hirsuta is not easily confused with other species
of Helıconıa, since it is of smaller stature and possesses hirsute
inflorescences. Heliconia hirsuta var. hırsuta has a very pubescent
perianth, and the var. rubiflora possesses a slightly to moderately
pubescent perıanth. The two varieties are similar in habit, and ın
occupying similar habitats.

Key to Varieties of Heliconia hirsuta

Plant 1-3 m. tall; perianth yellow with green tip, hirsute.........
var. hirsuta

Plant ca. 1 m. tall, perianth brick-red; slightly to moderately
hirsute........   var. rubiflora

Helıconıa hirsuta L. f. var. hirsuta

    Branch-bracts 5-9, red to yellow, lower branch-bracts 9-11 cm.
long. Upper branch-bracts 3-4 cm. long. Flowers 4-10 per branch-
bract, pale yellow with greenish tips, hirsute, 3.5-4.5 cm. long;
perianth 2.5-2.8 cm. long, hirsute; pedicel 1.0-1.8 cm. long, hirsute-
tomentose.

    Dıstribution: Extends throughout northern South America into
Panama.

Helıconıa hirsuta L. f. var. rubiflora R. R. Smith var. nov.

    Planta ca. 1 m. alta. Lamina 15-25 cm. longa, 5-6 cm. lata.
Bracteae 4-5, rubrae; bracteae infernae 6-11 cm. longae. Flores
ın bractearum axillis 4-8, rubri; perıanthium 2.5-3.0 cm. longum,
puberulum ad sparse hirsutum.

Plant ca. 1 m. tall. Leaf-blades 15-25 cm. long, 5-6 cm. wide.
Branch-bracts 4-5, red; lowest branch-bracts 6-11 cm. long, upper-
most branch-bracts 2.5-4.0 cm. long; all branch-bracts 0.4-0.7 cm.
side width. Flowers 4-8 per branch-bract, red, 3.5-4.0 cm. long;
perianth 2.5-3.0 cm. long, puberulent to slightly hirsute with
short whitish hairs.

Type: BELIZE. DIST. STANN CREEK: swamp places, Stann Creek,
alt. 20 ft., 19 Sept. 1967, W. A. Schipp 357 (Holotype: MO;
Isotypes F, MICH, NY, UC).

Distribution: Has been found in only two Central American
countries, Belize and Nicaragua.

Representative Specimens:

BELIZE: DIST. STANN CREEK: growing in open flats, Stann
Creek, 6 Dec. 1931, W. A. Schipp 835 (F, MICH, MO, NY, UC).

NICARAGUA: DEPT. ZELAYA: Comarca del Cabo, Francis Sirpi,
between Waspan and Puerto Cabezas, elev. 0-100 m., 15 Mar. 1971,
J. T. Atwood 4793 (MO, SEY); near Esperanza, Rio Grande, alt.
0-15 m., 10 April 1949, A. Molina R. 2127 (F); Comarca del Cabo,
rain forest, Miguel Bikon, 13 July 1972, S. B. Robbins 5863 b
(MO, SEY); Commarca del Cabo, Cororia Bush, 40-45 km. s.w. of
Waspan, rain forest, elev. 10-100 m., 21 Jan. 1970, F. C.
Seymour and J. T. Atwood 3725 (GH, MO, SEY); Comarca del Cabo,
swamp near Bilwaskarma, elev. 0-100 m., 14 Mar. 1971, F. C.
Seymour 4702 (SEY).

Heliconia schiedeana K., Linnaea 20:463. 1847. Synonyms: Bihai
schiedeana (kl.) O. Ktze., Rev. Gen. Pl. 2:685. 1891. H. hirsuta
Cham. and Schlect. in Linnaea 6:57. 1831. (non H. hirsuta L. f.).
H. pochutlensis Conzatti, Fl. Tanon Mex. 3:129. 1947.

Moderately stout plant, 1.5-3.0 m. tall, musoid habit.
Inflorescence erect, pubescent, 25-30 cm. long. peduncled; rachis
slightly flexuosed to straight, densely tomentose pubescent,
spiralled. Branch-bracts 6-12, dark-red, red-green yellow to
orange, shallow boat-shaped, may become strongly reflexed while
maturing. Floral-bracts 4-8, long ovate to wide lanceolate, 3-4
cm. long, 0.5-1.0 cm. wide, often pubescent on outer surface.
Flowers 6-21 per branch-bract, yellow densely villous; perianth
2.5-5.0 cm. long, villous; pedicel 1.0-1.5 cm. long, villous-
pilose. Fruit dark, subglobose to 3-sided, villous, ca. 1 cm.
diameter.

Heliconia schiedeana is not closely related to other species

of <u>Heliconia</u>. The outstanding characteristics of the species are
the pubescence of the inflorescence and the showy floral-bracts
extending out over the long thin branch-bracts. The fruits are
exposed on long, hairy pedicels, and not hidden within the branch-
bracts.

Key to the varieties of <u>H</u>. <u>schiedeana</u>

Branch-bracts red or red-green; perianth 3.5-5.0 cm. long........
var. <u>schiedeana</u>

Branch-bracts yellow or yellow-green; perianth 2.5-3.0 cm. long..
var. <u>spissa</u>

<u>Heliconia schiedeana</u> Kl. var. <u>schiedeana</u>

Inflorescence erect, pubescent, 15-45 cm. long, rachis
slightly to moderately flexuose. Branch-bracts 6-15, dull-red to
red-green. Flowers 6-21, pale yellow 3.0-6.5 cm. long; perianth
villous.

Distribution: Steep, moist slopes of ravines and mountains.
Endemic to Mexico.

<u>Heliconia schiedeana</u> Kl. var. <u>spissa</u> (Griggs) R. R. Smith, var nov.

Synonyms: <u>Heliconia</u> <u>spissa</u> Griggs, Bull. Torr. Bot. Club 30:
652. 1903. <u>Bihai</u> <u>spissa</u> (Griggs) Griggs, Bull. Torr. Bot. Club 31:
445. 1904.

Lamina oblongo-ovata, 60-75 cm. longa, 20 cm. lata, apice acuta,
basi roundata; petiolus longus. Inflorescentia erecta, 15-35 cm.
longa, pubescenti. Bractae 9-12, luteae ad aurantiacae. Flores in
bractearum axillis 10-17, lutei, 4 cm. longi; perianthium 2.5-3.0 cm.
longum, villosum.

Leaf-blade oblong-ovate, 60-75 cm. long, 20 cm. wide acute
apex, rounded base; petiole long. Inflorescence erect, pubescent,
15-35 cm. long; rachis red, nearly straight. Branch-bracts 9-12,
bright yellow to orange; middle branch-bracts 5-7 cm. long.
Flowers 10-17 per branch-bract, yellow, 4 cm. long; perianth 2.5-3.0
cm. long, villous. Fruit dark-yellowish color, subglobose to 3-
sided, 1 cm. diameter; seed 0.8 cm. long, 0.7 cm. wide; seedcoat
sclerified, rough somewhat wrinkled.

Type: GUATEMALA: DEPT. ALTA VERAPAZ: near finca Sepacuite,
30 March 1902, <u>O. F. Cook and R. R. Griggs 359</u> (US).

Distribution: Rain forest, thickets near river. s. Mexico to Nicaragua.

Representative Specimens:

BELIZE: DIST. BELIZE: in high forest, churchyard on Sibum River, Jan.-June 1936, <u>C. L. Lundell 6950</u> (MICH, NY). DIST. EL CAYO: between El Cayo and Benque Viejo, 15 Feb. 1931, <u>H. H. Bartlett 11511</u> (MICH).

GUATEMALA: DEPT. VERAPAZ: thicket along river Pantin, below Tamahu, alt. ca. 600 m., 5 April 1939, <u>P. C. Standley 70571</u> (F); damp forested slopes, along road between San Crestobal Verapaz and Chexoy, alt. 1200-1300 m., 19 Feb. 1942, <u>J. A. Steyermark 43899</u> (F). DEPT. PETEN: Tikal, 12-15 April 1931, <u>H. H. Bartlett 12638</u> (GH, MICH); Santa Teresa, Subin River, 13 April 1933, <u>C. L. Lundell 2890</u> (F, GH, MICH); Cerro Ceibal, alt. 75-150 m., 30 April 1942, <u>J. A. Steyermark 46089</u> (F).

HONDURAS: DEPT. COMAYAGUA: Quebrada Montanuelas, alt. 1400 m., 18 July 1962, <u>A. Molina R. 10873</u> (F); Palm Grove, San Louis, alt. 2500 ft., 5 May 1933, <u>J. B. Edwards 598</u> (F, GH). DEPT. EL PARISO: Montana Cifuentes, 15 March 1963, <u>A. Molina R. 11404</u> (F).

MEXICO: STATE CHIAPAS: near Laguna Ocotal Grande, elev. ca. 950 m., 20 July - 20 Aug. 1954, <u>R. L. Dressler 1584</u> (GH).

NICARAGUA: DEPT. NEUVA SEGOVIA: vicinity of Jalapa, 23 Dec. 1973, <u>J. T. Atwood, S. A. Marshall, and D. A. Neill 6773</u> (SEY).

ACKNOWLEDGEMENTS

The author wishes to thank Mr. Frank Seymour and Dr. Gilbert Daniels for reading the manuscript and offering valuable comments. Additional gratitude is extended to Dr. Daniels for providing the photograph of the holotype of Heliconia longiflora.

LITERATURE CITED

Smith, Robert R. 1968. A Taxonomic Revision of the Genus Heliconia in Middle America. Ph.D. Thesis. University of Florida. Gainesville, Fla.

Seymour, Frank C. 1973. Cassia in Nicaragua. Phytologia Vol. 27, No. 5:330-348.

BOOK REVIEWS

Alma L. Moldenke

"FOOD AND AGRICULTURE" edited by Dennis Flanagan and the Board of
Editors of "Scientific American Books", iv & 154 pp., illus.,
W. H. Freeman & Company, San Francisco, California 94104.
1976. $9.00 hardcover, $4.95 softcover.

The 12 chapters in this valuable, timely book originally appeared as articles in the September 1976 issue of "Scientific
American". It is the 27th issue published annually and separately
on a single focus. As in all of these excellent works, the illustrations are of very great value. In this one there are 102
impressive diagrams and special computer-enhanced, multispectral
scanning LANDSAT earth-resources satellite photographs.

The chapters are written by different authors with scientific
and practical experience. Jean Mayer indicates that about 1/3rd
of mankind, mainly in Asia and Africa, is undernourished. Agriculture in the United States, which is really technological
agribusiness so advanced that one farmer's efforts feeds 50
others, offers some realistic prospects for feeding some of the
3 billion additional people who will probably join the earth's
population by the end of the 20th century. Important!

"THINKING LIKE A MOUNTAIN: Aldo Leopold and the Evolution of an
Ecological Attitude Toward Deer, Wolves, and Forests" by
Susan L. Flader, xxv & 284 pp., illus., University of Mis-
souri Press, Columbia, Missouri 65201. 1974. $12.50.

When young, Leopold "thought that because fewer wolves meant
more deer, that no wolves would mean hunters' paradise" until he
watched that first old wolf he had shot die. Years later he was
able to conclude "Only the mountain has lived long enough to lis-
ten objectively to the howl of a wolf." Leopold used the expres-
sion "thinking like a mountain to characterize objective or ecological thinking: it should not be viewed as a personification".

Aldo Leopold (1887—1948) has endeared himself to the many
naturalist-oriented readers with "A Sand Country Almanac" (1949)
and will continue to do so. He has been a professional conserva-
tionist, wildlife manager, etc. in our southwestern national
forests and in Wisconsin. He "insisted that the evolution of a
'land ethic' was an intellectual as well as emotional process"
involving the ecological leaders and the general public in a
philosophy of a natural self-regulating system.

The author has carefully studied available source material to
produce this interestingly oriented professional biography.

"LE CONTRÔLE DE L'ALIMENTATION DES PLANTES CULTIVÉES" Volumes I
and II, edited by Pál Kozma with the collaboration of D.
Polyák & E. Hervay, 1014 pp., illus., Akadémiai Kiadó, Budapest V, Hungary. 1975. $60.00.

In the fall of 1972 the Third European and Mediterranean Collo-
quium on this topic met in Budapest with 44 specialists from seven
socialist countries, 31 from eleven western countries and 63 from
the host country. About 100 basically well documented papers in
Russian, German, French, Spanish, Italian, and English, each with
its own language summary and one in at least one of the other languages, are grouped in the following sections: (1) general problems
and methods, (2) grain, forage and industrial crops, (3) vegetable,
ornamental and forest plants, (4) mediterranean subtropical cultivation, (5) viticulture, and (6) fruit culture. Throughout the
role of macro- and micro-minerals in plant nutrition, their absorp-
tion rhythms, dilution, migration, accumulation, effects and sympt-
oms of excessive or of insufficient amounts, in photosynthesis, in
enzyme and in protein synthesis are considered analytically and
experimentally.

Much valuable information is contained within the covers of these
two fine volumes.

"MALAYAN WILD FLOWERS — DICOTYLEDONS" by Murray Ross Henderson,
ii & 478 pp., illus., Malayan Nature Society, Kuala Lumpur,
Malaysia. Reprint 1974.
"MALAYAN WILD FLOWERS — MONOCOTYLEDONS" by Murray Ross Henderson,
357 pp., illus., Malayan Nature Society, Kuala Lumpur, Malay-
sia. Reprint 1974. The 2-volume set is available outside of
Asia through Otto Koeltz Science Publishers, D-624 Koenigstein/
Taunus, West Germany. 86 DM.

Much of this material was first published serially in 1949, 1950
and 1951 in the Malayan Nature Journal. Since the first non-
technical work on the local flora was Corner's "Wayside Trees of
Malaya", the author of these volumes instead concentrates upon
the "majority of the smaller plants to be found not only in the
forest, but by the roadsides, on the seashores and in the waste
spaces of Malaya." There are 625 figures of line drawings, often
multiparted, that illustrate accurately. The keys are readily
workable and the text well written by a highly competent author.
Since this work was prepared over 25 years ago, it could not record recent nomenclatural changes.

"FAUNA AND FLORA OF THE BIBLE - Helps for Translators" Volume II
by the Committee on Translations of the United Bible Societies, xv & 207 pp., illus., United Bible Societies, London
EC4V 4BX, England. 1972 paperbound.

After an explanatory introduction and giving a standard set of

abbreviations the biblical animals and then the biblical plants
are arranged alphabetically by their English names, their scientific names, transliterations of the textual Hebrew and Greek names,
problems of specific identifications, features of behavior and
appearance especially in regard to symbolic and figurative usage,
and the corresponding scriptural references.

There are several "Difficult Passages" discussed, as in case
of the identity of the "behemoth". Ficus sycomorus L. has long
been known from the Bible land area and it has been realized for
a couple of centuries that the King James translation as "syca-
more" is incorrect as the authors of this work carefully explain.
It would have been helpful if they would have recommended
"sycomore" or "sycomore fig" as the common name. The book is
well indexed. This study should prove of real value to many more
folks than just biblical translators.

"A GARDEN OF PLEASANT FLOWERS" or "PARADISI IN SOLE: PARADISUS
    TERRESTRIS" by John Parkinson, x & 626 pp., illus., Facsimile
    Edition of Dover Publications, Inc., New York, N. Y. 10014.
    1976. $25.00 oversize.

This is an unabridged republication of the 1629 first edition
of which there can be only a few precious copies extant. A single
page Publisher's Note has been added giving pertinent data about
this famous book, including the parody on the author's name shown
at the bottom of the Garden of Eden* cartouche. PARADISI IN SOLE =
Park in Sun = Parkinson.

The text in readable English even today discusses about 1000
plants from all over the then-known world that could be grown in
English gardens and has 108 full-page plates illustrating 812 of
these plants. The bulk of the book comprises a 'Garden of
Pleasure' with fragrant herbs, attractively flowering herbaceous
plants and shrubs. There is a 'Kitchen Garden' of culinary herbs
and vegetables. There is also an 'orchard' with its trees, shrubs
and vines producing edible fruit.

There is an index of Latin plant names (pre-Linnean, of course),
one of English plant names, and a table of adjunct medicinal uses.

How fortunate it is that many, many people can now join the
English Queen Henrietta Maria (to whom this work is dedicated) in
perusing at leisure in many private and public libraries this
historically and horticulturally significant book.

*The publishers have used much of this drawing as a cover for their
recent reprint of "Bible Plants for American Gardens".

PHYT

Designed to expedite botanical publication

l. 36   August 1977   No 4

CONTENTS

Published by Harold N. Moldenke and Alma L Moldenke

303 Parkside Road
Plainfield, New Jersey 07060
USA

Price of this number, $3, per volume, $9.75 in advance or $10.50 after close of the volume, 75 cents extra to all foreign addresses, 512 pages constitute a volume; claims for issues lost en route must be made immediately after receipt of the next issue.

UNA NUEVA ESPECIE DE
CROTALARIA (LEGUMINOSAE) DEL VALLE DE MEXICO

Judith Espinosa G.
Laboratorio de Botánica Fanerogámica
Escuela Nacional de Ciencias Biológicas
Instituto Politécnico Nacional
México 17, D.F.

Al hacer la revisión del género Crotalaria de la
familia Leguminosae para la Flora del Valle de México,
se encontró un ejemplar determinado como Crotalaria
angulata Mill., cuyas características no corresponden
a las de esta especie, actualmente considerada como
sinónimo de C. rotundifolia var. vulgaris Windler.

Después de revisar el trabajo de Windler*, llega-
mos a la conclusión de que dicho ejemplar no corresponde a ninguna de las especies ahí descritas, por lo que
pensamos que se trata de una nueva especie.

Crotalaria rzedowskii Espinosa sp. n.

Planta herbacea, annua, erecta, 20-40 cm alta; ra-
dix palaris, 1.5-3 mm diametro in parte crassissima;
caulis cylindricus, 2 mm diametro, e basi ramosus, pubescens, trichomatibus patentibus, 1-1.5 mm longis;
stipulae una vel duae per plantam, aliquod nullae, fo-
liaceae, decurrentes, lobis nullo modo acutis nec patentibus sed brevibus et rotundatis; folia simplicia,
elliptico-oblonga, 1.5-4.5 cm longa, 1-2 cm lata, apice
rotundato, marginibus integris, basi cuneata, utrinque
pubescentia, trichomatibus 1-1.5 mm longis, petiolo 2
mm longo; inflorescentiae oppositifoliae, pedunculo 2-
8 cm longo, floribus 4-5 per racemum; bracteae lineari-
lanceolate, 3-5 mm longae, 0.5 mm latae; pedicellus 3-
4 mm longus, pilosus; bracteolae lineares, 3-5 mm longae, 0.5 mm latae; calyx 10-14 mm longus, pilosus, trichomatibus adpressis 1 mm longis, tubo 3-4 mm longo;

---

* Windler, D. R. Systematic studies in Crotolaria
sagittalis L. and related species in North America
(Leguminosae). Phil. D. Thesis. Dept. of Botany.
University of North Carolina. Chapel Hill. 1970.

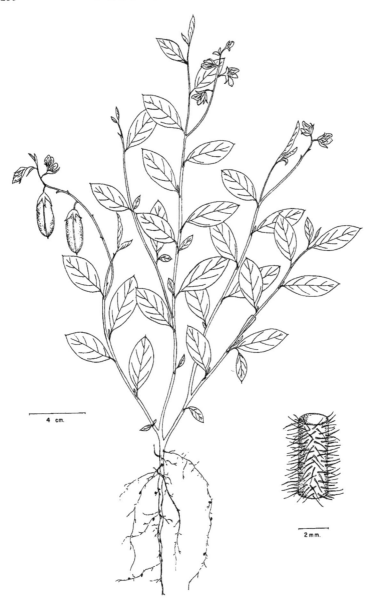

4 cm.

2 mm.

Fig. 1.- Crotalaria rzedowskii Espinosa sp. n.

corolla lutea, vexillum vittis vinosis, 10 mm longum,
parte superiore 5 mm latum; stamina 10, 5 antheris 1 mm
longis, cetera 5 antheris 0.5 mm vel minus longis; sty-
lus 7 mm longus; legumen 2-3 mm longum, 1 cm diametro,
glabrum, juventute viride-flavidum, demum nigrum.

Planta herbácea, anual, erguida, de 20 a 40 cm de
altura; raíz pivotante de 1.5 a 3 mm de diámetro en su
parte más gruesa; tallo cilíndrico, de 2 mm de diámetro, ramificado desde la base, pubescente, con los tricomas extendidos, de 1 a 1.5 mm de largo; estípulas
una o dos en toda la planta, a veces ausentes, foliá-
ceas, decurrentes y sin lóbulos agudos extendidos, sino breves y redondeados, hojas simples, elíptico-oblon-
gas, de 1.5 a 4.5 cm de largo por 1 a 2 cm de ancho,
ápice redondeado, borde entero, base cuneada, haz y envés pubescente con tricomas de 1 a 1.5 mm de largo, pe-
cíolo de 2 mm de largo; inflorescencias opositifolias
con el pedúnculo de 2 a 8 cm de largo, flores 4 a 5 en
cada racimo; bracteas linear-lanceoladas, de 3 a 5 mm
de largo por 0.5 mm de ancho; pedicelo de 3 a 4 mm de
largo, piloso; bracteolas lineares de 3 a 5 mm de largo por 0.5 mm de ancho; cáliz de 10 a 14 mm de largo,
piloso, con los tricomas adpresos de 1 mm de largo, tubo de 3 a 4 mm de largo; corola amarilla, estandarte
listado de rojo-guinda, de 10 mm de largo por 5 mm de
ancho en la parte superior; estambres 10, 5 de ellos
con las anteras de 1 mm de largo y los otros 5 con las
anteras de 0.5 mm de largo o menos; estilo de 7 mm de
largo; legumbre de 2 a 3 cm de largo por 1 cm de diámetro, glabra, de color verde-amarillento cuando no está madura y negra después. Semillas de color verde ama-
rillento, brillantes, de 2 mm de largo. Florece de sep-
tiembre a diciembre y fructifica de noviembre a febre-
ro.

Habitat: Es una planta escasa en laderas húmedas
del sur del Valle de México con vegetación de encinar,
en las que se ha colectado entre 2400 y 2700 m de al-
titud.

Tipo: MEXICO: DISTRITO FEDERAL: DELEGACION DE
TLALPAN: Cerca de Xicalco, sobre la carretera México-
Cuernavaca; alt. 2700 m; en ladera andesítica con vegeta
ción de encinar; 25 IX 1966; J. Rzedowski 23205 (ENCB).

Otros ejemplares examinados y tomados en cuenta
para la descripción: MEXICO: DISTRITO FEDERAL: DELEGA-
CION DE TLALPAN: Cerca del Mirador, a un lado de la ca
rretera vieja México-Cuernavaca; alt. 2700 m; encinar
perturbado; 4 VIII 1972; J. Espinosa 1041 (ENCB).
DELEGACION DE XOCHIMILCO: Santa Cecilia; alt. 2550 m;
bosque de encinos; 3 X 1976; A. Ventura 2226 (ENCB).
Cerro de Santa Cecilia; alt. 2500 m; en ladera con bos
que de encino; 6 XI 1976; A. Ventura 2372 (ENCB).
Rancho del Conejo; alt. 2400 m; en bosque de encino;
30 I 1977; A. Ventura 2552 (ENCB).

Esta especie pertenece a la subsección Iocaulon
de la sección Calycinae del género Crotalaria (Windler
op. cit.). La presencia de inflorescencias opositifolias y de pubescencia formada de tricomas extendidos,
la aproximan a C. sagittalis L., pero sobre todo a C.
rotundifolia var. vulgaris Windler. De la primera di-
fiere en la escasez de estípulas y la forma de las mis
mas. De la segunda, por ser una planta anual erguida y
no perenne y debumbente.

MANCOA ROLLINSIANA, UNA ESPECIE NUEVA DE CRU-
CIFERAS ENCONTRADA EN EL VALLE DE MEXICO

Graciela Calderón de Rzedowski[+]
Laboratorio de Botánica Fanerogámica
Escuela Nacional de Ciencias Biológicas
Instituto Politécnico Nacional
México 17, D.F.

En junio de 1975 colectamos, en los alrededores
de la Presa Jaramillo, municipio de Pachuca, Hidalgo,
una pequeña planta semi-rastrera, perteneciente a la
familia de las Crucíferas y que no habíamos encontra-
do con anterioridad. Al estudiarla con detenimiento,
se llegó a la conclusión de que corresponde al género
Mancoa, pero, al no coincidir con las descripciones
de ninguna de las especies conocidas, se sospeché que
podría tratarse de una nueva entidad. Posteriormente
se envié una muestra de la planta al Dr. R. C. Rollins, director del Gray Herbarium de la Universidad
de Harvard, quien tuvo la amabilidad de examinar el
material y coincidió con esta opinión.

Mancoa rollinsiana Calderón sp. n.

Herba perennis, semirastrera, 5-20 cm diametro,
caules fructiferi usque ad 15 cm alti, pubescentia
trichomatibus stellatis et simplicibus sparsioribus;
caules plures e basi oriundi; foliorum laminae ambito

[+]Trabajo parcialmente subvencionado por el Consejo Nacional de Ciencia y Tecnología, en el marco del proyecto "Flora y vegetación del Valle de México".

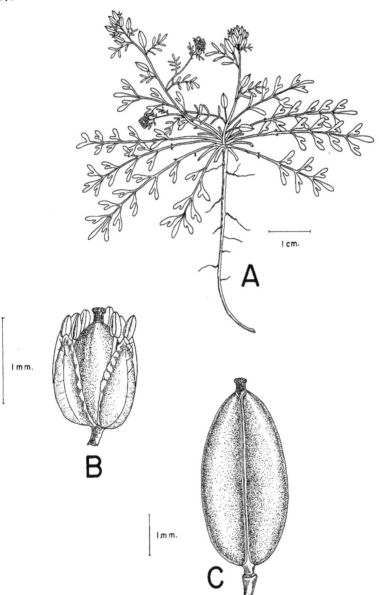

Mancoa rollinsiana Calderón. A. Aspecto general de
la lanta. B. Flor. C. Fruto.

generali oblongae, pinnatisectae usque ad bipinnati-
sectae, lobulis ovatis usque ad linearibus, folia
basalia usque ad 3.5 cm longa petiolo incluso, usque
ad 1 cm lata, caulina similia sed minora, sessilia
et plerumque auriculata; inflorescentiae densae, co-
rymbiformes; flores albidi; sepala oblonga, 1.2 mm
longa, viridula vel rubella, margine scariosa, decidua; petala nulla vel lineari-oblanceolata, usque ad
1.2 mm longa, alba, inconspicua et decidua; infruc-
tescentiae elongatae, racemosae, 3-8 cm longae; si-
liquae ascendentes, glabrae, oblongae, 5 mm longae,
1.5-2 mm latae; semina 35-40 per loculum, 0.5 mm
longa, funiculi filamentosi, cum septo post dehis-
centiam persistentes.

Hierba baja, perenne, más bien semi-rastrera,
extendiéndose radialmente (de 5 a 20 cm de diámetro),
aunque las inflorescencias con tendencia a elevarse
del suelo hasta unos 15 cm de alto, con pubescencia
principalmente de tricomas estrellados, combinados
con escasos pelos simples; raíz pivotante larga y
delgada; tallos varios, saliendo de la base, de entre las hojas basales; láminas de contorno oblongo
en general, las hojas basales pinnatisectas a bipi-
nnatisectas con los lóbulos por lo general ovados a
lineares, hasta de 3.5 cm de largo (inclusive el pe-
ciolo que mide alrededor de la mitad del largo total), por un máximo de 1 cm de ancho, semejantes a
las caulinas, pero estas últimas son sésiles, con
frecuencia auriculadas y de dimensiones menores
(1.5 cm de largo por 0.8 cm de ancho); inflorescencias, densas, corimbiformes; flores blanquecinas,
frecuentemente con una bráctea foliácea en la base;
sépalos oblongos, con el ápice algo agudo, inflexo
en el botón, poco pubescentes, de 1.2 mm de largo,
verdosos o algo rojizos, con el borde escarioso, deciduos; pétalos ausentes o linear-oblanceolados, in-
conspicuos y deciduos (a veces hasta de 1.2 mm de
largo), blancos; infrutescencias alargadas, en raci-
mos de 3 a 8 cm de largo, más bien ascendentes, silicuas glabras, oblongas, angostándose hacia el ápice, de unos 5 mm de largo por 1.5 a 2 mm de ancho,
infladas pero a la vez algo comprimidas en el senti-

do contrario al septo, estilo de 0.3 mm de largo por
0.3 mm de ancho; semillas 35 a 40 en cada lóculo,
ovadas a anchamente elípticas, de color café claro,
de 0.5 mm de largo por 0.3 mm de ancho, insertas sobre un funículo filamentoso, tortuoso, más largo que
la propia semilla y que persiste al igual que el
septo después de la dehiscencia.

Tipo: MEXICO: HIDALGO: Alrededores de la Presa
Jaramillo, cerca de Cerezo, municipio de Pachuca; en
suelo húmedo en pradera de la cola de la presa; alt.
2800 m; 22-VI-1975; Rzedowski 33312 (ENCB; isotipo
en GH). Ejemplares en flor y fruto.

Mancoa es un grupo afín a Capsella y algunas de
sus especies fueron inicialmente descritas bajo este
último nombre genérico. La distribución conocida de
Mancoa se restringe al Nuevo Mundo y Rollins (Contr.
Dudley Herb. Vol. 3, No. 6: 191-196. 1941) reconoce
4 especies sudamericanas y 3 norteamericanas. De las
últimas, todas existen en México y la que se describe aquí extiende más al sur el área conocida del género en Norteamérica.

Evidentemente M. rollinsiana está más relaciona-
da con M. mexicana Gilg & Muschler, de San Luis Potosí, que con cualquier otra especie. Estos dos taxa
se distinguen entre sí principalmente en que:

1.- Mancoa mexicana es planta más densamente pubescente y con los tricomas mayores que los propios
de M. rollinsiana.

2.- Los frutos de M. mexicana son densamente pubescentes y más anchos que los de M. rollinsiana,
cuyas silicuas son totalmente glabras.

3.- El estilo en el fruto de M. mexicana se
aprecia más angosto y largo que en el de M. rollinsiana.

Tales características se resumen en el siguiente
cuadro:

	Pubescencia de la planta	Diámetro de los tricomas (mm)	Fruto		Estilo Largo y ancho (mm)
			Ancho (mm)	Pubescencia	
M. mexicana	densa	± 0.4	2-3	densa	0.4x0.2
M. rollinsiana	medianamente densa	± 0.2	1.5-2	nula	0.3x0.3

El nombre de la especie está dedicado al Dr. Reed C. Rollins, quien por años se ha ocupado del estudio de la familia Cruciferae incluyendo muchos elementos de la flora mexicana y con quien estamos en deuda por su ayuda en la resolución de varios problemas que se han presentado al preparar las Crucíferas para la Flora del Valle de México.

ADDITIONAL NOTES ON THE GENUS VERBENA. XXVI

Harold N. Moldenke

VERBENA OFFICINALIS L.

Additional bibliography: Sweet, Hort. Brit., ed. 1, 1: 325 (1826) and ed. 3, 553, 1839; Moldenke, Phytologia 36: 221, 235, & 247--250. 1977.

Additional vernacular names reported for this species are "èrbo dé mèrvèyo", "èrbo dé Nouéstro-Damo", "erculania", "erva de la mivuza" [=jaundice-herb], "escalaurus", "eseberus", "exuperans", "exupra matricalis", "eysemchrawl", "ferria", "flegwurt", "fleur de madame", "gerabotamm", "gerobotanis", "grünkraut", "hardiizer", "herba sagminalis", "herba sancti Johannis", "herba sanguinalis", "herba verminata", "herbe à la croix", "herbe à tous maux", "herbe au chat", "herbe de l'effort", "herbena", "herculana", "hiera", "hierobotamm", "hraetelwyrt", "iiserhard", "iiserkruid", "isanina", "isarna", "iserenhard", "isinina", "isnwurze", "Juno's teares", "kerckkruid", "kroazik", "lanzaouenn ar groaz [=herb of the cross], "lerepontrina", "lirobotamm", "lustago", "menthe de chat", Mercurie's moistblood", "militaris", "mormèno", "nymphaea", "palumbaris", "peltoclotis", "peltodotes", "pijounièro", "qanna-biyyé", "recia", "riz' el hhamân" [=pigeon's food], "sabiarella", "sacra frondis", "sacralis", "sacratimen", "sagmen"m "sagmina", "sanguinaria", "thiabsenti", "trixago", "varlenn", "varvan-na", "varvègne", "varvéna", "varvéne", "varvèno", "vèlvône", "vènèré", "vèrbèna", "verbana columbina", "verbena mascula", "verbena recta", "vèr bèn ày'", "verbene", "verbien", "verbigena", "vèrléne", "vèrlin-ne", "vermaine", "vèrmèno", "vermina", "verminaca", "verminacia", "verminatio", "verpidion", "vertipedium", "vertipo-dum", "vertiroedum", "verveine droicte", "vèrvéle", "vèrvèlo", "vervenn", "vervine", "vèrvin-ne", "vèrvouin-ne", "virvoni", "vorouéne", "vouorvéno", "vratour", "vrèvin-na", "werbinádj", "ysendeck", and "yzerne". Quite a few of these names go back to classical times and to the Middle Ages. This brings to 421 the number of vernacular names thus far listed by me for this plant.

Don (1830) calls V. officinalis the "official vervain", V. sororia the "sister vervain", and V. spuria the "spurious vervain" — the first of these he admits as native to England, the second he says was introduced into English gardens from Nepal in 1824, and the third as introduced from "N. Amer." in 1731. Sweet calls the same three taxa the "common", "sister", and "jagged-leaved" vervain and gives 1823 as the date of introduction of the second.

Rolland (1910) also tells us, quoting various (mostly old) authors, that "cette herbe sacrée était cueillie à Rome, sur le Capitole par les Féciaux. Celui qui la portait était appelé 'verbenarius'." Of the name, "fleur de madame", he notes that this is in contradistinction to "fleur de monsieur", which is applied to Sisymbrium officinale. He continues: "L'érbo de la vérméno Fo coulá lou sanc

274

séns doubri la véno = la verveine fait couler le sang sans qu'on
ouvre la veine.....La verveine passe pour guérir nombre de mala-
dies et en biens des endroits pour garantir des sorciers celui qui
la porte sur soi. On emploie cette plante comme détersive, on dit
qu'elle mang le sang.....Prenes chaque matin, pendant neuf jours,
un morceau de pain et une gousse d'ail; puis allez les déposer
devant un pied de verveine. Cela fait vous aures soin d'uriner sur
le tout; c'est un remède infaillible contre les fièvres quartes....
Pisser sur la verveine porte bonheur.....Pour morsure de serpent,
ardez la vervainne et en faites poudresi la metez sur la morsure,
si garira. Et qui portera la vervainne sus soy jamais serpent ne
le grevera......Les vieilles femmes vont chercher la verveine dont
elles se serviront comme remède, au clair de la lune et en marchant à reculons....

"Le sorcier, qui veut savoir quelle maladie a celui qui vient le
consulter, cueille, en décours de la lune, trois branches de verveine qu'il laisse macérer pendant trois jours et troi muits dans
du vinaigre. Au lever de la lune, il reconnait, à la manière dont
les feuilles sont antrelacées, de quelle maladie son client est
atteint.....Sur les vertus médicinales de la verveine, voyez encore:
'Dialogue des créatures' (1482)....Pour n'estre point las en allant,
prens verveine cueillie la veille de la Saint-Jehan et la porte sur
toy.....Pour qu'un cheval ne se lasse pas en courant attaches lui la
grosse dent du loup en marchant et pendes lui au cou des racines de
verveine et d'armoise.....Pour escalader facilement les montagnes,
les armaillis et surtout les garçons de chalet mettent à leur jar-
retière un rameau de verveine qu'ils appellent 'vérvéna à corre' =
verveine à courir.....Pour faire dix lieues à l'heure vous appli-
ques sur la rate un amplâtre composé de divers ingrédients parmi
lesquels sept feuille de verveine......

"La verveine donne de la force aux lutteurs....Au Sabbat les
diablotins se font des jarretières de verveine pour marcher sans
fatigue.....Pour que les pommiers produisent beaucoup, on met un
brin de verveine dans les branches.....Si on fromme les poules avec
de la verveine sauvage, on est assuré de les vendre un bon prix...
La femme qui a perdu son lait, doit, pour le recouvrer, cueillir
trois sommités de branches de verveine, en récitant le 'Pater nos-
ter' el l'Ave Marie' et les porter sur soi....."un brin de verveine
porté constamment sur soi, rend chaste.....Cueillie et brûlée le
jour de la Saint-Jean la v.guérit les gernies. Nos paysans disent
ironiquement que, dans sa cendre, se trouve le 'gekkensteen'
(=pierre des fous)......

"Les femmes blanches, apparitions nocturnes, présentent aux per-
sonnes qu'elles rencontrent une branche de chêne ou d'herbe de la
croix (verveine). Si l'on accepte ce talisman, on sera doué d'aut-
ant d'années de puissance et de joie que la branche a de feuilles;
mais au bout de ce temps, votre âme appartiendra au démon...Quand
on veut acquérir l'affection d'une personne, on lui présente de la
verveine......

"Conjuration pour se faire aimer. Il faut par trois vendredis
à duit heures du matin faire autour d'une verveine trois tours à
rebours et bénédiction de la main gauche; et le dernier vendredi

l'arracher de la main gauche et en la cueillant il faut dire ces
mots: 'o pega vervena, o pega, o pega, Lucia vervena, Lucia vervena, Lucia, o Luna, Luna'. Puis it faut faire poudre de cette
verveine, en disant: 'Je te conjure aux noms de Vénus et de Cupi-
don, du Soleil et de la Lune que celle de toi je toucherai ne
puisse nul autre aimer que moi et n'aime comme toi même'. Puis
en touchant la fille, dites: 'Audi filia (ici le nom de la fille)
et inclina aurem tuam et obliviscere populum tuum et domum patris
tui et sequere me'.....

"Autre incantation pour se faire aimer: Le premier vendredi de
la nouvelle lune, il faut avoir un couteau neuf et aller cueillir
une verveine. Il faut se mettre à genoux, la face tournée vers
le soleil levant et, coupant la date herbe avec le couteau, dire:
'Sara isquina safos; je te cueille, herbe puissante, afin que tu
me serves à ce que je voudray'. Puis vous vous léverez sans re-
garder derrière vous. Étant dans votre chambre vous la ferez
sécher et pulvériser et vous ferez avaler cette poudre à la personne.....

"Autre incantation pour se aimer: Prenez de la verveine que
vous pilerez et en frotterez le talon de votre main gauche, puis
avec cette main, vous formerez un signe de croix à votre front,
ensuite à celui de la fille, en disant: 'Cathos, que ton désir
seconde au mien comme celui de Saint-Joseph avec Marie!'.....

"Le verveine prise le soleil étant en Aries, avec de la graine
de pivoine d'un an, mises en poudre, si la poudre est mise entre
deux amants, aussitôt ils auront querelle......

"Language des fleurs. — La verveine signifie: ne me laisses
pas pour une autre....Un brin de verveine sauvage, offert à une
fille, équivant à une déclaration....La verveine symbolise les
enchantements....La verveine symbolise l'inspiration, la poésie."

Gibbs (1974) reports cyanogenesis absent from the leaves and
seeds of this species, leucoanthocyanin absent from the leaves,
and syringin absent from the stems; in addition, the Ehrlich test
gave negative results in the leaves.

Keys (1976) says that in China the leaves are used medicinally.
The taste is bitter. The "Plant contains an essential oil (com-
prising citral, geraniol, limonene, verbenone), invertin, a bitter
principle, verbenalol ($C_{11}H_{14}O_5$; long prismatic needles; m.p. $133°$;
soluble in ether, acetic acid; poorly soluble in water) and the
glycoside verbenalin ($C_{17}H_{25}O_{10}$; bitter needles; m.p. $178°$; freely
soluble in water; slightly soluble in alcohol, acetone; insoluble
in chloroform, ether). Verbenalin in frogs produces mucosal ex-
coriation. Prescribed as emmenagogue. Dose, 19—18 gm."

Sutherland refers to V. officinalis as "frequent" on steep
slopes in Ethiopia; the corollas are described as having been "purple" on his no. 309, "bluish-violet" on Westra & Rooden 189, and
"blue" on Beach 5110. Beach describes the plant as "a vigorous
much-branched weed". On Sebald 381 the corollas were "lilac" in
color when fresh.

The Fitter work cited in the bibliography of this species is
dated "1974", but was not actually issued until February 17, 1975.

It is also worth noting here that although Bock (1522) uses the name, Verbena foemina, in his text, his accompanying illustration very plainly is meant to depict V. officinalis.

Stokes (1812) gives some interesting sidelights about this species, telling us that Linnaeus believed the species "to have been wafted by the sea to Sweden". He quotes Miller as saying that it is "Never found above a quarter of a mile from a house". He states that Curtis, Lightfoot, and Scopoli classified it "after Mentha at the end of the order Gymnospermia of the class Didynamia, considering as a genus connecting the Personatae to the Verticillatae", while Hudson placed it "between Nepeta and Mentha, apparently regarding it with Ray as really belonging to the Verticillatae, in which disposition he has been followed by Walcott, Sibthorpe, and Smith.....where it is acknowledged to be erroneously placed in the order Gymnospermia of the class Didynamia." He affirms that Ray described the corollas as pale-blue, while Bauhin called them pale-purplish. He goes on to say: "Root perennial according to Ray and Curtis, Sibth. Salisb. W. and Smith; biennial according to Huds, Hort. kew. and Donn; annual according to Boerh. Linn. and Lightf. Are we hence to conclude that it is sometimes perennial, sometimes biennial and sometimes annual, or to learn how little dependence in investigating plants is to be placed on the marks even of cultivating botanists, when we observe Curtis, Auton and Boerhaave holding opposite opinions?"

Loessner (1912) cites Seler 3460 from Atascosa County, Texas, as "V. officinalis L. forma?", but the plant represented is almost certainly V. halei. The Curtiss 677 and Rugel 121, cited below, were previously regarded by me as V. domingensis Urb., but I feel now that they are probably a form of V. officinalis (or even perhaps V. halei Small). This may prove true of most, if not all, of the Cuban material previously cited under V. domingensis, a species very likely endemic to Hispaniola.

The Herb. Zuccarini s.n. [Hort. bot. Monac.], cited below, is a mixture with xV. baileyana Moldenke. The Meebold 12840, distributed as V. officinalis, actually is V. brasiliensis Vell., McNeal 925 is V. californica Moldenke, Eggers 2175 is V. domingensis Urb., Karwinski s.n. is V. ehrenbergiana Schau., D. Dickinson s.n. [June 8, 1918], Lindheimer 1076 & s.n. [Galveston, May 1843], Pickett & Bot. Class 60, Thomas & Gremillion 2487, and Wurzlow s. n. [Sept. 15, 1917] are V. halei Small, Mukherjee s.n. [12.9.74] is xV. hybrida Voss, M. E. Jones 2215 is V. lasiostachys Link, Clemens 42170, Meebold 7818, 8304, & 12839, and Schlieben 7691 are V. litoralis H.B.K., Pringle 8534 is V. menthaefolia Benth., Repton 1298 is V. officinalis var. natalensis Hochst., J. Z. Weber 2294 is V. supina f. erecta Moldenke, and Robertson s.n. [June 5, 1899] is V. xutha Lehm.

Additional citations: CUBA: Havana: Curtiss 677 (Cm, Es, Es, Mu,

N, Vt, W—522300). Matanzas: Rugel 121 [856] (C). CHILE: Valdiv-
ia: Hollermayer 607 (Mu). MACARONESIA: Gran Canaria: Kunkel 11247
(Mu), 11429 (Mu). GREAT BRITAIN: England: Harz 183 (Mu). SPAIN:
J. Kraft JK.39 (Mu). GERMANY: Berger s.n. [München] (Mu—335);
Brixle s.n. [Herb. Merxmüller 14339] (Mu); Herb. Mus. Bot. Landis-
huth s.n. (Mu—333); Herb. Schmiedel 93 (Mu—329), s.n. (Mu—328);
Herb. Schreber s.n. [Marlofstein, 1784] (Mu—330); Herb. Univ.
Maximil. s.n. (Mu—331); Herb. Zuccarini s.n. (Mu—338, Mu—339).
ITALY: Gröbner s.n. [3.6.1968] (Mu); Zollitsch 4626 (Mu). MALTA:
Westra & Rooden 189 (Ld). YUGOSLAVIA: Micevski 57110 (Mu). MO-
ROCCO: Rauh 426 (Mu). ALGERIA: Doppelbaur 107 (Mu). EGYPT: Sisi
s.n. [El Giza, 24.5.1973] (Mu). ETHIOPIA: Sebald 381 (Mu); Suth-
erland 309 (Ws). UNION OF SOCIALIST SOVIET REPUBLICS: Karachayeva-
Cherkesskaya: Vekhor s.n. [Teberda, VII.1863] (Mu). Republic un-
determined: Herb. Grimm s.n. [Novogorod] (Mu—344). IRAN: Redding
4 (Mi). AFGHANISTAN: Beach 5110 (Ln—192056); Podlech 11391 (Mu),
18248 (Mu), 18654 (Mu), 19914 (Mu); K. H. Rechinger 19229 (Mu).
PAKISTAN: Baluchistan: K. H. Rechinger 30270 (Mu). Northwestern
Provinces: Brandis 1608 (Mu—1120). SIKKIM: J. D. Hooker s.n.
[alt. 6000 ped.] (Mu—349). INDIA: Assam: Jenkins s.n. [Assam]
(Mu—350); Watt 10362 (Mu). East Punjab: T. Thomson s.n. [alt. 1-
4000 ped.] (Mu—348). Uttar Pradesh: Wallich 1825/4 (Mu—1264).
State undetermined: Hügel s.n. [Mapuri, Ind. sup.] (Mu—353).
BURMA: Upper Burma: Huk s.n. [Chin hills, June 1892] (Mu—3802);
Luxburg s.n. [24.2.1903] (Mu). CHINA: Fukien: En 2022 (Mu), 2689
(Mu). FORMOSA: Tanaka & Shimada 11032 (Mu). THAILAND: Larsen &
Larsen s.n. (Ac). JAPAN: Kyushu: Oldham 619 (Mu). AUSTRALIA:
Queensland: M. S. Clemens 42796 (Mi). CULTIVATED: Germany: Herb.
Schreber s.n. [1789] (Mu—345); Herb. Zuccarini s.n. [Hort. bot.
Monac.], in part (Mu). Mexico: Hendricks 590 (Ws). Sweden: Col-
lector undetermined s.n. [20 Aug. 1835] (Ac).

VERBENA OFFICINALIS var. GAUDICHAUDII Briq.
    Additional bibliography: Moldenke, Phytologia 28: 364 & 443
(1974) and 34: 602. 1976.
    The Burke 55, previously cited by me as this variety, actually
proves better regarded as representing var. natalensis Hochst.
    Additional citations: AUSTRALIA: State undetermined: F. v.
Mueller s.n. (Mu—1574).

VERBENA OFFICINALIS var. MACROSTACHYA (F. Muell.) Benth.
    Additional bibliography: Moldenke, Phytologia 24: 27. 1972.
    Additional citations: AUSTRALIA: Queensland: F. v. Mueller s.n.
[Peake Downs] (Mu—1571—isotype, Z—photo of type).

VERBENA OFFICINALIS var. NATALENSIS Hochst. ex F. Krauss, Flora
    28: 68, hyponym. 1845.
    Bibliography: F. Krauss, Flora 28: 68. 1845; Moldenke, Résumé
Suppl. 4: 17. 1962; Moldenke, Phytologia 10: 198 & 213. 1964; Moldenke, Fifth Summ. 2: 687. 1971; Moldenke, Phytologia 28: 364
(1974) and 34: 261, 262, & 279. 1976.
    Although unaccompanied by a formal description, this variety is
clearly based on <u>Krauss 151</u> from "ad fluv. Umlaas, Natal, Dec."
The unnumbered Krauss collection in the Munich herbarium is probably a part of the type collection.
    Recent collectors speak of this plant as an erect perennial
herb, 3 feet tall, and have found it growing in waste ground and
"locally frequent" along old roads, at altitudes of 1300—1580 m.,
flowering in August, October, and December, and fruiting in August.
The corollas are said to have been "purple" on <u>Repton 1298</u>, "mauve"
on <u>E. A. Robinson 5596</u>, and "pale mauve" on <u>Acocks 20990</u>.
    Material of this taxon has most generally been identified and
distributed in herbaria as typical <u>V. officinalis</u> L. and many of
the southern African specimens cited by me in previous installments of these notes may prove, on re-examination, to represent
this variety. The <u>Burke 55</u>, cited below, was incorrectly cited by
me in my 1974 work as var. gaudichaudii Briq.
    Citations: ZAMBIA: <u>E. A. Robinson 5596</u> (Mu). RHODESIA: <u>Fries,
Norlindh, & Weimarck 4002</u> (Mu); <u>Morris 355</u> (Mu). SOUTH AFRICA:
Cape Province: <u>Krauss s.n.</u> [=151?] (Mu—352—isotype?). Orange
Free State: <u>Acocks 20990</u> (Mu). Transvaal: <u>Burke 55</u> (Pd); <u>Repton
1298</u> (Z).

VERBENA OFFICINALIS var. PROSTRATA Gren. & Godr.
    Additional bibliography: Moldenke, Phytologia 28: 362, 364, 392,
& 427 (1974) and 34: 260. 1976.
    Additional citations: GERMANY: <u>Schultes s.n.</u> (Mu—332). ETHI-
OPIA: <u>Schimper 145</u> (Mu—347).

VERBENA OFFICINALI-VENOSA Paxt.
    ddiional bibliography: Moldenke, Phytologia 28: 364—365 &
464A 1974.

xVERBENA OKLAHOMENSIS Moldenke
    Additional bibliography: Moldenke, Phytologia 28: 365 & 457.
1974.

VERBENA ORCUTTIANA Perry
    Additional bibliography: Moldenke, Phytologia 28: 365. 1974.
    Moran encountered this plant at 1500—1680 meters altitude,
flowering and fruiting in July and August, referring to it as
"common", "common in meadows and common for miles in open pine
forests", and "common with <u>Artemisia</u> tridentata in openings in
Jeffrey pine forests. He describes the corollas as "blue" on his
<u>nos. 16439</u> & <u>16479</u>.

Additional citations: MEXICO: Baja California: R. V. Moran 16439 (Sd—71882), 16479 (Sd—71870), 18135 (Sd—76425).

VERBENA ORIGENES R. A. Phil.
Additional bibliography: Moldenke, Phytologia 30: 161. 1975.
Additional citations: CHILE: Coquimbo: Werdermann 225 (Mu).

VERBENA ORIGENES var. GLABRIFLORA Moldenke
Additional bibliography: Moldenke, Phytologia 24: 29. 1972.
Zöllner has found this plant growing at 4000 meters altitude, flowering and fruiting in January.
Additional citations: CHILE: Antofagasta: Zöllner 8309 (Ac).

xVERBENA OSTENI Moldenke
Additional bibliography: Moldenke, Phytologia 28: 365. 1974.
Recent collectors have encountered this plant "em afloramento rochoso", flowering in October and November. The corollas are described as having been "white" on both of the Brazilian collections cited below.
Additional citations: BRAZIL: Rio Grande do Sul: Lima, Vianna, Ferreira, & Irgang ICN.20982 (Ut—320453); Lindeman, Irgang, & Valls ICN.8484 (Ut—320454). URUGUAY: Herter 1000 [Herb. Herter 82763] (Mu).

VERBENA OVATA Cham.
Additional & emended bibliography: Buek, Gen. Spec. Syn. Candoll. 3: 495. 1858; Bolkh., Grif, Matvej., & Zakhar., Chrom. Numb. Flow. Pl., imp. 1, 717 (1969) and imp. 2, 717. 1974; Moldenke, Phytologia 30: 161. 1975.
The corollas are said to have been "violet" on Schinini & Carnevali 10297 and the plant was found growing "en pantanos".
Additional citations: PARAGUAY: Fiebrig 502 (Mu—4040). ARGENTINA: Corrientes: Schinini & Carnevali 10297 (Ld).

VERBENA PARODII (Covas & Schnack) Moldenke
Additional & emended bibliography: Bolkh., Grif, Matvej., & Zakhar., Chrom. Numb. Flow. Pl., imp. 1, 715 (1969) and imp. 2, 715. 1974; Moldenke, Phytologia 30: 162 (1975) and 31: 388. 1975.
Recent collectors have found this plant in fruit in November (in addition to the months previously reported by me). The corollas on Fabris & Schwabe 5022 are said to have been "lilac" in color when fresh.
Additional citations: ARGENTINA: Buenos Aires: Fabris & Schwabe 5022 (Mu). Mendoza: G. Dawson 3252 (Mu).

VERBENA PARVULA Hayek
Additional bibliography: López-Palacios, Revist. Fac. Farm. Univ. Los Andes 14: 23 (1974) and 15: 90 & 93. 1975; Moldenke, Phytologia 30: 162 (1975) and 34: 257 & 258. 1976; Soukup, Biota 11: 19. 1976; Moldenke, Phytologia 36: 30, 33, 52, 226, & 236. 1977.

Taylor has encountered this plant "in low forest and wet upland pastures on mountain slopes" in Costa Rica. López-Palacios (1974) cites López-Palacios 2552 from Mérida, Venezuela, deposited in the Universidad de Los Andes herbarium. In his 1975 work he comments that "En los pocos ejemplares venezolanos (López-Palacios 2552 y Ruiz-Terán & López-Figueiros 2377), fuera de su hábito reducido no le encuentro diferencia alguna con la _V. litoralis_, y a mi modo de ver creo que esta especie no pasa de ser una forma enana de aquella, debida a las condiciones ecológicas desfavorables de clima y altura." He then quotes Macbride (1960) and concludes "En el primer párrafo simplemente he expuesto mi opinión para si algún día llega a tenerse en cuentra por quien se enfrente con la revisión del género."

Additional citations: COSTA RICA: Heredia: J. Taylor 17625 (N).

VERBENA PARVULA var. GIGAS Moldenke

Additional bibliography: Moldenke, Phytologia 24: 31. 1972; López-Palacios, Revist. Fac. Farm. Univ. Los Andes 15: 93. 1975; Soukup, Biota 11: 19. 1976; Moldenke, Phytologia 34: 257 & 258 (1976) and 36: 33. 1977.

Recent collectors have encountered this plant at altitudes of 2100—3600 meters. In addition to the months previously reported by me, it has been found in anthesis in May and in fruit in May and June.

López-Palacios describes it as "hierba erecta de ca. 60 cm. de aspecto similar a la _V. litoralis_, ca. de la cual crece, pero de hojas más pequeñas y más pilosas" and "hierba más o menos postrada, espigas algo cilíndricas, fl. violado-lilas". He found it flowering and fruiting in December and January.

Additional citations: ECUADOR: Chimborazo: Herb. Univ. Cent. Quito 2345 (Mu). Loja: López-Palacios 4164 (Ld). Pichincha: Herb. Univ. Cent. Quito 2340 (Mu), 2343 (Mu), 2346 (Mu); López-Palacios 4198 (Z). BOLIVIA: La Paz: W. Forster s.n. [8.I.1954] (Mu).

VERBENA PARVULA var. OBOVATA Moldenke, Phytologia 36: 52. 1977.

Bibliography: Moldenke, Phytologia 36: 52. 1977.
Citations: ECUADOR: Pichincha: López-Palacios 4250 (Z—type).

VERBENA PERAKII (Covas & Schnack) Moldenke

Additional & emended bibliography: Bolkh., Grif, Matvej., & Zakhar., Chrom. Numb. Flow. Pl., imp. 1, 715 & 717 (1969) and imp. 2, 715 & 717. 1974; Moldenke, Phytologia 30: 162. 1975.

VERBENA PERENNIS Wooton

Additional & emended bibliography: G. W. Thomas, Tex. Pl. Ecolog. Summ. 78. 1969; Bolkh., Grif, Matvej., & Zakhar., Chrom. Numb. Flow. Pl., imp. 1, 717 (1969) and imp. 2, 717. 1974; Moldenke, Phytologia 30: 162 (1975) and 36: 244. 1977.

Higgins encountered this plant in gravelly to sandy or sandy

clay-loam soils in pinyon-juniper association, while the Corrells
encountered it in "rocky soil in small mountains". Chiang and
his associates found it growing in calcareous gravel in "isotal o
matorral" on limestone hills, associated with Dasylirion, Quercus,
and Rhus. Hess & Stickney refer to it as a "common perennial, 60
cm. tall", with "blue-violet" corollas, and found it on slopes of
low limestone hills in pinyon-juniper-grassland with Agave, Dasylirion, Opuntia, Nolina, Rhus, Berberis, etc. The corollas are
said to have been "purple" on Correll & Correll 30882. Thomas
(1969) calls it "perennial verbena".

Additional citations: TEXAS: Brewster Co.: Hess & Stickney 3406
(N). Culberson Co.: Correll & Rollins 23897 (N). Pecos Co.: Cor-
rell & Correll 30882 (N). NEW MEXICO: Eddy Co.: Higgins 9228 (N).
Guadalupe Co.: Higgins 8996 (N). MEXICO: Coahuila: Chiang C.,
Wendt, & Johnston 9178 (Ld).

VERBENA PERENNIS var. JOHNSTONI Moldenke

Additional bibliography: Moldenke, Phytologia 28: 367. 1974.

Henrickson found this plant to be "frequent perennial on rocky
limestone slopes", at 4500 feet altitude, growing in association
with Acacia, Dalea, Dasylirion, Hilaria, Mimosa, and Yucca, flowering and fruiting in July. The corollas on his no. 11366 are
said to have been "purple-blue" when fresh.

Additional citations: MEXICO: Coahuila: Henrickson 11366 (Ld).

xVERBENA PERPLEXA Moldenke

Additional bibliography: Moldenke, Phytologia 24: 33 (1972)
and 36: 147. 1977.

Reeves encountered what appears to be this hybrid at an altitude of 3500 feet, flowering and fruiting in August, but misidentified it as V. ciliata Benth.

Additional citations: ARIZONA: Santa Cruz Co.: T. Reeves R.
1166 (W--2737258).

xVERBENA PERRIANA Moldenke

Additional synonymy: Verbena x perianna Perkins, Estes, &
Thorp, Bull. Torrey Bot. Club 102: 197, in textu. 1975.

Additional bibliography: Perkins, Estes, & Thorp, Bull. Torrey Bot. Club 102: 194 & 197. 1975; Moldenke, Phytologia 30:
162 (1975), 34: 249, 250, & 279 (1976), and 36: 221. 1977.

The Iltis, Bell, Melchert, Patman, & Witt 12361, cited below,
was previously incorrectly cited by me as V. hastata L. It was
collected in flower and fruit in October.

Additional citations: ILLINOIS: Cass Cc.: Geyer s.n. [Beards-
town, July 1842] (Mu--416--cotype). WISCONSIN: Marquette Co.:
Iltis, Bell, Melchert, Patman, & Witt 12361 (Ws). MISSOURI:
Saint Louis: Engelmann s.n. [St. Louis] (Mu--417--cotype, Mu--
1678--cotype).

VERBENA PERUVIANA (L.) Britton

Additional synonymy: Verbenella chamaedryfolia Juss. ex Spach, Hist. Nat. Veg. Phan. 9: 238. 1840. Verbena chamaedrufolia Robledo, Bot. Med. 392, sphalm. 1924. Verbena lindleyi Paxt., in herb.

Additional & emended bibliography: Loud., Hort. Brit., ed. 2, 552. 1832; Sweet, Hort. Brit., ed. 3, 553. 1839; Spach, Hist. Nat. Veg. Phan. 9: 238—239. 1840; Baxt. in Loud., Hort. Brit. Suppl. [3]: 655. 1850; Buek, Gen. Spec. Syn. Candoll. 3: 494—496. 1858; Vilm., Fl. Pleine Terr., ed. 1, 939 (1865), ed. 2, 2: 976 (1866), ed. 3, 1: 1200 (1870), and ed. 4, 1067. 1894; Gilg in Engl., Syllab. Pflanzenfam., ed. 7, 314, fig. 413 D—F (1912) and ed. 8, 318, fig. 413 D—F. 1919; Knoche, Fl. Balear., imp. 1, 1: 59. 1921; Gilg in Engl., Syllab. Pflanzenfam., ed. 9 & 10, 339, fig. 418 D—F. 1924; Robledo, Bot. Med. 382. 1924; Pittier, Man. Pl. Usual. Venez. 395 & 450. 1926; Diels in Engl., Syllab. Pflanzenfam., ed. 11, 339, fig. 432 D—F & L. 1936; Bolkh., Grif, Matvej., & Zakhar., Chrom. Numb. Flow. Pl., imp. 1, 715—717. 1969; R. E. Harrison, Handb. Bulbs & Perenn. S. Hemisph., ed. 3, 266. 1971; Knoche, Fl. Balear., imp. 2, 1: 59. 1974; Kooiman, Act. Bot. Neerl. 24: 464. 1975; López-Palacios, Revist. Fac. Farm. Univ. Los Andes 15: 89. 1975; Moldenke, Phytologia 30: 139, 150, & 162—164 (1975), 31: 409—411 (1975), 34: 270 & 279 (1976), and 36: 122, 142, 153, & 228. 1977.

Additional & emended illustrations: Gilg in Engl., Syllab. Pflanzenfam., ed. 7, fig. 413 D—F (1912), ed. 8, 318, fig. 413 D—F (1919), and ed. 9 & 10, 339, fig. 418 D—F. 1924; Diels in Engl., Syllab. Pflanzenfam., ed. 11, 339, fig. 432 D—F. 1936; Melchior in Engl., Syllab. Pflanzenfam., ed. 12, 436, fig. 184 E & L (in part). 1964.

The corollas of this plant are described as "red" on Cristóbal & al. 1141 and on Schinini & Miranda 9576 and "very red" on Turner 9176. Gillanders and his associates (1973) recommend this plant for "sunny places and walls". Sweet (1830) says that this species was introduced into English gardens from Buenos Aires, Argentina, in 1827; he calls it the "scarlet-flowered vervain". Spach calls it the "verbénelle a feuilles de germandrée". Knoche (1921) reports it cultivated in the Balearic Islands, where it is called "carmelitana", a name also used for xV. hybrida Voss.

The Herb. Herter 82941, Herter 1057, Herzog 1217, and Pflanz 951, distributed as V. peruviana, are actually V. incisa Hook., while Hort. Parag. 11782 and T. Rojas 3395 are V. phlogiflora Cham., Block s.n. is V. rigida Spreng., and Torgo s.n. [Herb. Brad. 21257] is V. selloi Spreng.

Additional citations: URUGUAY: Herter 19 [Herb. Herter 71313] (Mu—4369). ARGENTINA: Buenos Aires: Cabrera & Fabris 14743 (Mu); Fabris 2707 (Mu). Córdoba: Lorentz 13 (Mu—1560); Lossen 10 (Mu—4368). Corrientes: Cristóbal, Quarín, Schinini, & Mirandi 1141 (Ld); Schinini & Mirandi 9576 (Ld). La Pampa: Fortuna s.

n. [15.XII.43] (Ut—330574B). Santa Fé: Turner 9176 (Ld). Tucu-.
mán: Meyer, Vaca, & Gómez 22305c (N). CULTIVATED: Belgium: Martens s.n. [h. b. lov.] (Mu—390). California: Germer s.n. [Los
Angeles, 1876] (Mu—1279, Mu—1280). France: Weinkauff s.n.
[Jard. des plant.] (Mu—1250). Germany: Herb. Kummer s.n. [Hort.
bot. Monac.] (Mu—1249); Herb. Schwasgrichen s.n. [Hort. Lipsi-
ensis] (Mu—1266); Herb. Staatsherb. München s.n. [18.X.1957]
(Mu); Herb. Zuccarini s.n. [Hort. bot. Monac. 1834] (Mu—301), s.
n. [Hort. Monac. 1846] (Mu—302). India: Herb. Hort. Bot. Calcutt. s.n. (Mu—1119).

VERBENA PERUVIANA f. ROSEA Moldenke

Additional bibliography: Anon., N. Y. Times D.41, April 6.
1975; Moldenke, Phytologia 30: 163—164. 1975.
Additional illustrations: Anon., N. Y. Times D.41, April 6,
1975.
It seems probable, from the illustration and description,
that Stern's "Pink Princess" Trailing Verbena may well represent
this form. In the advertisement cited in the bibliography above
(1975) the plant is described as a "full-flowering Trailing Verbena", giving "a refreshing flowerfall of exquisite two-toned
pink blooms all summer long" when grown in a hanging basket. It
is said to bloom "continuously from spring 'til frost. Just fill
flowerpots, planters, boxes or baskets with our brilliantly
blooming 'Pink Princess (T)[=Trade name] Trailing Verbena and
hang in full or part sun in your windows, but preferably outdoors
on porch or patio. With a little loving care and plenty of water-
ing, you can enjoy magnificent two-toned pink sprays all summer
and fall! Pinch back stems for continual heavy blooming and
lovely symmetrically shaped plants. Keep indoors over the winter
and next spring enjoy another fantastic flowerfall"...."If you
plant our Trailing Verbena 6" to 12" apart soon delicate pink
blooms cover the ground in closely clustered waves of color to
cover bare spots [as a groundcover]....create beautiful borders..
form a spectacular sea of pink under gladioli, lilies and other
tall plants. And they keep producing flowers right up to the
heavy frost! Not hardy — so treat as Geraniums or Petunias".

VERBENA PHLOGIFLORA Cham.

Additional synonymy: Verbenella tweediana Hook. ex Spach, Hist.
Nat. Veg. Phan. 9: 239. 1840' Verbena phlogiflora ♂ vulgaris
Schau. in A. DC., Prodr. 11: 537. 1847. Verbena phlogiflora ♂
vulgaris [Cham.] ex Buek, Gen. Spec. Syn. Candoll. 3: 495. 1858.
Verbena decemloba Mart. ex Moldenke, Phytologia 34: 278, in syn.
1976.
Additional & emended bibliography: G. Don in Loud, Hort. Brit.
Suppl. 1: 680. 1832; Baxt. in Loud., Hort. Brit. Suppl. 2: 680.
1839; Sweet, Hort. Brit., ed. 3, 553. 1839; Spach, Hist. Nat. Veg.
Phan. 9: 239. 1840; Baxt. in Loud., Hort. Brit. Suppl. [3]: 655.

1850; Buek, Gen. Spec. Syn. Candoll. 3: 494—496. 1858; Vilm., Fl.
Pleine Terr., ed. 1, 939 (1865), ed. 2, 2: 976 (1866), ed. 3, 1:
1200 (1870), and ed. 4, 1067. 1894; Sonohara, Tawada, & Amano, Fl.
Okin. 132. 1952; Bolkh., Grif, Matvej., & Zakhar., Chrom. Numb.
Flow. Pl., imp. 1, 715 & 717 (1969) and imp. 2, 715 & 717. 1974;
León & Alain, Fl. Cuba, imp. 2, 2: 282. 1974; Moldenke, Phytologia
30: 139, 140, 150, & 164 (1975), 31: 411 (1975), 34: 278 & 279
(1976), and 36: 47, 229, & 240. 1977.

Hatschbach has encountered this plant repent on dry rocky campos and in abandoned cultivated ground, while Schinini & Carnevali
found it "en bajo pantanoso entre matas de gramineas y ciperaceas".
Others have encountered it "in wet drainage zone of campo at margin
of capão or wood island" and in a "small campo on flat hilltop, an
herb to 1 m. tall, living in forest margin". Bornmüller reports it
from 400 m. altitude in Brazil; Quarín found it "en pantano" in
Argentina.

The corollas on Hatschbach 35201 and on Schinini & Carnevali
10300 are said to have been "purple", those on Hatschbach 35172 &
35640 were "violet", those on Pabst 6093 and E. Fereira 6266 were
"light-violet", and those on Quarín 2863 were lilac-violet". Lindeman & Haas describe the corollas on their no. 3998 as "purple
(1/2P3/8)" and on their no. 1120 as "purple 10P7/6, above 10P5/10,
center 10PB6-7/6".

Spach (1840) calls the species "verbénelle de Tweedie", while
Sonohara and his associates (1952) call it "bijozakura". Sweet
(1839) tells us that it was introduced into English gardens from
South America before 1839.

Material of V. phlogiflora has been misidentified and distributed in some herbaria as V. chamaedryfolia Juss. On the other hand,
the Rodrigues 530 [Herb. Inst. Miguel Lillo 31561], previously reported by me and distributed as V. phlogiflora, actually is V.
spectabilis Moldenke, while Bornmüller 143 is V. incisa Hook.

Additional citations: BRAZIL: Minas Gerais: Duarte 6309 [Herb.
Brad. 16885] (Mu); Widgren s.n. [1845] (Mu—1577). Paraná:
Hatschbach 35172 (Ld), 35201 (Ld), 35640 (Ld), 39150 (Ld); Linde-
man & Haas 1120 (Ut—320418), 3998 (Ld). Rio Grande do Sul:
Bornmüller 160 (Mu—4294). Santa Catarina: Pabst 6093 [E. Perei-
ra 6266; Herb. Brad. 22512] (Mu). São Paulo: Martius s.n. [Ypa-
nema, Jan. 1818] (Mu—365), s.n. [Prov. St. Pauli] (Mu—366).
PARAGUAY: T. Rojas 3395 [Herb. Parag. 11782] (Mn), 4736 (Mu). AR-
GENTINA: Corrientes: Quarín 2863 (Ld); Schinini & Carnevali 10300
(Ld).

VERBENA PINETORUM Moldenke

Additional bibliography: Moldenke, Phytologia 28: 372 (1974)
and 36: 244. 1977.

The Walker collection cited below matches the type of the species very well, although the bracts are relatively longer than they
are in some other collections referred here.

Additional citations: MEXICO: Chihuahua: S. Walker 76037 (N).

VERBENA PINNATILOBA (Kuntze) Moldenke
    Additional bibliography: Moldenke, Phytologia 24: 40—41. 1972.
The corollas are said to have been "blue" on Ibarrola 542.
    Additional citations: ARGENTINA: Corrientes: Ibarrola 542 (Ut—
3305818).

VERBENA PLATENSIS Spreng.
    Additional bibliography: Loud., Hort. Brit., ed. 2, 552. 1832;
G. Don in Loud., Hort. Brit. Suppl. 2: 704. 1839; Sweet, Hort.
Brit., ed. 3, 553. 1839; Baxt. in Loud., Hort. Brit. Suppl. [3]:
655. 1850; Buek, Gen. Spec. Syn. Candoll. 3: 495 & 496. 1858;
Vilm., Fl. Pleine Terr., ed. 1, 938—939 (1865), ed. 2, 2: 976
(1866), ed. 3, 1: 1199—1200 (1870), and ed. 4, 1066—1067 & 1070.
1894; Bolkh., Grif, Matvej., & Zakhar., Chrom. Numb. Flow. Pl.,
imp. 1, 715 & 717 (1969) and imp. 2, 715 & 717. 1974; El-Gazzar,
Egypt. Journ. Bot. 17: 75 & 78. 1974; Molina R., Ceiba 19: 96.
1975; Moldenke, Phytologia 30: 150 & 164—165 (1975), 31: 392 &
412 (1975), and 34: 270. 1976.
    Additional illustrations: Voss in Vilm., Fl. Pleine Terr., ed.
4, 1066. 1894.
    Molina R. (1975) records this species as cultivated in Hondu-
ras, but I suspect that he may actually be referring to white-
flowered forms of xV. hybrida Voss which would be far more likely
to be found there.
    Sweet (1839), calling V. platensis the "germander-like ver-
vain", asserts that it was introduced into English gardens from
Buenos Aires, Argentina, in 1837. The corollas on the Cabrera &
al. collections and on Lossen 117, cited below, are said to have
been "white" when fresh.
    Additional citations: URUGUAY: Herter 750a [Herb. Herter
85298] (Mu). ARGENTINA: Buenos Aires: Cabrera & Fabris 16531
(Mu). Córdoba: Lossen 117 (Mu—4376). Jujuy: Cabrera, Arambarri,
Cabrera, & Malacalza 17276 (Mu). CULTIVATED: Germany: Herb. Reg.
Monac. s.n. [Hort. Monac. 1849.16.VIII] (Mu—393); Herb. Zuccar-
ini s.n. [hort. monac. 1840] (Mu—394), s.n. [hort. Monac. 1846]
(Mu—395); Prince Paul of Wurtemberg s.n. [Hort. Mergentheim 1840]
(Mu—1583).

VERBENA PLATENSIS f. IVERIANA (Bosse) Moldenke
    Additional bibliography: Moldenke, Phytologia 30: 164—165
(1975) and 31: 392 & 412. 1975.

VERBENA PLICATA Greene
    Additional & emended bibliography: G. W. Thomas, Tex. Pl. Eco-
log. Summ. 78. 1969; Bolkh., Grif, Matvej., & Zakhar., Chrom.
Numb. Flow. Pl., imp. 1, 717 (1969) and imp. 2, 717. 1974; E. H.
Jordan, Checklist Organ Pipe Cactus Natl. Mon. 7. 1975; Moldenke,
Phytologia 30: 156 & 165 (1975) and 36: 244. 1977.
    The Spellenbergs found this plant growing "in large open grassy
areas with mesquite and Ephedra in a low swale, densely grassy and

weedy, with much tobosa and desert-willow". Other recent collectors have encountered it in calcareous gravel in "matorral desértico inerme on limestone hills and north-facing fans", growing in association with <u>Agave lecheguilla</u>, <u>Cordia parvifolia</u>, <u>Larrea tridentata</u>, and <u>Orusonia</u>. Higgins encountered it "on gravelly exposed limestone outcrops in <u>Acacia-Prosopis-Sophora</u> association" and "in sandy to gravelly limestone soils in <u>Larrea-Acacia-Prosopis</u> association".

The corollas on <u>Chiang, Wendt, & Johnston 9137</u> are said to have been "lavender" in color when fresh, while those on <u>Spellenberg & Spellenberg 3823</u> were "pale blue-lavender". Mrs. Jordan (1975) calls this the "ribbed vervain".

The Reeves <u>R.1198</u>, distributed as <u>V. plicata</u>, actually is <u>V. neomexicana</u> (A. Gray) Small, while <u>R. V. Moran 16893</u>, <u>20727</u>, <u>20748</u>, & <u>21749</u> are <u>V. neomexicana</u> var. <u>hirtella</u> Perry.

Additional citations: OKLAHOMA: Jackson Cc.: <u>Waterfall 11980</u> (Mi). TEXAS: Presidio Cc.: <u>Higgins, Higgins, & Higgins 9867</u> (N), Val Verde Co.: <u>Higgins, Higgins, & Higgins 9964</u> (N). NEW MEXICO: Hidalgo Cc.: <u>Spellenberg & Spellenberg 2823</u> (Ld, N). MEXICO: Coahuila: <u>Chiang, Wendt, & Johnston 9137</u> (Ld).

VERBENA PLICATA var. DEGENERI Moldenke

Additional bibliography: G. W. Thomas, Tex. Pl. Ecolog. Summ. 78. 1969; Moldenke, Phytologia 24: 46. 1972.

Thomas (1969) calls this the "Degener verbena". It seems most likely to me now that this taxon is not worthy of more than form rank.

VERBENA POGOSTOMA Klotzsch

Additional bibliography: Buek, Gen. Spec. Syn. Candoll. 3: 495. 1858; Moldenke, Phytologia 24: 46. 1972; Soukup, Biota 11: 19. 1976.

VERBENA PORRIGENS R. A. Phil.

Additional bibliography: Moldenke, Phytologia 30: 165. 1975.

The Werdermann <u>789</u>, distributed as <u>V. porrigens</u>, actually is <u>V. atacamensis</u> Reiche.

VERBENA PULCHELLA Sweet

Emended synonymy: <u>Glandularia pulchella</u> Sweet ex Spach, Hist. Nat. Veg. Phan. 9: 240. 1840; Troncoso & Cabrera, Fl. Prov. Buenos Aires 137. 1965.

Additional & emended bibliography: Sweet, Hort. Brit., ed. 2, 419. 1830; G. Don in Loud., Hort. Brit. Suppl. 1: 680. 1832; Loud., Hort. Brit., ed. 2, 552. 1832; Baxt. in Loud., Hort. Brit. Suppl. 2: 680. 1839; Sweet, Hort. Brit., ed. 3, 553. 1839; Spach, Hist. Nat. Veg. Phan. 9: 240—241. 1840; Schau. in A. DC., Prodr. 11: 552. 1847; Baxt. in Loud., Hort. Brit. Suppl. [3]: 655. 1850; Buek,

Gen. Spec. Syn. Candoll. 3: 495. 1858; Vilm., Fl. Pleine Terr.,
ed. 1, 937—938 (1865), ed. 2, 2: 975 (1866), ed. 3, 1: 1198
(1870), and ed. 4, 1066. 1894; Moldenke, Phytologia 30: 166 & 172
(1975), 34: 279 (1976), and 36: 139, 153, 231, & 240. 1977.

Additional illustrations: Voss in Vilm., Fl. Pleine Terr., ed.
4, 1066. 1894.

Vilmorin (1863) calls this species "verveine délicate", while
Spach (1840) calls it "glandularia élégant". Sweet (1830) calls
it the "pretty vervain" and avers that it was introduced into
English gardens from Buenos Aires, Argentina, in 1827.

The Hiendlmayr s.n. [Hort. Lipsiensis], Herb. Zuccarini s.n.
[h. Monac. 1836], and Herb. Kummer s.n. [h. Paris.], distributed
as V. pulchella, actually are V. tenuisecta Briq.

Additional citations: URUGUAY: Herter 1805 [Herb. Herter 96556]
(Mu).

VERBENA PULCHELLA f. COROLLA-ALBIDA Paxt.
Additional bibliography: G. Don in Loud., Hort. Brit. Suppl. 1:
680. 1832; Baxt. in Loud., Hort. Brit. Suppl. 2: 680. 1839; Sweet,
Hort. Brit., ed. 3, 556. 1839; Moldenke, Phytologia 28: 375. 1974.

VERBENA PUMILA Rydb.
Additional synonymy: Verbena pumila "Rydb. in Small" ex G. W.
Thomas, Tex. Pl. Ecolog. Summ. 78. 1969.
Additional & emended bibliography: G. W. Thomas, Tex. Pl. Eco-
log. Summ. 78. 1969; Bolkh., Grif, Matvej., & Zakhar., Chrom.
Numb. Flow. Pl., imp. 1, 717 (1969) and imp. 2, 717. 1974; Hinton
& Rzedowski, Anal. Esc. Nac. Cienc. Biol. 21: 111. 1975; Moldenke,
Phytologia 30: 166 (1975), 31: 392 (1975), 34: 279 (1976), and 36:
141 & 147. 1977.
The D. Wright 23, distributed as V. pumila, actually is V. bi-
pinnatifida Nutt.
Additional citations: OKLAHOMA: Comanche Co.: Waterfall 11938
(Mi). TEXAS: Comal Co.: Lindheimer 1075 (Mu—4087). Young Co.:
Correll & Johnston 22126 (N). County undetermined: Lindheimer
IV.501 (Mu—268).

VERBENA PUMILA f. ALBIDA Moldenke
Additional bibliography: Moldenke, Phytologia 30: 166 (1975)
and 31: 392. 1975.

VERBENA QUADRANGULATA Heller
Additional & emended bibliography: G. W. Thomas, Tex. Pl. Ecolog. Summ. 78. 1969; Bolkh., Grif, Matvej., & Zakhar., Chrom.
Numb. Flow. Pl., imp. 1, 717 (1969) and imp. 2, 717. 1974; Moldenke, Phytologia 30: 166 (1975), 34: 252 (1976), and 36: 147.
1977.
Thomas (1969) calls this the "four-angle verbena". Recent collectors have encountered it in calcareous gravelly loam "in izotal
in canyon through limestone plateau", in calcareous rocky loam "in

crasirosulifolio espinosos on limestone hills", in sandy alluvium
in "materral desértico inerme on sandy fans", and in gypsiferous
clay-loam on limestone ridges in association with Agave lecheguilla, A. falcata, A. asperrima, Feuquieria splendens, Yucca filifera, Acacia rigidula, Opuntia leptocaulis, O. rufida, Hechtia,
Larrea, Dasylirion, Nolina, Quercus, and Prosopis.

Additional citations: TEXAS: Duval Co.: Croft s.n. [San Diego,
1885] (Mu). Nueces Co.: A. A. Heller 1388 (Ln—70254—isotype).
MEXICO: Coahuila: Johnston, Wendt, & Chiang C. 10602a (Ld). Nuevo León: Johnston, Wendt, & Chiang C. 10209a (Ld), 10226e (Ld),
10248h (Ld).

VERBENA RACEMOSA Eggert
Additional & emended bibliography: G. W. Thomas, Tex. Pl. Ecolog. Summ. 78. 1969; Bolkh., Grif, Matvej., & Zakhar., Chrom.
Numb. Flow. Pl., imp. 1, 717 (1969) and imp. 2, 717. 1974; Moldenke, Phytologia 30: 167 (1975) and 36: 244. 1977.

Strohmeyer describes this plant as having "stems several from
taproot, apparently annual base, without old stems, the taproot
only 3 mm. across at thickest point", and encountered it on a
river floodplain, flowering and fruiting in April. The corollas
on Meebold 26696 are described as having been "white" when fresh.

Additional citations: TEXAS: Brewster Co.: Strohmeyer s.n. [4
April 1975] (Ld, Z). NEW MEXICO: Dona Ana Co.: Meebold 26696 (Mu).

VERBENA RADICATA Moldenke
Additional & emended bibliography: G. Don in Loud., Hort. Brit.
Suppl. 1: 680. 1832; Sweet, Hort. Brit., ed. 3, 768. 1839; Bart. in
Loud., Hort. Brit. Suppl. 2: 680 (1839) and [3]: 655. 1850; Buek,
Gen. Spec. Syn. Candoll. 3: 495. 1858; Bolkh., Grif, Matvej., &
Zakhar., Chrom. Numb. Flow. Pl., imp. 1, 715 & 717 (1969) and imp.
2, 715 & 717. 1974; Hocking, Excerpt. Bot. A.23: 290 & 291. 1974;
Moldenke, Phytologia 28: 372 & 377. 1974.

Sweet (1839), calling this the "rooting vervain", avers that it
was introduced into English gardens from Chile in 1832.

VERBENA RADICATA var. GLABRA (Hicken) Moldenke
Additional & emended bibliography: Hocking, Excerpt. Bot. A.23:
290 & 291. 1974; Moldenke, Phytologia 28: 377. 1974.

VERBENA RECTA H.B.K.
Additional bibliography: Buek, Gen. Spec. Syn. Candoll. 3: 495.
1858; Moldenke, Phytologia 28: 377—378. 1974; Hinton & Rzedowski,
Anal. Esc. Nac. Cienc. Biol. 21: 111. 1975; Moldenke, Phytologia
36: 145. 1977.

Beaman encountered this species in small open grassy meadows
surrounded by Abies forest.

Additional citations: MEXICO: Michoacán: Beaman 4353 (Ln—171723).
Oaxaca: Pringle 4769 (Ln—70253, Mu). State undetermined: Karwin-

ski s.n. (Mu—300).

VERBENA RECTILOBA Moldenke

Additional bibliography: Moldenke, Phytologia 28: 378 & 440. 1974; Hocking, Excerpt. Bot. A.26: 4. 1975.

VERBENA REGNELLIANA Moldenke

Additional bibliography: Moldenke, Phytologia 24: 127 (1972) and 36: 231. 1977.

Martius s.n., cited below, was collected in anthesis in February and the corollas are said to have been "lilac" in color when fresh. It was misidentified and distributed as V. erinoides Lam.

Additional citations: BRAZIL: Minas Gerais: Martius s.n. [ad S. João d'El Rey, Febr.] (Mu—309).

VERBENA RIBIFOLIA Walp.

Additional bibliography: Buek, Gen. Spec. Syn. Candoll. 3: 495. 1858; Moldenke, Phytologia 28: 378. 1974.

VERBENA RIBIFOLIA var. LONGAVINA (R. A. Phil.) Acevedo de Vargas

Additional bibliography: Moldenke, Phytologia 24: 127—128. 1972.

Additional citations: CHILE: Province undetermined: Philippi s. n. [Hacienda de Cauquenes] (Mu—1579).

VERBENA RIGIDA Spreng.

Additional synonymy: Verbena venosa var. aspera Lorenz ex Moldenke, Phytologia 34: 279, in syn. 1976.

Additional & emended bibliography: G. Don in Loud., Hort. Brit. Suppl. 1: 602 & 680. 1832; Loud., Hort. Brit., ed. 2, 552 & 602 (1832) and ed. 3, 602. 1839; Sweet, Hort. Brit., ed. 3, 553. 1839; Baxt. in Loud., Hort. Brit. Suppl. 2: 680 (1839) and [3]: 655. 1850; Buek, Gen. Spec. Syn. Candoll. 3: 494—496. 1858; Vilm., Fl. Pleine Terr., ed. 1, 937 (1865), ed. 2, 2: 974 (1866), ed. 3, 1: 1198 (1870), and ed. 4, 1065. 1894; Pittier, Man. Pl. Usual. Venez. 398 & 450. 1926; Ewan in Thieret, Southwest. La. Journ. 7: 11 & 42. 1967; Bolkh., Grif, Matvej., & Zakhar., Chrom. Numb. Flow. Pl., imp. 1, 717. 1969; G. W. Thomas, Tex. Pl. Ecolog. Summ. 78. 1969; R. E. Harrison, Handb. Bulbs & Peren. S. Hemisph., ed. 3, 266. 1971; Healy, Gard. Guide Pl. Names 225. 1972; Williamson, Sunset West. Gard. Book, imp. 11, 437. 1973; Bolkh., Grif, Matvej., & Zakhar., Chrom. Numb. Flow. Pl., imp. 2, 717. 1974; R. D. Gibbs, Chemotax. Flow. Pl. 3: 1752—1755 (1974) and 4: 2295. 1974; S. B. Jones, Castanea 39: 137. 1974; León & Alain, Fl. Cub., imp. 2, 281 & 282, fig. 121 B. 1974; Duncan & Foote, Wildfls. SE. U. S. 150, [151], & 295. 1975; Kooiman, Act. Bot. Neerl. 24: 464. 1975; Lopez-Palacios, Revist. Fac. Farm. Univ. Los Andes 15: 93. 1975; Moldenke, Phytologia 30: 134 & 167—168 (1975), 31: 374—376, 387, 392, & 409 (1975), and 34: 20, 250, 260, & 279. 1976; Park Seed Cc., Park Seeds Fls. & Veg. 1976: 90. 1976; Moldenke, Phytologia 36: 47, 131, 141, 226, & 228. 1977.

Additional illustrations: Voss in Vilm., Fl. Pleine Terr., ed.
3, 1: 1198 (1870) and ed. 4, 1065. 1894; León & Alain, Fl. Cuba,
imp. 2, 281, fig. 121 B. 1974; Duncan & Foote, Wildfls. SE. U. S.
[151] (in color). 1975.

Fittier (1926) records the popular name, "virginia", for this
species in Venezuela, while Thomas (1969) calls it "veiny verbena"
and Walker (1976) records "bijo-zakura" [bijo, a beautiful woman,
zakhura, flowering cherry]. The corollas on Correll & Rollins
21061 are said to have been "deep reddish-purple".

López-Palacios (1975) comments that this species "Es originar-
ia del sur del Continente, pero ha sido difundida ampliamente en
cultivos de jardines botánicos y particulares, aunque en Venezuela aún no es muy abundante. Desde el punto de vista taxonómico,
es inconfundido con las restantes especies venezolanas y fácil-
mente distinguible por sus hojas de ancha base semiabrazadoras o
subcordadas."

Balakrishnan refers to this plant as a "0.5 m. tall shrub with
mauve flowers", but this is certainly either an error in observation [since the plant is not a shrub] or a case of mixed labels.
If the label accompanying the collection is correct in other details, he encountered the species in open places along railroad
tracks [a very typical habitat], at 2400 meters altitude, flowering in December. Griffiths refers to it as a "fleshy rhizomatous
perennial, 6—15 inches tall, single- to multi-stemmed; stems
green, basally and apically flushed purplish-red, square, well-
foliaged; leaves semi-erect, decussate, dark-green; inflorescence
dense, near-capitate, compound, those on major peduncles more or
less well separated; flowers deep bluish-purple, richly colorful".

Other recent collectors describe the stems as scabrous, 4 mm.
in diameter, and the corolla slightly longer than the calyx. They
have encountered it in well-drained soil in association with
Baptisia and in sandy clay in oak-pine flatwoods. Spindler refers to it as "infrequent in pastures or sandy soil". Godfrey
speaks of the plant as "locally abundant" in old fields. Mrs.
Clemens found it on the "margin of grasslands" in Queensland,
while Codd refers to it as "occasional patches on roadsides" in
the Transvaal. Ghazal found it to be "widespread" in Louisiana.
My wife and I found it locally and sporadically very abundant
on grassy road shoulders and along fencerows in North and South
Carolina, Georgia, and Florida, although not as widespread as V.
tenuisecta in those states. Thomas reports it naturalized in a
Louisiana cemetery.

Sweet (1839) calls this species the "strong-veined" and "wrinkled" vervain, introduced into English gardens from Buenos Aires,
Argentina, in 1837.

The corollas on Ardoin 21, Hatschbach HH.10300, Herb. Brad.
30320, Kummrow 775, and E. Fereira 7683 are described as having
been "violet" in color when fresh, while those on Codd 10208 were
"dark-purple", those on C. M. Allen 658, Curry, Martin, & Allen
273, Darus 124, and Lemmon 1100 were "purple", those on Augustin

37 and Horst JWH.B.281 were "deep-purple", on Vann 22 "bright-
purple", on Bougere 2113 "lavender-purple", on Gonella 24 "purple
to violet", on Webster & Wilbur 3352 "lavender", and those on E.
Pereira 5187 "roxo-claro".

Gibbs (1974) reports leucoanthocyanin absent from the leaves of
this species, cyanogenesis absent from the stems and leaves, syringin absent from the stems, and the Ehrlich test giving negative
results in the leaves.

The Herb. Kummer s.n. [Hort. bot. Monac.] collection, cited below, is a mixture with V. hispida Ruís & Pav., while Hatschbach
HH.10300, Herb. Brad. 30320, and E. Pereira 7682 are mixtures with
V. rigida var. obovata (Hayek) Moldenke rather than var. glandulifera Moldenke as distributed by the collectors.

Additional citations: NORTH CAROLINA: Harnett Co.: Moldenke &
Moldenke 30003 (Gz). SOUTH CAROLINA: Allendale Co.: Moldenke &
Moldenke 29961 (Ac, Gz, Ld, Tu). Clarendon Co.: Moldenke & Mol-
denke 29974 (Ac, Ld, Tu). GEORGIA: Carlton Co.: Moldenke & Mol-
denke 29279 (Ld). Coweta Co.: Moldenke & Moldenke 29304 (Ac).
Lee Co.: Moldenke & Moldenke 29344 (Kh), 29360 (Gz). Meriwether
Co.: Moldenke & Moldenke 29305 (Gz). Schley Co.: Moldenke & Mol-
denke 29330 (Kh). Sumter Co.: Moldenke & Moldenke 29339 (Ld).
Talbot Co.: Moldenke & Moldenke 29317 (Tu). FLORIDA: Hernando
Co.: Moldenke & Moldenke 29495 (Tu). Leon Co.: Godfrey 72332
(W—2734291). ALABAMA: Sumter Co.: J. A. Churchill s.n. [10 May
1955] (Ln—204151); R. Kral 26416 (Lc). MISSISSIPPI: Rankin Co.:
Webster & Wilbur 3352 (Mi). LOUISIANA: East Baton Rouge Par.:
Beck 453 (Lv); E. A. Bessey s.n. [VII.14.09] (Lv, Lv, Lv); Block
s.n. (Lv); C. A. Brown 1054 (Lv)' 17096 (Lv); Correll & Correll
10449 (Lv); Gonella 24 (Lv); Horst JWH.B.281 (Lv); Lawrence &
Crawford s.n. [Oct. 4, 1916] (Lv); N. E. Petersen s.n. [March 17,
1909] (Lv), s.n. [May 7, 1909] (Lv); W. D. Phillips s.n. [March
1904] (Lv); Ross 67 (Lv). East Feliciana Par.: C. A. Brown 18587
(Lv); Darus 124 (Lv); Ghasali s.n. [Idlewild Exp. Sta.] (Ld).
Lafayette Par.: B. E. Lemmon 1100 (Lv). Livingston Par.: Beck
166 (Lv); Montz 1895 (Lv); Vann 22 (Lv). Ouachita Par.: Thomas &
Reid 24091 (Kl—21812). Pointe Coupee Par.: M. Chaney 393 (Lv).
Saint Helena Par.: C. M. Allen 658 (Lv). Saint Landry Par.: Ar-
doin 21 (Lv). Saint Mary Par.: C. A. Brown 6569 (Lv); Bynum, In-
gram, & Jaynes s.n. [Apr. 13, 1933] (Lv); J. A. Churchill s.n. [2
May 1955] (Ln—204157); Correll & Correll 9346 (Lv). Saint Tam-
many Par.: Arsène 11080 (Lv); Bougere 2113 (Lv); C. A. Brown
17736 (Lv); N. F. Petersen s.n. [Oct. 9, 1909] (Lv). Tangipahoa
Par.: Kirby 116 (Lv); H. R. Wilson 82 (Lv). Terrebonne Par.:
Arceneaux 35a (Lv). Union Par.: R. D. Thomas 23465 (Kl—21955).
Washington Par.: C. A. Brown 18429 (Lv). West Feliciana Par.:

Augustin 37 (Lv); Curry, Martin, & Allen 273 (Lv); Spindler 82
(Lv). TEXAS: Waller Co.: Correll & Rollins 21061 (N). BRAZIL:
Paraná: Kummrow 775 (Ld); E. Pereira 5187 [Herb. Brad. 13596] (Mu),
7683, in part [Hatschbach HH.10300; Herb. Brad. 30320] (Mu). AR-
GENTINA: Buenos Aires: Lorenz 2289 (Mu—1833). Province undeter-
mined: Freiland s.n. [Aug. 1902] (Mu). SOUTH AFRICA: Transvaal:
Codd 10208 (Mu). State undetermined: Penther 1764 [Komgha] (Mu).
SRI LANKA: Balakrishnan NBK.413 (W—2721204). AUSTRALIA: New
South Wales: Meehold 3187 (Mu). Queensland: M. S. Clemens 42111
(Mi), 42206 (Mi). CULTIVATED: California: A. Griffiths 4515
[LASCA 55-E-935] (Sd—90062). Germany: Herb. Kummer s.n. [Hort.
Nymphenbergensis 1840] (Mu—1677), s.n. [Hort. bot. Monac.] (Mu);
Herb. Schwaegrichen s.n. (Mu—1275); Herb. Von Schönau s.n. (Mu);
Herb. Zuccarini s.n. [Hort. bot. Monac. 1836] (Mu—418).

VERBENA RIGIDA var. GLANDULIFERA Moldenke
    Additional synonymy: Verbena rigida var. glandulos Moldenke,
Phytologia 34: 279, in syn. 1976. Verbena rigida var. glandulosa
Moldenke, Phytologia 34: 279. 1976.
    Additional bibliography: Moldenke, Phytologia 24: 132 (1972),
34: 279 (1976), and 36: 47. 1977.
    Hatschbach refers to this plant as ruderal and found it in
flower and fruit in December. The corollas on his no. 37899 are
said to have been "lilac" in color when fresh.
    The Hatschbach HH.10368, Herb. Brad. 30089 & 30956, Pabst 7898,
and E. Pereira 7752 & 8623, distributed as V. rigida var. glandulifera, actually are var. obovata (Hayek) Moldenke, while Hatschbach HH.10300, Herb. Brad. 30320, and E. Pereira 7683 are mixtures
of V. rigida Spreng. and V. rigida var. obovata.
    Additional citations: BRAZIL: Paraná: Hatschbach 37899 (Ld).

VERBENA RIGIDA var. LILACINA (Benary & Bodger) Moldenke
    Additional synonymy: Verbena lilacina Harrison, Handb. Bulbs &
Peren. S. Hemisph., ed. 3, 266. 1971.
    Additional bibliography: R. E. Harrison, Handb. Bulbs & Peren.
S. Hemisph., ed. 3, 266. 1971; Moldenke, Phytologia 28: 381 (1974),
31: 387 & 392 (1975), and 34: 279. 1976.
    Ferreira encountered this plant growing at the edge of railroad
tracks, flowering and fruiting in October, describing the corolla
color as "lilac". It is very probable that some of the collections
cited under typical V. rigida, whose flowers are described by their
collectors as "lila" may belong here, but until it is definitely
ascertained what collectors mean by "lila" it is, perhaps, best to
leave them under the typical form.
    Additional citations: BRAZIL: Paraná: L. F. Ferreira 117 (Ld).

VERBENA RIGIDA var. OBOVATA (Hayek) Moldenke
    Additional bibliography: Moldenke, Phytologia 24: 220. 1972.

Porto & Oliveira describe this plant as an herb to 50 cm. tall
and the corolla color as "roxas 5P4/11", while Lindeman and his
associates refer to the corolla color as "roxas 4P5/8" and found
the plant growing in "terrena baldio". The corollas on Herb.
Brad. 30956, Pabst 7898, and E. Pereira 8623 were "roxas", while
those on Bornmüller 204 are said to have been "hellrosa" [=light
pink] when fresh.

Hatschbach HH.10300, Herb. Brad. 30320, and E. Pereira 7683,
cited below, are mixtures of this variety and typical V. rigida
Spreng.

Material of this variety has been misidentified and distributed in some herbaria as V. rigida var. glandulifera Moldenke.

Additional citations: BRAZIL: Paraná: E. Pereira 7683, in part
[Hatschbach HH.10300; Herb. Brad. 30320] (Mu), 7752 [Hatschbach
HH.10368; Herb. Brad. 30089] (Mu). Rio Grande do Sul: Bornmüller
162 (Mu—4292), 204 (Mu—4293); Lindeman, Irgang, & Valls ICN.
8775 (Ut—320450); E. Pereira 8623 [Pabst 7898; Herb. Brad. 30956]
(Mu); Porto & Oliveira ICN.9584 (Ut—320449). ARGENTINA: Corrientes: Krapovickas, Cristóbal, Schinini, & González 25650 (Ld).

VERBENA RIGIDA f. PARAGUAYENSIS Moldenke, Phytologia 34: 20. 1976.
    Bibliography: Moldenke, Phytologia 34: 20 & 260. 1976.
    Citations: PARAGUAY: T. Rojas 3407 [Hort. Parag. 11794] (Mu—
type, Z—isotype, Z—photo of type).

VERBENA RIGIDA var. REINECKII (Briq.) Moldenke
    Additional bibliography: Moldenke, Phytologia 24: 133. 1972.
    Additional citations: BRAZIL: Rio Grande do Sul: Reineck &
Czermak 68 (Mu—3492).

VERBENA RINCONENSIS Moldenke
    Additional bibliography: Moldenke, Phytologia 24: 133 (1972)
and 34: 252. 1976.

    Henrickson reports this plant "infrequent in open desert" and
"infrequent annual among Larrea on rocky limestone hillside with
Chihuahuan Desert vegetation", associated with Berberis, Bouvar-
dia, Dasylirion, Dyssodia, Fouquieria, Opuntia, Psatrophe, Yucca,
grasses, etc., at 2200—2365 m. altitude, flowering and fruiting
in August and September.

    Additional citations: MEXICO: Zacatecas: Henrickson 6295 (Ld),
6698 (Ld).

VERBENA RIPARIA Raf.
    Additional bibliography: Moldenke, Phytologia 28: 381, 399, &
464 (1974) and 30: 159. 1975.

    The Monts 2294, distributed as V. riparia, actually is V. xu-
tha Lehm.

    Additional citations: VIRGINIA: Smyth Co.: J. K. Small s.n.
[July 1, 1892] (Lv).

VERBENA ROBUSTA Greene

Additional synonymy: Verbena prostrata var. glandulosa Dunkle, in herb.

Additional bibliography: R. D. Gibbs, Chemotax. Flow. Pl. 3: 1752 & 1755 (1974) and 4: 2295. 1974; Moldenke, Phytologia 30: 168 (1975) and 36: 221 & 232. 1977.

Moran reports this plant "occasional by streams" in Baja California, while Thorne found it growing in association with Baccharis glutinosa, Ambrosia psilostachya, etc., along streams on Santa Catalina island.

Dunkle's V. prostrata var. glandulosa seems to be based on his no. 8548 from a dry stream bed on Santa Cruz island in the Channel Islands of California, where he collected it in flower and fruit in August.

Gibbs (1974) reports leucoanthocyanin absent from the leaves of V. robusta, saponin also absent, but tannin probably present.

The Blakley 5394, distributed as V. robusta, is actually V. bracteata Lag. & Rodr.

Additional citations: CALIFORNIA: Alameda Co.: Hillebrand 1863 (Mu). CHANNEL ISLANDS: Santa Catalina: Thorne 36670 (Sd—69572). Santa Cruz: Dunkle 8548 (N). MEXICO: Baja California: Meebold 20234 (Mu); R. V. Moran 16655 (Sd—73068).

VERBENA RUFIFLORA Rojas

Additional & emended bibliography: Krapovickas, Bol. Soc. Argent. Bot. 11: Supl. 269. 1970; Moldenke, Phytologia 30: 168 (1975) and 31: 388. 1975.

VERBENA RUNYONI Moldenke

Additional bibliography: G. W. Thomas, Tex. Pl. Ecolog. Summ. 78. 1969; Moldenke, Phytologia 28: 382. 1974; D. S. & H. B. Correll, Aquat. & Wetland Pl. SW. U. S., imp. 2, 2: 1397, 1400, & 1775. 1975; Moldenke, Phytologia 34: 153. 1976.

Thomas (1969) calls this the "Runyon verbena".

VERBENA RUNYONI f. ROSIFLORA L. I. Davis

Additional bibliography: Moldenke, Phytologia 28: 382. 1974; D. S. & H. B. Correll, Aquat. & Wetland Pl. SW. U. S., imp. 2, 2: 1397, 1400, & 1775. 1975.

xVERBENA RYDBERGII Moldenke

Additional bibliography: R. A. Davidson, State Univ. Iowa Stud. Nat. Hist. 20 (2): 77. 1959; Moldenke, Phytologia 30: 145 & 168 (1975), 31: 377 (1975), 34: 270 (1976), and 36: 29. 1977.

Davidson (1959) records this hybrid from Louisa and Muscatine Counties, Iowa, where he found it in a swamp and on sandy slopes above a bog.

Additional citations: MISSOURI: Saint Louis: Engelmann s.n. [Banks of the Mississippi, July 1842] (Mu—361). CULTIVATED: Germany: Herb. Staatsherb. Münch. s.n. [Hort. bot. Monac. 28.7.53] (Mu).

VERBENA SAGITTALIS Cham.
    Additional bibliography: Buek, Gen. Spec. Syn. Candoll. 3: 495.
1858; Moldenke, Phytologia 28: 383. 1974.

VERBENA SANTIAGUENSIS (Covas & Schnack) Moldenke
    Emended synonymy: Glandularia santiaguenensis Covas & Schnack
ex Bolkh., Grif, Matvej., & Zakhar., Chrom. Numb. Flow. Pl., imp.
1, 715, sphalm. 1969.
    Additional & emended bibliography: Bolkh., Grif, Matvej., &
Zakhar., Chrom. Numb. Flow. Pl., imp. 1, 715 (1969) and imp. 2,
715. 1974; Moldenke, Phytologia 28: 383 & 457—458 (1974) and 30:
178. 1975.
    Recent collectors refer to this as a decumbent plant and have
found it growing on low campos, flowering in November and December. The corollas are described as having been "purple" on Quarín, Schinini, & González 2437 and as "violet" in color on Krapovickas, Schinini, & Quarín 26701.
    Additional citations: ARGENTINA: Corrientes: Quarín, Schinini,
& González 2437 (Ld). Salta: Krapovickas, Schinini, & Quarín
26701 (Z).

VERBENA SANTIAGUENSIS (Covas & Schnack) Moldenke x V. PERUVIANA
    (L.) Britton
    Additional bibliography: Moldenke, Phytologia 28: 383, 457, &
458. 1974.

VERBENA SANTIAGUENSIS f. ALBIFLORA Moldenke
    Additional bibliography: Moldenke, Phytologia 24: 138. 1972.
    Troncoso states that in her opinion the type sheet of this
form is what she calls Glandularia pulchella var. gracilior
Troncoso.

VERBENA SCABRA Vahl
    Additional bibliography: Sweet, Hort. Brit., ed. 1, 1: 325.
1826; G. Don in Loud., Hort. Brit., ed. 1, 247 (1830) and ed. 2,
247. 1832; Loud., Hort. Brit., ed. 2, 552. 1832; G. Don in Loud.,
Hort. Brit., ed. 3, 247. 1839; Sweet, Hort. Brit., ed. 3, 553.
1839; Buek, Gen. Spec. Syn. Candoll. 3: 495. 1858; G. W. Thomas,
Tec. Pl. Ecclog. Summ. 78. 1969; Fong, Trojánkova, Trojánek, &
Farnsworth, Lloydia 35: 147. 1972; León & Alain, Fl. Cuba, imp.
2, 2: 281, fig. 121A. 1974; D. S. & H. B. Correll, Aquat. & Wetland Pl. SW. U. S., imp. 2, 2: 1396—[1398] & 1775, fig. 654
a—f. 1975; Moldenke, Biol. Abstr. 59: 6926. 1975; Moldenke,
Phytologia 30: 140, 168—169, & 176 (1975), 31: 410 (1975), and
34: 250. 1976; Lakela, Long, Fleming, & Cenelle, Pl. Tampa Bay,
ed. 3 [Bot. Lab. Univ. S. Fla. Contrib. 73:] 116 & 182. 1976;
Long & Lakela, Fl. Trop. Fla., ed. 2, 741, 742, & 961. 1976;
Moldenke, Phytologia 36: 126 & 145. 1977.
    Additional illustrations: León & Alain, Fl. Cuba, imp. 2, 2:
281, fig. 121A. 1974; D. S. & H. B. Correll, Aquat. & Wetland
Pl. SW. U. S., imp. 2, [1398], fig. 654 a—f. 1975.

Curiously, Don (1830) places this species in his section Trifidae (with leaves trifid) even though its leaves certainly are
not normally trifid. He calls it the "scabrous vervain" and says
that it was introduced into English gardens from Mexico in 1825.
Sweet (1830) calls it the "rough vervain" and gives its introduction date as 1822. Thomas (1969) calls it "harsh verbena" and
reports that it is also known locally as "white vervain". Churchill found it growing in sloughs in Florida; Adams found it as
the edges of pastures. Brumbach encountered it in disturbed
ground and reports the corolla color as "lavender-purple". Other
recent collectors describe the plant as having fibrous roots,
stems 1—2.5 m. tall, and "corollas as long as the calyx", and
have found it growing in sandy soil, marshy land, and on spoil-
banks of bayous in cypress-tupelo swamps. Lakela and her associates (1976) assert that in the Tampa Bay area [Florida] it inhabits "woods, margins, glades, [and] waste places", flowering in
spring and fall.

The corollas are said to have been "purple" on Curry, Martin, &
Allen 449, "lavender" on C. A. Brown 3892, "pale mauve-pink with
darker eye" on C. D. Adams 11408, "whitish" on C. A. Brown 4084a,
and "white" on Stam 60.

Material of V. scabra has been misidentified and distributed in
some herbaria as "V. polystachia H.B.K." On the other hand, the
Arceneaux 383, distributed as V. scabra, actually is V. urticifolia L., while J. A. Churchill s.n. [22 August 1970] is V. urticifolia var. leiocarpa Perry & Fernald and Herb. Zuccarini s.n.
[Hort. bot. Monac.] is a mixture of V. officinalis L. and xV.
baileyana Moldenke.

Additional citations: FLORIDA: Collier Co.: J. A. Churchill
s.n. [7 May 1969] (Ln—230044); Moldenke & Moldenke 29608 (Ac,
Gz, Ld, Tu). Levy Co.: Meebold 26897 (Mu). Monroe Co.: Meebold
27567 (Mu). Sanibel Island: Brumbach 8786 (N, W—2773210).
LOUISIANA: Catahoula Par.: Thomas & al. 10859 (Kl—11561). Pointe
Coupee Par.: C. A. Brown 3892 (Lv). Saint James Par.: Stam 60
(Lv). Saint Tammany Par.: C. A. Brown 4084a (Lv). Tangipahoa
Par.: Correll & Correll 9239 (Lv). Terrebonne Par.: Wurzlow s.
n. [June 20, 1912] (Lv). West Feliciana Par.: Curry, Martin, &
Allen 449 (Lv). Grand Isle: Cangemi & Andrus 80 (Lv). TEXAS:
Comal Co.: Lindheimer 1077 (Mu—4090). CALIFORNIA: San Bernar-
dino Co.: Parish & Parish 1043 (Mu—1578). JAMAICA: C. D. Adams
11408 (Mu, Mu); Wullschlägel 968 (Mu—299). HISPANIOLA: Domini-
can Republic: Bertero s.n. [St. Domingo] (Mu—415). PUERTO RICO:
Eggers 996 (Mu—3894); Sintenis 767 (Mu—1584).

VERBENA SCABRA f. ANGUSTIFOLIA Moldenke
Additional bibliography: Moldenke, Phytologia 28: 384. 1974;
D. S. & H. B. Correll, Aquat. & Wetland Pl. SW. U. S., imp. 2,

2: 1396, 1397, & 1775. 1975.

VERBENA SCABRA var. TERNIFOLIA Moldenke
    Additional bibliography: Moldenke, Biol. Abstr. 59: 6926. 1975;
Moldenke, Phytologia 30: 169 (1975) and 31: 378. 1975.

xVERBENA SCHNACKII Moldenke
    Additional bibliography: Moldenke, Phytologia 28: 384 & 457.
1974.

VERBENA SCROBICULATA Griseb.
    Additional bibliography: Moldenke, Phytologia 30: 169. 1975.
The corollas on Legname & Cuezzo 10450c are said to have been
"purple" when fresh.
    Additional citations: ARGENTINA: Jujuy: Legname & Cuezzo
10450c (N).

VERBENA SEDULA Moldenke
    Additional bibliography: Balgooy, Pacif. Pl. Areas 3: 245.
1975; Moldenke, Phytologia 28: 384 (1974) and 36: 236. 1977.
    López-Palacios describes this species as "hierba de 0.6—1.5
m., hojas anchas, superiores relativamente estrachas, espigas
largas y delgadas", the corollas on his no. 4294 as "lila", on
no. 4297 as "morado lila", and on no. 4296 as "rosado morada
claro", and found it in flower and fruit in February.
    Additional citations: GALAPAGOS ISLANDS: Indefatigable: López-
Palacios 4294 (Ld), 4296 (Ld), 4297 (Ld).

VERBENA SEDULA var. FOURNIERI Moldenke
    Additional bibliography: Moldenke, Phytologia 24: 141. 1972.
Schimpff encountered this plant at 180 meters altitude and
distributed it as V. litoralis H.B.K. Werff states that it is
"a common plant in the fern-sedge zone" on Chatham island and
that "the plant is somewhat spreading and the flowers are pale-
blue". He found it in flower in October. His nos. 1481 & 1508
are placed here tentatively, awaiting the publication of his
full study of Galápagos vervains; of no. 1481 he notes "leaves
close together; flowers pale-blue, almost white".
    Additional citations: GALAPAGOS ISLANDS: Chatham: Schimpff 132
(Mu); Werff 1481 (Z), 1508 (Z), 2205 [1505] (Ld).

VERBENA SELLOI Spreng.
    Additional bibliography: Buek, Gen. Spec. Syn. Candoll. 3:
496. 1858; Moldenke, Phytologia 30: 152 & 169 (1975) and 36: 231.
1977.
    Troncoso feels that A. santiaguensis f. albiflora Moldenke is
synonymous with what she calls Glandularia pulchella var. gracilior Troncoso and which I regard as V. selloi. Araujo refers to
V. selloi as a very frequent heliophilous herb with "red" flowers.
    Material of V. selloi has been misidentified and distributed

in some herbaria as V. chamaedrifolia Juss.

Additional citations: BRAZIL: Rio Grande do Sul: Araujo 1312 [Herb. FEEMA 12258] (Pf); Torgo s.n. [Herb. Brad. 21258] (Mu). URUGUAY: Herter 181 [Herb. Herter 79174] (Mu).

VERBENA SESSILIS (Cham.) Kuntze

Additional bibliography: Buek, Gen. Spec. Syn. Candoll. 3: 496. 1858; Moldenke, Phytologia 28: 385 & 464 (1974), 34: 260 (1976), and 36: 237. 1977.

Pedersen refers to this plant as having weak, leaning stems, and encountered it on campos, fruiting in November. The corollas on Pedersen 7171 are said to have been "purple" when fresh.

Additional citations: PARAGUAY: Fiebrig 4031 (Mu—4141). ARGENTINA: Entre Ríos: Pedersen 7171 (N).

VERBENA SIMPLEX Lehm.

Additional & emended bibliography: Sweet, Hort. Brit., ed. 1, 1: 325. 1826; G. Don in Loud., Hort. Brit., ed. 1, 246 (1830) and ed. 2, 246. 1832; Loud., Hort. Brit., ed. 2, 552. 1832; G. Don in Loud., Hort. Brit., ed. 3, 246. 1839; Sweet, Hort. Brit., ed. 3, 553. 1839; Buek, Gen. Spec. Syn. Candoll. 3: 494—496. 1858; Conard, Pl. Iowa 44. 1951; R. A. Davidson, State Univ. Iowa Stud. Nat. Hist. 20 (2): 77. 1959; Cooperrider, State Univ. Iowa Stud. Nat. Hist. 20 (5): 70. 1962; Barker, Univ. Kans. Sci. Bull. 48: 571. 1969; Bolkh., Grif, Matvej., & Zakhar., Chrom. Numb. Flow. Pl., imp. 1, 716 & 717 (1969) and imp. 2, 716 & 717. 1974; E. T. Browne, Castanea 39: 183. 1974; El-Gazzar, Egypt. Journ. Bot. 17: 75 & 78. 1974; Rousseau, Géogr. Florist. Qué. [Trav. Doc. Cent. Étud. Nord 7:] 376—377, 480, 505, 643, & 788, map 827. 1974; Duncan & Foote, Wildfls. SE. U. S, 150, [151], & 295. 1975; Lópes-Palacios, Revist. Fac. Farm. Univ. Los Andes 15: 90. 1975; Moldenke, Phytologia 30: 169—170 (1975), 31: 376 & 377 (1975), and 34: 247—249 & 279. 1976; Grimé, Bot. Black Amer. 191. 1976; Moldenke, Phytologia 36: 29, 134, 145, 221, & 244. 1977; F. H. Montgomery, Seeds & Fruits 202, fig. 1, & 230. 1977.

Additional illustrations: Duncan & Foote, Wildfls. SE. U. S. [151] (in color). 1975; F. H. Montgomery, Seeds & Fruits 202, fif. 1. 1977.

Don (1830) calls this the "narrow-leaved vervain" and says that it was introduced into English gardens from "N. Amer." in 1802. Montgomery (1977) describes the seeds as "Nutlets as in V. bracteata, 2.5 x 0.8 x 0.6 mm." Browne (1974) records this species from Stone County, Arkansas. Churchill found it growing on road embankments; Harold encountered it in grassy woodlands. Demaree records it as "common on clay banks" at 275 feet altitude in Arkansas. Cooperrider (1962) found it to be frequent in low sandy ground along rivers in Clinton and Jones Counties, Iowa, while Davidson (1959) refers to it as infrequent in dry sandy soil in Louisa and Muscatine Counties in the same state. Barker (1969) records it from Butler, Chase, Cowley, Lyon, and

Wabaunsee Counties in Kansas, where, he says, it is "Occasional
in rocky upland prairie, on rocky prairie slopes, [and] along roadsides and right-of-ways".

Muehlenbach considered his no. 1525 to represent a natural
hybrid of this species with V. bracteata Lag. & Rodr., but I see
no evidence of this supposed hybridity. I regard his collection
as merely representing the edaphically erect form of V. bracteata.

The corollas on F. R. Fosberg 35716 are said to have been
"blue-lavender" when fresh.

Material of V. simplex has been misidentified and distributed
in some herbaria as V. carolina L. The Curtiss 1955 collection
is a mixture with var. eggerti Moldenke. The Gattinger s.n. [Nash-
ville, June–August] and Tans 1360, distributed as typical V. simplex, are better regarded as var. eggerti, while V. Keller s.n.
[Aug. 24, 1917] is V. halei Small and J. A. Churchill s.n. [25 May
1954] is V. stricta Vent.

Additional citations: NEW YORK: County undetermined: Schoepf
s.n. [Ad New York] (Mu—249). NEW JERSEY: Atlantic Co.: J. A.
Churchill s.n. [19 June 1957] (Ln—230284). County undetermined:
Hillebrand s.n. (Mn); Herb. Zuccarini s.n. (Mu—254). VIRGINIA:
Fauquier Co.: F. R. Fosberg 35916 (N). ILLINOIS: Will Cc.: Mee-
bold 19175 (Mn). Winnebago Co.: Bebb s.n. [Fountaindale] (Mu).
INDIANA: Jefferson Co.: Frazee s.n. [June 30, 1885] (Lc). KEN-
TUCKY: County undetermined: Herb. Kummer s.n. (Mu—1241). TEN-
NESSEE: Blount Co.: Curtiss 1955, in part (Mu—1278). Morgan
Co.: J. A. Churchill s.n. [8 May 1955] (Ln—204158). KANSAS: Ly-
on Co.: Harold s.n. [6/22/60] (Lc). Miami Co.: Hauser & Brooks
2887 (N, N). MISSOURI: Cass Co.: Meebold 24221 (Mu). Dent. Co.:
Meebold 25368 (Mu). Jackson Co.: Meebold 25199 (Mu), 26766 (Mu).
Ripley Co.: Morizot & Whatley s.n. [30 June 1968] (Kl—11866).
Saint Louis: Mühlenbeck 1208 (Mu). ARKANSAS: Izard Co.: R. D.
Thomas 15366 (Kl—11812). Randolph Co.: Demaree 69646 (Ld).
UNITED STATES: State undetermined: Hooker s.n. [S. States] (Mu—
253). NORTH AMERICA: Locality undetermined: Harz s.n. [N. Am.]
(Mu). CULTIVATED: Germany: Herb. Mus. Bot. Landishuth s.n. (Mu—
248); Herb. Schwaegrichen s.n. [Hort. Lipsiensis] (Mu—1260), s.
n. (Mu—1247); Herb. Zuccarini s.n. [H. b. M. 1819] (Mu—251),
s.n. [Hort. bot. Monac. 1830] (Mu—252).

VERBENA SIMPLEX var. EGGERTI Moldenke
    Additional synonymy: Verbena multicaulis Raf., Herb. Raf. 65,
nom. nud. 1833.
    Additional & emended bibliography: Raf., Herb. Raf. 65. 1833;
E. D. Merr., Ind. Raf. 205 & 295. 1944; Moldenke, Résumé Suppl.
7: 2, 3, & 10. 1963; Moldenke, Phytologia 11: 473 & 480. 1965;
Moldenke, Fifth Summ. 1: 18, 38, 44, 46, & 66 (1971) and 2: 695,

793, 917, & 920. 1971; Moldenke, Phytologia 23: 373 (1972), 24:
147 (1972), 34: 248, 249, & 279 (1976), and 36: 244. 1977.

Tans encountered this variety in roadside gravel on a sunny
moraine ridge, growing with Ambrosia psilostachya, Andropogon
gerardi, Asclepias verticillata, Carduus nutans, Euphorbia corol-
lata, Desmodium canadense, and D. illinoense, and describes the
corollas as light-blue.

Curtiss 1955 is a mixture of this variety and typical V. simplex Lehm. Material of the variety has been found growing along
roadsides and has generally been distributed in herbaria as V.
angustifolia Michx.

Additional citations: TENNESSEE: Blount Co.: Curtiss 1955, in
part (Mu—1278). Davidson Co.: Gattinger s.n. [Nashville, June-
Aug.] (Mu). WISCONSIN: Walworth Co.: Tans 1360 (Ts).

VERBENA SPECTABILIS Moldenke
Additional bibliography: Moldenke, Phytologia 24: 147 (1972)
and 34: 259 & 260. 1976.

The corollas on Porto & Oliveira ICN.9586 are described as
having been "vermelhas 5R2/12", while those on Rodriguez 530
were "violet". The latter collection was in flower in August
and has previously been incorrectly cited by me as V. phlogiflora Cham.

Additional citations: BRAZIL: Rio Grande do Sul: Porto & Oliveira ICN.9586 (Ut—320461). ARGENTINA: Misiones: D. Rodriguez
530 [Herb. Inst. Miguel Lillo 31561] (N).

VERBENA SPHAEROCARPA Perry
Additional bibliography: Moldenke, Phytologia 28: 387. 1974.

Felger encountered this plant on sunny slopes and the slopes
of volcanic cones, at 600—1000 meters altitude, flowering and
fruiting in March. The corollas are said to have been "bluish"
on his no. 15758 and "lavender" on no. 15822.

Additional citations: MEXICAN OCEANIC ISLANDS: Socorro: Felger
15758 (Sd—83127), 15822 (Sd—83196).

VERBENA STACHYS Raimondi
Additional bibliography: Moldenke, Phytologia 24: 147—148.
1972; Soukup, Biota 11: 19. 1976.

VERBENA STELLARIOIDES Cham.
Additional & emended synonymy: Verbena stellarioides α decurrens Cham. ex Schau. in A. DC., Prodr. 11: 541. 1847. Glandularia stellatioides (Cham.) Covas & Schnack ex Bolkh., Grif, Matvej.,
& Zakhar., Chrom. Numb. Flow. Pl., imp. 1, 715, sphalm. 1969.

Additional & emended bibliography: Buek, Gen. Spec. Syn. Can-
doll. 3: 496. 1858; Bolkh., Grif, Matvej., & Zakhar., Chrom. Numb.
Flow. Pl., imp. 1, 715 (1969) and imp. 2, 715. 1974; Moldenke,
Phytologia 28: 368, 387, 457, & 464 (1974) and 36: 48 & 153. 1977.

The Fiebrig 4031, distributed as V. stellarioides, seems better regarded as representing V. sessilis (Cham.) Kuntze, a very
closely related taxon.

VERBENA STEWARTII Moldenke
    Additional bibliography: Moldenke, Phytologia 24: 148—149.
1972; Balgooy, Pacif. Pl. Areas 3: 245. 1975; Moldenke, Phytologia 36: 33. 1977; Van der Werff, Bot. Notiser 130: 96—97.
1977.
    López-Palacios describes this plant as "hierba de hojas y es-
pigas delgadas, fl. morado claro" and found it in flower and
fruit in February.
    Van der Werff (1977) is of the opinion that this taxon is con-
specific with a very variable V. townsendii Svenson.
    Additional citations: GALAPAGOS ISLANDS: Narborough: López-
Palacios 4300 (Z).

VERBENA STOREOCLADA Briq.
    Additional bibliography: Moldenke, Phytologia 28: 388 & 465.
1974.

VERBENA STRICTA Vent.
    Additional & emended bibliography: Sweet, Hort. Brit., ed. 1,
1: 325. 1826; G. Don in Loud., Hort. Brit., ed. 1, 246 (1830) and
ed. 2, 246, 1832; Loud., Hort. Brit., ed. 2, 552. 1832; G. Don in
Loud., Hort. Brit., ed. 3, 246. 1839; Sweet, Hort. Brit., ed. 3,
553. 1839; Buek, Gen. Spec. Syn. Candoll. 3: 494—496. 1858;
Kuntze, Rev. Gen. Pl. 2: 510. 1891; Conard, Pl. Iowa 44. 1951; E.
R. Spencer, Just Weeds, ed. 2, 201—204 & 332, fig. 65. 1957; R.
A. Davidson, State Univ. Iowa Stud. Nat. Hist. 20 (2): 77. 1959;
Hall & Thompson, Cranbrook Inst. Sci. Bull. 39: 74. 1959; Cooper-
rider, State Univ. Iowa Stud. Nat. Hist. 20 (5): 70. 1962; P. W.
Thompson, Cranbrook Inst. Sci. Bull. 52: 37. 1967; Barker, Univ.
Kans. Sci. Bull. 48: 571. 1969; E. R. Spencer, All About Weeds
201—204 & 332, fig. 65. 1968; Bolkh., Grif, Matvej., & Zakhar.,
Chrom. Numb. Flow. Pl., imp. 1, 717. 1969; G. W. Thomas, Tex. Pl.
Ecolog. Summ. 78. 1969; Scully, Treas. Am. Ind. Herbs 283. 1970;
Bolkh., Grif, Matvej., & Zakhar., Chrom. Numb. Flow. Pl., imp. 2,
717. 1974; El-Gazzar, Egypt. Journ. Bot. 17: 75 & 78. 1974; Rousseau, Géogr. Florist. Qué. [Trav. Doc. Cent. Étud. Nord 7:] 377,
465, 643, & 785, map 828. 1974; [Bard], Bull. Torrey Bot. Club
102: 431. 1975; Perkins, Estes, & Thorp, Bull. Torrey Bot. Club
102: 194—198. 1975; Moldenke, Phytologia 30: 170 (1975), 31: 415
(1975), and 34: 249 & 279. 1976; Anon., Biol. Abstr. 61: AC1.732
1976; Grimé, Bot. Black Amer. 191. 1976; Van Bruggen, Vasc. Pl.
S. Dak. 369, 520, & 536. 1976; Ziegler & Sohmer, Contrib. Herb.
Univ. Wisc. LaCrosse 13: 16. 1976; Moldenke, Phytologia 36: 29,
135, 150, 157, 217, 224, & 229. 1977; F. H. Montgomery, Seeds &
Fruits 202, fig. 2, & 230. 1977.
    Additional illustrations: E. R. Spencer, Just Weeds, ed. 2,
[203], fig. 65. 1957; E. R. Spencer, All About Weeds [203], fig.

65. 1968; F. H. Montgomery, Seeds & Fruits 202, fig. 2. 1977.

Don (1830) calls this species the "strict vervain" and says that it was introduced into English gardens from North America in 1802. Sweet (1826) calls it the "upright vervain". Thomas (1969) lists the name "mullenleaf verbain" (sic).

Churchill has found this species growing in dry open fields, on dry sandy knolls, and in gravelly disturbed soil around gravel pits in Michigan. Stephens & Brooks refer to it as "abundant with f. albiflora in dry, sandy, rocky, clay soil on prairie pasture hillsides. Montgomery (1977) describes the seeds as "Nutlets 2.8 x 0.7 x 0.6 mm., form as in V. bracteata, most of the dorsal surface coarsely reticulate, inner surfaces with a white pubescence or papillose".

Scully (1970) lists some Amerind uses for this plant, for which see under V. hastata L. in the present series of notes.

Perkins and his associates (1975) have studied "an association of V. stricta, V. bracteata Lag. & Rodr., V. halei Small, V. urticifolia L., and derived interspecific hybrids in southern Oklahoma. All hybrids but one involved V. stricta parentage. Of the four species, V. stricta produces the largest and most aggregated blossoms, the most accessible pollen, and the greatest quantity of nectar. The species is visited by the broadest spectrum of insects, including the following: Coleoptera: Anomala sp., Diptera: Atomosia melanopogon, Bacca sp. (with Verbena pollen on head), Exoprosopa sp. (with pollen on head), Neorhynchocephalus sp. (with pollen on head), Systropus sp. (with pollen on head), Homoptera: Cuerna sp., Hymenoptera: Apis mellifera (with pollen on head), Stictiella formosa, Bombus americanorum (with pollen on head), Ceratina shinnersi (with pollen on head), Dialictus sp. (with pollen on head), Megachile spp. (with pollen on head), Svastra atripes, Triepeolus sp., Xylocopa virginica, and Lepidoptera: Agraulis vanillae, Atalopedes campestris, Danaus plexippus, Euphyles vestris, Eurema lisa, E. mexicana, Grais stigmaticus, Hemiargus isola, Hylephila phyleus, Leptotes marina, Pieris protodice, Pholisora cattulus, Strymon melinus, Thorabes bathyllus, T. phylades, Vanessa virginiensis. Nine of these insects were also found visiting each of the other 3 species. Insect exclusion and artificial pollination experiments indicated that V. stricta is an outcrossing species with low seed-set through self-fertilization; V. halei and V. urticifolia are highly autogamous and V. bracteata less so. The liklihood that V. stricta will enter into interspecific hybridization is greater than that for the other 3 species. Number, frequency, and pollen-carrying ability of the interspecific pollen vectors favor crosses with V. stricta involving V. urticifolia and V. halei."

These authors note that V. stricta "occupies disturbed sites throughout the Central United States. In southern Oklahoma, it occurs within the ranges of V. bracteata.....V. halei..... and V.

urticifolia.....and hybrids frequently occur where the species
are sympatric....All four species and a series of putative hybrids
were found growing, almost to the exclusion of other species, on
a highly disturbed area, a hog lot.....Approximately 500—700
plants of V. halei and V. bracteata were present in the study
area. Verbena stricta and V. urticifolia were represented by only
50—60 individuals. Pure stands of V. stricta, V. urticifolia,
and V. bracteata occurred within the larger area, but in many sections at least three of the species, and at times all four, were
located within a 5 m. radius. All but one of the hybrids exhibited morphological affinities with V. stricta."

They found that 5 male fertile insect-visited plants of V.
stricta with 623 potential seeds had an 87.6 percent seed-set, 5
male sterile insect-visited plants with 747 potential seeds had
76.3 percent seed-set. Nine bagged plants yielded 1263 seeds and
a 7.5 percent seed-set; 3 geitonogamous artificially cross-
pollinated plants yielded 695 seeds and a 14.3 percent seed-set,
and 2 xenogamous artificially cross-pollinated plants yielded
433 seeds and a 78.5 percentage seed-set.

"All four species have two sterile stylar lobes, one or occasionally both of which extend as elongate lobes beyond the glandular stigmatic surface. Stigmatic receptivity appears to develop
first near these lobes and then to spread across the stigmatic
zone. Quantities of pollen were found in the axils of the lobes,
which may function as passive scrapers of pollen off the mouth-
parts of probing insects. In Verbena stricta dehiscence com-
mences the day before unfolding of the corolla lobes and continues
until the corolla falls away, a total of three days. Thus, self-
pollination could occur prior to exposure of the stigma; however,
autodeposition of pollen is unlikely: The anthers are borne ca 2
mm above the stigma, and the pollen remains at this level. The
corolla tube is inclined so that the distal third of the tube
which contains the anthers is nearly horizontal. Hairs in the
corolla also restrict the free fall of pollen to the stigma. Ad-
ditionally, the stigma is not receptive until the flower has
opened. Pollen was not observed on the stigma of bagged flowers.
Spatial separation of the anthers and stigma in the other three
species is less than in Verbena stricta."

They report the corolla color in V. stricta to be light-
purple and the corolla 9 mm. in diameter — the other species
have less showy and smaller flowers with the corolla-tube straight
(in V. stricta it is strongly curved below the anther level).

"Of the four species, Verbena stricta is probably the most attractive to insects. It is perhaps the tallest plant of the group
(mostly over 100 cm) and has 4—21 upright flowering branches
bearing dense aggregates of flowers.....The nectar....contains
38±2 percent dissolved solids. Assuming these solids to be sug-
ars, this percentage is approximately at the mid-range of a series
of sugar concentrations of nectar for 43 species of flowering
plants reported by Percival (1961). Although nectar exists in the

other three species, the amount was too slight for extraction.

"Several characteristics.....of the inflorescences and flowers of Verbena stricta indicate its probable attractiveness for a variety of anthophilous insects: (1) several dense, elongate spikes, each with a mass of light purple flowers forming a showy unit; (2) tubular corolla with a horizontal throat; (3) protected nectar source with a relatively high percentage of dissolved solids and the nectar sources massed.....; (4) a landing platform; and (5) a supply of easily accessible pollen at the throat of each corolla tube.....Verbena stricta was visited by the greatest diversity and number of potential pollinators.....

"All insect visitors appeared to probe the flowers of Verbena stricta for nectar with the exception of a small solitary bee (not captured), which gathered pollen only. The butterflies methodically probed several flowers on each dense branch and then flew to another branch on the same plant or an adjacent plant. Longer flights were infrequent. Foraging patterns of flies resembled those of butterflies, but the flies spent less time probing each flower and moved more rapidly from plant to plant. Bees tended to visit fewer flowers on each plant and to forage over greater distances, especially those which were gathering pollen.

"Most individual visitors were typically species specific. The only insect observed visiting indiscriminately both parental plants and their hybrid (V. stricta, V. urticifolia, and V. x desmii) was Systropus. A single Hemiargus isola was observed to visit V. bracteata and V. stricta successively....visitors to Verbena stricta and V. halei were fairly equally divided among the three major orders of specialized insect pollinators, Diptera, Hymenoptera, and Lepidoptera....it is obvious that pollen from the other three species is viable on the stigmas of Verbena stricta, but that interspecific crosses may produce fewer seeds than intraspecific, allogamous crosses.....Nevertheless, cross-pollination among these species involving V. stricta as the carpellate parent is likely and cross-fertilization feasible."

They point out that V. stricta sets significantly fewer seeds when bagged than when open-pollinated. "Although V. stricta and V. bracteata are capable of autogamous reproduction, each shows a tendency for cross-fertilization. This tendency is most pronounced in V. stricta. Artificial pollen transfer within a single plant of V. stricta also demonstrates considerably lower seed set. This species, then, is partially self-incompatible.

"Pollen fertility in Verbena stricta varies considerably with complete sterility evidence in 3 plants, about 14 percent fertility in 4 plants, under 50 percent in 8 other plants, and the remaining 17 plants sampled exhibiting 63—98.3 percent fertility. The 15 plants of V. stricta which exhibited less than 50 percent fertility also had indehiscent anthers. Yet even in the 3 completely sterile individuals seed set was 76 percent. A bagged, pollen sterile plant set no seed while bagged, but set full seed when exposed to pollinators. Seed set in pollen-sterile individuals approaches normalcy, and therefore such plants are ovule-fertile....pollen-sterile plants set seed only when pollen-fertile forms grow in close proximity and pollinators are readily available. Although the majority of the plants of V. stricta are capable of either outcrossing or selfing, a limited number in this locality function only as pistillate parents.

"On the basis of its reproductive biology and pollination system, Verbena stricta seems more likely to enter into interspecific hybridization than the other three species...for the following reasons: (1) V. stricta is primarily allogamous, whereas in V. halei and V. urticifolia, approximately 50 percent of the gametes are committed to autogamy; (2) V. stricta shares taxa of pollinators with each of the other three species; (3) pollen washings also indicate that interspecific crosses should occur most frequently among the triad V. stricta, V. halei and V. urticifolia. Visitors shared among V. stricta, V. halei and V. urticifolia were primarily bees and flies. The most common visitors shared by V. stricta and V. urticifolia were butterflies, which may be less dependable pollinators.

"Only one hybrid plant of V. stricta x V. bracteata parentage was found. The two outcrossing species are the least likely to hybridize. The hybrids most frequently found involved Verbena stricta as one of the parental species. Interspecific cross-pollinations, therefore, appear to be skewed in favor of V. stricta, the most attractive to the greatest number and variety of insects. In conditions where the four species are sympatric, hybrids would be expected to include V. stricta as one of the parental types, with crosses involving V. halei and V. urticifolia the most likely to occur and V. bracteata the least likely to do so."

These workers have done a tremendously important piece of research and deserve the highest praise! Many more such studies are urgently needed in this genus (and in other genera of this family) to help us to understand the taxonomic and systematic position of the many groups of puzzlingly variable or puzzlingly similar taxa.

Hall & Thompson (1959) record V. stricta from Oakland County, Michigan, where they say it is occasional in fields, especially in sandy soil; Thompson (1967) lists it for Leelanau County. In Iowa Davidson (1959) records it as a common "Weed of roadsides, railways, pastures, and other open, often sandy, places", while Cooperrider (1962) also refers to it as common in sandy roadsides and other open sandy places in that state. In Kansas Barrell notes that it is "Common, in upland prairies, on prairie slopes, along roadsides and railroad right-of-ways. Occurs throughout the area." Barber also found it "occasional in grassy lightly grazed uplands with Chrysopsis villosa" in Kansas.

Grimé (1976) says "Vervain [is] a cure for rheumatism. The following notes from Cameron Mann, a well known botanist of Kansas

City, are worth recording: 'The following remedy for rheumatism
was recently communicated to me by a respectable lady, who had
found it quite successful in her own case, and also with several
of her friends. It was given her by an old negress who claimed
many years ago to have been told of it by an Indian. The medicine
is made by boiling the root and part of the stalk of one of the
blue vervains in vinegar for twelve hours, and then rubbing the
decoction upon the afflicted parts. Which of the species was
used I could not tell from the species [specimens] shown me, as
they consisted merely of root and stalk, with, fortunately one
stalk bearing withered flowers. The latter identified the plant
as a Verbena, but there being no leaves, I could not tell whether
it was Verbena angustifolia, Verbena hastata, Verbena stricta, [or]
Verbena bracteosa, all four being native here."

The Herb. Zuccarini s.n. [Hort. bot. Monac.] collection, cited
below, is a mixture with xV. moechina Moldenke.

Material of V. stricta has been misidentified and distributed
in some herbaria as V. bracteosa Michx. and as V. simplex Lehm.
On the other hand, the Bynum, Ingram, & Jaynes s.n. [Apr. 13,
1933], distributed as V. stricta, actually is V. rigida Spreng.

Additional citations: NEW YORK: Suffolk Co.: Kandler s.n. [Up-
ton, 5.8.1956] (Mu). ILLINOIS: Adams Co.: Purpus 117 (Mu—4289).
Winnebago Co.: Bebb s.n. [Fountaindale] (Mu—4252). County unde-
termined: Herb. Zuccarini s.n. (Mu—373). INDIANA: Tippecanoe Co.:
Bresinsky s.n. [2.7.1967] (Mu). IOWA: County undetermined: Hille-
brand s.n. [Iowa, 1871] (Mu). MICHIGAN: Antrim Co.: T. H. P. Mar-
shall 1738 (Ln—127640). Macomb Co.: J. A. Churchill s.n. [18
July 1954] (Ln—204161). Oakland Co.: J. A. Churchill s.n. [25
May 1954] (Ln—203432). WISCONSIN: La Crosse Co.: Doppelbaur &
Doppelbaur 963 (Mu). MINNESOTA: Clay Co.: O. A. Stevens 1271 (Mi).
SOUTH DAKOTA: Lawrence Co.: N. F. Petersen s.n. [Spearfish] (Lv).
Pennington Co.: O. M. Clark 6044 (Mu). Stanley Co.: Stephens &
Brooks 33954 (Sd—74497). KANSAS: Barber Co.: Barrell 38-72 (W—
2802774). Lyon Co.: Hauser & Brooks 3058 (N); F. L. Nelson s.n.
[June 29, 1960] (Lc); A. V. Weber s.n. [June 28, 1959] (Lv). MIS-
SOURI: Cass Co.: Meebold 24220 (Mu). Cole Co.: Meebold 25461 (Mu).
Ralls Co.: J. Davis 4488 (Mu). Saint Louis: Engelmann s.n. [St.
Louis] (Mu); Mühlenbeck 145 (Mu, Mu). County undetermined: Mar-
tens s.n. (Mu—574). ARKANSAS: Izard Co.: R. D. Thomas 15346 (Kl—
11810). Phillips Co.: Demaree 15224 (Lv). NEBRASKA: Pierce Co.:
N. F. Petersen s.n. [Aug. 11, 1910] (Lv). CULTIVATED: France:
Herb. Schreber s.n. [Ex horto Parisii] (Mu—370). Germany: Herb.
Schreber s.n. [Hortus Erlangensis 1805] (Mu—371); Herb. Zuccarini
s.n. [Hort. bot. Monac.] (Mu—375, Mu—376).

VERBENA STRIGOSA Cham.
    Additional bibliography: Buek, Gen. Spec. Syn. Candoll. 3: 496.

1858; Moldenke, Phytologia 30: 170 (1975) and 36: 226. 1977.

Recent collectors describe this plant as a decumbent herb, 50 cm. tall, found it growing along roadsides, and report that the flowers "attract many bees". They have found it in flower in September and both in flower and fruit in December.

The corollas on Lindeman & Haas 2475 are described as having been "blue-purple (10PB6/6)", while on Lourteig 2143 they were "lilac". Handro says of the plant "planta do campo, formando grandes touceiras; flores roxo-azuis".

The Reitz & Klein 17616, cited below, was previously erroneously cited by me as V. hirta var. gracilis Dusén. The Dusén 242, distributed as V. strigosa, actually is V. lobata Vell.

Additional citations: BRAZIL: Paraná: Hatschbach 4216 (Mu); Lindeman & Haas 2475 (Ld); Reitz & Klein 17616 (Ac, N, W—2548333). Santa Catarina: Lourteig 2143 (N). São Paulo: Handro 607 [Herb. Inst. Bot. S. Paulo 55438] (W—2748270).

VERBENA SULPHUREA D. Don

Additional synonymy: Glandularia sulphurea D. Don ex Spach, Hist. Nat. Veg. Phan. 9: 241. 1840. Verbera diceras Bert. ex Moldenke, Phytologia 36: 48, in syn. 1977.

Additional & amended bibliography: G. Don in Loud., Hort. Brit. Suppl. 1: 680. 1832: Loud., Hort. Brit., ed. 2, 553. 1832; Baxt. in Loud., Hort. Brit. Suppl. 2: 680. 1839; Sweet, Hort. Brit., ed. 3, 553. 1839; Spach, Hist. Nat. Veg. Phan. 9: 241. 1840; Baxt. in Loud., Hort. Brit. Suppl. [3]: 655. 1850; Buek, Gen. Spec. Syn. Candoll. 3: 494 & 496. 1858; Bolkh., Grif, Matvej., & Zakhar., Chrom. Numb. Flow. Pl., imp. 1, 717 (1969) and imp. 2, 717. 1974; Moldenke & Neff, Orig. & Struct. Ecosyst. Techn. Rep. 74-18: 57, 75, 116, & 118. 1974; Moldenke, Phytologia 30: 171 (1975), 34: 274 (1976), and 36: 40 & 48. 1977.

Spach (1840) calls this species the "glandularia jaune", while Loudon (1832) calls it the "sulphur-coloured vervain" and says that it was introduced into English gardens from Chile in 1834.

Additional citations: CHILE: Aconcagua: Dessauer s.n. [Aconcagua '97] (Mu). Coquimbo: Grau & Grau 1663 (Mu). Valparaíso: Behn s.n. [16 Oktobr. 1929] (Mu); Bertero 1392 (Mu—377), 1809 (Mu—377); Buchtien s.n. [21.IX.1895] (Mu—1839). Province undetermined: Dessauer s.n. [Fishermans] (Mu); Frömbing s.n. [1886] (Mu—1787). CULTIVATED: Germany: Herb. Kummer s.n. [Hort. bot. Monac. 1846] (Mu—1268).

VERBENA SULPHUREA var. FUSCORUBRA Skottsberg

Additional bibliography: Moldenke & Neff, Orig. & Struct. Ecosyst. Techn. Rep. 74-18: 118. 1974; Moldenke, Phytologia 24: 227. 1972.

[to be continued]

Keys to the Flora of Florida -- 3, Boraginaceae [1]

Daniel B. Ward and Paul R. Fantz
Department of Botany, Agricultural Experiment Station
University of Florida, Gainesville, Fla.

ABSTRACT: A key is provided to the 10 genera of Boraginaceae native and naturalized in the state of Florida, U.S.A. The genera, with the number of included species, are: *Bourreria*, 3; *Buglossoides*, 1; *Cordia*, 2; *Cynoglossum*, 2; *Heliotropium*, 7; *Lithospermum*, 3; *Mallotonia*, 1; *Myosotis*, 1; *Onosmodium*, 1; and *Tournefortia*, 2. Amplified keys are presented to the species within each genus. The keys are supplemented with discussion of nomenclature and justification of generic placement and specific delimitation in *Bourreria*, *Buglossoides*, *Heliotropium*, and *Mallotonia*. *Buglossoides arvensis*, *Cynoglossum furcatum*, *Heliotropium procumbens*, and *Myosotis macrosperma* are newly reported for Florida. *Heliotropium europaeum*, *Myosotis virginica*, and *Borago officinalis* are excluded.

BORAGINACEAE Juss.   Borage Family

The Boraginaceae well typify the southeastern family as treated by J. K. Small. He placed its species also in two segregate families, Ehretiaceae and Heliotropiaceae, which are not now generally recognized. The numerous papers of I. M. Johnston are most important in determining generic alignments and the nomenclature of many species.

1. Shrubs, small trees, or woody vines.

   2. Leaves linear-spatulate, succulent, with dense silky-gray pubescence.   Mallotonia

   2. Leaves broad, membranous, variously pubescent to glabrous but not silky-gray.

      3. Shrubs or small trees; styles divided toward apex.

         4. Styles once forked; flowers small, white, in open cymes.   Bourreria

---

[1] This paper is Florida Agricultural Experiment Station Journal Series No. 398.

309

4. Styles twice forked; flowers large and orange *or* small and in dense heads.

3. Woody vines or rarely low shrubs; styles wholly united.   Tournefortia

1. Annual or perennial herbs.

   5. Flowers in condensed terminal cymes.

     6. Annual; corolla white to bluish. Buglossoides

     6. Perennial; corolla pale yellow to orange-yellow.
                                    Lithospermum

   5. Flowers in elongate cymes usually showing pronounced scorpioid curvature.

     7. Style long-exserted from corolla.
                                  Onosmodium

     7. Style short, included in corolla.

       8. Nutlets with stout hooked (grapnel-like) bristles; cauline leaves clasping.   Cynoglossum

       8. Nutlets without hooked bristles; cauline leaves cuneate to petiolate.

         9. Flowers sessile; style at apex of ovary.
                                  Heliotropium

         9. Flowers short-pedicellate; style arising between ovary lobes.   Myosotis

BOURRERIA

The Strongbacks have not been given sufficient attention by previous workers to permit confident naming of all Florida plants. Johnston almost wholly omitted study of *Bourreria*, leaving a revision by Schulz (in Urban, Symbolae Antillanae 7:45-70. 1911) the only comprehensive analysis of the genus. The Rough Strongback, a shrub or small tree of the lower Florida Keys, has from an early date been distinguished by its strigose leaves from the much more common smooth-leaved Bahama Strongback, *B. ovata* Miers, yet doubt remains as to its correct scientific name.

The Rough Strongback was well described from Florida
materials at least as early as 1860 by Chapman (Flora of the
Southern United States, ed. 1) under the name *Ehretia radula* Poir.
Small (Flora of the Southeastern United States, ed. 1. 1903) and
others soon named the plant *Bourreria radula*, in recognition of
its correct generic placement. The epithet "radula," or *rasp*,
seemed a felicitous term for the distinctive abrasive leaves of
this plant.

Loss of this mnemonic word was a consequence of the 1911
revision by Schulz. He saw the rough-leaved Florida plant as
falling within *Bourreria revoluta* HBK. and made that in turn a
variety (var. *revoluta* (HBK.) O. E. Schulz) of the generic type,
B. succulenta Jacq. (1760). Assignment of the Florida plant to
revoluta (although at the specific level) was accepted by Britton
& Wilson (Sci. Survey of Porto Rico and the Virgin Islands, 1930)
and by Small (Manual of the Southeastern Flora, 1933), and more
recently has been employed by Little (Checklist of Native and
Naturalized Trees, 1953). Long & Lakela (Flora of Tropical
Florida, 1971) conformed more closely to Schulz, treating
revoluta as a variety of *B. succulenta* (although incorrectly
attributing this last species to C. E. Stahl, a twelve year old
boy on the centennial of Jacquin'a actual description).

The epithet *revoluta*, however, whether used at the level of
species or variety, cannot refer to the Florida plant. *Bourreria
revoluta* was based by Kunth (Nov. Gen. 3:67. 1818) on materials
from Regla, Hidalgo, Mexico, at an elevation above 2000 m., and
was described as glabrous. In habitat and form it is unlikely to
be the rough-leaved Florida plant, and in date it is preceded by
more probable Antillean names.

The name that most certainly applies to the Florida rough-
leaved plant, although not necessarily the earliest, is *Bourreria
radula* (Poir. in Lam.) G. Don, based upon *Ehretia radula* Poir.
(in Lamarck, Encyc. suppl. 2:2. 1811). Although typified by
materials from Hispaniola, it was described by Poiret in terms
that were both detailed and apt for the plants of the Florida
Keys; the leaves, in part, were seen as "toutes couvertes en
dessus de points blancs tres-rudes," an unmistakable reference to
the distinctive stout white-based hairs of the upper leaf surface.
Although Schulz considered *B. radula* to be synonymous with a
second Hispaniolan species, *B. tomentosa* (Lam.) G. Don, that name
has not previously been applied to Florida materials, and the
tomentose lower leaf surface of *B. tomentosa* may preclude such
application. Until the much needed further study of this genus
is undertaken, it seems both prudent and satisfying to return to
the epithet long ago employed by Chapman, with the Rough Strongback again known as *Bourreria radula*.

In recent literature regarding the Florida species of
Bourreria, the common name often employed has been "Strongbark,"
surely a corruption of the Bahamian "Strongback," in reference to
the practice (still continued, as observed by George Avery) of
using the leaves in a tea which will give one "strength to work
all day."

<u>Bourreria</u> P. Browne   Strongbacks

1. Leaves narrowly obovate to spatulate, hispidulous above, small
   (blade to 2.5 cm. long, 0.8 - 1.2 cm. wide; petiole to 3 mm.
   long); flowers white; drupes orange, 7 - 8 mm. broad; small
   twiggy shrub, under favorable conditions to 2.5 m. tall but
   usually much lower; very local and now nearing extinction
   through habitat destruction, pinelands of Big Pine Key, Monroe
   County, and Long Pine Key of the Everglades National Park,
   Dade County. July - September.
   LITTLE STRONGBACK.   B. <u>cassinifolia</u> (A. Rich.) Griseb.

1. Leaves broadly obovate, of medium size (blade 2.5 - 6 cm.
   long, 1.4 - 4 cm. wide; petiole 3 - 18 mm. long); drupes 8 -
   11 mm. broad.

   2. Leaves strigose above, the individual hairs stout and at
      length white-based; petioles 3 - 7 mm. long; cymes usually
      few (5 - 30) - flowered; flowers white; drupes orange;
      small tree; originally in dry hammocks of Key West, now
      precariously persisting in yards, with a few individuals
      on adjacent keys. Nearly all year. [*B. revoluta* HBK.,
      misapplied]
      ROUGH STRONGBACK.   <u>B</u>. <u>radula</u> (Poir. in Lam.) G. Don

   2. Leaves smooth above; petioles 10 - 18 mm. long, rarely
      less; cymes often many (20 - 80) - flowered; flowers white;
      drupes yellow to orange-red with increasing maturity; small
      to occasionally fair-sized tree; frequent in hammocks, at
      times somewhat adventive in cut-over areas, throughout the
      Keys, sparingly at Bear Lake and elsewhere near the edge of
      Florida Bay. June - October.
      BAHAMA STRONGBACK.   <u>B</u>. <u>ovata</u> Miers

BUGLOSSOIDES

This genus has been segregated by I. M. Johnston (Jour.
Arnold Arb. 35:38-46. 1954) from *Lithospermum* primarily on the
basis of its well developed insect guide lines in the corolla
throat, with a number of its other floral features (apiculate
anthers, lobed sterile tip of style, small cylindric pollen, and
blue as a common corolla color) occurring only sporadically else-

where in *Lithospermum* or allied genera and never with the same
degree of uniformity as in *Buglossoides*. The desirability of
adopting this ineuphonious name has not impressed itself upon
most recent floristic writers (cf. Hommersand in Radford et al.,
Manual of the Vascular Plants of the Carolinas, 1968), but Johnston's rationale for its separation from *Lithospermum* is clearly
expressed and seems well founded and has been concurred in by
such modern writers as Correll & M. C. Johnston (Manual of the
Vascular Plants of Texas, 1970) and Fernandes (in Tutin et al.,
Flora Europaea, 1972).

The one Florida species, *Buglossoides arvensis*, is based on
two remote collections which appear to be the first legitimate
records of this plant for the state. Fernald (Gray's Manual of
Botany, 1950) reported for the species (as *Lithospermum arvense*)
to extend "s. to Fla." However, no documenting specimen presently
exists at the Gray Herbarium, nor is there other indication of the
basis for Fernald's statement.

Buglossoides Moench   Gromwells

1. Flowers small, corolla scarcely exceeding calyx, white with
   tube purplish-blue below; nutlets gray with the surface elabo-
   rately wrinkled and pitted; leaves oblong-lanceolate, small;
   annual herb; weedy elsewhere, known in Florida only from two
   collections: open pineland, Tampa, Hillsborough County [*S. C.
   Hood 4147*, 26 Apr 1951, FLAS]; vacant lot, Pensacola, Escambia
   County [*J. R. Burkhalter 3473*, 6 Mar 1976, FLAS]. March –
   April. [*Lithospermum arvense* L.]
   CORN GROMWELL.   **B. arvensis** (L.) I. M. Johnst.

CORDIA

The two Florida species of *Cordia* have very similar ranges
but greatly different structure, a fact recognized by Small
(Manual of the Southeastern Flora, 1933) who placed them in different genera. Studies by Johnston (Jour. Arnold Arb. 30:85-104.
1949; 30:111-127. 1949) are helpful in placing these species in
proper relation to each other and to their allies.

Cordia L.   Cordias

1. Corolla orange, 3 - 4 cm. long; fruit tightly enclosed by
   fleshy white calyx, 2 - 2.5 cm. long at maturity; leaves
   entire, broadly ovate, the blade 15 - 20 cm. long, on a 3 - 5
   cm. petiole, occasionally much smaller; shrub or small tree;
   infrequent but conspicuous in flower, hammocks, rocky areas,

roadsides, often in cultivation; Florida Keys and (more
rarely) southern Dade County.   June - October.   [*Sebesten
Sebestena* (L.) Britt.]
GEIGER-TREE   C. sebestena L.

1. Corolla white, about 0.7 cm. long; fruit a red naked one-
   seeded drupe, 0.3 - 0.5 cm. long; leaves coarsely serrate,
   ovate, 2 - 3 cm. long; weak-stemmed shrub; occasional, in
   hammocks on Florida Keys, rarely on north shore of Florida
   Bay. [All Florida materials have been determined as var.
   humilis (Jacq.) I. M. Johnston.]   May - August (December).
   [*Varronia globosa* Jacq.]
   VARRONIA.   C. globosa (Jacq.) Kunth

CYNOGLOSSUM

   Cynoglossum furcatum Wall. in Roxb., an Indian species not
previously reported for continental North America, has been collected near the south shore of Lake Okeechobee (see key, below).
Its occurrence elsewhere in the New World has been noted only by
Liogier (Rhodora 67:349. 1965) who records it for Maricao, Puerto
Rico. In 1932 it was grown as a cultivated ornamental in Gainesville (*Watkins*, FLAS), but at the present time it does not seem
to be known in Florida horticulture. It is not known whether the
Okeechobee plants were derived from this or other intentional introductions, or were truly adventive. Recent collection activities in the Okeechobee area have not encountered the species.

   The second species of this genus, *Cynoglossum virginianum*
L., although an unquestioned native, is a northern plant that has
only been recognized in recent years to extend southward into
Florida. Neither Chapman nor Small appear to have been aware of
its presence on the Apalachicola River bluffs. This may have
been a consequence of its occurrence only on the geologically
older portions of the bluffs (now in Torreya State Park) away
from the meanders of the river; the steamboat landings and early
road connections were only at points where the bluffs and river
were in proximity so that construction would not be swept away by
the yearly floods.

        Cynoglossum L.   Wild Comfreys

1. Leaves large (to 30 cm. long, 8 cm. wide), basal ones long-
   petiolate, cauline ones clasping; pubescence coarsely hirsute;
   flowers white (in ours; often light blue elsewhere); style in
   fruit shriveled and very inconspicuous; nutlets 6 - 8 mm.
   long; coarse perennial herb; a single Florida station: wooded

ravines and slopes, Torreya State Park, Liberty County.
March - April. [*Cynoglossum* "*virginicum*," Chapman, Small, in
error.]
WILD COMFREY.   C. virginianum L.

1. Leaves small (4 - 6 cm. long, 1 - 1.5 cm. wide), cauline ones
   clasping; pubescence appressed-silky; flowers blue; style in
   fruit prominent, 2 mm. long; nutlets 3 mm. long; perennial?;
   a native of India, collected once in Florida: South Bay, s.
   of Lake Okeechobee, Palm Beach County [*W. M. Buswell*, 1 May
   1942, FLAS]. April - May.

                                    C. furcatum Wall. in Roxb.

HELIOTROPIUM

    The only significant studies of the Florida Heliotropes are
those of Johnston (Cont. Gray Herb. 81:21-23. 1928; Jour. Arnold
Arb. 30:133-138. 1949), and even those are only incidentally
applicable to Florida plants. A recent short note by Long
(Rhodora 72:32-33. 1970) has proposed a new varietal combination
that thése earlier papers would suggest is unneeded.

    Long, as others have been, was impressed with the great sim-
ilarity among three supposed Florida species -- *Heliotropium
polyphyllum* Lehm., *H. leavenworthii* Torr., and *H. horizontale*
Small -- similarities that Small (Manual of the Southeastern
Flora, 1933) managed very largely to obscure. Outside of the
literature influenced by Small, this grouping has long been known
by the first of these three names. Long saw these names as representing two varieties: var. *polyphyllum* (including *H. leavenworthii*) was erect, often strictly so, while var. *horizontale*
(Small) Long, his new combination, included plants that were
spreading-decumbent to prostrate. He found these two varieties
to be readily distinguishable. In contrast, he found plants with
white or with yellow corollas to be unworthy of taxonomic distinction; *H. leavenworthii* had been separated from *H. polyphyllum*
primarily on its yellow flowers.

    Johnston's papers were not cited by Long and in part compel
a different interpretation. The type for Lehmann's *H. polyphyllum* was from coastal northern South America, an area in which
Johnston (1928) made particular study of *Heliotropium*. It thus
becomes untenable to treat the type of the species as erect, as
does Long, in the face of Johnston's observation that "all South
American materials referrable to *H. polyphyllum* appears to be
prostrate-spreading...."

    Johnston does express qualifications as to the Florida
plants so to make the reciprocal of his statement not necessarily
true: the prostrate Florida plants may themselves not be

identical with the South American type. Resolution of this will
come only with further study. But if varietal distinction is
desired for the upright plants that predominate in Florida, *H.
polyphyllum* var. *leavenworthii* Gray (1874) is available and unam-
biguous. (It is a curious circumstance that "*Heliotropium
Leavenworthii* Torr.," so widely used on specimens of the distinctive yellow-flowered Florida plant, appears not to be properly
recorded as to authorship. This epithet was published by Gray in
varietal status, with only indirect attribution to Torrey. Seemingly Small (Flora of the Southeastern States, 1903) was the
first to give it legitimate specific status; as such it becomes
H. leavenworthii (Gray) Small.)

     The color variations must not be too lightly dismissed.
Collectors in Dade County and on the Florida Keys have repeatedly
observed that plants growing in close proximity may differ only
in the white or yellow colors of their corollas. Yet in most
areas of Florida one color or the other predominates or, more
commonly, is exclusive. This distribution is approximately given
in the accompanying key. The significance of this pattern is unknown, although a reasonable interpretation might be that Florida
plants are descended from relatively few introductions of different genic systems further south in the Caribbean. In conformity
with both Johnston and Long, these color forms are not here given
formal designations.

     Equal uncertainty accompanies the level of distinction to be
accorded the prostrate to decumbent Florida plants. Small named
these *Heliotropium horizontale* (Bull. New York Bot. Gard. 3: 435-
436. 1905) on the basis of yellow-flowered collections obtained
in the rocky pinelands north and west of Homestead, Dade County.
He continued with this habit and range in his later publications.
Long, although including Small's types under the new var. *horizontale*, termed the plant a "maritime ecotype," expanded the
range northward to Palm Beach on the east coast and Pinellas
County on the west coast, included white-flowered specimens
(*Perkins*, GH), and increased the permissible height to 2 feet
(*Rehder 853*, GH). Defined in this fashion, var. *horizontale* cannot be distinguished from the preponderant Florida form. Yet *H.
horizontale*, in the original restricted usage of Small, is worthy
of some degree of recognition, at least in its most extreme prostrate form. Since it is surely close to, if not identical with,
the typical *H. polyphyllum* of northern South America, it is here
tentatively termed var. *polyphyllum*, with var. *leavenworthii* re-
served for the much more common upright specimens.

                    Heliotropium L.   Heliotropes

1. Plant completely glabrous, very succulent, usually somewhat
   glaucous; leaves linear-spatulate, fleshy; stems prostrate

with ascending shoots; corollas white with yellow eye; perennial herb; a frequent plant of coastal shores, usually just
above high tide, or on spoil banks or waste areas; Florida
Keys north Tampa Bay and to Cape Canaveral, occasionally
inland (Putnam, Seminole counties), but there only on brackish
soils. March - August.
SEASIDE HELIOTROPE.   H. curassavicum L.

1. Plant hairy, not decidedly succulent, never with a waxy bloom.

    2. Delicate annual, not above 2 dm. tall; some flowers sub-
    tended by foliaceous bracts 2 - 3 times length of flower;
    corolla minute, white; rare, open hammock, Key West (for-
    merly), Sugarloaf Key. August. [*H. phyllostachyum* Lehm.]
                                                            H. fruticosum L.

    2. Perennial or annual, if annual, lacking foliaceous bracts.

        3. Perennial; leaves sessile, linear to narrowly ovate;
        stems prostrate to erect.

            4. Flowers purple with yellow eye; inflorescence a clus-
            ter of 2 - 5 prominently scorpioid cymes; leaves to
            7 cm. long, 1.5 cm. wide; carpels 2, remaining intact
            at maturity, each with 2 seeds; plants sprawling to
            erect; sporadic, but often abundant locally, dry soil
            of roadsides and waste area, occasionally a lawn
            weed, north and central peninsular Florida, less
            often in the Panhandle. March - August. [*H.*
            anchusaefolium Poir. in Lam.; *Cochranea anchusaefolia*
            (Poir.) Guerke]
            WILD HELIOTROPE.   H. amplexicaule Vahl

            4. Flowers white, white with yellow eye, or yellow;
            inflorescence largely unbranched, straight or apical-
            ly gently curving; leaves to 2 cm. long, 0.2 - 0.3
            cm. wide; carpels 2, each splitting at maturity, to
            form 4 separate nutlets; stems prostrate to erect;
            frequent, marl prairies, moist pinelands, savannas,
            brackish shores, and roadbanks; south Florida, north
            along coasts to Taylor County on west and Volusia
            County on the east. [This species exhibits complex
            and as-yet unexplained variations of flower color and
            habit. In Dade County and on the Florida Keys both
            white and yellow flowers occur. White predominates
            from Collier County northward to beyond Tampa Bay,
            and inland to Highlands County. Only yellow occurs
            north of Hernando County, in the Pinecrest area of
            Monroe County, and along the east coast north of Dade
            County. Particularly near Homestead, Dade County,
            but occasionally elsewhere, prostrate plants occur;

these may be termed var. *polyphyllum*. They intergrade
with much more widespread ascending to erect plants,
which may be distinguished as var. *leavenworthii*
Gray.] All year. [incl. *H. Leavenworthii* Torr.; *H.
horizontale* Small; *H. polyphyllum* var. *horizontale*
(Small) Long]

H. polyphyllum Lehm.

3. Annual; leaves petiolate, variously broad, not linear;
stems erect.

5. Flowers light purple with white or yellow eye; fruits
2 - 2.5 mm. long, with sharp longitudinal ridges,
separating into two 2-seeded nutlets; leaves broadly
ovate, the blades 3 - 5 cm. across; weedy herb; uncommon and sporadic, floodplains of Choctawhatchee
River and Apalachicola River (Washington, Jackson,
Calhoun, Gadsden and Liberty counties). August –
September. [*Tiaridium indicum* (L.) Lehm.]

H. indicum L.

5. Flowers white or with yellow eye; fruits 1 - 1.5 mm.
long (excluding persistent style base if present),
without prominent ridges; leaves narrowly ovate, elliptic, to spatulate, 1.5 - 2.5 (- 3.5) cm. across.

6. Fruits very much broader than long, separating
into two 2-seeded nutlets; style lacking on mature
fruits; leaves becoming blackened upon drying;
common, hammocks, waste areas, shell mounds,
citrus groves (where often weedy), roadsides, and
saline shores; south Florida, north in coastal
counties to Tampa Bay on the west and Volusia
County (Turtle Mound) on the east. All year. [*H.
parviflorum* L.; *Schobera angiosperma* (Murr.)
Britt.]
SCORPION-TAIL   H. angiospermum Murr.

6. Fruits scarcely broader than long, separating into
four 1-seeded nutlets; style persisting on mature
fruits as a sharp dark beak; leaves remaining
green upon drying; rare, an erratic introduction
on the wooded floodplain of the Apalachicola
River, Calhoun County. August - September. [This
species, although newly discovered in Florida (*R.
K. Godfrey 75520*, 14 Sept 1976, FLAS, FSU) is familiar along the Gulf Coast westward, where it has
been erroneously known as *Heliotropium europaeum*
L. (see Excluded Species).]

H. procumbens Mill.

Excluded Species

Heliotropium europaeum L.   This species was collected on
"waste ground, Pensacola, Florida," in August 1901 [*A. H. Curtiss
6864* (GH), fide C. E. Wood].   Recent active collectors in the
Pensacola area have not encountered it, and it is assumed not to
have persisted in the state.   Its white to bluish flowers,
minutely pubescent verrucose nutlets to 3 mm. long, and hirsute
stems permit separation from the white (drying yellowish)
flowers, strigose but otherwise smooth nutlets to 1.5 mm. long,
and appressed-pubescent stems of the closely related *H. procumbens* Mill.

LITHOSPERMUM

This genus has been well treated by Johnston (Jour. Arnold
Arb. 33:299-363. 1952) except for his near-total omission of detailed distributional data.   *Lithospermum incisum* is more extensively adventive than has been previously recognized.   This
species is also characterized by the presence of numerous very
small cleistogamous flowers, from which most of the nutlets are
produced, following withering of the conspicuous and large
chasmogamous flowers.

The authorship of *Lithospermum caroliniense* has by now been
fully argued (Wilbur, Jour. Elisha Mitchell Sci. Soc. 78:125-132.
1962; Ward, Rhodora 64:87-92. 1962).

Lithospermum L.   Puccoons

1. Stems arising from a cluster of basal leaves very much larger
   than the cauline ones; leaves obovate to elliptic; corollas
   small (to 6 mm. long), yellow or infrequently cream; roots
   non-purpling, fusiform and fascicled; perennial herb; local,
   on moist hardwood slopes and calcareous bluffs, north Florida,
   disjunct in area of upper Apalachicola drainage (Jackson to
   Liberty counties) and east of Suwannee River (Suwannee to
   Alachua counties).   March – April.
                                     L. tuberosum Rugel ex DC.

1. Stems without basal leaves or basal leaves not appreciably
   larger than the cauline ones; leaves lanceolate to linear (or
   if ovate, reduced toward base); corollas of chasmogamous
   flowers large (15 – 35 mm. long); root often causing purpling
   of pressing and mounting papers.

   2. Stems arising from a stout vertical root which is almost
      always broken in collecting; root strongly purpling col-
      lecting papers, the stain characteristically penetrating

sheets upon which plant is mounted; corollas bright orange-
yellow, the lobes entire; stout perennial herb; frequent,
dry pinelands, western panhandle Florida, east to Apalachicola River drainage. March - May (August). [*Batschia
caroliniensis* (Walt.) Gmel.]
PUCCOON.   L. caroliniense (J. F. Gmel.) MacMill.

2. Stems arising from slender taproot, often fully collected;
   root lightly purpling contiguous papers; corollas bright
   yellow, the lobes erose; perennial herb; infrequent and
   sporadic, adventive along sandy roadsides and railroads,
   occasionally on limestone ledges, north Florida, south to
   Hernando and Seminole counties. March - April. [*Batschia
   linearifolia* (Goldie) Small]

                                          L. incisum Lehm.

MALLOTONIA

    Florida has one species of this genus, *Mallotonia gnaphal-
odes*, the Sea-lavender. It is often placed within *Tournfortia*,
although its maritime habitat and linear-spatulate gray-pubescent
leaves at least superficially set it sharply apart from the
species allied to *T. hirsutissima*, the type of that genus.
Johnston, the foremost recent student of this group, has not been
consistent in its assignment. Initially (Cont. Gray Herb. 92:66-
89. 1930) he saw the Sea-lavender as within *Tournefortia*, then he
segregated it with two closely related Old World species to form
Messerschmidia (Jour. Arnold Arb. 16:161-166. 1935), and still
later returned it to *Tournefortia* (Jour. Arnold Arb. 30:129-133.
1949).

    Messerschmidia had been described originally by Linnaeus
(Mantissa 42, 1767; as "*Messersmidia*") as applying to one of the
Old World species (*M. sibirica*, of Asia) included by Johnston.
The name has undergone various interpretations by later authors,
as well as several spellings. Johnston (1930) discussed these
aspects rather fully, concluding, "...the confusion that has
attended the history of the name is quite sufficient to warrant
its rejection as a nomen confusum, at least as far as our American plants are concerned." Later Johnston (1935) reversed field,
finding that *Messerschmidia* was indeed an acceptable generic name.

    Britton (Ann. Missouri Bot. Gard. 2:47. 1915) transferred
the Sea-lavender without comment to form the monotypic *Mallotonia*,
and has been followed by Small (Manual of the Southeastern Flora,
1933), Gooding et al. (Flora of Barbados, 1965), Adams (Flowering
Plants of Jamaica, 1972), Gillis (Rhodora 76:111. 1974), and
others. The rationale for this placement seems never to have
been fully discussed, although Johnston (1935), in recognizing

Messerschmidia, called attention not only to its species' wide departure in general appearance from the species of *Tournefortia*, sensu stricto, but to anatomical differences in their corky bark and pubescence structure.

Nor has there been adequate discussion -- or perhaps not adequate realization -- that *Messerschmidia* was used by Linnaeus so many years earlier for a plant that is surely congeneric. It is only by acceptance (by implication, if not by overt intent) of Johnston's argument, which he later abandoned, of the invalidity of *Messerschmidia*, that *Mallotonia* can be seen as the correct segregate name. Perhsps ultimately, if the consensus remains firm that the Sea-lavender (with its two Old World allies) merits generic segregation from *Tournefortia*, conservation of *Mallotonia* via the International Code will provide more certain stability.

<u>Mallotonia</u> Britt.   Sea-lavender

1. Leaves densely clustered at ends of twigs, linear-spatulate, 4 - 9 cm. long, covered with silky gray pubescence; inflorescence a long-peduncled very congested one-sided cyme; flowers white with pink tinge in throat, small (4 - 5 mm. long); fruit a dry brown 2-seeded drupe; small erect shrub, to 2 m. tall; infrequent, on front line of dunes, outer edge of salt flats, always fronting on ocean; not found on quiet bays or other low-energy coasts; Florida Keys, northward only along east coast, to Cape Canaveral. December - March. [*Tournefortia gnaphalodes* (L.) R. Br. ex R. & S.]
SEA-LAVENDER   <u>M. gnaphalodes</u> (L.) Britt.

MYOSOTIS

The specific separation of *Myosotis macrosperma* from *M. virginica* is adequately supported on the basis of available materials. The merits of such a separation have been discussed by Fernald (Rhodora 41:558. 1939; 43:637. 1941), while the contrary view has been presented by Steyermark (Flora of Missouri, 1963).

Myosotis virginica (L.) BSP. (=*M. verna* Nutt.) has not been collected in Florida. The basis of its report for this state by Fernald (Gray's Manual of Botany, 1950) is a collection from Chattahoochee (*A. H. Curtiss*, GH). The collection has been examined by C. E. Wood who believes it to be *M. macrosperma*. Since the alluvial river edge at Chattahoochee is one of the Florida stations for *M. macrosperma*, there is no hesitancy in accepting Dr. Wood's determination.

Myosotis L.   Forget-me-nots

1. Corolla small, white; calyx covered with hooked hairs, the two
   lower lobes appreciably longer than the three upper ones; leaves
   oblong to spatulate; annual soft-pubescent herb; very local,
   alluvial deposits along bank of Apalachicola River, Gadsden and
   Liberty counties. March. [*M. virginica* (L.) BSP. var. *macrosperma* (Engelm.) Fern.]

                                        M. macrosperma Engelm.

ONOSMODIUM

     This small genus has been discussed in considerable detail by
Johnston (Contr. Gray Herb. 70:17-18. 1924; Jour. Arnold Arb. 35:18-
24. 1954).

          Onosmodium Michx.   False Gromwells

1. Corolla 8 - 12 mm. long, exceeding calyx, cream at base with the
   lobes yellow-green; style undivided, long-exserted (to twice
   length of corolla), persisting on young fruits; nutlets light
   gray, smooth and shining, only one maturing per flower; leaves
   elliptic to obovate, harshly pubescent; perennial herb; frequent,
   dry open sandy woods and roadsides, north Florida, south to
   Hillsborough and Highlands counties. March - April.
   FALSE GROMWELL.   O. virginianum (L.) A. DC.

TOURNEFORTIA

     Tournefortia volubilis has long been recognized as a member of
the Florida flora. Small (Manual of the Southeastern Flora, 1933)
admitted a second very closely allied species, *T. poliochros* (under
his segregate name, *Myriopus poliochros*). Although recent West
Indian treatments recognize these two as specifically distinct, the
differences are almost inconsequential as exemplified by Florida
collections, and are not here maintained.

     All Florida plants of this complex are more or less pubescent.
In southern Florida all collections are very lightly appressed
sericeous on the lower leaf surface, only occasionally developing
a gray cast. A single series of collections from Green Mound,
south of Daytona Beach (*Small, Small, DeWinkeler 10726*, 7 Sept 1922;
FLAS) appears to be the only basis for the white-canescent form that
has been called *T. poliochros*.

Tournefortia L.   Tournefortias

1. Leaf blade 3 - 5 cm. long, ovate to elliptic lanceolate; pubescence fine, closely appressed, scant to hoary gray on lower leaf surface; drupe white with small black spots, 1 - 4 seeded, each seed forming a separate lobe under the tightly stretched flesh; corolla dark yellow to greenish white, the lobes subulate; climbing and scrambling woody vine, occasionally free-standing as a low shrub; hammocks, thickets, shell mounds; frequent in Florida Keys and southern Dade County, disjunct on west to Hillsborough County (Cockroach Bay), rare along east coast (Merritt Island, Brevard County) and north to now-extirpated station at Green Mound, Volusia County (this station is the basis for *T. poliochros* in Florida, a variant with leaves hoary gray below). (December) March - August. [*Myriopus volubilis* (L.) Small; *T. poliochros* Spreng. in L.; *Myriopus poliochros* (Spreng.) Small] SOLDIER-BUSH.   T. volubilis L.

1. Leaf blade 7 - 18 cm. long, elliptic; pubescence coarse, spreading, particularly abundant on young stems; drupe uniformly white, spherical, usually 4-seeded; corolla white, the lobes ovate; robust scrambling vine; occasional, tropical hammocks; southern Dade County, Monroe County (but absent from the Florida Keys), north to the Fahkahatchee Slough of Collier County. March - May. HAIRY TOURNEFORTIA.   T. hirsutissima L.

EXCLUDED GENERA

    'Borago officinalis L. Borage. Two Florida collections of this Mediterranean species have been made (FLAS) -- both in Alachua County. Both are believed to represent cultivated plants. Borage in Europe has long been cultivated as a flavoring for beverages; although introduced and sparingly escaped in eastern North America it seems scarcely adapted to independent survival here.

INDEX TO THE RUBIACEAE BY JULIAN A. STEYERMARK

IN THE BOTANY OF THE GUAYANA HIGHLAND

BY B. MAGUIRE AND COLLABORATORS

Compiled by Joseph H. Kirkbride, Jr.[1].

Julian A. Steyermark has now completed his treatment of the Rubiaceae for the Botany of the Guayana Highland. It appeared as four installments in the Memoirs of The New York Botanical Garden: 10(5): 186-278, 17 Feb 1964, 12(3): 178-285, 10 Sep 1965; 17(1): 230-436, 22 Dec 1967; 23: 227-832, 30 Nov 1972. The importance of this work and its structure make it imperative that a complete index to it be available.

This index includes all taxa that appear in the foregoing papers. New taxa, new combinations, new names, new statuses, and emendations are underlined. Names in synonymy are indicated by underlined page numbers. Illustrations are indicated by the letter 't' preceding the page number.

Acanthaceae 23: 232
Alibertia 12: 211, 222, 223;
  23: 227, 346, 356
  acuminata 12: 222, 223, 224,
    225
    var acuminata 12: 222,
      223
    var obtusiuscula 12: 222,
      223
  bertierifolia 12: 222, 223
  dolichophylla 23: 355, 356
  edulis 12: 219, 222, 223,
    224, 225, 226; 23: 356
  granulosa 12: 225
  hispida 12: 211, 23: 356
  latifolia 12: 222, 223,
    224, 225

  var latifolia 12: 225
  var paragueniana 12: 225
  var parvifolia 12: 225
  longistipulata 12: 224
  myrciifolia 12: 222, 226,
    227
    var myrciifolia 12: 226
    var tepuiensis 12: 226
  obidensis 12: 221
  panamensis 12: 224
  stenantha 23: 356
  surinamensis 23: 355
  triflora 12: 222, 223, 226
  triloba 12: 222, 227;
    23: 356
  trinitatis 12: 222

  1. Smithsonian Institution, U. S. National Museum, Washington, D. C. 20560.
  Errors or omissions in this index are the responsibility of the compiler.

324

Coffea 17: 354
  calycina 17: 360; 23: 385
  crassiloba 411
  didymocarpa 23: 517
  flavicans 17: 395
  guianensis 17: 380
  herbacea 17: 380
  laurifolia H.B.K. 17: 423, 424
  laurifolia Salisb. 17: 423
  occidentalis 17: 384
  sessilis 23: 715
  spicata 23: 590, 591
  (?) stipulacea 17: 421
  subsessilis 23: 713
  tenuiflora 17: 357; 23: 386
  truncata 17: 374
Commianthus 12: 227
  concolor 12: 229
  discolor 12: 244
  pilosus 12: 232
  schomburgkii 12: 240
  speciosus 12: 243
Compositae 23: 655
Condalia
  lanceolata 17: 306
  repens 17: 300, 301, 302
Condaminea
  utilis 12: 187
Condamineae 12: 178, 185
Conosiphon 23: 322
  aureus 23: 322, 323
  polycarpon 23: 324
  striiflorus 23: 322
Cordia
  poeppigii 23: 370
Cordiera
  acuminata 12: 222, 225
  latifolia 12: 225
  myrciifolia 12: 226
  surinamensis 17: 359;
    23: 386
  triflora 12: 226
Coryphothamnus 12· 263, 264
  auyantepuiensis 12: 264,
    t265
Cosmibuena 23: 227, 295
  arborea 23: 295
  grandiflora 23: 295
    var grandiflora 23: 295
    var latifolia 23: 295,
      296
  latifolia 23: 295, 296

obtusifolia 23: 295, 296
  var latifolia 23: 296
quinqueflora 23: 295, 296
triflora 23: 295
Coupoui
  aquatica 12: 210
  brasiliensis 12: 209
  martiniana 12: 209
  micrantha 12: 205
Coussarea 17: 360, 396, 412;
    23: 227, 451, 489, 493
  amapaënsis 17: 362, 366
  benensis 17: 365
  bernardii 17: 361, 371
  brevicaulis 17: 361, 364,
    365
  capitata 17: 366
  fanshawei 17: 361, 365
  froelichia 17: 367
  grandis 17: 360, 363, 23: 389
  hallei 23: 387
  hirticalyx 17: 362
    var glabrior 17: 360, 363
    var hirticalyx 17: 362,
      363
  hyacinthiflora 17: 369
  klugii 17: 361, 363
  lasseri 17: 361, 366
  leptoloba 17: 360, 362
    var leptoloba 17: 362
    var mutisii 17: 362
  leptophragma 17: 361, 368, 370;
    23: 489, 495
  longiflora 17: 361, 364, 365
    var benensis 17: 364, 365
    var longiflora 17: 364
  mapourioides 17: 361, 367
  martini 17: 367
  micrococca 17: 361, 370
  moritziana 17: 361, 370, 371
  mutisii 17: 362
  paniculata 17: 361, 367
  pentamera 17: 370
  pittieri 17: 361, 366
  racemosa 17: 362, 369
  revoluta 17: 361, 368
  scalaris 17: 387, 388
  schomburgkiana 17: 368
  sprucei 17: 368
  surinamensis 17: 362, 369
  violacea 17: 360, 361, 362,
    368, 369

rogaguana 23: 790
sarmentosa 23: 789, 791,
   792
setigera 23: 801, 802, 803,
   804
simplex 23: 789
spicata 23: 789, 790
surinamensis 23: 794, 795,
   797
teres 23: 788, 798
   subsp angustata 23: 799
      var angustata 23: 799
         f angustata 23: 798,
            799
         f latior 23: 798,
            799
   subsp prostrata 23: 799
      var prostrata
         f latifolia 23: 798,
            800
         f leiocarpa 23: 798,
            800
         f prostrata 23: 798,
            800
   subsp teres 23: 798
virginica 23: 788, 790
Dioicodendron
   dioicum 12: 185
Duggena
   grisea 23: 313
   hirsuta 23: 313, 314
   incanescens 23: 313, 314
Duidania 17: 230, 232; 23:
   232, 233
   montana 17: 230, t231
Duroia 12: 198, 211; 23:
   227, 346, 356
   amapana 12: 198, 209
   aquatica 12: 198, 210;
      23: 343
   bolivarensis 12: 200, 206
   duckei 12: 199, 201;
      23: 345
   eriopila 12: 198, 199,
      201, 204, 210, 218, 219
      var brevidentata 12:
         201, 203
      var eriopila 12: 203
         f eriopila 12: 203
         f glabra 12: 201, 203
      var tafelbergensis 12: 203
   fusifera 12: 199, 204, 205,

207, 218
genipoides 12: 200, 205,
   207, 218
gransabanensis 12: 199, 205
hirsuta 12: 198, 200, 201,
   218
kotchubaeoides 12: 198, 201,
   t202; 23: 356
longiflora 12: 200, 209
longifolia 12: 210, 211, 218
macrophylla 12: 198, 209;
   23: 343
maguirei 23: 343, t344
   var macrocarpa 23: 345
   var maguirei 23· 345
   var patentinervia 23: 345
martiniana 12: 198, 209
merumensis 12: 200, 207
nitida 12: 200, 208
oocarpa 12: 211, 218
palustris 12: 199, 204
paraensis 12: 200, 207
paruensis 12: 200, 206;
   23: 356
petiolaris 12: 199, 204,
   218; 23: 356
plicata 12: 198, 210
prancei 23: 345
retrorsipila 12: 200, 208
saccifera 12: 198, 210, 218
sprucei 12: 199, 200, 201,
   205, 207
steinbachii 12: 211
stenophylla 12: 211; 23: 356
strigosa 12: 199, 204
surinamensis 12: 203
trichocarpa 12: 211
triflora 12: 199, 204; 23:
   345, 356
velutina 12: 198, 209, 219
Einsteinia
   sericantha 10: 214
   speciosa 10: 214
Elaeagia 12: 185, 186, 192;
   23: 306
   alterniramosa 12: 187
   asperula 12: 186, 191
   barbata 12: 186, 191
   brasiliensis 12: 180, 192
   cuatrecasasii 12: 186, 191
   cubensis 12: 187, 189
   ecuadorensis 12: 186, 191

Additions to the Flora of Futuna Island,
   Horne Islands
         Pacific Plant Studies 33

              Harold St. John
      B. P. Bishop Museum, Honolulu, Hawaii

   In 1971 the first account of the flora of Futuna
Island was printed by St. John and Smith (1971). At
that time the known flora was 152 species. This
total can be segregated into 2 endemics; 91
indigenous species, including the 22 ferns; 14
adventives; and 45 cultivated ornamentals and
crop plants.
   Now, there has come to hand a new collection
from the island, made in June and July 1974, by
the anthropologist Patrick V. Kirch, of the
Bishop Museum. It consists of 168 specimens.
Besides the species already known on the island,
the Kirch collection adds 1 new endemic; 12
indigenous; 9 adventives; and 26 ornamentals or
crops.
   The additional and significant species are here
listed. The native species are printed in italic.

                 Pteridophyta
                 Adiantaceae
Antrophyum reticulatum (Forst. f.) Kaulf. Nuku
   Singave, on vertical rock face in humid forest,
30-40 m alt., Kirch 150.
                 Phanerogamae
                 Monocotyledones
                 Gramineae
Bambusa vulgaris Schrad. ex Wendl. "kofe fiti."
   Matutufu, Singave, stream bank, 10 m alt.,
   Kirch 52.
Echinochloa colonum (L.) Link. Nuku, Singave,
   pondfield borders, common, weed, Kirch 114.
Ischaemum rugosum Salisb. "vao papalangi."
   Nuku, Singave, weed, border of pondfields,
   Kirch 113.
Miscanthus floridulus (Labill.) Warb. "u."
   Maunga, Nuku, Singave, 2nd growth, 175 m alt.,
   Kirch 100.

Oplismenus compositus (L.) Beauv. "mutie." Aloalo,
  Nuku, Singave, weed, dry garden zone, 20 m alt.,
  Kirch 71.
Paspalum conjugatum Bergius. "mutie." Aloalo,
  Nuku, Singave, dry garden zone, weed, 20 m
  alt., Kirch 66.
Schizostachyum glaucifolium (Rupr.) Munro.
  "kofe Futuna." Leava Valley, Singave, cookhouse
  zone, Kirch 63.

Cyperaceae

Cyperus odoratus L. Nuku, Singave, pondfield
  embankment, weed, 25 m alt., Kirch 136.
Eleocharis geniculata (L.) R. & S. "kutu." Kirch 109.
E. orostachys Steud. "kutu." Nuku, Singave, abundant in pondfield, 10 m alt., Kirch 110.

Araceae

Caladium bicolor (Ait.) Vent. "kape faka teuteu."
  ornamental, cookhouse zone, Kirch 49.
Colocasia esculenta (L.) Schott, var. antiquorum
  (Schott.) Hubb. & Rehd. "talo uli," a new name,
  Singave, Kirch 54.
Cyrtosperma Chamissonis (Schott) Merr. "pulaka."
  Aloalo, Nuku, Singave, dry garden zone, Kirch 11.
Epipremnum pinnatum (L.) Engl. Aloalo, Nuku,
  Singave, on trees, dry garden zone, Kirch 69.
Xanthosoma atrovirens C. Koch & Bouché, "talo
  fiti uli." Aloalo, Nuku, Singave, cult., dry
  garden zone, Kirch 2.

Dioscoreaceae

Dioscorea alata L. "ufi fakasoa." abandoned hill
  garden, 50 m alt., Kirch 124.
D. esculenta (Lour.) Burkill, with three cultivars:
  "ufi lei," Aloalo, Nuku, Singave, dry garden
    zone, 20 m alt., Kirch 5.
  "ufi lei vai," Aroa-Vele uplands, Alo Dist.,
    75-100 m alt., Kirch 17.
  "ufi lei lotuma," ditto, Kirch 18.

Cannaceae

Canna indica L. "fanagana." Nuku, Singave, cookhouse zone, 10 m alt., Kirch 57.

Piperaceae

Peperomia pallida (Forst. f.) A. Dietr., var.
  tuamotensis (F. Br.) Yuncker, on rock face of
  irrigation ditch, 10 m alt., Kirch 168.

Moraceae

Artocarpus altilis (Parkins. ex Z) Fosb. "mei aveave." Nuku, Singave, cookhouse zone, cult., Kirch 62.

Urticaceae

Elatostema Yenii St. John. "lole." Nuku, Singave, irrigation ditch embankment, 20 m alt., Kirch 137. New island record.

Maoutia australis Wedd. Maunga, Nuku, Singave, 2nd growth, 150-175 m alt., Kirch 82; ditto 83.

Amaranthaceae

Alternanthera sessilis (L.) R. Br. "vao." Aloalo, Nuku, Singave, dry garden zone, weed, Kirch 70; ditto, 111.

Leguminosae

Caesalpinia pulcherrima (L.) Sw. "oai." Nuku, Singave, cookhouse zone, 10 m alt., Kirch 58.

Pueraria lobata (Willd.) Ohwi. "aka." Asoa-Vele uplands, Alo Dist., in swidden garden, Kirch 22; 23; 41.

Vigna marina (Burm.) Merr. "fue." fallow pondfield, 10 m alt., Kirch 117.

Rutaceae

Citrus sinensis (L.) Osbeck. "moli." Nuku, Singave, 2nd growth, 50 m alt., Kirch 162.

Euphorbiaceae

Codiaeum variegatum (L.) Bl. "lafakau kula." Nuku, Singave, cookhouse zone, ornamental, Kirch 36.

Tiliaceae

Grewia crenata (J. R. & G. Forst.) Schinz & Guillem. "iti." Maunga, Nuku, Singave, 2nd growth, 150-175 m alt., Kirch 87.

Malvaceae

Hibiscus tiliaceus L. "fau fatu." New vernacular name. Fata-asau, Nuku, Singave, 2nd growth, 100 m alt., Kirch 97.

Sterculiaceae

Melochia vitiensis Gray. "ito." Asoa-Vele uplands, Alo Dist., 2nd growth, 75-100 m alt., Kirch 24.

Passifloraceae

Passiflora maliformis L. "pasio." Aloalo, Nuku, Singave, dry garden zone, cult. Kirch 74.

Melastomataceae

Medinilla racemosa sp. nov.
  M. samoensis sensu St. John & A. C. Smith,
    Pacif. Sci. 25: 335, 1971, non (Hochr.)
    Christophersen, Bishop Mus., Bull. 154:
    30, 1938.   Fig. 1.
    Diagnosis Holotypi: Liana lignosa in arbor-
ibus scandens est, ramulis 3-7 mm diametro
subcarnosis, cortice stramineo laevi sed in sicco
cun fugis longitudinalibus, internodis 1.5-3.3
cm longis, foliis oppositis glabris, cicatricibus
    foliorum 5 mm latis lunatis, cicatricibus
    fascicularum 7 in ordini circulari, petiolis
    5-30 mm longis, laminis 4-12 cm longis 3-8.5
    cm latis late ellipticis basi cuneata 5-pli-
    nervatis supra obscure viridibus infra viridibus
    et cum nervis elevatis, racemis 3-5 cm longis
    fere 9-13-floriferis puberulentis cum pilis
    badiis scabris, bracteis ex axilibus 5-7 mm
longis spatulatis, pedicellis 7 mm longis
1-floriferis et proxima calycem cum 2 bracteis
2.7 mm longis anguste lanci-ellipticis foliaceis,
calycibus cum tubo 6.2 mm longo hemisphaerico
firmo pallido sed lobis nullis, 5 petalis 11 mm
longis albis sed roseitinctis obovatis nigri-apic-
ulatis, staminibus 8, filamentis 6 mm longis
ligulatis membranaceis, antheris 5 mm longis 1.8
mm latis cum 2 lobis basilibus tomaculiformatis
divergentibus obscuris firmis, sacis fertilibus
1.8 mm longis erectis cylindraceis obtusis et a
foramini terminali dehiscentibus, ovario 2.5 mm
diametro subgloboso, stylo 6 mm longo filiformi,
stigmate vadose convexo, (fructibus incognitis).
    Diagnosis of Holotype: Woody vine, climbing on
trees; branchlets 3-7 mm in diameter, somewhat
fleshy, with smooth stramineous bark, when dried
with longitudinal ridges; internodes 1.5-3.3 cm
long; leaveˢopposite, glabrous; leaf scars 5 mm
wide, lunate; bundle scars 7, in a circular pattern;
petioles 5-30 mm long; blades 4-12 cm long,
3-8.5 cm wide, broadly elliptic, the base cuneate,
5-plinerved, above dark green, below green with
outstanding nerves; racemes 3-5 cm long, about
9-13-flowered, puberulent with brown scabrous
hairs; axillary bracts 5-7 mm long, spathulate;

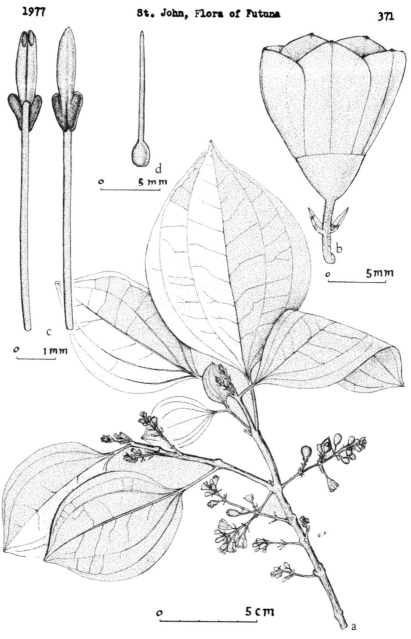

Fig. 1. **Medinilla racemosa**

pedicels 7 mm long, 1-flowered, and just below
the calyx with 2 bracts 2.7 mm long, narrowly
lance-elliptic, foliaceous; calyx tube 6.2 mm long,
hemispheric, firm, pale, without distinct lobes;
petals 5 and 11 mm long, white, tinged with pink,
obovate, darkly apiculate; stamens 8; filaments
6 mm long, ligulate, membranous; anthers 5 mm long,
1.8 mm wide, the 2 basal lobes 0.4 mm long,
sausage-shaped, dark, firm, divergent, the fertile
sacs 1.8 mm long, erect, cylindric, obtuse,
dehiscent by a terminal pore; ovary 2.5 mm in
diameter, subglobose; style 6 mm long, filiform;
stigma low convex, (fruit unknown).

Holotypus: Polynesia, Horne Islands, Futuna
Island, Nuku, Singave, humid forest, climbing on
trees, 100 m alt., July 12, 1974, Patrick V.
Kirch 130 (BISH).

Specimens Examined: Polynesia, Horne Islands,
Futuna Island, ditto, Kirch 129 (BISH); pentes
sud de Mt. Puke, restes de forêt humide sur terrain
volcanique, 500-600 m alt., 27.X.1968, Mackee
19,852 (BISH).

Discussion: M. racemosa is most closely related
to M. samoensis (Hochr.) Christophersen, of
Upolu, Tutuila, and Savai'i in Samoa, a species
with the young shoots brown hirsutulous with scabrous hairs, these more or less persistent;
inflorescences umbellate, hirsutulous; petals 5 mm
long. M. racemosa has the leaves glabrous, and
the young stems nearly so; inflorescences racemose,
puberulent; and the petals 11 mm long.

The new epithet is the Latin adjective racemosa,
with a raceme, and it refers to the inflorescence
of the species.

Onagraceae

Ludwigia octivalvis (Jacq.) Raven, subsp octivalvis.
"ta'ekana." Nuku, Singave, taro terraces, weed,
Kirch 115.

Convolvulaceae

Ipomoea brasiliensis (L.) Sweet. "fue kau kula."
Matatufu, Nuku, Singave, second growth, abandoned swidden garden, 20 m alt., Kirch 53.

Merremia peltata (L.) Merr. "pulupulou." New
vernacular name. Kirch 88.

Apocynaceae

<u>Alyxia bracteolosa</u> Gray. "maile kulu." New vernacular name. Fata-asau, Nuku, Singave, 2nd growth, 100 m alt., Kirch 93.

Labiatae

Coleus scutellarioides (L.) Benth. Matufu, Nuku, Singave, houseyard, weed, Kirch 43.

Solanaceae

Capsicum annuum L. "polo." Matufu, Nuku, Singave, cookhouse zone, Kirch 45.

Rubiaceae

<u>Canthium Merrillii</u> (Setchell) Christophersen. "funa." Nuku, Singave, humid forest, 30-40 m alt., Kirch 160.

Cucurbitaceae

Luffa cylindrica (L.) Roem. "timo vao." Nuku, Singave, 2nd growth, feral vine, 25 m alt., Kirch 138.

Compositae

Synedrella nodiflora (L.) Gaertn. "petelo." Aloalo, Nuku, Singave, dry garden zone, weed, Kirch 73.

Literature Cited

St. John, Harold and Albert C. Smith, 1971. The Vascular Plants of the Horne and Wallis Islands. Pacif. Sci. 25: 313-348.

The Flora of Niuatoputapu Island, Tonga
Pacific Plant Studies 32

Harold St. John
Bishop Museum, Honolulu, Hawaii, 96818

Geography

Niuatoputapu Island is almost at the northern end
of the Tongan chain and is very close to Samoa (140
miles to Upolu Island). Its closest neighbors are
the volcanic ash cone of Tafahi, only 5 miles to
the north, and Niuafo'ou (Tin Can) Island, some
120 miles to the west.

Niuatoputapu is of volcanic origin, with a central ridge of bedded tuff, breccia, and lavas
which rises to a maximum height of 165 meters.
Surrounding this central ridge is a terrace,
probably wave-cut in origin and Pleistocene in
age, now covered with a thick and fertile deposit
of clay soils. Falling away in a relatively steep
bluff, this terrace is in turn surrounded by an
apron-like plain of uplifted recent marine sedi-
ments: sand, coral cobbles, and other reef detri-
tus. This apron of low-lying terrain has been
uplifted as a result of tectonic activity within
the period of Polynesian occupation of Niutopu-
tapu. The island is 6.8 kilometers long and
4.5 kilometers wide.

The vegetation of Niuatoputapu has been extensively modified as a result of some 3,000 years
of Polynesian occupation. The present Tongan-
speaking population practices a form of shifting
cultivation (or bush-fallow rotation) in which the
principal garden type is the mixed yam-aroid
swidden (refer to P. V. Kirch, "Indigenous agriculture in Uvea" in press, Economic Botany) for
a description of a highly similar West Polynesian
agricultural system. The principal cultivated
yams are Dioscorea alata and D. esculenta;
among the cultivated aroids are Colocasia
esculenta, Alocasia macrorrhiza, and Xanthosoma
sagittifolium. Bananas, particularly diploid
hybrids of the Musa section, are extensively
planted in swiddens following the harvest of
yams and aroids. Arboriculture also is a sig-

nificant part of the indigenous agricultural
system, with Artocarpus altilis being the dominant crop. Inocarpus faqifer is also widely cultivated.

Exploration

The first plant collection on Niuatoputapu
was made by Dr. H. Hürlimann, of Basel, between
Dec. 5, 1951, and Jan. 2, 1952. He collected 79
species, and the specimens are deposited in the
herbaria of the University of Zürich, and of
DePauw University.

A recent collection was made by the anthropologist Dr. Patrick V. Kirch, of the Bishop Museum.
He collected between July and December, 1976,
and gatered 138 species of vascular plants, many
of which duplicated and confirmed those of Hürlimann, but 59 others were new locality records.

Flora

The flora as now known total 10 ferns and fern
allies and 201 phanerogams. These are classed as:

Indigenous	91
Endemic	1
Ornamentals	19
Crops	53
Weeds	37
	201 total flora

Comparisons have been made with the published
flora of Tonga by Yuncker (1959).

Many of the Pacific Islands have notable endemism in their floras, but it is not so with Niuatoputapu. Although politically placed in Tonga,
the island is closest to Samoa, which is 140
miles to the northeast. Futuna is 290 miles to
the northwest. Taveuni in Fiji is 400 miles to the
west. Niue is 320 miles to the southeast; and
Rarotonga is 1,020 miles to the southeast. Of
the larger islands of Tonga which lie to the south,
Vavau is 160 miles distant.

When the flora is compared with that of these
adjacent islands, it is found that all of the
indigenous species are also found in Samoa, except
10. These are: Pandanus Mbalawa, P. odoratissimus,
var. Setchellii, Ficus Storckii, Santalum Yasi,

Canavalia ssricea, Dysocylum Forsteri, Eugenia
dealata, Burckella Richii, Ipomoea macrantha,
and Solanum amicorum.

When compared with the flora of Fiji, it is
found that all of the indigenous species also
occur there, except 11. They are: Pandanus
odoratissimus, var. Setchellii, Rhus taitensis,
Melochia aristata, Eugenia clusiaefolia, Eugenia
dealata, Burckella Richii, Planchonella torricellensis, Hoya chlorantha, Solanum amicorum, and
Psychotria insularum.

When compared to Vavau, the nearest high
Tongan Island to the south, all of the indigenous
species are also there except: Psilotum nudum,
Nephrolepis biserrata, N. exaltata, Pandanus
angulosus, P. Mbalawa, P. turritus, Fimbristylis
pycnocephala, Taeniophyllum fasciola, Trema
orientalis, var. viridis, Ficus scabra, Sesuvium
Portulacastrum, Canavalia sericea, Vigna marina,
Suriana maritima, Securinega samoana, Terminalia
Catappa, Eugenia dealata, E. neurocalyx, Burckella
Richii, Planchonella torricellensis, Diospyros
elliptica, Geniostoma insulare, forma insulare,
Hoya chlorantha, and Ipomoea macrantha.

Likewise, less afinity is shown to the flora
or other islands. Niue Island lacks 24 of the
species; and Rarotonga lacks 44.

Hence, it is evident that the flora of Niuatoputapu is very closely, and almost equally, related to that of Samoa and of Fiji, and much less
so to the large islands of southern Tonga.

The following species are additions to the
known flora of Tonga: Amorphophallus campanulatus (Roxb.) Bl., "lena," a cultivated food plant;
Planchonella torricellensis (K. Schum.) H. J. Lam,
"kalaka," an indigenous tree, common in Samoa;
Hoya chlorantha Rech., a native vine, also known
in Samoa; and Gardenia rotumaensis St. John, a
cultivated ornamental tree, called "siale lotuma,"
(= Gardenia from Rotuma). It is a native to
Rotuma Island, to the north of Fiji.

List of the Vascular Flora

The indigenous species are distinguished by italic type.

PTERIDOPHYTA
Psilotaceae

Psilotum nudum (L.) Griseb. On tree trunks, Kirch 240; on limestone, Hürlimann 205.

Lycopodiaceae

Lycopodium sp. An epiphyte similar to L. phyllanthum, observed by Kirch.

Pteridaceae

Pteris tripartita Sw. "hulufe." Climax forest, Kirch 287.

Davalliaceae

Davallia solida (Forst. f.) Sw. "lau fare," and "lau fale." On tree trunk, Kirch 176; in shade, Kirch 182.

Nephrolepis biserrata (Sw.) Schott. "hulufe." In shade, Kirch 181.

Nephrolepis exaltata (L.) Schott. "lau fale." Coastal strand, Kirch 245.

Aspidiaceae

Cyclosorus invisus(Forst. f.) Copel. "hulufe." Mesophilous forest, Hürlimann 342.

Aspleniaceae

Asplenium nidus L. "katafa." Second growth on ridge, Kirch 224.

Polypodiaceae

Phymatodes Scolopendria (Burm.) Ching. In shade, Kirch 183. Like the Malayan plants, it has the sori in double rows on each side of the midrib.

Pyrrosia adnascens (Sw.) Ching. On Cocos trunks, Kirch 310.

SPERMATOPHYTA
MONOCOTYLEDONES
Cycadaceae

Cycas circinalis L. Cultivated. Hürlimann 616.

Pandanaceae

Pandanus angulosus St. John. "fa." Regenerated forest near ridge summit, Kirch 188.

Pandanus Mbalawa St. John. "fafa," and "fa tea." Offshore islet, Kirch 279. New for Tonga, growing also on Fulanga Island, Fiji.

Pandanus odoratissimus L. f., var. Setchellii

Humid alkaline sands, Hürlimann 388.

Pandanus turritus Martelli. "fala hola." Only
one tree, cultivated by a house, Kirch 266.
Fruit used for "kahora." Native to Samoa.

Pandanus spp.
"kie." Kirch 300. Leaves used for plaiting.
"tofua tapahina." Kirch 301. Leaves used for
    plaiting.
----- Kirch 328. Leaves used for plaiting.

                    Gramineae

Bambusa vulgaris Schrad. "pitu." Second growth
forest, Kirch 345.

Cenchrus echinatus L. "hefa." Coastal strand,
Kirch 252.

Centotheca latifolia (L.) Trin.
    C. lappacea (L.) Desv. Forest clearing,
    Hürlimann, 377; edge of plantation, Hürlimann
    635; grassland, Kirch 211.   r

Cynodon Dactylon (L.) Pers. Roadside, Hülimann
634.

Eragrostis tenella (L.) Beauv. ex R. & S.
"musie." Roadside on sandy soil, Hürlimann
378; coastal strand, Kirch 253.

Miscanthus floridulus (Labill.) Warb. "kaho."
Dry forest on volcanic soil, Hürlimann 349.

Paspalum conjugatum Berg. "mohuku." Roadside,
Hürlimann 633

Paspalum distichum L. (P. vaginatum Sw.).
Alkaline mud, Hürlimann641.

Saccharum officinarum L. "to." Second growth
swidden, Kirch 315.

Sporobolus elongatus R. Br. "musie." Coastal
strand, Kirch 249.

Sporobolus indicus (L.) R. Br. Grass lawn,
Hürlimann 327; roadside, Hürlimann 647.

Urochloa ambigua (Trin.) Pilger. Roadside,
Hürlimann 620.

Zea Mays L. Observed by Kirch.

                    Cyperaceae

Cyperus brevifolius (Rottb.) Hassk. "musie."
Grassland, Kirch 184.

Cyperus compressus L. "pakopako." Roadside,
Hürlimann 394.

Cyperus javanicus Houtt. "mahelehele." and
"musie." Wet sandy muddy soil, Hürlimann 384;

swampy pond edge, Kirch 341.

Cyperus rotundus L. "pakepake.: Grassland,
　　Hürlimann 326.

Fimbristylis autumnalis (L.) R. & S., var.
　　complanata (Retz.) Kuekenth. "takataka."
　　Wet sandy muddy soil, Hürlimann 383.

Fimbristylis pycnocephala Hbd. "pakopako,"
　　and "pako fae lolo."

　　F. cymosa R. Br., var. pycnocephala(Hbd.)
　　　　Kuekenth., f. pycnocephala, and f.
　　　　paupera Hürlimann, Bauhinia 3: 193, 1967.
　　Wet soil, Hürlimann 387; sandy soil,
　　Hürlimann 671; tidal swamp, Kirch 308.

Scleria lithosperma (L.) Sw. "musie." Grassland,
　　Kirch 185.

Scleria polycarpa Boeck. "mahelehele," and
　　"musie."

　　S. margaritifera sensu Hürlimann, non Gaertn.
　　Dry forest, Hürlimann 356; grassland, Kirch 185.

Palmae

Cocos nucifera L. "niu." Secondary forest,
　　Hürlimann 632; observed by Kirch.

Pritchardia pacifica Seem. & Wendl., var. pacifica.
　　Second growth forest, Kirch 239. Fruit edible;
　　leaves used for umbrellas. "piu."

Araceae

Alocasia macrorrhiza (L.) Schott. "kape." Swidden
　　garden, Kirch 321.

Amorphophallus campanulatus (Roxb.) Bl. "lena."
　　In second year swidden, Kirch 191.

Colocasis esculenta (L.) Schott, var. antiquorum
　　(Schott) Hubb. & Rehd. "talo Tonga." The culti-
　　vars are:　:
　　"talo Niue." Second year swidden, Kirch 194.
　　"talo uli." Second year swidden, Kirch 200.
　　"talo vahe." Second year swidden, Kirch 199.
　　"lau ila." Second year swidden, Kirch 193.

Cyrtosperma Chamissonis (Schott) Merr. "via."
　　Observed by Kirch.

Epipremnum pinnatum (L.) Engler. Vine on coconut
　　tree, Kirch 220.

Xanthosoma sagittifolium (L.) Schott. "talo
　　Futuna." Second year swidden, Kirch 196; and
　　"talo tea." Kirch 195.

Bromeliaceae

Ananas comosus (Stickm.) Merr. "fainā." Second
  year swidden, Kirch 314.

Commelinaceae

Rhoeo spathacea (Sw.) Stearn. "faina faitoka."
  Coastal strand, Kirch 250.

Liliaceae

Cordyline terminalis (L.) Kunth, var. terminalis.
  "si." In second growth, Kirch 186. Formerly
  roots eaten; leaves used as ornaments, Kirch 186.

Amaryllidaceae

Crinum asiaticum L. Escaped, Hürlimann 399
Zephyranthes candida (Lindl.) Herb. Grassland,
  Hürlimann 398.

Taccaceae

Tacca Leontopetaloides (L.) Ktze. "mahoa'a."
  Secondary forest, Hürlimann 363; coastal strand,
  Kirch 241.

Dioscoreaceae

Dioscorea alata L. "ufi." The cultivars are:
  "kahokaho." Second year swidden, Kirch 187.
  "kaumeile." Second year swidden, Kirch 203.
  "kulo." First year swidden, Kirch 198.
  "tuaata." Second year swidden, Kirch 197.
Dioscorea bulbifera L. "hoi." Mesophilous coastal
  forest, Hürlimann 650; third year swidden,
  Kirch 205; in second growth, said to be
  poisonous, Kirch 347.
Dioscorea esculenta (Lour.) Burkill. "ufi lei
  lotuma " (=Rotuma). Second year swidden,
  Kirch 201; Kirch 202.
Dioscorea nummularia Lam. "palai." Observed by
  Kirch.
Dioscorea pentaphylla L. "lena." In garden
  complex, Kirch 333.

Musaceae

Musa, hybrids. "hopa." Observed by Kirch.

Zingiberaceae

Hedychium coronarium Koenig in Retz.
  Cultivated, Kirch 325.
Zingiber Zerumbet (L.) Roscoe in Sm. Around
  garden house, Kirch 324.

Cannaceae

Canna indica L. Abandoned garden, Kirch 216.
Orchidaceae
Dendrobium Tokai Rchb. f. "pipini." On tree trunk
in tidal swamp, Kirch 313.
Taeniophyllum fasciola (Forst. f.) Rchb. f. On
trunks of Cocos at edge of tidal swamp,
Kirch 309.
Vanilla planifolia Andrews. "vanilla." Cultivated,
Observed by Kirch.
DICOTYLEDONES
Casuarinaceae
Casuarina litorea Stickm. (C. equisetifolia L.)
"toa." Dry forest on ridge, volcanic soil,
Hürlimann355; and observed by Kirch.
Piperaceae
Piper methysticum Forst. f. "kava." Second growth
swidden, Kirch 232. Roots used to make a beverage.
Ulmaceae
Trema orientalis (L.) Bl., var. viridis Lauterb.
"mangele." Secondary forest on ridge, volcanic
soil, Hürlimann 637.
Moraceae
Broussonetia papyrifera (L.) Vent. "hiapo."
Observed by Kirch.
Artocarpus altilis (Parkins. ex Z) Fosb. "mei."
Cultivated tree in garden, Kirch 327.
Ficus obliqua Forst. f. "ovava." Mesophilous
forest, Hürlimann 372; coastal strand,
Kirch 337.
Ficus scabra Forst. f. "masi." Mesophilous forest
on volcanic soil, Hürlimann 338; second growth
on ridge, Kirch 222.
Ficus Storckii Seem. "masi." Second growth,
Kirch 294.
Ficus tinctoria Forst. f. "masi." Coastal tidal
swamp, Kirch 307.
Urticaceae
Pipturus argenteus (Forst. f.) Wedd., var. lanosus
Skottsb. Dry forest on volcanic soil, Hürlimann
364; second growth on ridge, Kirch 230.
Santalaceae
Santalum Yasi Seem. "ahi." Forest on ridge,
volcanic soil, Hürlimann 350; second growth
forest with Myrtaceae, Kirch 339.

Amaranthaceae

Achyranthes aspera L. "tamatama." Third year
swidden, Kirch 206; clearing in climax forest,
Kirch 299.

Nyctaginaceae

Boerhavia diffusa L. Weed, Hürlimann 630.

Pisonia grandis R. Br. "pukavai." Large tree in
climax forest, Kirch 295.

Aizoaceae

Sesuvium Portulacastrum (L.) L. On coral apron,
Kirch 269.

Annonaceae

Cananga odorata (Lam.) Hook. f. & Thoms. "honolulu."
"mohokoi." Cultivated shrub, Kirch 312; 342.

Lauraceae

Cassytha filiformis L. "fatai." Parasitic on
Tarenna, Hürlimann 390; coastal strand,
Kirch 304.   Hernandiaceae

Hernandia nymphaeifolia (Presl) Kubitzki. "fotulona;"
"puko vili." H. ovigera Stickm. H. peltata Meissn.
Mesophilous forest, Hürlimann 373; coastal strand,
Kirch 264; 305.

Cruciferae

Lepidium virginicum L. Ruderal, Hürlimann 645.

Crassulaceae

Kalanchoe pinnata (Lam.) Pers. "te'e kosi."
coastal strand, Kirch 243.

Rosaceae

Parinarium glaberrimum Hassk. "pipi." Second
growth forest, Kirch 331. Timber used for
building.

Leguminosae

Abrus precatorius L. "matamoho." Forest on
mountain ridge, Hürlimann 353.

Acacia simplex (Sparrm.) Pedley. "tatangia."
A. simplicifolia (L. f.) Druce
Coastal forest, Hürlimann 370; coastal strand,
Kirch 256.

Adenanthera pavonina L. "lopa." Second growth on
ridge, Kirch 231. Fruit edible.

Caesalpinia major (Medic.) Dandy & Exell.
C. crista sensu Hürlimann, non L. "talatala
'amoa." Forest on ridge, volcanic soil,
Hürlimann 351.

<u>Canavalia sericea</u> Gray. "fue kula." Offshore
coral islet, Kirch 271.

Crotalaria retusa L. Coastal bush, calcareous
soil, Hürlimann375.

Delonix regia (Boj.) Raf. " 'ohai." Cultivated,
Hürlimann 379.

<u>Dendrolobium umbellatum</u> (L.) Benth., f. <u>hirsutum</u>
(DC.) Ohashi. "lala."
<u>Desmodium umbellatum</u> (L.) DC.
Edge of dry forest on ridge, volcanic soil,
Hürlimann 358; second growth, Kirch 285.

Derris elliptica (Roxb.) Benth. "kava." Second
growth forest, Kirch 238.

<u>Erythrina variegata</u> L., var. <u>orientalis</u> (L.)
Merr. "ngatae." Second growth forest, Kirch 233.

Indigofera suffruticosa Mill. " 'akau veli."
Second growth forest, Kirch 284.

Inocarpus fagifer (Parkins. ex Z) Fosb. "ifi."
Secondary forest on mountain ridge, volcanic
soil, Hürlimann357; cultivated or protected
near village, Kirch 283.

Mimosa pudica L., var. unijuga (Duchass. & Walp.)
Griseb. Second growth, Kirch 318.

Pueraria lobata (Willd.) Ohwi. "aka," "hakataha."
P. Thunbergiana (Sieb. & Zucc.) Benth.
Dry forest on ridge, volcanic soil, Hürlimann
354; mesophilous forest on coastal plain,
calcareous soil, Hürlimann 640; second growth,
Kirch 235. Tuber edible.

Tephrosia purpurea (L.) Pers. "kavahuhu." Grassy
edge of ridge, volcanic soil, Hürlimann 344;
exposed locality, Kirch 180.

Uraria lagopodioides (L.) Desv. "iku'ipuai."
Roadside, Hürlimann 367.

<u>Vigna marina</u> (Burm.) Merr. "fue lau tolu." Shore,
on coral sand, Hürlimann 617; offsore islet.
Kirch 270.

                    Rutaceae
Citrus Aurantium L. "moli," "moli kai." By road
in forest, Hürlimann 631; around garden house,
Kirch 322.

Citrus limonia Osbeck. Observed by Kirch.
                  Simaroubaceae
<u>Suriana maritima</u> L. "nyengie." Coral shore,
Hürlimann 628.

Meliaceae

Aglaia saltatorum A.C. Sm. "langakali." Cultivated, Hürlimann 649.

Dysoxylum Forsteri (Juss.) C. DC. "mo'ota."
Forest, Kirch 223; offshore islet, Kirch 282;
climax forest, Kirch 291.

Melia Azedarach L. "sita." Second growth, Kirch 340.

Euphorbiaceae

Acalypha boehmerioides Miq. "hongohongo." Roadside,
Hürlimann 396.

Aleurites moluccana (L.) Willd. "tuitui." Cultivated, Kirch 268.

Bischofia javanica Bl. "koka." Second growth forest,
Kirch 237. Bark used for a brown dye.

Breynia disticha Forst. f., forma nivosa (W. G. Sm.)
Croizat. Cultivated, and escaped, Hürlimann 400.

Euphorubia Atoto Forst. f. On sand at shore,
Hürlimann 497.

Euphorbia hirta L. "sakisi." Grassland, Hürlimann
328. E. pilulifera L.

Glochidion ramiflorum Forst f. "malolo," and
"mahame." Secondary forest, volcanic soil,
Hürlimann 333.

Homalanthus nutans (Forst. f.) Pax. "fenua malala."
Mesophilous forest, Hürlimann 332; coastal
forest, Hürlimann 651; second growth foest on
steep ridge, Kirch 219; 348.

Jatropha Curcas L. "fiki." Roadside, Hürlimann 644.

Macaranga Harveyana (Muell.-Arg.) Muell.-Arg.
"loupata." Kirch 229; coastal strand, Kirch 251;
second growth forest, Kirch 289.

Manihot esculenta Crantz. "manioka," and "mata
ki eua," and "falaoa." Cultivated, Kirch 213; 215.

Phyllanthus Niruri, sensu Hürlimann, non L. Weed,
Hürlimann 329.

Phyllanthus simplex Retz. Roadside, Hürlimann 397.

Securinega samoana Croizat. "poumuli." Mesophilous
forest, volcanic soil, Hürlimann 339; second
growth on ridge, Kirch 228. Best timber for
house posts.

Anacardiaceae

Mangifera indica L. "mango." Cultivated, Kirch 316.

Pleiogynium timoriense (DC.) Leenhouts. Regenerated
forest near ridge summit, Kirch 187.

Rhus taitensis Guillem. "tavahı." Dry forest on
  ridge, volcanıc soil, Hürlimann 346; regenerated forest near ridge summit, Kırch 175.
  Wood used in construction.
Spondias dulcis Parkins. ex Z. "vı." Observed
  by Kırch.

Sapindaceae

Cardiospermum Halicacabum L. Weed, Hürlimann 626.
Elatostachys falcata (Seem.) Radlk. "ngatata."
  Clımax forest, Kırch 297.
Pometia pinnata J. R. & G. Forst. "tava."
  Cultıvated, Hürlimann 377; Kırch 349.

Rhamnaceae

Alphitonia zizyphoides (Soland.) Gray. Ridge
  summit, Kırch 173.
Colubrina asiatica (L.) Brongn. "fiho'a."
  Coastal forest, Hürlimann 382; second growth
  forest, Kırch 234; coastal strand, Kırch 254;
  offshore islet, Kırch 278.

Tiliaceae

Triumfetta Bartramia L. "mo'osipo." Coastal strand,
  Kırch 248; weed, Kırch 207.
Triumfetta procumbens Forst f. "mo'osipo."
  coastal strand, Kırch 303.

Malvaceae

Hibiscus Manihot L. "pele." Second year swıdden,
  Kırch 192. Leaves eaten as greens. The blades
  are broadly cordate, and only shallowly lobed,
  instead of being deeply cleft. This variety is
  also in cultıvation in Fiji and on Niue (Sykes:
  113).
Hibiscus rosa-sınensis L. "kaute." Cultıvated,
  Kırch 267.
Sıda acuta Burm. f. "te'chosı." Weed, Hürlimann
  643.
Sıda rhombıfolia L. "te'hoosı." Weed, Kırch 208.
Thespesia populnea (L.) Soland. ex Correa. "mılo."
  Offshore coral islet, Kırch 277. Wood used for
  house construction, and spear handles.

Sterculiaceae

Melochia aristata Gray. "mako." Secondary forest
  on ridge, volcanıc soil, Hürlimann 362; mesophilous coastal forest, Hürlimann 627; second
  growth forest, Kırch 217.

Guttiferae

Calophyllum Inophyllum L. "feta'u." Coastal strand,
  Kirch 260. Wood used for bowls.

Flacourtiaceae

Flacourtia Rukam Zoll. & Mor. in Mor. "filimoto."
  Mesophilous forest at base of ridge, volcanic
  soil, Hürlimann 336; second growth forest,
  Kirch 343.

Passifloraceae

Passiflora maliformis L. "vaini." Second growth
  forest, Kirch 330. Fruit eaten.

Thymeliaceae

Phaleria acuminata (Gray) Gilg. "huni." Coastal
  strand, Kirch 242.

Caricaceae

Carica Papaya L. "lesi." Observed by Kirch.

Lythraceae

Pemphis acidula J. R. & G. Forst. "ngingi'e."
  Sandy coast, Hürlimann 386; offshore islet,
  Kirch 273. Trunks used for posts and firewood.

Barringtoniaceae

Barringtonia asiatica (L.) Kurz. "futu." Mesophilous forest, Hürlimann 374; coastal strand,
  Kirch 336. Fruits used as fish poison.

Combretaceae

Terminalia Catappa L. "telie." Forest on ridge
  top, volcanic soil, Hürlimann 393; coastal strand,
  Kirch 334; 335.

Terminalia litoralis Seem., var. tomentella
  Hemsl. "telie." Offshore islet, Kirch 281.

Myrtaceae

Eugenia clusiaefolia Gray. "mafua."
  Syzygium clusiaefolium (Gray) C. Mueller.
  Mesophilous forest on ridge, volcanic soil,
  Hürlimann 340; ridge summit, Kirch 171. Wood
  used for house posts.

Eugenia corynocarpa Gray. "heahea."
  Syzygium corynocarpum (Gray) C. Mueller.
  Second growth forest, Kirch 346. Fruit edible.

Eugenia dealata Burkill. "mafua ai lulu."
  Syzygium dealatum (Burkill) A. C. Sm.
  Climax forest, Kirch 298. Fruit eaten.

Eugenia malaccensis L. "fekika."
  Syzygium malaccense (L.) Merr. & Perry
  Cultivated, Kirch 329.

Eugenia neurocalyx Gray. "fekika vao."
  Syzygium neurocalyx (Gray) Christoph.
  Mesophilous coastal forest, Hürlimann 648.
Psidium Guajava L. Second growth forest, Kirch 317.
                Sapotaceae
Burckella Richii (Gray) H. J. Lam. "kau." In
  village, Kirch 338.
Planchonella torricellensis (K. Schum.) H. J. Lam.
  "kalaka." Climax forest, Kirch 292.
                Ebenaceae
Diospyros elliptica (J. R. & G. Forst.) P. S.
  Green, var. elliptica. "mapa." Mesophilous
  coastal forest, Hürlimann 345; climax forest on
  ridge top, Kirch 169. Fruit edible.
Diospyros samoensis Gray. "koka uli." Dry forest
  on ridge, volcanic soil, Hürlimann 352; 361;
  second growth forest on ridge summit, Kirch 172.
                Oleaceae
Jasminum didymum Forst. f. "tutu'ila." Forest on
  mountain ridge, volcanic soil, Hürlimann 365.
                Loganiaceae
Geniostoma insulare A. C. Sm. & Stone, f. insulare.
  "te'e pilo a Maui." Dry forest on mountain
  ridge, volcanic soil, Hürlimann 366; offshore
  islet, Kirch 275.
                Apocynaceae
Alyxia stellata (Forst. f.) R. & S. "maile."
  Forest on mountain ridge, volcanic soil,
  Hürlimann 347; 368; Kirch 174. Used for
  personal adornment.
Cerbera Manghas L. Calcareous sandy coastal
  forest, Hürlimann 624; second growth forest,
  Kirch 320.
Neisosperma oppositifolia (Lam.) Fosb. & Sachet.
  "fao." Ochosia oppositifolia (Lam.) K. Schum.
  Coastal sandy forest, Hürlimann 391; coastal
  strand, Kirch 262; offshore islet, Kirch 280.
  Wood for house construction.
                Asclepiadaceae
Hoya chlorantha Rech. Vine on rocks, Kirch 177.
                Convolvulaceae
Ipomoea Batatas (L.) Poir. "kumala." Observed
  by Kirch.
Ipomoea indica (Burm.) Merr. Second growth on
  ridge, 100 m. alt., Kirch 226.

Ipomoea macrantha R. & S. "pula."
  Ipomoea tuba (Schlecht.) G. Don
  Clearing in climax forest, Kirch 290.
                    Boraginaceae
Cordia subcordata Lam. "milo." Coastal
  strand, Kirch 258; 259.
Messerschmidia argentea (L. f.) Johnston. "tohuni."
  Offshore islet, Kirch 276.
                    Verbenaceae
Clerodendrum inerme (L.) Gaertn. Dry forest on
  ridge, volcanic soil, Hürlimann 359; second
  growth forest, Kirch 306.
Lantana Camara L. "talatala." In clearing,
  Hürlimann 369; on rocks, Kirch 179.
Premna taitensis Schauer, var. rimatarensis
  F. Br. "volovalo." Forest on ridge, volcanic
  soil, Hürliman 360; second growth forest,
  Kirch 236; coastal strand, Kirch 247; second
  growth, Kirch 319.
Stachytarpheta urticaefolia (Salisb.) Sims.
  Roadside, Kirch 293.
                    Labiatae
Coleus sp.   "kaloni." cultivated, Hürlimann 623.
Pogostemon Cablin (Blanco) Benth. Near garden
  house, Kirch 326.
                    Solanaceae
Lycopersicon esculentum Mill. "temata." Observed
  by Kirch.
Nicotiana Tabacum L. "tapaka Tonga." Observed
  by Kirch.
Physalis minima L. "ku'uai." Roadside, Hürlimann
  395.
Solanum amicorum Benth. "polo Tonga." Strand,
  Kirch 302.
Solanum nigrum L. "polo kai." In garden, Kirch
  190; in clearing, Kirch 296. Eaten like pepper.
                    Scrophulariaceae
Lindernia crustacea (L.) F. Muell. Weed by path,
  Hürlimann 393.
                    Rubiaceae
Borreria laevis (Lam.) Griseb. "mohuku." Weed,
  Kirch 244.
Gardenia rotumaensis St. John. "siale Lotuma."
  cultivated, Kirch 344.
Geophila herbacea (Jacq.) Ktze. In shade of Cocos,

Kirch 212.

Guettarda speciosa L. "puapua." Coastal strand,
Kirch 261.

Morinda citrifolia L. "nonu." Mesophilous coastal
forest, Hürlimann 392; coastal strand, Kirch
218; 257.

Psychotria insularum Gray. "ola vai."ᵣMesophilous
secondary forest, volcanic soil, Hürlimann 331;
climax forest, Kirch 221; 227; 286.

Randia cochinchinensis (Lour.) Merr. "ola."
Mesophilous forest on ridge, volcanic soil,
Hürlimann 330; climax forest on summit, Kirch
170; 209; coastal strand, Kirch 255.

Tarenna sambucina (Forst. f.) Durand. "manonu."
Saline coast, Hürlimann 389; Kirch 225; climax
forst, Kirch 288.

Cucurbitaceae

Momordica Charantia L. "vaini ai kumā." Weed in
third year swidden, Kirch 204.

Goodeniaceae

Scaevola Taccada (Gaertn.) Roxb., var. Taccada.
"ngahu." Coastal strand, Kirch 263.

Scaevola Taccada (Gaertn.) Roxb., var. sericea
(Vahl) St. John. "ngahu." Offshore islet,
Kirch 274.

Compositae

Ageratum conyzoides L. Garden weed, Kirch 214.

Vernonia cinerea (L.) Less., var. parviflora
(Reinw. in Bl.) DC. Weed, third year swidden,
Kirch 210.

Wedelia biflora (L.) DC. " 'ate." Offshore islet,
Kirch 272.

Bibliography

Sykes, W. R. 1970. Contributions to the Flora of
Niue. New Zealand Dept. Sci. Ind. Res., Bull.
200: 1-321, figs. 1-45.

Yuncker, T. G. 1959. Plants of Tonga. B. P.
Bishop Mus., Bull. 220: 1-283, figs. 1-17.

. ADDENDA
Pandanaceae
Pandanus Whitmeeanus Martelli. "paongo." Grove by
  roadside between Hihifo and Matavai, Kirch 265.
  Leaves used for mats; fruit reddish orange,
  fragrant, used for garlands.

Malvaceae
Hibiscus tiliaceus L. "fau." Common, observed
  by Kirch

Bombacaceae
Ceiba pentandra (L.) Gaertn. "vavae." In Hihifo
  village, semi-cultivated, Kirch 350.

NOTES ON THE CLUSIACEAE - CHIEFLY OF PANAMA. I.

Bassett Maguire [1]

The preparators of the Flora of Panama have been good enough to give me the privilege of examining the large series of Clusiaceae more recently collected in Panama in the field survey of the interesting and complex flora of that country, which bridges those of Central America to the north and continental South America to the south. Besides yielding many endemic taxa of its own, the flora partakes to a great extent of much of that of the two neighboring regions.

Therefore, any phytotaxonomic or phytogeographic study of Panama requires an examination of affinities northward into Mexico, southward through the Andean countries to Bolivia, and eastward into Venezuela. Such geographic connections will be abundantly demonstrated in this projected short series of papers which will make names and nomenclatural changes available for the Flora of Panama.

There is no intended significance in the order of presentation of taxa. Comments on the genus Clusia, being the largest of the family and the most difficult, would be offered last.

The genera Symphonia and Mammea are both generally treated as monospecific in America. Two species of the first, including the American S. globulifera, are commonly recognized in tropical Africa, while perhaps 20 species are variously recognized for Madagascar.

Mammea americana L , the type of that genus, is indigenous in the Antilles, Central America, and northern South America, and is commonly cultivated throughout the tropical world because of the large delicious fruit which it produces in abundance. There appear to be two species native to tropical Africa, and, if transfers from Ochrocarpos are correctly made, then numerous species in Madagascar, Malesia, and Pacifica. Further comment is offered under Mammea.

These observations are made because of the historic and geographical significance of such distributions, a pattern for which is found among other genera of the Clusiaceae.

---

[1] The New York Botanical Garden.

Symphonia globulifera L. f. Suppl. 302. 1781.

> Symphonia fasciculata (Pl. & Tr.) Vesque, DC. Monogr. Phan.
>   8: 232. 1893.
> S. coccinea (Aubl.) Oken. Allg. Naturgesch. 3^2: 431. 1841,
>   = Moronobea coccinea Aubl. Hist. Pl. Guiane Fr. 2:
>   789 (excl. Fig . a-j which are of Symphonia globulifera
>   L. f.). 1775.

Type. Linnaean Herbarium No. 853. Linnaean Society.
London.

In America as well as in Madagascar Symphonia is variable
and has been responsive to ecologic gradient. High mountain
forms, for instance, tend to have larger flowers and reduced
cymes. Vegetative modification may be parallel. These modifications have not seemed, possibly in large part because of inadequacy of materials, to be correlatable with geography. For
these reasons I have made no attempt to initiate any systematic
review of such infraspecific variation within Symphonia globulifera, and have not attempted to assess the two recent propo-
sitions of R E. Schultes, viz., S. microphylla[2] and S. utilis-
sima[2].

Now, because of the excellent and convincing series of
collections of Symphonia globulifera centered around Cerro
Santa Rita, Colón, Panama, an area known for a selective narrow endemism, I am required to recognize the local narrow-
leafed form as the var. angustifolia. This action will
stimulate inquiry into the composition of the remainder of
the species.

---

[2]

Schultes, R. E. Bot. Mus. Leafl. Harvard Univ. 17: 20-22.
1955.

<u>Symphonia</u> <u>globulifera</u> L. f. var <u>angustifolia</u> Maguire, var nov

Folia peranguste lanceolata, 8-12(18) mm lata, 4-6(8) cm longa; sepalis submembranaceis, minute fimbriatis, jugis exterioribus acutiusculis; arboribus parvis.

Type. Santa Rita Ridge east of transisthmian highway, alt 300-400 meters, tropical wet forest, tree 4 meters, flowers red, Prov. Colón, Panama, 16 Dec 1972, <u>Alwyn Gentry</u> <u>6557</u> (holotype MO, isotype NY).

Distribution. Apparently a narrow endemic confined to the Santa Rita Ridge as indicated by the following citations:

PANAMA. Prov. Colón, summit of Cerro Santa Rita: shrub 5 ft, latex yellow, flowers pink, 1200-1500 ft alt, 13 Sept 1947, <u>Paul H. Allen</u> and <u>Dorothy O. Allen</u> <u>5101</u> (NY, MO); shrub to 15 ft, fruit red, 19 km from main highway, Jan 1968, <u>J. D. Dwyer</u> <u>8580</u> (MO, sheet 1); shrub to 10 ft, buds rose colored, 19 km from transisthmian highway, 28 Jan 1968, <u>Dwyer 8580</u> (MO, sheet 2); arbol delgado de 7 m, flores rojas, látex amarillo, zona maderera de Santa Rita, 10 Oct 1968, <u>M. D. Correa y R. L. Dressler</u> <u>1085</u> (MO, 2 sheets); arbolito de 7 m, flores coral, frutos chocolates, látex amarillo, camino maderero de Santa Rita, 20 Mar 1969, <u>Correa y Dressler</u> <u>1205</u> (MO); shrub 3 m, flowers red, ovary yellowish, fruit pink, plants in vegetative condition appear to be abundant in the area, near Agua Clara rainfall station, 23 Apr 1970, <u>Robin Foster</u> <u>1738</u> (MO); slender tree ca 4 m high, buds rose-colored, along road ca 1 mi from Boyd-Roosevelt Highway, 9 Jul 1971, <u>Croat</u> <u>15337</u> (MO, NY); tree 4 m, flowers red, east of transisthmian highway, 300-500 m alt, tropical wet forest, 16 Dec 1972, <u>Gentry</u> <u>6557</u> (holotype MO, isotype NY); tree 3 m, petals rose-red, sap yellow, 15 km from Boyd-Roosevelt Highway, 450 m alt, in wet forest, 14 Mar 1975, <u>Mori & Kallunki</u> <u>5057</u> (MO). Prov. Panama: 5-6 mi north of El Llano, near San Blas border, 1300 ft alt, tree 10 meters, flowers red, 8 Sept 1972, <u>Gentry</u> <u>5814</u> (MO).

Mammea L.

As indicated here above, the genus Mammea is represented
historically by a single American species, M. americana L.,
and a single African species, M. africana D. Don.   Should
Mammea, especially the American species, be treated as conspecific with Ochrocarpos (Ochrocarpos africana Oliver), the
broader genus then would consist of some 50 species.   I
attempt no generic evaluation here.

Our attention to the genus Mammea, in the restricted
sense, is occasioned by a recent collection, Mori, Kallunki
& Gentry 4699, made in interior Panama, which I immediately
associated with M. africana, in the assumption that its occurrence in Panama is the result of introduction.

However, Doctor Mori (personal communication) is of the
very strong opinion that this tree in question is established
as a well defined population in primary woodland and does not
bear evidence of introduction.   Mori has offered the view that
the tree is indigenous.

Should this be the case, then, as for Symphonia globulifera, we should be confronted with a second amphi-Atlantic
species representing the two indicated species, or that we have
here a narrow endemic American species (the second for the
Western Hemisphere), but one closely allied to Mammea africana
D. Don.

The Mori-Kallunki-Gentry collection is in mature fruiting condition.   To resolve the problem here presented, it
will be necessary to reexamine the population in the field so
as to reach a further considered opinion as to its possible
indigenous character, and more especially to collect an adequate series of specimens in flowering, young fruiting, and
mature fruiting condition.   Fluid-preserved specimens
should be obtained for cytological and morphological examination.

PANAMA.   Prov. Panama: tree 30 m tall, fruit brown,
warty, 4-locular, 4-seeded, wet forest at 350 m alt, El
Llano-Carti Road, 12.7 km from Inter-American Highway, 15
Feb 1975, Mori, Kallunki & Gentry 4699 (MO).

Chrysochlamys Poeppig

The separation of the four genera, Tovomita, Tovomitopsis, Chrysochlamys, and possibly Balboa, is at best maintained on tenuous morphologic grounds. It may at some date
of more advanced knowledge of them be required to bring these
names together under a single generic designation. To do so,
however, if even-handed application of criteria is applied,
would also require a larger sweeping performance among other
clusioid genera. This I would be loathe to make.

Chrysochlamys clusiaefolia Maguire, Bol. Soc. Venez. Cien.
    Nat. 25: 225. 1964.

To the present time Chrysochlamys clusiaefolia had been
known only from the region of the coastal Andes of Venezuela.
Now a collection of undoubted assignment to the species has
been made in the considerably disjunct Serranía del Darien,
along the Panamanian-Colombian frontier. Other comparable
geographical disjuncts are well known.

The Panamanian collection is faithful to the facies and
form of the Venezuelan plants except that its flowers (male)
lack the closely subtending outer pair of sepals (? bracts)
which are characteristic of all of the original specimens.
The subspecies offered here is based upon that distinction.

Chrysochlamys clusiaefolia Maguire subsp. clusiaefolia

Small or median-sized trees of wet primary forests above
1000 m altitude, Venezuelan coastal Cordillera: Estados
Aragua, Yaracuy, Falcon, and Dist. Federal.

Chrysochlamys clusiaefolia Maguire, subsp panamaensis Maguire,
    subsp nov

Subspeciei clusiaefoliae similis, jugis parvis sepalis
exterioribus deficientibus exceptis.

Type. Tree 10 m, buds green, lower montane wet forest,
1400 m alt, base camp, Cerro Mali, Serranía del Darien,
Panama, Colombian frontier, 21 Jan 1975, A. Gentry & S. Mori
13769 (holotype MO).

Known only by the type collection.

<u>Chrysochlamys</u> sp nov, aff <u>Ch</u>. <u>clusiaefolia</u> Maguire

The specimen cited below is not determinable.

PANAMA.   Prov. Coclé: small tree, fruits green in axillary and extra-axillary clusters, latex scanty, greenish-
white, elfin forest, Cerro Caracoral, 1000 m alt, 19 Jan 1968,
<u>Duke & Dwyer</u> <u>15102</u> (MO).

Obviously more of this entity should be sought.

Standley and Williams have described three species of
<u>Chrysochlamys</u> from Panama, viz <u>Ch</u>. <u>pauciflora</u> Standley, Ceiba
3: 214. 1953, and <u>Ch</u>. <u>eclipes</u> L. Wms. and <u>Ch</u>. <u>standleyana</u> L.
Wms., Trop. Woods 111: 15-16. 1959.   I have not had opportu-
nity to form judgment on these three proposals.

<u>Tovomitopsis</u> Planchon & Triana

As indicated earlier in this paper, the significant morphologic qualities of the genera <u>Tovomita</u>, <u>Tovomitopsis</u>, and <u>Chrysochlamys</u> are inadequately investigated. L. O. Williams[3], careful student of Central American botany, while admitting the inadequacy of morphologic, anatomic, and bio- logic evidences, has nonetheless brought <u>Tovomita</u> and <u>Tovo- mitopsis</u> together under the former name.

This action, while understandable, I feel is somewhat premature. More detailed study should give sounder basis for such taxonomic and nomenclatural adjustment.

Repeatedly the question has been put to me as to the distinction of the three generic groupings. I am unable to give matured judgment on the matter at this time, as said above, because of the inadequacy of information of struc- tural flowering and fruiting details. I thus offer here below a tentative generalized table which purports to separate the three entities, and with which my own work must suffice until further detailed study is accomplished.

---

[3]
  Williams, L. O. Guttiferae from Middle America.   Tropical Woods 111: 15. 1959.

TABLE I.
Tentative Differentiation of Genera

	Sepals	Petals	Stamens	Ovary	Aril	Inflorescence
Tovomita	Sepals 2-4, the outer valvate, exceeding and enclosing other flower parts.	Petals 4-(5-8), decussate to somewhat imbricate.	Stamens numerous, free or united at the base.	Ovary 4-(5) locular, locules 1-ovulate.	Exarillate, outer integument carnose, ariliform.	Inflorescence axillary and/or terminal.
Tovomitopsis	Sepals 4, decussate, outer smaller and shorter than inner, not exceeding or enclosing other flower parts.	Petals 4-5, decussate.	Stamens numerous, free or united at base.	Ovary 4-5-locular, locules 1-2-ovulate.	Exarillate, integuments (1 or both) carnose, sacciform.	Inflorescence axillary and/or terminal.
Chrysochlamys	Sepals 5, imbricate.	Petals 5, imbricate.	Stamens numerous, free or congested in center.	Ovary 5-locular, locules 1-ovulate.	Exarillate, integuments (? 1 or both) carnose, sacciform.	Inflorescence cauliflorous.

Tovomitopsis angustifolia Maguire, sp nov

Frutex parvus, dioecius, 2 m maximus; ramulis tenuibus, teretibus; latice albido; foliis appositis, anguste elliptico-lanceolatis, laminis vulgo 8-11 cm longis, 1.0-1.5 cm latis, apicibus basibusque anguste acuminatis; venis lateralibus paucis, vulgo 4-5, non-binatis, valde sursum arcuatis; petiolo 8-10 mm longo, tenui, anguste alato; inflorescentiis terminalibus, cymosis; masculinis. inflorescentiis ad 5 cm longis, vulgo 8-18-floribus; alabastris maturis globosis, 5-8 mm longis; sepalis decussatis, jugis exterioribus ovatis, obtusis, 3-4 mm longis; interioribus floribus includentibus; staminibus numerosis sed paucioribus quam 50, filamentis crassis, 3-4 mm longis; antheris oblongis, ca 0.5 mm longis, lateraliter dehiscentibus, 4-locularibus; granis pollinis tricolporatis, sphaeroideis, poris prominentibus, sulcis non-prominentibus, 22-24 u diam, sporodermate reticulato, foemineis: floribus mihi non visis; fructibus pyriformibus, stipitatis, ca 2 cm longis, 5-locularibus, loculo dispermis; placentatione axiali elongata; seminibus 5-6 mm longis, ca 1.5-2.0 mm latis, embryone non-differentationi, semine omnino in membrana involuta; gynobasi angusta, ca 5-6 mm longa, stylis 5, cornutis, ad bases connatis, ca 2 mm longis, stigmatibus sessilibus, anguste obovatis, ca 1 mm longis, distaliter introrsis.

Type. Shrub 2 m tall, petals white, filaments yellow, on road to Calovebora, along stream, NW of Santa Fe, 2.7 km from Escuela Agricola Alto de Piedra, Veraguas, Panama, 30 Mar 1975, S. Mori & J. Kallunki 5357 ♂ (holotype MO, isotype NY).

Distribution. Known certainly only from the Province of Veraguas above Santa Fe.

PANAMA. Prov. Veraguas: shrub 0.5-1.2 m tall, fruits green tinged with red, flowers white, Río Primero Braso, 2.5 km beyond Agriculture School Alto Piedra near Santa Fe, elev 700-750 m, 24 Jul 1974, Croat 25437 (MO, NY); much-branched shrub 1-2 m tall, forming clumps along stream edge, cut twigs exude milky sap, tropical wet forest, Atlantic Slope, 16 Nov 1974, Mori & Kallunki 3179 (MO, NY); 2.7 km from Escuela Agricola Alto de Piedra, 30 Mar 1975, Mori & Kallunki 5357 (holotype MO, isotype NY); streamside shrub to 2 m tall, very common, petals white, stamens yellow, NW of Santa Fe, 11 km from Escuela Agricola de Piedra, in valley of Río Dos Bocas, Atlantic slope, 450-550 m alt, 17 May 1975, Mori & Kallunki 6125 (NY). Gentry 8764, Río Guanche, 1-4 km upstream from Portobelo Road, 0-100 m alt, tropical wet forest, Colón, Panama, 10 Dec 1973, is doubtfully referred here. The specimen is inadequate for determination.

Tovomita Aublet

The genus Tovomita is the second largest of the American clusioid genera and, second to the largest genus, Clusia, possessive of the greatest range. Any consolidation of satellite genera with Tovomita would require that the Aublet name be retained.

Within these brief notes only two species will be given reference: Tovomita weddelliana, a long established and much fragmented species, and T. croatii, herein offered as new.

Because of its great range, extensive nomenclatural history, and diversified morphology, I here offer a rather full description of the species to include all here indicated segregates, some of which may indeed prove to represent geographic variants.

Tovomita weddelliana Planchon & Triana, An. Sc. Nat. Ser. 4.
                    Bot. 14: 277. 1860.

    Clusia oblanceolata Rusby, Desc. S. Am. Pl. 58. 1920.
        Type. Valparaiso, Santa Marta, Colombia, 20 Mar
        1899, H. H. Smith 1880 (NY).
    Clusia pithecobia Standl. & L. Wms. Ceiba 1: 244. 1951.
        Rio Piedras Blancas, Prov. Puntarenas, Costa Rica,
        3 Aug 1950, Allen 5592 (F).
    Tovomita longicuneata Engl., Bot. Jahrb. 58. Beibl. 130:
        7. 1923. Type. Manzon, Huanuco, Peru, 900--1000 m,
        ♂, Apr 1904, Weberbauer 3446 (B).
    ?Tovomita sphenophylla Diels, Notizblatt 14: 32. 1938.
    Tovomita rhizophoroides Cuatr., An. Inst. Biol. Mex. 20:
        101. 1949. Type. Rio Naya, Valle, Colombia,
        Cuatrecasas 14280 (F).
    Tovomita ligulata Cuatr., An. Inst. Biol. Mex. 20: 99.
        1949. Type. Rio Digua, Valle, Colombia, 27 Aug 1943,
        Cuatrecasas 14949 (F).
    Tovomita lanceolata Cuatr., An. Inst. Biol. Mex. 20: 102.
        1949. Type. Rio Calima, Valle, Colombia, Cuatre-
        casas 21278 (F).
    Tovomita glossophylla Cuatr., Rev. Acad. Col. Cien. 8:
        62. 1950. Type. Quebrada del Caquetá, Caquetá,
        Colombia, 6 Apr 1940, Cuatrecasas 9194 (US, F).
    Tovomita angustata Steyerm., Fieldiana Bot. 88: 399, fig.
        82. 1952. Type. Kavanayén, Bolívar, Venezuela, 23 Nov
        1944, Steyermark 60475 (holotype F, isotype NY).

Dioecious tree or often the flowers hermaphrodite, to
15 m high, branchlets terete, latex moderate milky or cream-
colored; leaves opposite, subcoriaceous, narrowly oblanceolate, (6)10-25(28) cm long, 2-5 cm broad, the apex acute or
short acuminate, gradually drawn to a narrow abrupt base;
petiole 3-5 mm long; midrib prominent, especially on the lower
surface; primary veins prominulous, 1.0-1.5 mm apart, rising
at a 10⁰ angle; inflorescence cymose, terminal or axillary,
staminate inflorescence open, seldom conferted, multiflorous,
to 20(25) cm long, primary branches to 8 cm long; flower buds
globose, when mature commonly 6 mm long, pistillate buds
larger than the staminate; sepals 2 pairs, decussate, subor-
bicular, concave, the external pair in the bud longer than
and enclosing the interior pair and the other flower parts,
the margins subconnivent or the first somewhat overlapping the
second; petals 4, decussate, oblong to obovate, commonly 6-8
mm long; male flowers: buds commonly 4 mm long, stamens numerous, filaments free, ca 2 mm long, thickened at the base,
borne on a corona ca 0.5 mm high, anthers 2-lobed, ovoid, ca
0.5 mm long, connective narrow; pollen tricolporate, broadly
oblong-elliptic, to 45 u in equatorial view, rounded, 25-30
u in polar view, exine minutely granular; stamens or stamino-
dia numerous, when fertile similar to those of ♂ flowers,
ovary 5-6-locular, ovule solitary in each loculus, gynobase
prominent, placenta axillary, linear, ca 10 mm long, central;
embryo erect, little differentiated, stigmas 5-6, sessile;
fruit strongly pyriform, commonly 4-5 cm long, the base narrowed forming a gynobase to 2 cm long, above the middle enlarged, obovate, stigmas sessile, obovate, 2.5-3.0 mm long,
seed solitary in each loculus, seed linear, 15-16 mm long,
5-6 mm thick, somewhat 3-angled; seed coat 0.2-0.3 mm thick,
indurated; seed enveloped in a fleshy mass possibly arillar in
nature.

Type. Bolivie septentrionale, valée de Tipuani, province de Larecaja, Weddell ann. 1851 (holotype B).

Distribution. Tovomita weddelliana becomes a tree to 15
m tall, is a forest dweller or may occur in more open ecotone
habitats. It ranges from Costa Rica to Peru and Bolivia along
the Andean axis; in the Venezuelan Andes to the Federal District, and in the eastern Guayana Highland.

It is difficult to assign a center of distribution.
Leaves are greatly variable in size and form (the chief
basis of segregation), as noted in the discription.
Evaluation of any of the several varietal designations
must await more adequate collection.

Some 72 collections have been examined in this review
of Tovomita weddelliana. Only those of Panama, numbering
22, are cited herein:

PANAMA. Prov. Colón: shrub 2 m (tree at maturity),
Santa Rita Ridge, 1 Mar 1971, Croat 13896 (MO); 7 meter
tree, 10 mi SW of Puertobelo, 2-4 mi from coast, 10-200 m
alt, 24 Mar 1973, Liesner 1058 (MO); tree 8 m tall, Río
Guanche, near Portobelo, tropical wet forest, 25 m alt,
24 Mar 1975, Mori & Kallunki 5216 (MO). Darien: stilted
tree, 8" dbh, fairly common in elfin forest, latex not obvious, fruits green, Cerro Pirre, 2500-4500 ft, 9-10 Aug
1967, Duke & Elias 13733 (NY). Panamá: tree 15 ft tall,
fruits green, Cerro Jefe, 14 Feb 1968, Correa & Dressler 721
(MO); arbol 10 m alto, látex, fruto rojo-verde, camino de
Llano a Cartí, altura ± 400 m, 20 Feb 1973, Correa, Dressler
et al 1855 (MO); small tree to 4 m, leaves leathery, sap
white, 200-500 m alt, 19 km above Pan-American Highway on road
from El Llano to Carti-Tupile, 20-21 Feb 1973, Kennedy 2516
(MO); tree 10 m tall, 15 cm dbh, premontane wet forest along
El Llano-Carti road, 16-18-1/2 km by road N of Pan American
Highway at El Llano, alt 400-450 m, 28 Mar 1974, Nee & Tyson
10973 (MO); tree 7 m tall, 10 cm dbh, wet forest, 350 m alt,
El Llano-Carti Road, 17.5 km from Inter-American Highway,
14 Feb 1975, Gentry, Mori & Kallunki 4597 (MO). San Blas:
primary forest, along newly cut road from El Llano to Carti-
Tupile, Continental Divide to 1 mi from Divide, 300-500 m
alt, 30 Mar 1973, Liesner 1269 (MO). Veraguas: tree, stilt
roots, 60 ft, milky latex, flowers white with yellow stamens,
vicinity of Santa Fe, forested slopes of Cerro Tute, 3000 ft
alt, 24 Mar 1947, P. H. Allen 4352 (MO, NY); tree 4 m,
flowers in bud, white, fruits green tinged with purple, along
Río Dos Bocas, ca 12 km beyond Santa Fe, 450 m alt, 25 Jul
1974, ♂, Croat 25781 (MO); tree 6 m, flowers white, stamens
and style yellow, fruits purplish, along road between Escuela
Agricola and Alto Piedra (above Santa Fe) and Río Dos Bocas
ca 5-8 km from Escuela, 730-770 m elev, 26 Jul 1974, Croat
25961 (MO, 2 sheets); tree 10 m, flowers white, Valley of Río
Dos Bocas, 11 km from Escuela Agricola Alto Piedra (Above

Santa Fe) on the road to Calovebora; primary forest along river, 450 m alt, 30 Aug 1974, Croat 27492 (MO); tree 4 m, Valley of Río Dos Bocas along road between Escuela Agricola Alto Piedra and Calovebora, 15.6 km northwest of Santa Fe, primary forest, along trail to Santa Fe, steep forested hill east of river, 450-550 m alt, 31 Aug 1974, Croat 27664 (MO); shrub 4 m tall, fruit red, along stream, NW of Santa Fe, 2 6 km from Escuela Agricola Alto de Piedra, 23 Feb 1975, Mori & Kallunki 4761 (MO).

Panama specimens with more obtuse or rounded apexes are: Prov. Panamá: tree 25 m, flowers white, in forest about 1 mi upstream from Frizzel's Vinca Indio, on slopes of Cerro Jefe, 9 Sept 1970, Foster & Kennedy 1832 (MO); epiphytic shrub, top of Cerro Jefe, 1 Apr 1972, Gentry 4877 (MO, NY); tree 15 m, old fruit reddish, Cerro Jefe, 22 Sept 1972, Gentry 6146 (MO).

Panama specimens with smaller shorter leaves in the manner of Tovomita rhizophoroides: Prov. Coclé: shrub 3 m, La Messa above El Valle, in forest on both sides of junction with road to Cerro Pilon, elev ca 800 m, 21 Jul 1974, Croat 25422 (MO, NY). Prov. Panamá: arbol, altura de 10 m, con raices fulcreas y leche color crema, Cerro Azul, 600 m alt, 15 Abr 1971, E A. Lao & L A. Holdridge 31 (MO).

Tovomita croatii Maguire, sp nov

Frutex ad 5 m altus; ramis teretibus; foliis sessilibus, subamplecticaulibus, chartaceis, ellipticis vel late
oblanceolatis, acuminatis, vulgo 7-10 cm latis, 17-28 cm
longis, integris, pinnivenatis, venis primariis in nervo
marginale collectis, glabris, subtus valde vittatis, con-
spicue squamoso-punctatis; inflorescentiis terminalibus cymosis, multifloribus, ramulis ad 10 cm longis, bracteis con-
spicuis oblanceolatis vel lanceolatis vel lineari-lanceolatis,
ad 15 mm longis; floribus masculinis: sepalis 4, decussatis,
2 exterioribus valvatis, 2 interioribus involventibus et ex-
cedentibus, ad 12 mm longis, convexis, vittatis; staminibus
plus minusve 20-30, 3-4 mm longis, filamentis teretibus,
liberis, antheris bilobatis, ca 0.6 mm latis, 0.4 mm longis,
hippocrepiformibus, lobis lateralibus; polline parvo, tri-
colpis, ca 22-25 u diam; floribus foemineis: non visis;
fructibus elliptico-ovalibus, ca 18-20 mm longis, 5-loculari-
bus, loculis dispermis, stylis partibus liberis, ca 0.5 mm
longis, stigmatibus terminalibus, orbicularibus; seminibus
linearibus, in membranis (tegumentorum ?) longis, subcarnosis,
involutis, axe subbasi affixis; endospermio relative magno,
embryone erecto, tenuiter lineari involuto.

Type. El Valle, behind Club Campestre, shrub 2 m,
flowers white, sap viscid, cloudy, Prov. Coclé, Panama,
12 Apr 1971, Thomas B. Croat 14268A (holotype MO).

Distribution. PANAMA. Prov. Bocas del Toro: shrub 1-2
m tall, bracts mauve and maroon, between Q. Gutierrez and east
slope of La Zorra, headwaters of Río Mali, Chiriqui Trail,
18 Apr 1968, Kirkbride & Duke 732 (MO, 2 sheets). Prov.
Coclé: shrub 1.5 m, buds pinkish, Cerro Caracoral, in the rain
forest below the elfin forest, 24 Apr 1968, Kirkbride 1122 (MO,
2 sheets); shrub to 15 ft, fruits cream flushed with red,
seeds 7 (as seen in X-sect), elongate, testa red, pulp of
fruit colorless, sticky, 25 Jul 1968, Dwyer & Correa 8913 (MO);
shrub to 10 ft, leaves not glandular-punctate, bracts and axes
of infl. rose-red, as are fruits, 25 Jul 1968, Dwyer & Correa
8855 (MO); shrub to 2 m high, buds dull pink, Cerro Pilón,
hill below summit, above El Valle de Antón, 2000-2700 ft alt,
in rain forest, 28 Mar 1969, Dwyer, Durkee, Croat & Castillon
4556 (MO); shrub 2 m, flowers white, sap viscid, cloudy, El
Valle, behind Club Campestre, 12 Apr 1971, Croat 14268A (holotype MO); shrub 2 m, flowers white, sap viscid, cloudy, Cerro
Pilón (above El Valle de Anton), 13 Apr 1971, Croat 14338
(MO); 1.5 m tall, terrestrial, in tropical wet forest, Cerro
Pilón, 900-1173 m, 16 Mar 1973, Liesner 783 (MO).

Tovomita croatii bears the most visibly conspicuous
character of the genus, that of the external pairs of essentially connivent sepals exceeding and enclosing the remaining organs of the flower, i e, inner pair of sepals,
petals, androecium and/or gynoecium, as they may be present.

The morphologic distinction among the three closely
related genera, Chrysochlamys, Tovomita and Tovomitopsis,
of the arillar structure is yet to be systematically investigated. The locules of the ovary in the three genera
are said to be uniovulate. Yet, at least in Tovomitopsis
some of the locules may produce two ovules, and therefore become bispermic. In the present species all locules seem
constantly to bear two ovules and, hence, become two-seeded.

In our species, the carnose envelopes of the twin seeds
appear to be of chalazal attachment. Which, or what parts of
the teguments are involved, is not here determined. However, the envelopes are complete and seem to be separate for
each seed.

Tovomita coriacea Maguire, sp nov

Tovomita weddellianae affine.

Arbor 10-20 m alta, latice albida; foliis appositis,
confertis, coriaceis, anguste oblanceolatis, apicibus obtusis vel rotundatis, basibus angustis, petiolo ca 5 mm
longo; venis lateralibus, 1.0-2.0 mm apartis, adscendentibus a 10°-20° angulo; floribus et staminibus non visis;
floribus pistillatis: alabastris ovato-globosis, ca 5 mm
longis, sepalis 4, decussatis, concavis, marginibus non-
marginatis, jugis exterioribus in alabastro connatis; petalis 4; staminodiis destitutis; ovario 5-loculari, sessili,
loculis uniovulatis; ovulis axe lateraliter affixis; fructu
globoso ca 2 cm lato, sine gynobasi; stigmatibus 5, sessilibus.

Type. Common, pistillate tree 10-20 m, leaves at
summit of branches, erect, latex creamy-white, flower buds
white, fruit maroon, obovoid-ovoid, rounded at the summit,
Cerro de Humo, bosque nublado virgen, 12 km norte del pueblo
de Río Grande arriba, 1273 m alt, Peninsula de Paría, Estado Sucre, Venezuela, 2 Mar 1966, Julian A. Steyermark
94884 (holotype NY, isotype VEN).

Known only by the type collection, but ascribed by
the collector to be common.

Tovomita coriacea is clearly related to Tovomita weddelliana Pl. & Tr., as is demonstrated by the very similar
leaves, but differs from that species in the coriaceous
quality of the leaves and the rotund non-stipitate fruit,
the prominent gynobase being lacking.

The species is published here, although of Venezuelan
origin, because of its affinity to the widespread Tovomita
weddelliana. I wish to acknowledge the kindness of Doctor
Steyermark, collector, for permitting me to do so.

Clusiella elegans Pl. & Tr., An Sc. Nat Ser. 4. Bot. 14: 254. 1860.

  These apparently are initial records for Panama. The center of distribution of the species lies in Colombia. Its range extends into South American Panama, Venezuela and Ecuador.

  PANAMA. Prov. Darien: epiphytic shrub, the stems rooting at nodes and becoming fastened at several points; flowers white, fruits green, vicinity of upper gold mining camp of Tyler Kittredge on headwaters of Río Tuquesa, ca 2 air km from Continental Divide, in recently cleared primary forest, 26 Aug 1974, Croat 27213 (MO) Prov. Panamá: tree fallen, unknown height estimated over 4 m, leaves leathery, calyx green, petals white, glandular area around ovary yellow, stigma white, in tropical wet forest, 16 km above Pan-American Highway on road from El Llano to Carti-Tupile, 13 Feb 1973, Helen Kennedy, R. L. Dressler & Anne Mahler 2394 (MO); same data, 13 Feb 1973, Kennedy, Dressler & Mahler 2397A (MO); 13 Mar 1973, Croat 22894 (MO). Prov. San Blas: 13 Mar 1973, Liesner 658 (MO); 26-27 Mar 1973, Liesner 1208 (MO); 12.7 km Llano-Carti Road, 15 Feb 1975, Mori, Kallunki & Gentry 4701 (MO); same, Mori, Kallunki & Gentry 4705 (MO). Prov. Veraguas: Santa Fe, 20 Dec 1974, Mori, Kallunki et al 3875 (MO).

ADDITIONAL NOTES ON THE GENUS AVICENNIA. X

Harold N. Moldenke

AVICENNIA L.

Additional synonymy: Bontia "L. ex Loefl." apud Soukup, Biota
11: 6, in syn. 1976. Bontia "P. Br. ex Airy Shaw in Willis" apud
Soukup, Biota 11: 6, in syn. 1976. Upata "Rheede ex Adans." apud
Soukup, Biota 11: 6, in syn. 1976.

Additional & emended bibliography: Jacq., Select. Stirp. Amer.,
imp. 1, 177—178, pl. 112, fig. 1 & 2. 1763; Sweet, Hort. Brit.,
ed. 2, 419. 1830; G. Don in Loud., Hort. Brit., ed. 1, 247 (1830)
and ed. 2, 247. 1832; Loud., Hort. Brit., ed. 2, 554. 1832; G.
Don in Loud., Hort. Brit., ed. 3, 247. 1839; Sweet, Hort. Brit.,
ed. 3, 554. 1839; A. DC., Prodr. 11: 701. 1847; Buek, Gen. Spec.
Syn. Candoll. 3: 46. 1858; Crozet, Voy. Tasmania [transl. Ling
Roth] 5: 36. 1891; Estores Anzaldo, Marañon, & Ancheta, Philip.
Journ. Sci. 86: 236 & 239. 1958; Puri, Indian Forest Ecol. 1: 31
(1960) and 2: 223—227 & 229—232. 1960; Golley, Odum, & Wilson,
Ecology 43: 9—19. 1962; Gaussen, Viart, Legris, & Labroue, Trav.
Sect. Scient. Techn. Inst. Franç. Pond., Hors Ser. 5: 25 & 26.
1965; Burns & Rotherham, Austral. Butterflies 104. 1969; Gill,
Forest Sci. 17: 462—465. 1971; Jacq., Select. Stirp. Amer., imp.
2, 177—178. 1971; Moore, Miller, Albright, & Tiessen, Photosyn-
thetica 6: 393. 1972; Ewel & Whitmore, U. S. Dept. Agr. Forest.
Serv. Res. Pap. ITF-18: 16. 1973; Tomlinson & Gill in Meggers,
Ayensu, & Duckworth, Trop. Forest Ecosyst. Afr. & S. Am. 129—
133, & 142. 1973; Chai, Malays. Forest. 38: 188, 204—205, & 207.
1975; Lugo, Evink, Brinson, Broce, & Snedaker in Golley & Medina,
Ecolog. Stud. 11: 336, 338, 339, 344, & 345. 1975; Occhioni
Martins, Leandra 5: 138. 1975; Anon., Biol. Abstr. 61: AC1.559.
1976; Anon., Biores. Index 12 (11): B.75. 1976; M. F. Baker, Fla.
Wild Fls., ed. 2, imp. 2, 190. 1976; Bultman & Southwell, Biotrop-
ica 8 (2): 76 & 92. 1976; Cambie, Journ. Roy. Soc. N. Zeal. 6
(3): 333. 1976; Corner, Seeds Dicot. 1: 276. 1976; Felger & Lowe,
Nat. Hist. Mus. Los Angeles Co. Contrib. Sci. 285: 5 & 50. 1976;
Fleming, Ganelle, & Long, Wild Fls. Fla. 15 & 41. 1976; Follmann-
Schrag, Excerpt. Bot. A.26: 503. 1976; F. R. Fosberg, Biol. &
Geol. Coral Reefs 3, Biol. 2: 272. 1976; F. R. Fosberg, Rhodora
78: 112. 1976; Hocking, Excerpt. Bot. A.26: 422 (1976) and 28:
259 & 260. 1976; Lakela, Long, Fleming, & Genelle, Pl. Tampa Bay,
ed. 3 [Bot. Lab. Univ. S. Fla. Contrib. 73:] 115, 149, & 150.
1976; Laurence & Mohammed, Journ. Agr. Soc. Trin. & Tob. 76: 345.
1976; Long & Lakela, Fl. Trop. Fla., ed. 2, 17, 732, & 930. 1976;
Lugo, Inst. Forest. Latinoam. Invest. Bull. 50: 49 & 54. 1976;
Moldenke, Phytologia 34: 18, 70—94, 167—203, 247, 248, 252—256,
261—263, 265—269, 271, 278, 485, 499, 504, 507, & 509 (1976)
and 35: 13. 1976; Moldenke & Sm. in Reitz, Fl. Ilust. Catar. I
Erioc: 97. 1976; Raven, Evert, & Curtis, Biol. Pl., ed. 2, 427 &

670, fig. 20-12. 1976; Rahm, Environ. Conserv. 3: 47—57. 1976;
Rogerson & Becker, Bull. Torrey Bot. Club 103: 145 & 277. 1976;
Soukup, Biota 11: [1], 6, 21, & 22. 1976; Jiménez & Liogier, Moscosoa 1 (2): 17. 1977; Moldenke, Biol. Abstr. 63: 2451—2452.
1977; Moldenke, Phytologia 35: 507 (1977) and 36: 31, 32, 34, 38,
39, 46, & 47. 1977; A. L. Moldenke, Phytologia 36: 88. 1977; Rogerson, Becker, & Prince, Bull. Torrey Bot. Club 104: 82. 1977;
Ward, Phytologia 35: 409. 1977.

Golley and his associates (1962) aver that mangroves cover a-
bout 1/3 of the coastlines of tropical America. Sweet (1830,
1839) places the genus _Avicennia_ in the Myoporinae [Myoporaceae].
Burns & Rotherham (1969) report that the larvae of the Copper
Jewel butterfly (Hypochrysops apelles), attended by small black
ants, live in the dead or folded leaves of _Avicennia_ in Australia. When feeding, the larvae eat the epidermis of the leaves,
leaving a network of veins which become dry and assume a scorched
appearance.

It is perhaps worth noting here that Cambie (1976), Lakela,
Long, Fleming, & Genelle (1976), Fleming, Genelle, & Long (1976),
and Lakela & Long (1976) also accept the family Avicenniaceae as
a separate family from the Verbenaceae and Myoporaceae.

AVICENNIA AFRICANA P. Beauv.
    Additional bibliography: Buek, Gen. Spec. Syn. Candoll. 3: 46.
1858; Anon., Biol. Abstr. 61: AC1.559. 1976; Hocking, Excerpt.
Bot. A.28: 259 & 260. 1976; Moldenke, Phytologia 33: 239, 250, 252,
259, 261, & 262 (1976) and 34: 74, 202, & 203. 1976.

AVICENNIA ALBA Blume
    Additional bibliography: Buek, Gen. Spec. Syn. Candoll. 3: 46.
1858; Fedde & Schust. in Just, Bot. Jahresber. 44: 253. 1922; Lamberti, Univ. São Paulo Fac. Filos. Bol. 317 [Bot. 23]: 120, 150,
155, 160, & 165. 1969; Chai, Malays. Forest. 38: 188, 205, & 207.
1975; Anon., Biol. Abstr. 61: AC1.559, 1976; Moldenke, Phytologia
34: 70—72, 75, 76, 80, 84, 85, 90, 91, 93, 94, 167, 170, 172,
179, 180, 185—188, 190, 194, 195, 197, 198, 200—203, 262, 265,
267, 268, & 271 (1976) and 36: 38. 1977.

Recent collectors describe this species as a large tree or a
small shrub-like tree, 2—30 m. tall, often divided at the base
into 2 main stems, the trunks to 20 cm. in diameter at breast
height, with many pneumatophores, 10—30 cm. tall, in a radius of
4 m. from the base of the tree, the trunk and branches black, the
bark dark-brown, more or less smooth, the leaves dark-green above,
grayish-green or whitish beneath, and the fruit light-green. They
have found it growing in "open muddy mangrove forests" and "very
common on seashores and on mud flats in moderately firm soils or
soft mud". Van der Kevie describes the corollas as "yellow to
orange", while Foreman & Katik refer to them as simply "yellow".
They have found it in anthesis in June and in fruit in August.
McCusker describes the corollas as "bright-yellow" and encountered
the species "on both seaward and landward edges of mangrove swamps"

Chai (1975) has studied this tree extensively in Malaysia and describes it there as follows: "Small to huge tree to 70 feet tall, 7 ft. girth. Bark dark brown to black. No buttresses but may develop slender, soft stilt roots. Leaves lanceolate or elliptic-obovate with tapering base, lower leaf surfaces whitish, salt being excreted from this surface. Fruit glaucous green, leech-shaped. Another pioneer species colonising newly formed mud flats as Sonneratia alba. Often gregarious along low convex banks of the rivers near the sea but later replaced by Rhizophora apiculata and Bruguiera parviflora. Rare inland." In his key he distinguishes it from the two other species of the genus known there [A. marina and A. officinalis] as follows: "Medium to large tree to 70 ft. tall; bark dark gray to black, often with white patches, not flaky; leaves oblong-elliptic, whitish below; soft mud." He calls it "api-api hitam".

Additional citations: THAILAND: Maxwell 75-918 (Ac); Van der Kevie 1 (Ac). MALAYA: Selangor: McCusker 303 (Ld). BISMARK ARCHIPELAGO: Manus: Foreman & Katik LAE.59275 (Mu).

AVICENNIA ALBA var. LATIFOLIA Moldenke
Additional bibliography: Anon., Biol. Abstr. 61: AC1.559. 1976; Moldenke, Phytologia 34: 71, 91, 198, & 265. 1976.

AVICENNIA BICOLOR Standl.
Additional bibliography: Anon., Biol. Abstr. 61: AC1.559. 1976; Hocking, Excerpt. Bot. A.26: 422. 1976; Moldenke, Phytologia 33: 240. 1976.

Pohl and Davidse, identifying this as A. germinans (L.) L., refer to it as "one of the dominants in mangrove swamps". They found it in fruit in June.

The H. Kennedy 2281, distributed in some herbaria as A. bicolor, actually represents A. germinans var. guayaquilensis (H.B.K.) Moldenke.

Additional citations: COSTA RICA: Guanacaste: Pohl & Davidse 10588a (W—2774496).

AVICENNIA ELLIPTICA Holm
Additional bibliography: Buek, Gen. Spec. Syn. Candoll. 3: 46. 1858; Anon., Biol. Abstr. 61: AC1.559. 1976; Moldenke, Phytologia 34: 71, 202, & 278 (1976) and 36: 32 & 34. 1977.

The Dias da Rocha 108, F. C. Hoehne s.n. [Herb. Inst. Bot. S. Paulo 24908], Löfgren s.n. [Herb. Inst. Bot. S. Paulo 15596], Usteri s.n. [Herb. Inst. Bot. S. Paulo 15598], and Herb. Com. Geogr. & Geol. 3062, distributed as A. elliptica, actually represent A. schaueriana f. candicans instead.

Additional citations: BRAZIL: Bahia: Lanna 716 [Castellanos 25468; Herb. FEEMA 4562] (Ld), 747 [Castellanos 25497; Herb. FEEMA 4561] (Z).

AVICENNIA ELLIPTICA var. MARTII Moldenke

Additional bibliography: Moldenke, Phytologia 32: 438, 454, & 455 (1975) and 33: 255, 262, & 269. 1976; Anon., Biol. Abstr. 61: AC1.559. 1976; Moldenke, Phytologia 36: 32 & 34. 1977.

The Drouet 2442, previously cited by me as A. elliptica var. martii, now seems to me to be only one of the many forms of A. germinans var. guayaquilensis (H.B.K.) Moldenke; the same is true of Ducke 5407, Lanjouw & Lindeman 301, and Smith & Smith 546.

AVICENNIA EOCENICA Berry

Additional bibliography: Moldenke, Phytologia 32: 455. 1975; Anon., Biol. Abstr. 61: AC1.559. 1976.

AVICENNIA EUCALYPTIFOLIA Zipp.

Additional synonymy: Avicennia marina var. australasica (Walp.) Moldenke, Phytologia 34: 72, in syn. 1976.

Additional bibliography: Fedde & Schust. in Just, Bot. Jahresber. 44: 253. 1922; Anon., Biol. Abstr. 61: AC1.559. 1976; Hocking, Excerpt. Bot. A.28: 259. 1976; Moldenke, Phytologia 34: 71—72, 84, 85, 91, 93, 94, 177, 188, 268, & 271. 1976.

Stoddart and Thom refer to this species as it occurs on the Great Barrier Reef as a shrub or tree, 1—3 meters tall, with vertical pneumatophores (and "roots"), the leaves often with prominent leaf-galls, and found it "occasional" on the seaward side of Rhizophora colonies and on shingles, shingle ramparts, and shingle or mangrove cays, flowering and fruiting already (although apparently very sparingly) at 1—3 meters height.

Additional citations: GREAT BARRIER REEF ISLANDS: Hampton: Thom 4211 (W—2744309). Howick: Thom 4203 (W—2744287). Low: Stoddart 4336 (W—2744301). Lowrie: Stoddart 4998 (W—2744204). Low Wooded: Stoddart 4524 (W—2744316). Sand: Stoddart 4210 (W—2744311). Three: Stoddart 4499 (W—2744321). West Hope: Stoddart 4405 (W—2744299).

AVICENNIA GERMINANS (L.) L.

Additional & emended bibliography: Jacq., Select. Stirp. Amer., imp. 1, 177—178, pl. 112, fig. 1 & 2. 1763; Sweet, Hort. Brit., ed. 2, 419. 1830; G. Don in Loud., Hort. Brit., ed. 1, 247 (1830), ed. 2, 247 (1832), and ed. 3, 247. 1839; Schau. in A. DC., Prodr. 11: 699—700. 1847; Buek, Gen. Spec. Syn. Candoll. 3: 46. 1858; Pat., Ill. Biol. Monog., ser. 2, 4: 482. 1916; Sydow in Just, Bot. Jahresber. 44: 595. 1923; Fedde in Just, Bot. Jahresber. 44: 1377. 1927; M. F. Baker, Fla. Wild Fls., ed. 2, imp. 1, 190. 1938; Lamberti, Univ. São Paulo Fac. Filos. Bol. 317 [Bot. 23]: 40, 41, 45, 46, 120, 149, & 165. 1969; Jacq., Select. Stirp. Amer., imp. 2, 177—178, pl. 112, fig. 1 & 2. 1971; Ewel & Whitmore, U. S. Dept. Agr. Forest Serv. Res. Pap. LTF-18: 16. 1973; Tomlinson & Gill in Meggers, Ayensu, & Duckworth, Trop. Forest Ecosyst. Afr. & S. Am. 129—133 & 141—142. 1973; Lugo, Evink, Brinson, Broce, & Snedaker in Golley & Medina, Ecolog. Stud. [Jacobs, Lange, Ol-

son, & Wieser, Ecol. Stud. 11:] 335—339, 344, & 345. 1975; Anon.,
Biol. Abstr. 61: AC1.559. 1976; Anon., Biores. Index 12 (11): B.
75. 1976; M. F. Baker, Fla. Wild Fls., ed. 2, imp. 2, 190. 1976;
Felger & Lowe, Nat. Hist. Mus. Los Angeles Co. Contrib. Sci. 285:
5 & 50. 1976; Fleming, Genelle, & Long, Wild Fls. Fla. 15 & 41.
1976; F. R. Fosberg, Rhodora 78: 112. 1976; Lakela, Long, Fleming,
& Genelle, Pl. Tampa Bay, ed. 3 [Bot. Lab. Univ. S. Fla. Contrib.
73:] 115 & 149. 1976; Laurence & Mohammed, Journ. Agr. Soc. Trin.
& Tob. 76: 345. 1976; Long & Lakela, Fl. Trop. Fla., ed. 2, 17,
732, & 930. 1976; Lugo, Inst. Forest. Latinoam. Invest. Bull. 50:
49 & 54. 1976; Moldenke, Phytologia 34: 72—76, 84—86, 91, 93,
94, 170, 172, 177, 180, 199—203, 248, 252—256, & 271. 1976;
Raven, Evert, & Curtis, Biol. Pl., ed. 2, 427 & 670, fig. 20-12.
1976; Rehm, Environ. Conserv. 3: 47—57. 1976; Soukup, Biota 11:
6. 1976; Jiménez & Liogier, Moscosoa 1 (2): 17. 1977; Moldenke,
Phytologia 36: 31, 34, & 41. 1977; A. L. Moldenke, Phytologia 36:
88. 1977; Ward, Phytologia 35: 409. 1977.

Additional & emended illustrations: Jacq., Select. Stirp. Amer.,
imp. 1, pl. 112, fig. 1 & 2 (1763) and imp. 2, pl. 112, fig. 1 &
2. 1971; Fleming, Genelle, & Long, Wild Fls. Fla. 41 (in color).
1976; Lugo, Inst. Forest. Latinoam. Invest. Bull. 50: 54. 1976;
Raven, Evert, & Curtis, Biol. Pl., ed. 2, 427, fig. 20-12. 1976.

Some of the leaves on Sachet 459, from the Cayman Islands, resemble those seen on typical var. guayaquilensis; still, it seems
rather plain that such leaves are far more common in populations
of northern South America. Possibly both the typical form and
var. guayaquilensis occur in that southern West Indies - northern
South American area, a condition not at all unusual; certainly
some specimens of Asplund 16588, Budowski 25, Chapin 1129, Fournier 81, Hagen 8 & 809, Haught 4855, Pittier 11011, Romero-Castañeda
7275, D. H. Knight 1032, H. H. Smith 1937, Snodgrass & Heller 368,
Stewart 3267, T. W. J. Taylor TT.91, Wiggins 18310, and Wiggins &
Porter 517 seem to indicate this. Possibly var. guayaquilensis
would better be regarded as a form rather than a variety.

The corollas are said to have been "whitish" on Ventura A. 5226
and the label accompanying this collection makes the remarkable
claim that the specimen was collected at "40 m." altitude — doubtless an error.

Ewel & Whitmore (1973) say that "Some of the low alluvial areas
on the south coast of Puerto Rico contain saline soils, such as the
Santa Isabel series, and the vegetation on these sites is dominated
by Prosopis juliflora.....Mangrove forests....form parts of the
coastal associations in this life zone, but the development of tall,
luxuriant mangrove forests may, in some locations, be limited by
the scarce surface runoff, which can result in higher salinities
and lower nutrient inputs than would be the case along coasts with
more rainfall." Lugo and his associates (1975) have compared the
net daytime photosynthesis, nighttime respiration, and their ration
in Rhizophora, Avicennia, Laguncularia, and Conocarpus, noting that
only Avicennia among these genera comprises salt-excreting plants.
                                        [to be continued]

BOOK REVIEWS

Alma L. Moldenke

"FLORIDA WILD FLOWERS — An Introduction to the Florida Flora" by
Mary Francis Baker, xiii & 269 pp., illus., Reprint Edition
by Horticultural Books, Inc., Stuart, Florida 33494. 1976.
$5.95 paperbound.

This work was first published by the Macmillan Company in 1926
which also released the New Edition of 1938. The "Flowering Tree
Man of Florida", the indefatigable Dr. Edwin Menninger, has used
the latter for his Horticultural Books replication. "Eight hundred of our more common and more interesting herbs, shrubs, and
trees" are described and keyed. There are 49 black/white plates
of the author's plant photographs.
It is nice to have this old time favorite available again.

"ARK II Social Response to Environmental Imperatives" by Dennis C.
Pirages & Paul R. Ehrlich, x & 344 pp., illus., W. H. Freeman
& Company, San Francisco, California 94104. 1974. $3.95
paperbound.

"Prologue: Noah had ample warning from a respected authority to
build his ark and he used his time to good advantage. Skeptics
laughed, ridiculed, and drowned — but Noah, the original prophet
of doom, survived. We too have been warned that a flood of problems now threatens the persistence of industrial society, but this
time the [new institutional]....ark must ensure our survival by re-
designing the political, economic and social institutions of industrial society".
Dr. Pirages, a social scientist, prepared this text while serving
as a research associate with Prof. Ehrlich at Stanford University
1970—1973. The analyses and suggestions offered are logical, hon-
est, and radical (in the literal sense), but implementing them or
other valid substitutes (not palliative gestures) will be so very
difficult to achieve in time. More thoughtful reading of this book
could help.

"DNA SYNTHESIS" by Arthur Kornberg, ii & 399 pp., illus., W. H.
Freeman & Company, San Francisco, California 94104. 1974. $18.

In 1972 this Nobel-laureate biochemical authority in this in-
triguing complicated field presented the Robbins lectures at Pomona
College. His preparations have since been developed into this outstanding graduate level text for courses in molecular genetics,
biophysics, biochemistry, etc., for the student with a beginning

413

interest in DNA synthesis and for those working directly or in-
directly in this field.

The main topics are: structure, functions and precursors of
DNA, — DNA polymerases, of *Escherichia coli*, of phage induction
into bacteria and of eukaryotic cells, — replication of DNA
viruses, repair, recombination and restriction, — RNA polymerases,
transcription, and synthesis of genes. The book closes with an
intelligent analysis of social concerns about genetic chemistry
or what is suspiciously labeled as "genetic engineering" by some.

The text is written in direct, clear language. The 194 excellent illustrations consist mainly of interpretive diagrams
(149 in color) and electron micrographs and 69 tables.

"WATER — A Primer" by Luna B. Leopold, xvi & 172 pp., illus., W.
     H. Freeman & Company, San Francisco, California 94104. 1974.
     $4.95 clothbound, $2.95 paperbound.

This publication is part of the Series of Books in Geology
edited by James Gilluly and A. O. Woodford and it covers very
clearly and logically the general principles of hydrology for beginning students interested in a variety of environmental problems. Those readers seeking further explanations or applications
are referred to the Professional Papers and the Water Supply
Papers of the United States Geological Survey prepared by assorted
hydrological specialists. Since the principles of hydrology are
non-controversial this excellent brief introduction to them could
(on reading, of course) only result in a better informed citizenry
capable of wiser choices about the nation's development of its
water resources involving flood controls, irrigation, pollution
control, etc.

Did you know that "In some of the arid parts of western United
States water is being pumped that fell as rain during the ice age,
at least 10,000 years ago"?

"CHEMISTRY FOR THE LIFE SCIENCES" by J. G. Dawber & A. T. Moore,
     xiii & 426 pp., illus., McGraw-Hill Book Company, New York,
     N. Y. 10020. 1973. $11.95 paperbound.

This text, published in England, presumes an introductory
course in chemistry and is "designed for students taking courses
in Applied Biology and Medical Laboratory Sciences.....[and] as a
text for ancillary chemistry courses required in degree courses
in biological sciences and medicine." The chapter headings are:
Atomic structure and chemical bounding; Structural theory of organic chemistry; Physical aspects of chemical reactions including
thermodynamics and kinetics of chemical reactions; Reactions of
organic molecules by making and breaking covalent bonds mainly;
Physical chemistry of liquids and solutions including surface and
membrane phenomena and electrolytic solutions; Chromatography and

spectroscopy including magnetic resonance methods; Structure and properties of such natural organic compounds as amino acids, lipids, carbohydrates, nucleosides, vitamins and steroids; Structure and properties of biopolymeres and enzymic catalysis; and Chemical reactions in living organisms.

This book makes an excellent ancillary text in the United States for courses in biochemistry, molecular biology, and related topics and for personal study and review.

"ANNUAL REVIEW OF ECOLOGY AND SYSTEMATICS" Volume 5 edited by Richard F. Johnston with Peter W. Frank and Charles D. Michener, ii & 448 pp., illus., Annual Reviews, Inc., Palo Alto, California 94306. 1974. $15.00 U.S.A., $15.50 foreign.

Herein are 18 carefully presented, currently interesting articles provided with detailed bibliographies and often with tables and diagrams starting with H. J. Baker's "Evolution of Weeds" and ending with J. L. Harper's and J. White's "Demography of Plants". Between these excellent papers there are others on the ecology of mangroves, of macroscopic marine algae, of secondary successions, of island biogeography re equilibrium theory, etc. Many readers will be interested in J. Cracraft's "Continental Drift and Vertebrate Distribution" as well as I. Noy-Meir's "Desert Ecosystems: Higher Trophic Levels" and R. D. Alexander's "Evolution of Social Behavior".

"LECTURES ON THE PHENOMENA OF LIFE COMMON TO ANIMALS AND PLANTS" by Claude Bernard, Volume I, translation by Hebbel E. Hoff, Roger Guillemin & Lucienne Guillemin, xxxv & 288 pp., illus., Charles C. Thomas, Publishers, Springfield, Illinois 62717. 1974. $12.95.

This famous work first appeared in 1878 within months after the author had corrected the final proofs on his deathbed in Paris. It was intended as the initial segment of a Cours de Physiologie Général du Muséum d'Histoire Naturelle. The translators' introduction is followed by the original foreword and two funeral orations emphasizing not only Claude Bernard's principal discoveries and great influence on experimental physiology and experimental medicine, but also those personal traits that endeared him to his confrères and students.

The ten lectures follow, each preceded by a topical outline, and conclude that "Determinism remains the great principle of the science of physiology.....with no difference between the sciences of inanimate objects and the science of living bodies".

The translated English at times is stilted, but no original meaning is altered.

"TREES OF THE BERKELEY CAMPUS" Revised Edition by Robert A. Cock-
   rell & assisted by Frederick F. Warmke, vi & 97 pp., illus.,
   Division of Agricultural Sciences, University of California,
   Richmond, California 94804 or Berkeley, California 94720.
   1976. $5.90 paperbound oversize.

   Campus trees and larger shrubs are listed alphabetically by
scientific names, then common names, descriptions and locations.
Appendix I lists most of the species surrounding the buildings
and landmarks on the campus. Since Quercus agrifolia, the coast
live oak, is the most common native tree growing in the Berkeley
Hills with one or more visible from almost any point on campus,
it is not listed under each of these sites. Appendix II indexes
the common names and is followed by a glossary and selected
references. The illustrations include a keyed diagram of the
Berkeley campus, some plantings as the redwood grove and a photograph of Woodbridge Metcalf. "Woody", to whom this publication
deservedly is dedicated, was a "friendly man, and ardent conser-
vationist, and an outstanding teacher".

"CHINESE HERBS — Their Botany, Chemistry and Pharmacodynamics"
   by John D. Keys, 388 pp., illus., Charles E. Tuttle Company, Inc., Bunkyo-ku, Tokyo, & Rutland, Vermont 05701.
   1976. $15.00.

   Just as surely as this is a 'work of abiding love', it is also a thorough study botanically, historically, culturally and
pharmaceutically, requiring the past 20 years in preparation. As
a lad, the author became intrigued with the many herb-shops in
San Francisco's Chinatown. This continuing interest led to the
study of Chinese, the translation of over 20 books (mostly phar-
maceutical or culinary) into English, and the study of Japanese
and French.
   For over 250 Chinese character labeled and illustrated herbs
are given the scientific name and family, common English name,
botanical description, habitat, synonymy, pharmacodynamic inves-
tigations, Chinese therapeutic usage, dosage, incompatible drugs,
and related plants used for the same purpose. Appendices include
tables of (I) Supplementary Botanical Drugs, (II) Mineral Drugs,
(III) Drugs of Animal Origin, (IV) Collection of Chinese Prescrip-
tions, (V) Table of Toxic Herbs, and a glossary and list of references.
   It is interesting to find that Laminaria japonica Aresch. was
prescribed by ancient Chinese for goiter control. Eriocaulon
sieboldianum is now usually reduced to the synonymy of the wide-
ranging E. cinereum R. Br. or classified as a variety of it.
   This is a fascinating fine study which should interest many
different folks.

Designed to expedite botanical publication

| Vol 36 | August 1977 | No 5 |

CONTENTS

Published by Harold N Moldenke and Alma L Moldenke

303 Parkside Road
Plainfield, New Jersey 07060
U S A.

Price of this number, $3, per volume, $9 75 in advance or $10 50 after close of the volume, 75 cents extra to all foreign addresses, 512 pages constitute a volume, claims for numbers lost in the mail must be made immediately after receipt of the next following number.

HISTORY and DISTRIBUTION of EURASIAN WATERMILFOIL
in UNITED STATES and CANADA

Clyde F. Reed

Reed Library and Herbarium
Baltimore, Maryland

Watermilfoils have frequently been problems in ponds, lakes
and canals throughout North America in the past, but in the late
1950's and 1960's, they became serious ecological problems in much
larger bodies of water, as in the Potomac River, Chesapeake Bay
and TVA reservoir. From the early 1800's Myriophyllum spicatum L.
had been listed as the species causing the problem. In 1919 Fernald separated all North American plant-specimens from the Eurasian specimens, calling them M. exalbescens Fern. Since 1919 most
authors of floras, including Fernald, have used M. exalbescens for
all American material, ignoring M. spicatum completely. However,
both do exist in North America, typical M. spicatum being the species which caused the recent problem.

The purposes of this paper are to establish the fact that
both Myriophyllum spicatum L. and M. exalbescens Fern. exist in
United States and Canada, to review the literature for usage of
M. spicatum L. in North America in the past, to account for the
recent explosive growth of M. spicatum L. in Eastern United States
in the 1950's and 1960's, and to cite the herbarium specimens
collected from that time up to date (mainly in the Reed Herbarium),
showing the extent of the spread of M. spicatum L. as an economi-
cally important and ecologically dangerous waterweed.

Taxonomic considerations

Eurasian watermilfoil (Myriophyllum spicatum L.) was described
by Linnaeus (Sp. Pl. 2: 992. 1753), based on specimens from Europe,
from quiet waters, as a perennial.

In North America, Asa Gray (1848, p. 140) listed M. spicatum
L. from northern United States. Tatnall (1860. p. 98) suggested M.
spicatum L. probably could be found in New Castle Co., Dela., without any definite record of specimens. Again, Gray (1867, 1880
and 1887) and Robinson & Fernald (1908, p. 604) give M. spicatum
L. from northeastern United States, this time stating it as being
introduced from Europe. Probably all these references should be
considered the plant now called M. exalbescens Fern. However, some
of these early specimens have turned out to be M. spicatum L.

Ward (1881, pp. 24, 80, 160) was the first to cite definite specimens (No. 303) of M. spicatum L. from the Potomac River (below Alexandria and opposite Ft. Foote), presumably from the Virginia side of the river. Ricker (1906, p. 84) also cited specimen No. 303 as M. spicatum L. from Potomac River. Hitchcock and Standley (1919) added Hunting Creek (Va.) for this species, noting "widely distributed in North America, Europe and Asia", thus placing typical M. spicatum with all other North American material, most of which was later to be placed in M. exalbescens.

Fernald (1919) clearly indicated that American plants differed in several aspects from Eurasian specimens, and named the American material M. exalbescens Fern., without indicating that some of the material could be typical M. spicatum L. Hulten (1947) regarded M. exalbescens as a subspecies, namely M. spicatum subsp. exalbescens (Fern.) Hult. Löve (1961) discussed the situation and found that both had 2n=42 (hexaploid) chromosomes, but suggested the two names be retained for the different populations. (Löve originally (1948) recorded from Icelandic specimens 2n= ca. 36 for M. spicatum L.; later (1954) be corrected it to 2n= 28; and finally 'with better fixed material from Iceland' decided 2n=42, which is the same number as M. exalbescens Fern. from Lake Manitoba, Canada).

In giving the details of morphological differences between the two species, Löve says nothing about the plants of the Chesapeake Bay region, Potomac River or Tennessee Valley region. Plants from these areas do fit his description of M. spicatum L. and not for M. exalbescens Fern. This is important since Fernald (1950), Gleason (1952), Love (1961) and many others since then have assumed that all plants in North America, except some from Alaska and the Aleutian Islands, are M. exalbescens, noting 'M. spicatum of Amer. auth., not L.'.

Gleason (1952) added, 'perhaps better subordinated to the Eurasian M. spicatum as var. exalbescens (Fern.) Jeps. Patten (1950) noted 'there exists a possibility that M. exalbescens Fern. and M. spicatum var. capillaceum Lange are the same, since both of these description were based, in part only in the former instance, upon material from Greenland. This would invalidate Fernald's name in the varietal category through precedence'. Therefore, M. spicatum subsp. exalbescens (Fern.) Hult. has been suggested.

Reed (1970) was the first to treat both species as being in North America. The annotated list of specimens below indicates those plants definitely identified as M. spicatum L. in North America, most of which have been collected since the Eurasian Watermilfoil explosion in the Chesapeake Bay, Potomac River and Tennessee River Valley regions.

Myriophyllum spicatum L.

Perennial, aquatic-rooted herb, reproducing by seeds, but very commonly and most efficiently spreading by rhizomes, fragmented stems, and axillary buds that occur throughout the year; stems long and branching, often from a depth of 5 m., most frequently to 1.6 m.), often forming extensive mats at the surface of the water, brick-red or olive-green in dried specimens; leaves whorled in 3's or 4's, to 35 mm. long, the principal leaves of the primary stems with 14-21 pairs of rigid slenderly linear divisions; bracts rhombic-obovate to elongate, the bractlets nearly round or kidney-shaped, broader than long, 0.5-0.8 mm. long; spikes terminal, 2.5-10 cm. long, often standing above the water level, after pollination then resubmerging; flowers (after emergence) with the stigmas ripening well in advance of the stamens (favoring cross-pollination); petals deciduous before ripening of the stamens; anthers linear, 1.8-2.2 mm. long; floral bracts longer than the fruits; schizocarp 4-locular, with 4 seeds; mericarps spherical, 4-angled, 2.5-3 mm. in diameter. Late July - September.

In fresh and saline waters, on muck to hard-packed sand; most common and a nuisance, especially to sportsmen. Native of Eurasia and parts of Africa. In many distinct areas of Eastern and Central United States, as far west as Wisconsin and Texas; distinct area in west-central California.

Myriophyllum exalbescens Fern.

Perennial aquatic herb, reproducing by seeds, running rhizomes and fragments of the stems; stems simple or forking, purplish, when dry becoming white, up to 1 m. in length; leaves whorled, in 3's and 4's, 1.2-3 cm. long, with 6-11 (-14) pairs of capillary flaccid or slightly stiffish divisions, the primary leaves submersed, 1-5 cm. long, 1.4 cm. broad; spikes almost naked, terminal, with the flowers in whorls, the lower flowers pistillate, the upper staminate; bracts persistent, rarely equaling the fruit, spatulate-obovate or oblong-shell-shaped, 0.8-1.8 mm. long, the lower serrate, the upper entire; bracteoles ovate, entire, 0.7-1 mm. long; petals oblong-obovate, concave, 2.5 mm. long; anthers 1.2-1.8 mm. long; schizocarp nearly globose, very slenderly 4-sulcate, 2.3 3 mm. long; mericarp rounded on the back, smooth or roughened. July - September.

Lakes, ponds, pools, and quiet waters, often brackish or calcareous; especially troublesome around edges of lakes. Native, throughout the northern part of the United States, south to Delaware on the east coast and to the Mexican border on the west; south into northwestern Mexico and north into Canada, from Newfoundland and Labrador to Alaska; Greenland.

Myriophyllum spicatum L., Eurasian watermilfoil. A, Habit—× 0.5. B, Whorl of leaves—× 2, C, part of flower spike, with pistillate flowers below and staminate flowers above—× 4, D, immature fruits—× 4, E, mature fruit—× 4.

Patten (1954) made a comprehensive study of the floral characters for several populations of the M. spicatum-complex from lakes in New Jersey, and suggested that, in addition to the variations already known in Eurasian representative specimens noted by Hegi (1926), there were considerable intergradings between North American specimens of M. spicatum and M. exalbescens. Leaves of both vary from soft to stiff, the leaflets from straight to curved, either slumping together, as a feather, or remaining separate when taken from the water; stems of both vary from light green to pinkish or reddish, those of M. spicatum tending to be brighter green, and only occasionally whitened.

Introduction, Ecology and Spread of M. spicatum

Many aquatic plants have been and are still grown as aquarium plants. Eurasian Watermilfoil (M. spicatum L.), Parrotfeather (M. aquaticum (Vell.) Verdc. -- syn. M. brasiliense Cambess. and M. proserpinacoides Hook. & Arn.) and Anacharis canadensis (Michx.) Rich. (Elodea canadensis Michx.), are a few of the more common species used in aquaria. Over the years people wishing to dispose of unwanted aquarium contents, have dumped fish, snails and plants into various water sources, as old quarries, reservoirs, ponds, lakes, streams and rivers. For example, before they were filled in and built over, the auther collected all of these plants in the 1930's and 1940's from the limestone quarries near Cockeysville, Padonia and Texas, Baltimore County, Maryland. By the way, all these quarries drain into tributaries of the Big Gunpowder Falls which ultimately forms the Gunpowder River south of Joppatowne.

Barnes (1960) noted Eurasian Watermilfoil in the Chesapeake and Ohio Canal, near Cabin John and the Seven Locks area as early as 1945. Bertholdt (1958) wrote an article, stating "Your aquarium needs Myriophyllum -- plant of delicate beauty". Many aquarium magazines list Myriophyllum for sale. And, it is well-known that the vast growth in the TVA area in Tennessee was the result of plants introduced for the aquarium business.

The spotty distribution (see map) of authentic specimens of Eurasian Watermilfoil, in the past, would indicate independent introductions, probably from aquarium sources, from Canada to Florida, and occasionally westward.

The earliest published record for M. spicatum L. in the Potomac River is that of Lester Ward (1881), who wrote "Found in former years below Alexandria by Mr. Anton Zumbach. Probably still there" (p. 80). On p. 24, "Opposite Ft. Foote". Then in the Checklist following the Guide, on p. 116, "No. 303. Myriophyllum spicatum", a cited specimen. This reference would indicate the plant was in the Potomac River below Washington, D.C. previous to 1881.

Early Annotated Herbarium Specimens and Records

(1881 - 1953)

1881. Below Alexandria, Virginia. Anton Zumbach; opposite Ft.
Foote. (Potomac River). (Ward, 1881).

1902. In lake-like expanse of Sundrake's Creek, Gunpowder River
(Chesapeake Bay). Sept. 6, 1902. Geo. H. Shull 327. (US).

1915. Hunting Creek, Virginia. (Potomac River). Sept. 4, 1915.
W.L.McAtee 2340. (Patuxent Refuge Coll.).

1933. Creeks in Upper Potomac River. Francis M. Uhler. (U.S.
Fish & Wildlife Service (Unpubl. data), cited by Springer,
Beaven & Stotts (1961) and D. Haven (1962).

1937. Cecil Co., Maryland. (Chesapeake Bay, shore of Turkey
Point, Susquehanna Flats). June 18, 1937. J.B.Egerton &
C.F.Reed. (Reed).

1949. Montgomery Co., Maryland. C. & O. Canal. Well-established.
Oct. 1, 1949. F.H.Sargent. (Cath. U. -- annotated by C.F.
Reed in 1962). (M. aquaticum also well-established at this
time and collected).

1951. Montgomery Co., Maryland. C. & O. Canal between Locks
10 and 12, abundant. Aug. 10, 1951. E.P.Killip 41366. (US).

1952. Charles Co., Maryland. In Potomac River, along shore,
Chapel Point. Mar. 29, 1952. Reed 28087.

Ecology

     Most Myriophyllum species have definite affinities for
alkaline situations, and under high calcareous conditions in
ponds, lakes and quarries, they precipitate encrustations of
marl on their surfaces of the stems and leaves giving a white
appearance. This condition is found quite frequently on specimens of M. exalbescens throughout North America, as well as
on many of the specimens of M. spicatum from lakes and ponds
in both North America and Europe.

     Patton (1956) noted that "M. spicatum (var. exalbescens)
was instrumental in maintaining a high pH, and that the daily
pH cycle was so closely related to the physiological activities
of the plant that only a 3.8 unit fluctuation occurred, the
minimal and maximal of pH corresponding directly to photosyn-
thetic activities. During the period of highest pH and maximum
photosynthesis bicarbonate alkalinity dropped concomitant with
the appearance of several ppm. of carbonate, indicating photo-
synthetic utilization of half-bound carbon dioxide".

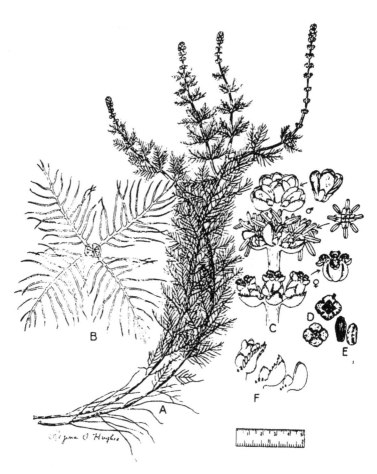

Myriophyllum exalbescens Fern. Northern watermilfoil A Habit—× 0.5, B, whorl of leaves
× 1.5 C flower spike with male and female flowers—× 5, D schizocarp—× 5, E mericarps—× 5, F bracts—
× 5

Although Eurasian Watermilfoil tolerates a broad range of
ecological conditions, it is absent from acid waters. Maximum
density of growth occurs in areas of fine organic ooze with a
muck or sandy-muck base, diminishing to near total absence on
pure sand. In grows in inland alkaline waters with a pH of about
8, and can tolerate tidal waters with salinity up to 16 ppt, or
about 46 percent of sea salinity; plants can retain growing tips
at 93 percent of sea salinity. Plants root best (experimentally)
at salinity of 3.5 ppt. (10 percent of sea salinity), and will
make good growth at 15 ppt. (43 percent of sea salinity).

Eurasian Watermilfoil is adaptable to rooting at varying
depths of water to 3 m. (rarely to 5 m.), with its long branching stems reaching the surface of the water level, but never extending beyond the water level, as they do in M. aquaticum. Most
plants are found attached in water 65-150 cm. in depth. Stems
and leaves are usually dark to medium bright green in deep water,
and both may become encrusted with algae and diatomaceous scum.
In winter the upper portions of the stems break up and float about
with the tides, each segment being capable to root and start a
new colony. By May or June free floating rooted segments are common in the water. Being perennial, the lower portions of the
plants remain green throughout the winter and send up new shoots
in the spring and summer. In mild winters and in more southerly
regions, floating stems can be seen year round (note dates on
annotated specimens). Summer growth is very rapid, measured growth
being 5-7 cm. per day. New Plants form readily by fragmentation
and long stems may take root anywhere along their length. Special
buds at the stem-tips are formed to produce new plants after the
natural breakup of the beds in winter. Plants also produce seeds
during the summer and early fall.

Explosive Growth and Distribution in Late 1950's

In 1962, in the Summary of the 1962 Interagency Research
Meeting on Eurasian Watermilfoil, Reed 'pointed out that in Maryland all heavily infested waters receive their runoff from limestone areas. This also is true of the TVA and other areas where
the plant is abundant. He felt that the presence of calcium ions
may be an important factor in milfoil abundance. He also noted
that the great reduction of coal-mining in Pennsylvania had
removed quantities of acid waters from the Susquehanna and that
contributions of calcium ions from that stream may have been
sharply increased during recent years, thus favoring milfoil
growth in the Upper Chesapeake".

When Eurasian Watermilfoil reached Cheaspeake Bay and the
Potomac River region, ideal conditions were present -- calcareous
waters (Shenandoah River drains off the limestone areas of north-
central Virginia into the Potomac, Hagerstown Limestone Valley
and along the Potomac and areas of Loudoun County, Virginia; large
areas of Pennsylvannia and New York drain limestone waters into
Susquehanna River, and then to Susquehanna Flats at the head of
Chesapeake Bay; Gunpowder River drain the rich limestone areas of
Cockeysville and Texas, Baltimore County, Maryland; Tennessee
Valley drains large calcareous areas of Virginia, Kentucky and
Tennessee) and calcareous beds (oyster beds along the Potomac
River and various areas along Chesapeake Bay), deep organic ooze
over the oyster beds, favorable pH ranges, warmer water tempera-
tures of the lower Potomac River drainage and Chesapeake Bay,
and the adaptability to the salinity of these regions. This combination of conditions led to the explosion in growth of M. spicatum in these regions in the 1950's and early 1960's.

    Most of these conditions had been present to some degree in
the Chesapeake Bay and lower Potomac River for many years, and M.
spicatum in one or anothers of its forms had been known and collected there since 1881.

    It is interesting to note here that rivers draining from
non-calcareous areas into Chesapeake Bay (as Patapsco, Patuxent,
Elk, Bohemia, Sassafras, Choptank and Wicomico (Wicomico Co.)
Rivers) do not have M. spicatum, even though each is tidal for a
for a portion of its distance. The acidity of these rivers, all
draining over acid rock or sand, probably prevents survival of
segments which may have been brought by tidal action.

    After 1960, M. spicatum spread down Chesapeake Bay into
Virginia as far as Princess Anne County (Virginia Beach) and
North Carolina (Currituck Sound), being found in areas with both
salinity and calcareous conditions similar to those further north.

    In a survey of Susquehanna Flats in 1957 by Robert E. Ste-
wart and Paul F. Springer, no plants of M. spicatum were reported;
in 1958, only one; but in 1959, it was found at 47 percent of
the hundred stations examined. Dense stands were seen along the
west side of Chesapeake Bay, some a mile long and one-eighth mile
wide.

    Preceding the growth explosion of M. spicatum in Chesapeake
Bay and the Potomac River, there had been a prolonged period of
natural ecological events -- hurricanes and tropical storms --
pushing more saline water up both more frequently and to greater
distances than normal tides would do, and then bringing downstream higher concentrations of calcium ions as the result of the
floods. The hurricanes, by raising the pH and the salinity at

the same time, especially in the upper reaches of the tidal portions of Chesapeake Bay (as Susquehanna Flats, mouth of Gunpowder
River and most of the estuaries of tidal rivers in the Upper
Chesapeake Bay) and the Potomac River up to Washington, D.C.,
and by pushing warmer waters into these areas (since most of the
hurricanes occurred from Juen to October), provided ideal conditions for M. spicatum to grow vigorously, and it did from about
1957. Additional incentives for milfoil to spread were the whip-
ping action of the hurricanes, the higher tides than normal, thus
keeping maximum growth of milfoil at maximum ecological conditions,
and man's intervention to control or irradicate it with mowing and
cutting devices.

  Recently, the National Hurricane Center, Coral Gables,
Florida, provided the author with a list of the hurricanes from
1945-1967 which directly affected the Chesapeake Bay-Potomac River
areas. They are listed below. Also many tropical storms, not listed as hurricanes, occurred during this time. For example, the
Guiness Book of World Records cites as a world record 1.23 inches
of rainfall in 1 minute at Unionville, Maryland at 3:23 P.M. on
July 4, 1956, the most intense recorded in modern times. This area
drains into the Potomac River.

Year	Date	Year	Date
1945	Sept. 18-19	1954	Oct. 15
1949	Aug. 28-29	1955	Aug. 12-13
1952	Aug. 31-Sept.1	1955	Aug. 17-19
1953	Aug. 13-14	1960	Sept. 12
1954	Aug. 30-31	1967	Sept. 16

  Hurricane Hazel (Oct. 1954) , tropical storms Connie and
Diane (Aug. 1955) and several flood-producing storms in 1956
probably set up ideal conditions on Susquehanna Flats snd along
the Potomac River for M. spicatum to thrive and spread rapidly
throughout the area from 1957 to 1959.

  Previous to this period of hurricanes and tropical storms
there had been several such flood-producing conditions in the late
1920's, up to the last big flood in 1936 (Hill, 1977). In the
1970's the most devastating hurricane and tropical storm in this
area were Hurricane Agnes (June 21, 1972) and Tropical Storm
Eloise (1975). Agnes wound up in Chesapeake Bay, and the declin-
ing salinity played havoc with the commercial seafood crops. Some
say thay the bay has yet to return to pre-Agnes days. (Hill,1977).

  Springer & Stewart (1960) made a study of the relation of
precipitation and chlorine content of surface water at Conowingo
(above tidal effects) and Turkey Point (near Susquehanna Flats),
and found that where the chlorine content was lower than 0.04-
0.07 percent of sea salinity, growth of milfoil was more abundant,
as at Carpenter Point, Perry Point and Fishing Battery.

Beaven (1960) indicated that at times temporary relief from the plant could be provided by cutting. Various modifications of power-driven sicklebars, rotary cutters, chain saws and sharp V-shaped drags were devised and used by local interests. However, since the sut-off portions continued to grow and take root, much like the normal after-winter fragmentation of plants, control by cutting tended to spread the plant to new areas.

Milfoil can be spread by entanglement on boat proleppors, anchors, nets and other gear moved from one body of water to another.

Following physical means to erradicate milfoil, chemical techniques were considered. In 1960 and 1961, Springer, Beaven and Stotts reported that formulations of three esters of 2,4-D, applied in dosages of 20 lbs. a.e. per acre, gave almost complete control of milfoil, especially at temperatures of 18°C during the period from mid-May through the first week of June, or until flowering was initiated.

Decline in growth of milfoil in the Upper Chesapeake Bay after 1962 was attributed to various pathogenic agents by Bayley et al. (1968 and 1970) and Bean et al. (1973.

Since Agnes and Eloise, growth has been becoming gradually increasing. This spring in May and June, the author has been able to collect Eurasian Watermilfoil in all the old familiar places where it had been collected 15-20 years ago, and in some places it is rather frequent.

At present it seems that M. spicatum is at a static or low rate of growth, only awaiting the next cycle of ideal ecological conditions to start another explosive growth cycle. The plants are there, the conditions are not quite optimal.

Illustrations and maps from Selected Weeds of the United States, U.S.D.A., Agric. Handb. No. 366. 1970.

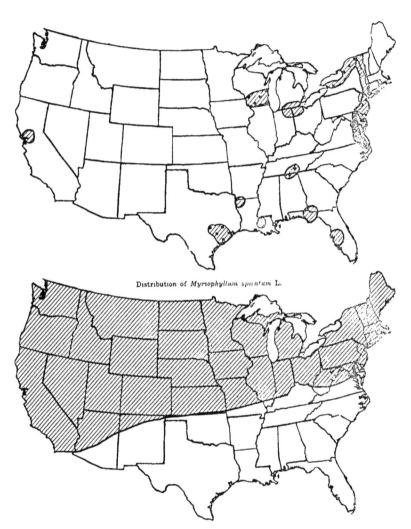

Distribution of *Myriophyllum spicatum* L.

Distribution of *Myriophyllum exalbescens* Fern.

Annotated Herbarium Specimens

(1954-1977)

MARYLAND

Hirzel (1962) stated 'more than 1oo,ooo acres of water in Maryland have been affected by milfoil. The plant has infested the Susquehanna".

Anne Arundel Co. -- Holly Beach at Rt. 50, Chesapeake Bay Bridge. June 15, 1970. Reed 88381 and 88383; shores of Little Magothy River at Cape St. Clair. June 15, 1970. Reed 88402. Covell (1961) stated "it has been found in the Magothy River, and in the Chesapeake Bay down to the Bay Bridge area".

Baltimore Co. -- Dundee Creek in very slightly brackish water. Aug. 10, 1960. F,M,Uhler et al. (US, F & WL); just E of Battery Point on west side of Gunpowder River. Aug. 25, 1960. N, Hotchkiss & G.H.Townsend 7670 (F. & WL); Galloway Cove, in 7 ft. of water. Aug. 15, 1960. G.H.Townsend. (F. & WL); Bird River near White Marsh. June 26, 1961. Reed 51126; Bird River Beach along Bird River. July 22, 1961. Reed 51410; along shores of Brown's Cove, branch of Middle River near Rocky Point and Cedar Point. Mar. 10, 1962. Reed 54154; along shore, Back River at Rocky Point. Mar. 10, 1962. Reed 54153; in shallow water at Rocky Point Cove, Back River. Mar. 10, 1962. Reed 54152; along Dundee Creek near Chase. Apr. 19, 1962. Reed 55784; Rocky Point on shore of Back River. July 8, 1963. Reed 64235; shores of Middle River at end of Kingston Road. July 8, 1963. Reed 64237; along Middle River at Wilson Point. July 8, 1963. Reed 64212 and 64213; along shore of Chesapeake Bay at Brown's Cove, near Breezy Point Park. Nov. 1, 1968. Reed; Dundee Creek near Chase. Oct. 24, 1970. Reed 87242; Dundee Creek, E of Chase, frequent. May 2, 1977. Reed 100560. Covell (1961) reported "it has flourished in Dundee and Saltpeter Creeks and is rapidly invading Middle River and other upper bay tributaries".

Calvert Co. -- Shore of Chesapeake Bay at Scientists Cliffs. Mar. 4, 1967. Reed 74496.

Cecil Co. -- Shores of Chesapeake Bay on Elk Neck at Hance Point. Aug. 23, 1962. Reed 58255; along beach at Charlestown. Nov. 23, 1962. Reed 59607; along shore of Elk River, Lewis Shore. May 25, 1963. Reed 62098; along Bohemia River at Rt. 213. May 25, 1963. Reed 62088; mouth of Northeast River at Charlestown, Susquehanna Flats. July 8, 1963. Reed 64335 and 64337; along Elk River off Elk Forest Road. May 25, 1963. Reed 62089; along Susquehanna River at Aiken, N of Perryville. July 8, 1963. Reed

64315; marsh along Bohemia River at Rt. 213 near Hanks Point.
Aug. 28, 1968. Reed 95406; shores, Susquehanna Flats, Charles-
town. June 15, 1968. Reed 77053; shallow water along shore,
White Crystal, mouth of Elk River. Aug. 28, 1968. Reed 79603;
White Crystal Beach, 6 mi W of Cecilton. June 18, 1977. Reed
100547; shore of Chesapeake Bay, Susquehanna Flats, Charlestown.
June 18, 1977. Reed 100549; frequent along shore of Susquehanna
River, head of Chesapeake Bay, Perryville. June 18, 1977. Reed
100550.

Charles Co. -- Covell (1961) reported " it forms dense beds in
Nanjemoy, Picowaxen and St. Patrick's Creeks, and in Port Tobacco and Wicomico Rivers, and in Neale and St. Catherine Sounds".
In Potomac River along shore, Chapel Point. Mar. 29, 1952. Reed
28087; West Hatton Point, Wicomico River. Oct. 22, 1957. R.E.
Stewart. (F. & WL); in Mattawoman Creek near Mason Springs Swamp
at Rts. 224-225. Oct. 27, 1963. Reed 64915; same area. May 18,
1969. F.M.Uhler. (Reed); exposed at low tide, Port Tobacco River,
cove at Brentland Wharf. May 30, 1969. F.M.Uhler (Reed).

Harford Co. -- Abington Beach on Bush River. July 8, 1963. Reed
64314; Otter Point on Bush River. July 8, 1963, Reed 64304;
shore of Chesapeake Bay, S of Harve de Grace. Dec. 11, 1965.
Reed 72271; Otter Point on Bush River. May 2, 1977. Reed 100561;
Spesutia Creek, attached to bottom, at Rt. US 40. May 14, 1977.
Reed 100532; shore of Chesapeake Bay, Harve de Grave. June 18,
1977. Reed 100550. Springer & Stewart (1959-1961) made an extensive study of Susquehanna Flats off Harve de Grace to Spe-
sutie Island.

Kent Co. -- Turner Creek off Sassafras River near Kennedyville.
Aug. 2, 1965. J.Stennis & Ted Stiles (Reed); frequent along
shore of Chesapeake Bay, mainly rooted fragments, Betterton.
June 18, 1977. Reed 100546.

Montgomery Co. -- Abundant in C. & O. Canal, 1 mi W of Cabin John.
Nov. 2, 1957. N. Hotchkiss 7594. (F. & WL.).

Queen. Annes Co. -- Flats in Northern Maryland, Maryland's
Eastern Shore as far south as Kent Island near the Chesapeake
Bay Bridge, and much of the western shore of Maryland from the
Upper Bay to the Potomac River.

St. Marys Co. -- Piney Point. 1959. Reported by Springer &
Stewart (1960).

VIRGINIA

Dexter Haven (1962) found M. spicatum in nearly every tri-
butary of the Virginia shore of the Potomac River, being most abundant in Hack, Lower Machodoc, Nomini, Popes, Mattox and Rossier's
Creeks.

Westmoreland Co. -- In 5-6 ft. of water, Cabin Point Cove, Lower
    Machodoc Creek. May 23, 1960. John Steenis, J. Gallagher & Jerry
    Townsend. (US); same locality. June 5, 1959. Dexter Haven;
    shore of Potomac River at Colonial Beach. Jan. 27, 1963. Reed
    60176; same locality. April 18, 1970. Reed 85653;

Princess Anne Co. -- (Virginia Beach). In marsh and ditches.
    Middle Creek Road, N of Back Bay. June 28, 1970. Reed 94127,
    94142, 94134; sandy waste, Back Bay Wildlife Refuge. June 30,
    1970. Reed 94146.

NORTH CAROLINA

Currituck Co. -- Edge of water, sound, Rt. 34 at Currituck. Aug.
    7, 1968. Reed 77484; Currituck Sound. Sept. 21, 1974. James
    A. Duke 17310 (Reed). John Steenis (1962) reported it in
    Pea Island Refuge on the outer banks.

Annotated Specimens of M. spicatum
Other localities of interest

California: Mountain Lake, San Francisco Co. June 1891. Michener
    & Bioletti. (Cath. U., Langlois).

Vermont: Franklin Co. Lake Champlain at St. Albans Bay, common
    here. July 9, 1965. Wm. D. Countryman1281. (F. & WL.).

Massachusetts: Fresh Pond, Cambridge. Aug. 11, 1879. Ex Herb.
    Thomas Morong. (Cath. U., Langlois).

Minnesota: Lake Collegeville. June 22, 1909. Jas. Hansen 281.
    (Cath. U., Langlois).

New Jersey: Hunterdon-Somerset Co., in flowing water a long Three
    Bridges Creek near Woodfern. July 12, 1962. Reed 57716 and 57723.

Pennsylvania: Common in shallow water along south shore of Lehigh
    River, opposite the foot of Jeter Island, Allentown. June 27,
    1963. N. Hotchkiss 7857. (US, F. & WL.).

New York: Paddy's Lake near Oswego. Sept. 1882. J.Herman Wibbe.
(Cath. U., Langlois).
  Monroe Co.: Riley's Pond, Cobb Hill Park, Rochester. July 11,
     1960. Ronald A. Ulrich. (F. & WL.).
  Seneca Co.: Seneca River at Waterloo. July 25, 1960. Ronald
     A. Ulrich. (F. & WL); north end of Seneca Lake, Rt. 5.
     Aug. 18, 1962. Reed 58277.
  Jefferson Co.: Westcott Beach, Lake Ontario, Thousand Island
     State Park. July 17, 1963. Reed 65823.
  Cayuga Co.: Along bay of Lake Ontario, Fairhaven. July 16,
     1963. Reed 65809.
  Schuyler Co.: Along canal between Weneta and Lamoka Lakes,
     near Weston. Sept. 7, 1963. Reed 64019; Waneta Lake, S of
     Wayne. Sept. 7, 1963. Reed 64002 and 64006.
  Wayne Co.: Occasional small plants, flowering on muck among
     scattered other plants, Sawmill Cove, SW side of Sodus Bay,
     Lake Ontario. Sept. 12, 1960. N. Hotchkiss 7684. (F. & WL.);
     abundant on east side of Sidus Bay, just N of US Rt. 104
     bridge, Lake Ontario. Sept. 12, 1960. N. Hotchkiss 7675
     (F. & WL.); in Lake Bluff on Lake Ontario. July 16, 1963.
     Reed 65801 and 65803; swamp, Bay Bridge, Rt. 104, W of
     Alton. July 16, 1963. Reed 65819.

Ohio: Trumbull Co., in pond along N side of Ohio Turnpike, 1.5
     mile SW of Lordstown. Sept. 4, 1962. P.F.Springer. (US, F. &
     WL.); Ottawa-Sandusky Co. line, abundant in gray silt in
     dike burrow pit on N side of Muddy Creek, SW of Port Clinton.
     Aug. 4, 1965. F.M.Uhler. (F. & WL.).

Tennessee: Rhea Co., Watts Bar Reservoir, Spring City. Nov. 1960.
     J.L.Frizzell. (US, F. & WL.).

Louisiana: Pointe Coupee Parish, False River near New Roads.
     July 1966. John W. Thieret. (F. & WL.).

Texas: Hays Co., occasional with dense other underwater vegetation
     in Aquarena Pond, San Marcos Spring, San Marcos. June 6, 1966.
     F.M.Uhler & N. Hotchkiss 8194. (F. & WL.); Burnet Co., common
     in hard-bottomed cove on E side of Inks Lake, Inks Lake State
     Park. June 8, 1966. Ulher & Hotchkiss 8204.(F. & WL.); Colorado-
     Fayette Co. line, filling a half-acre farm pond along Rt. 71,
     12 mi. NW of Columbus. June 10, 1966. Ulher & Hotchkiss 8208.
     (F. & WL.).

Cited Literature

Barnes, Irston R. Pest Threatening Waterfowl. Washington Post.
Jan. 17, 1960.

Beaven, G. Francis Water Milfoil invasion of Tidewater Areas.
Maryland Dept. Res. & Educ., Chesapeake Biol. Lab., Solomons,
Reference No. 60-28: 1-4. July 20, 1960.

Beaven, G.F. Summary of Myriophyllum test-plot observations in
Lower Machodoc Creek, Virginia. Maryland Dept. Res. & Educ.,
Chesapeake Biol. Lab., Solomons, Reference No. 60-34: 1-4.
(Mimeo). 1960.

Beaven, G.F. Water Milfoil studies in the Chesapeake Area. Maryland Dept. Res. & Educ., Chesapeake Biol. Lab., Solomons, Reference No. 60-52: 1-5. Oct. 10, 1960.

Beaven, G.F. (Summerizer). Summary of the 1962 Interagency Research
Meeting on Eurasian Watermilfoil. Nat. Res. Inst., Univ. Md.,
Chesapeake Biol. Lab., Solomons, Reference No. 62-15: 1-7. Feb.
23, 1962.

Beaven, G.F., C.K. Rawls and G.E. Beckett Field Observations upon
estuarine animals exposed to 2, 4-D. Contribution No. 192,
Nat. Res. Inst., Univ. Md., Chesapeake Biol. Lab., Solomons.
Proc. Northeastern Weed Control Conf., 16: 449-458. Jan. 3-5,
1962.

Bertholdt, W. Your aquarium needs Myriophyllum -- plant of delicate beauty. Aquarium Journ. 29(3): 106-107. 1958.

Chapman, A.W. Flora of the Southern United States, Ed. 2. Ivison,
Blakeman & Co., New York. 698. pp. 1889.

Covell, Charles Scientists Gunning for it: Renegade Plant Killing
Food of Waterfowl. The Evening Star (Washington). Nov. 21, 1961.

Fernald, M.L. Two new Myriophyllum and a species new to the United
States. Rhodora, 21: 120-124. 1919.

Fernald, M.L. Gray's Manual of Botany, Eighth Ed., 1652 pp. American Book Co., New York. 1950.

Gleason, H.A. The New Britton & Brown Illustrated Flora of the
Northeastern United States and adjacent Canada. New York Bot.
Gard. 2: 1-655. 1952.

Gray, Asa Manual of the Botany of the Northern United States
(from New England to Wisconsin and south to Ohio and Pennsylvania inclusive). Dedicated to John Torrey. 1848.

Gray, Asa Manual of the Botany of the Northern United States, including the District east of the Mississippi and north of
North Carolina and Tennessee. Fifth Ed. 1867; Eighth issue,
1880.

Gray, Asa Gray's Lessons in Botany and Manual of Botany. Rev.
Ed. 1887.

Grout, A.J. Notes on Myriophyllum. Bull. Torr Bot. Club, 23:
11. 1896.

Haven, Dexter Eurasian Water Milfoil in the Chesapeake Bay and
the Potomac River. Virginia Inst. Marine Science, Contribution
No. 108: 1-5. 1962. ("-- species first seen in creeks of the
upper Potomac in 1933 by Francis M. Uhler of the U.S. Fish and
Wildlife Service. (Unpubl. data)".

Hegi, G. Illustrierte Flora von Mittel-Europa. 5(2): 679-1562.
J.F.Lehmanns Verlag, Munchen.1926.

Hill, Michael It Wasnt Just Another Hurricane : Area Still Feel-
ing Effects of Agnes. The Evening Sun, Metro, p. Cl. June
20, 1977.

Hirzel, Donald A Growing Problem in Tidewater Maryland. Scientists Baffled by Sudden Spread of Exotic Water Plant. The Evening Sun (Baltimore), p. Bl. Mar. 2, 1962.

Hirzel, Donald Aquatic Enigma: Malignant Plant Choking Future of
Chesapeake Bay Area. The Sunday Star (Washington, D.C.).
Oct. 7, 1962.

Hitchcock, A.S. and Paul S. Standley Flora of the District of
Columbia and Vicinity. Contrib. U.S. Nat. Herb. 21: 214. 1919.

Horrocks, A.W. and R.F.Smith Control of Eurasian Watermilfoil,
Myriophyllum spicatum, in Lake Hopatcong, New Jersey. Proc.
Northeastern Weed Control Conf., 15: 549-557. 1961.

Hotchkiss, Neil Common Marsh, Underwater and Floating-leaved
Plants of the United States and Canada. 124 pp., over 750
illus. Dover Press. 1972.

Hulten, E. Flora of Alaska and Yukon, VII. Lund, Hakan Ohlasons
Boktryckeri. 1947.

Jepson, W.L. Manual of the Flowering Plants of California.
Assoc. Students Store, Univ. Calif., Berkeley. 1238 pp. 1925.

Kearney, T.H. and R.H. Peebles Flowering Plants and Ferns of
Arizona. U.S. Govt. Printing Office, 1069 pp. Wash., D.C.
1942.

Knupp, N.D. The Flowers of Myriophyllum spicatum L. Proc. Iowa
Acad. Sci. 18: 61-73. 1911.

Lange, J.M.C. Conspectus Florae Groenlandicae. Meddelelser om
Gronland udgivne af Kommissionen for ledelsen af de geologiske
og geografiske undersogelser i Gronland, 3: 237. 1887.

Linnaeus, C. Species Plantarum, 2d. 1. 2: 992. 1753.

Löve, Askell Some notes on Myriophyllum spicatum L. Rhodora, 63: 139-145. 1961.

Muencher, W.C. Aquatic Plants of the United States. Comstock Publ. Co., Ithaca, N.Y. 374 pp. 1944.

Patten Jr., Bernard C. Myriophyllum spicatum L. in Lake Musconetcong, New Jersey: its ecology and biology with a view toward control. M.S.Thesis, Rutgers Univ. 1954.

Patten Jr., Bernard C. The status of some American species of Myriophyllum as revealed by the discovery of intergrade material between M. exalbescens Fern. and M. spicatum L. in New Jersey. Rhodora, 56: 213-225. 1954.

Patten Jr., Bernard C. Germination of the seed of Myriophyllum spicatum L. Bull. Torr. Bot. Club,82(1): 50-56. Jan. 1955.

Patten Jr., Bernard C. Notes on the Biology of Myriophyllum spicatum L. in a New Jersey Lake. Bull. Torr. Bot. Club, 83(1): 5-18. Jan. 1956.

Peck, M.E. A Manual of the Higher Plants of Oregon. Binfords & Mort, Portland. 866 pp. 1941.

Perrot, E. Sur les organes appendiculaires des feuilles de certains Myriophyllum. Journ. de Bot. 14: 198-202. 1900.

Raciborski, M. Ueber die Inhaltskorper der Myriophyllum trichome. Ber. d. Deutsch. Bot. Gesellsch. 11: 348-351. 1893.

Reed, Clyde F. Selected Weeds of the United States. Agric. Handb. No. 366. USDA. 1970.

Ricker, P.L. A List of the Vascular Plants of the District of Columbia and Vicinity, p. 84. 1906.

Robinson, B.L. and M.L. Fernald Gray's New Manual of Botany, illustrated. 7th Edition. 1908.

Small, J.K. Manual of the Southeastern Flora (of United States). 1554 pp. 1933.

Smith, Gordon E. Eurasian Watermilfoil (Myriophyllum spicatum) in the Tennessee Valley. Meeting, Southern Weed Conference, Chattanooga, Tenn. 10 pp. (Mimeo). Jan. 17-19, 1962.

Springer, P.F. (Summarizer) Summary of Interagency Meeting on Eurasian Watermilfoil. Meeting, Patuxent Wildlife Research Center, 10 pp. Nov. 18, 1959.

Springer, P.F., G.F. Beaven and V.D.Stotts Eurasian Watermilfoil -- A rapidly spreading pest plant in Eastern waters. Presented at Northeast Wildlife Conference, Halifax, Nova Scotia. 6 pp. (Mimeo). June 11-14, 1961.

Springer, P.F. and R.E.Stewart Condition of Waterfowl feeding grounds on the Susquehanna Flats during the fall of 1959 with notes on the invasion of a serious pest plant. Administrative Rept., U.S. Dept. Interior, Bur. Sport Fisteries & Wildlife, Patuxent Res. Refuge, Laurel. 6 pp., map. (1960).

State Water Control Board, Richmond, Va. (R.V.D.) Purposeful Introduction of Chemicals into Virginia's Waters for the Control of Aquatic Vegetation or Animals. Jan. 29, 1964. Presented at Ann. Eurasian Water Milfoil Workshop, Annapolis, Md. Feb. 20, 1964.

Steenis, John H. and Gertrude M. King (Summarizers). Report of Interagency Workshop Meeting on Eurasian Watermilfoil. Annapolis, Md. 21 pp. Feb. 20, 1964.

Steenis, John H. and Vernon D. Stotts Progress Report on Control of Eurasian Watermilfoil in Chesapeake Bay. Proc. 15th Annual Meeting, Northeastern Weed Control Conf., 15: 566-570. Jan. 4-5, 1961.

Steenis, J.H., V.D. Stotts and Charles R. Gillette Observations on Distribution and Control of Eurasian Watermilfoil in Chesapeake Bay, 1961. Proc. Northeastern Weed Control Conf. (New York City), 16: 442-448. Jan. 3-5, 1962.

Stotts, V.D. (Summarizer). Summary of the Interagency Research Meeting on the Biology and Control of Eurasian Watermilfoil. Maryland Game and Inland Fish Commission, 7 pp. (Mimeo). 1961.

Tatnall, Eduard Catalogue of the Phanerogamous and Filicoid Plants of New Castle Co., Delaware. p. 98. 1860.

Tennessee Valley Authority Observations on mosquito production and control in Watermilfoil (Myriophyllum spicatum) Watts Bar Reservoir, 1961. Unpublished Rept., Vector Control Branch, 1961.

Tidestrom, Ivar Flora of Utah and Nevada. Contrib. U.S. Nat. Herb. 25: 1-665. 1925.

Tidestrom, Ivar and T. Kittell A Flora of Arizona and New Mexico. Cath. Univ. of Amer. Press, Wash., D.C. 896 pp. 1941.

Waesche, James F. and Richard Stacks (Photos). Worst Pest in Maryland Waters: Milfoil. Sunday Sun Magazine, pp. 10-12, 9 illus. July 19, 1964.

Ward, Lester F. Guide to the Flora of Washington and Vicinity. Bull. U.S. Nat. Mus. No. 26: 24, 80, 160. 1881.

Wiegand, K.M. and A.J.Eames The Flora of the Cayuga Lake Basin, New York. Cornell Univ. Agric. Expt. Sta., Mem. 92: 1-491.1925.

Harold N. Moldenke

ALOYSIA SCORODONIOIDES var. HYPOLEUCA (Briq.) Moldenke, comb. nov.
    Lippia scorodonioides var. hypoleuca Briq., Bull. Herb.
Boiss. 4: 338. 1896.

ALOYSIA SCORODONIOIDES var. LOPEZ-PALACII Moldenke, var. nov.
    Haec varietas a forma typica speciei laminis foliorum subtus
dense puberulentibus vel brevissime pubescentibus et configura
anguste ellipticis acutis parvis recedit.
    This variety differs from the typical form of the species
in leaves being very much smaller, the blades narrowly elliptic,
acute at both ends, only 2—2.5 cm. long and less than 1 cm.
wide, merely densely puberulent or only very shortly pubescent
beneath.
    The type of the variety was collected by Santiago López-
Palacios (no. 4249) -- in whose honor it is named — in cultivation in Quito, Pichincha, Ecuador, at 2800 meters altitude, on
February 4, 1977, and is deposited in my personal herbarium.
The collector describes the plant as a small tree, 3—5 m. tall,
cultivated from Ibarra to Tungurahma, with fragrant white flowers.
He notes that Acosta Solis (Recur. 2 (1): 76) refers to this
plant as Lippia ligustrina, with the vernacular name "verbena
persa", while Dr. Francisco Latorre has collected it as Aloysia
virgata. It bears no great resemblance to either of these taxa,
but its leaves, except in their pubescence, are almost identical
to those of var. parvifolia Moldenke.

ALOYSIA SCORODONIOIDES var. PARVIFOLIA Moldenke, var. nov.
    Haec varietas a forma typica speciei foliis multoties parvi-
oribus, laminis adultis anguste ellipticis plerumque 1—2.5 cm.
longis 5—14 mm. latis acutis subtus densissime tomentellis
recedit.
    This variety differs from the typical form of the species
chiefly in its very much smaller leaves, the blades of which
when adult are only 1—2.5 cm. long and 5—14 mm. wide, acute at
both ends. They differ from those of the previous variety in
being densely pubescent or tomentellous on the lower surface.
    The type of the variety was collected by Henry Hurd Rusby (no.
920) near La Paz, La Paz, Bolivia, at 10,000 feet altitude, in
October of 1885, and is deposited in the Princeton University
herbarium now on deposit at the New York Botanical Garden.

GLOSSOCARYA SCANDENS var. PUBESCENS Moldenke, var. nov.
    Haec varietas a forma typica speciei laminis foliorum subtus
dense pubescentibus differt.

This variety differs from the typical form of the species in having the lower surface of the leaf-blades densely pubescent.

The type of the variety was collected by A. H. M. Jayasuriya (no. 2038) in the jungle beside a rock outcrop south of Komari bridge, north of Pottuvil, Amaparai District, Sri Lanka, at low altitude, on May 4, 1975, and is deposited in the Britton Herbarium at the New York Botanical Garden. The collector describes the plant as a very scandent shrub, the branches reaching to 6 meters in length, and the corollas pure white.

PREMNA OBTUSIFOLIA f. SERRATIFOLIA (L.) Moldenke, stat. nov.
  Premna serratifolia L., Mant. 253. 1771.

STACHYTARPHETA ANGUSTIFOLIA var. BRITTONIAE Moldenke, var. nov.
  Haec varietas a forma typica speciei foliis anguste linearibus valde differt.

This distinct variety differs from the typical form of the species in its leaves being uniformly narrow-linear.

The type of the variety was collected by Nathaniel Lord Britton, Elizabeth Gertrude Britton, and Percy Wilson (no. 15709) in pinelands on the Isle of Pines, Cuba, between March 19 and 21, 1916, and is deposited in the Britton Herbarium at the New York Botanical Garden. It is dedicated to the two Brittons who, separately and together, contributed so very much to the advancement of systematic botany in the New World.

ADDITIONAL NOTES ON THE GENUS AVICENNIA. XI

Harold N. Moldenke

AVICENNIA L.
Additional bibliography: Engl., Syllab. Pflanzenfam., ed. 2,
189 (1898), ed. 3, 188 (1903), ed. 5, 193 (1907), and ed. 6, 198.
1909; Gilg in Engl., Syllab. Pflanzenfam., ed. 7, 314 & 315.
1912; R. T. Baker, Journ. Proceed. Roy. Soc. N. S. Wales 49: 257—
281. 1916; Gilg. in Engl., Syllab. Pflanzenfam., ed. 8, 319. 1919;
Fedde & Schust. in Just, Bot. Jahresber. 44: 253. 1922; Gilg in
Engl., Syllab. Pflanzenfam., ed. 9 & 10, 340. 1924; Kräusel in
Just, Bot. Jahresber. 44: 759—760. 1924; Fedde in Just, Bot. Jahresber. 44: 1377. 1927; Diels in Engl., Syllab. Pflanzenfam., ed.
11, 339. 1936; M. F. Baker, Fla. Wild Fls., ed. 2, imp. 1, 190.
1938; Lugo & Snedaker in R. F. Johnston, Ann. Rev. Ecol. Syst. 5:
40, 43, 45, 50, & 54. 1974; Gaussen, Legris, Meher-Homji, Fontale,
Pascal, Chandrahassan, Delacourt, & Troy, Trav. Sect. Scient.
Techn. Inst. Franç. Pond. Hors Ser. 14: 37 & 82. 1975; Von Erffa
& Geister, Mitt. Inst. Colombo-Aleman. Invest. Cient. Punta de
Betin 8: 176--179, pl. 7. 1976; Anon., Biol. Abstr. 63: 6131.
1977; Moldenke, Phytologia 36: 408—412. 1977.

AVICENNIA GERMINANS (L.) L.
Additional bibliography: Lugo & Snedaker in R. F. Johnston,
Ann. Rev. Ecol. Syst. 5: 40 & 45. 1974; Von Erffa & Geister, Mitt.
Inst. Colombo-Aleman. Invest. Cient. Punta de Betin 8: 176—179,
pl. 7. 1976; Anon., Biol. Abstr. 63: 6131. 1977; Moldenke, Phyto-
logia 36: 410—412. 1977.
Additional illustrations: Von Erffa & Geister, Mitt. Inst.
Colombo-Aleman. Invest. Cient. Punta de Betin 8: 178, pl. 7. 1976.
Von Erffa & Geister (1976) refer to remains of this species
having been found in the Holocene formations in Florida and in
Mexico. However, the remains found by them in Colombia are more
likely those of Rhizophora mangle.
The corollas on Dwyer & Pippin 10030 are said to have been
"white", taken from trees only "to 4 ft." tall, the leaves rather
more elongate than usual, on Correll 47404 they were "cream-color",
and on Correll & Correll 42342 "creamy-white".
Lakela and her associates (1976) assert that A. germinans in
Florida inhabits coastal lagoons in the mangrove belt, flowering
all year there. Laurence & Mohammed (1876) refer to the species
as a "prolific nectar yielder.....whose honey shows the two locally uncommon features of clarity and a tendency to crystallize" in
Trinidad & Tobago.
The Pohl & Davidse 10588a, distributed as A. germinans, actually is A. bicolor Standl., H. M. Curran 140 is A. germinans var.
cumanensis (H.B.K.) Moldenke, Budowski 98A-18, Curran 2032 &

2032-19, Fournier 81, Holm-Nielsen, Jeppesen, Løjtnant, & Øllgaard
7261, D. H. Knight 1032, Kennedy 2281, Kuntze s.n. [Trinidad, IV.
74], Romero-Castañeda 7275, and Simpson & Schunke V. 601 are A.
germinans var. guayaquilensis (H.B.K.) Moldenke, Britton & Britton
2595, Britton, Hazen, & Mendelson 541, Hahn 759, R. C. Marshall s.
n. [Herb. Trin. Bot. Gard. 12651], Nadeaud s.n. [Marais salé de
Victhersby, XI.1862], and Othmer s.n. are A. schaueriana Stapf &
Leechman, Killip & Cuatrecasas 38660 is A. tonduzii Moldenke, L.
M. Andrews 911 is Licania michauxii Prance, and Norris & Taranto
13327 is Laguncularia racemosa Gaertn.

Avicennia officinalis φ lanceolata Kuntze and A. officinalis
var. lanceolata Kuntze, previously included by me in the synonymy
of typical A. germinans, definitely belongs to that of var. guayaquilensis instead.

Additional citations: FLORIDA: Hillsborough Co.: Lakela 24616
(Ld, Ld). Monroe Co.: Perdue & Blum 4471a (Ld). Sanibel Island:
Brumbach 7904 (N). TEXAS: Cameron Co.: Webster & Wilbur 3035 (Mi).
Brazos Santiago Island: R. Runyon 2812 (Au—268808). Clark Is-
land: Lundell & Lundell 8760 (Ld). MEXICO: Sinaloa: Hernandez X.
676 (Ld). Veracruz: Ventura A. 5226 (Ld). Yucatán: Lundell &
Lundell 8140 (Ld). BELIZE: Dwyer & Pippin 10030 (Ac). BAHAMA
ISLANDS: Great Exuma: Correll & Correll 42342 (N). Inagua: D. S.
Correll 47404 (N). Long: S. R. Hill 2139 (N). CAYMAN ISLANDS:
Grand Cayman: Sachet 459 (W—2774762); Stoddart 7013 (W—2773944).
HISPANIOLA: Haiti: Ekman 8054 (Ld). PUERTO RICO: Stimson 3036
(Ld). COLOMBIA: Goajira: Romero-Castañeda 4496 (Ac). VENEZUELA:
Zulia: Budowski 25 (Ac); H. M. Curran 254 (N).

AVICENNIA GERMINANS var. CUMANENSIS (H.B.K.) Moldenke
     Additional bibliography: Moldenke, Phytologia 33: 255, 261,
262, 265—267, & 269 (1976) and 34: 256. 1976.

AVICENNIA GERMINANS var. GUAYAQUILENSIS (H.B.K.) Moldenke
     Additional synonymy: Avicennia officinalis φ lanceolata Kunt-
ze, Rev. Gen. Pl. 2: 502. 1891. Avicennia officinalis var.
lanceolata Kuntze ex Moldenke, Prelim. Alph. List Invalid Names 6,
in syn. 1940.
     Additional bibliography: Moldenke, Phytologia 34: 74, 75, 202,
203, 255, 256, & 271 (1976) and 36: 31, 34, 41, & 410—412. 1977.
     Recent collectors describe this plant as sometimes a bush or
shrub, flowering when only 0.5 m. tall, and at other times a low
or even tall tree, 2—25 m. tall, symmetrical in outline, with a
trunk diameter of 30 cm., the bark "nondescript gray-brown" or
gray, the leaves dark-green or very dark-green above, light-green
beneath, coriaceous, shiny above, dull beneath, the flowers
faintly fragrant, irregular, the calyx green, the corolla very
weakly zygomorphic, with 4 lobes, the fruit green or light yellow-

green, and the "seeds germinating in fruit".

They have found it growing in shallow water, at the margins of salt ponds, in saltwater estuaries, close to tideline, in small ponds back of beaches, in the mud at the head of lagoons, "in open areas where sand is only slightly above tideline and where subsand is moist", on mud banks, and in wet alluvial clay immediately above the high-tide zone, from sealevel to 2 m. altitude.

The corollas are said to have been "white" on Asplund 16588, Haught 4855, and Von Hagen 8, "pale-lemon" on Kramer & Hekking 2092, "light-cream, the throat rich yellow-gold" on Simpson & Schunke V. 601, "creamish-white" on Lasseigne 4409, and "yellowish-white" on Holm-Nielsen & al. 7261.

Additional vernacular names reported for this plant are "black mangrove", "jelly", "mangle negro", "negrita", and "parwa". Stewart reports it "common on beaches and around salt lakes" on Charles Island; Haught reports it "very abundant in sandy soil near sea" in Antioquia; Pittier found it "common in mangrove formation" in Miranda. On Indefatigable Island Taylor says "This is the less frequent of the mangroves which are almost all Rhizophora; this forms tall trees, 40--50 feet tall, which show up above the shrub vegetation of the adjacent dry belt -- the only other trees of any size are Opuntia and a few Piscidia and Erythrina." Kramer & Hekking found it "very common in saline silt forming extensive woods" in Surinam.

The Smith & Smith 546 collection, from St. Vincent island, is most interesting because of its geographic location. The collectors assert that the tree was "locally common on the windward side [of the island], rare on leeward side, near the coast in forest or secondgrowth".

Some leaves of Sachet 459, from the Cayman Islands, closely resemble those of typical A. germinans. The Breteler collections are especially interesting -- no. 4673 has typical quayaquilensis leaves; on 5178 they are blunt but narrow; on 5174 they are very long-pointed and very much like those of A. africana B. Beauv; those of 4677 are broad and very much like those of A. elliptica var. martii. Probably these collections were made deliberately because of the obvious differences in foliage characters. They are cited here tentatively. On Asplund 16588, Fournier 81, D. H. Knight 1032, Romero-Castañeda 7275, and some sheets of H. H. Smith 1937 they are much like those of typical A. germinans and probably should be so cited. The Drouet 2442 and Lanjouw & Lindeman 301 collections were previously cited by me as A. elliptica var. martii, but I now feel that they are better regarded as one of the many forms of A. germinans var. guayaquilensis; the same holds true for Ducke 9818.

Additional citations: WINDWARD ISLANDS: St. Vincent: Smith & Smith 546 (B, C). PEARL ISLANDS: San José: H. Kennedy 2281 (N). TRINIDAD AND TOBAGO: Trinidad: Kuntze s.n. [Trinidad, IV.74] (N).

COLOMBIA: Antioquia: Haught 4855 (N). Magdalena: Romero-Castañeda
7275 (Ac). VENEZUELA: Delta Amacuro: Budowski 98A-18 (Gz, N); H.
M. Curran 2032 (N), 2032-19 (Kh, N). State undetermined: Robert-
son & Austin 213 [Alcaballo] (Ld). GUYANA: Irwin 542 (Au—178000).
SURINAM: Kramer & Hekking 2092 (Ld); Lanjouw & Lindeman 301 (N, Ut-
17661b); Lasseigne 4409 (N). ECUADOR: Guayas: Asplund 18188 (Ld, N,
W—2652442); Fagerlind & Wibom 114 (Ld); Holm-Nielsen, Jeppesen,
Løjtnant, & Øllgaard 7261 (N). Manabi: Asplund 16588 (Ld, N, S, W—
2652442). GALAPAGOS ISLANDS: Charles: A. Stewart 3267 (N); Wiggins
& Duncan 517 (Ld). Indefatigable: Chapin 1120 (N); Fournier 81 (Ac);
D. H. Knight 1032 (Ac); T. W. J. Taylor T.T.91 (Gg—461253, N).
James: Wiggins & Porter 287 (Ld). Narborough: F. R. Fosberg 44703
(Ld); Wiggins & Porter 201 (Ld). Santa Cruz: Von Hagen 8 (N, N, N),
809 (N); Wiggins 18310 (Ld). PERU: Tumbes: Simpson & Schunke Vigo
601 (N, W—2799283). BRAZIL: Bahia: Belém & Pinheiro 3039 (Ld).
Ceará: Drouet 2442 (E—1110546, F—857471, F—949342, I, N, N, S,
Sp—37514, W—1594848). Pará: Ducke 9818 [Herb. Mus. Rio Jan. 5407]
(N).

AVICENNIA LANATA Ridl.
   Additional bibliography: Hocking, Excerpt. Bot. A.28: 260. 1976;
Moldenke, Phytologia 34: 75, 179, 201, 202, & 271. 1976.

AVICENNIA MARINA (Forsk.) Vierh.
   Additional synonymy: Avicennia marina (Forssk.) Vierh. ex Bult-
man & Southwell, Biotropica 8 (2): 76, sphalm. 1976.
   Additional bibliography: Buek, Gen. Spec. Syn. Candoll. 3: 46.
1858; Crozet, Voy. Tasmania [transl. Ling Roth] 5: 36. 1891; Fedde
& Schust. in Just, Bot. Jahresber. 54 (2): 746. 1934; Gaussen, Viart, Legris, & Labroue, Trav. Sect. Scient. Techn. Inst. Franç.
Pond. Hors Ser. 5: 25. 1965; Lamberti, Univ. São Paulo Fac. Filos.
Bol. 317 [Bot. 23]: 120, 155, 159, & 160. 1969; Chai, Malays. Forest.
38: 188, 205, & 207. 1975; Bultman & Southwell, Biotropica 8 (2):
76 & 92. 1976; Moldenke, Phytologia 34: 70—72, 75—94, 167—170,
172—174, 176, 177, 179, 180, 185—189, 193—195, 197—203, 261—
263, 265, 266, 268, & 271 (1976) and 36: 39, 410, & 411. 1977.
   Recent collectors in Thailand refer to this plant as a tree, 7 m.
tall, or a small shrub-like tree, 2—5 m. tall, the trunk gray, 30
cm. in diameter at breast height, the branches gray, the leaves
green above, gray- or grayish-green beneath, the pneumatophores to
15 cm. tall, the calyxes green, and the fruit light-green. The
corollas on Maxwell 75-917 are said to have been "orange" in color
when fresh, on Sumithraarachchi DBS.696 "yellow", and on Van der
Kevie 2 "yellow to orange". Maxwell refers to the plant as growing
in the "muddy mangrove zone along rivers" and Van der Kevie reports
it "very common on moderately firm soils of seashore".
   Chai (1975) has studied this mangrove intensively in Malaysia
and describes it as a "Shrublet (2 ft. tall) to medium-sized tree to

60 ft. tall. No buttresses, but slender, soft stilt roots may
develop. Bark reddish-brown, flaking off in irregular, thin pap-
ery flakes revealing green new bark surface. Leaves more or less
elliptic with blunt apices, lower surface glaucous, excreting
salt. Fruit glaucous-green, more or less heart-shaped, slightly
flattened. A pioneer species on new mud with a high proportion of
sand but does not seem to colonise pure mud. At the mouth of the
Bako river, it is slowly been replaced by A. alba. Found also a-
long sandy shores where it is seen to be in poor form and never
gregarious. Absent inland." In his key he distinguishes it from
A. alba and A. officinalis as follows: "Small shrub (3 ft. tall)
to big tree to 60 ft. tall; old bark grey pink or pinkish brown,
coming off in patches of irregular thin flakes revealing green new
bark; leaves elliptic, slightly whitish below; sandy soil". He
calls it "api-api merah".

Collectors in Tanzania describe the species there as a branched
tree, 9 meters tall, growing on the intertidal seashores at sealevel, the leaves about 12 cm. long and 7 cm. wide, the flowers
axillary, the corollas yellow or orange-yellow, the petals 4 (or 5),
and the stamens 4. They found it in flower there in December.

The Budowski 118-18, distributed as A. marina, seems to be A.
schaueriana Stapf & Leechman rather than A. germinans (L.) L. as
previously reported in this series of notes.

Additional citations: TANZANIA: Tanganyika: McCusker 221 (W—
2727153); Mwasumbi LBM.10438 (W—2727154). SRI LANKA: Bernardi
15299 (Mu); Davidse 7561 (Ld); Faden s.n. [Jayasuriya 2413] (Ld);
Sumithraarachchi DBS.696 (N, W—2804782). THAILAND: Larsen & Larsen 33784 (Ld); Maxwell 75-917 (Ac); Van der Kevie 2 (Ac).

AVICENNIA MARINA var. ACUTISSIMA Stapf & Moldenke
Additional bibliography: Moldenke, Phytologia 34: 167—169, 185,
194, 195, 201, & 262. 1976.

AVICENNIA MARINA var. RESINIFERA (Forst. f.) Bakh.
Additional bibliography: A. DC., Prodr. 11: 716. 1847; Buek, Gen.
Spec. Syn. Candoll. 3: 46. 1858; Crozet, Voy. Tasmania [transl. Ling
Roth] 5: 36. 1891; R. T. Baker, Journ. Proceed. Roy. Soc. N. S.
Wales 49: 257—281. 1916; Kräusel in Just, Bot. Jahresber. 44: 759—
760. 1924; Cambie, Journ. Roy. Soc. N. Zeal. 6 (3): 333. 1976; Mol-
denke, Phytologia 34: 169—180, 185, 195, 199—203, & 271 (1976) and
36: 39. 1977.

Dickson encountered this plant in salt marshes. Cambie (1976),
quoting an unpublished thesis by Cheeseman (1964) in the University
of Auckland library, reports the bark of this plant as containing
"taraxerol, taraxerone, tripertene hydrocarbons, taraxanthin, feru-
lic acid, caffeic acid, p-coumaric acid, [and] triacontane".

Additional citations: AUSTRALIA: Queensland: M. S. Clemens
42536a (Mi). NEW ZEALAND: Pollen Island: Dickson s.n. [10.3.71]
(Ac).

AVICENNIA MARINA var. RUMPHIANA (H. Hallier) Bakh.
  Additional bibliography: Moldenke, Phytologia 34: 170, 172,
178--180, 185, 200, 203, 266, & 271. 1976.

AVICENNIA OFFICINALIS L.
  Additional bibliography: Sweet, Hort. Brit., ed. 2, 419. 1830;
Loud., Hort. Brit., ed. 2, 554. 1832; Sweet, Hort. Brit., ed. 3,
554. 1839; Buek, Gen. Spec. Syn. Candoll. 3: 46. 1858; R. T. Baker,
Journ. Proceed. Roy. Soc. N. S. Wales 49: 257—281. 1916; Kräusel
in Just, Bot. Jahresber. 44: 759—760. 1924; Fedde in Just, Bot.
Jahresber. 44: 1377. 1927; Chai, Malays. Forest. 38: 188, 205, &
207. 1975; Moldenke, Phytologia 34: 167—172, 177, 179—203, 262,
267, 269, & 271 (1976) and 36: 410. 1977.
  Don (1830), Sweet (1830), and Loudon (1832) all list what is
probably this species (as "A. tomentosa") as cultivated in British
gardens in their day [in greenhouses?], introduced from India in
1793. They call it the "downy-leaved avicennia".
  Chai (1975) has studied this species intensively in Malaysia
and describe it there as a "Small to medium-sized tree to 55 ft.
tall. No buttresses but slender stilt roots may be present. Bark
surface brownish-grey to chocolate-brown, lenticellate, may be
narrowly cracked. Leaves spatulate or spoon-shaped, lower surfaces
very light brown; salt excretion from upper surface. Fruits more
or less heart-shaped, slightly flattened, beaked, covered in soft,
brown tomentum. Commonly inland but not gregarious, along river
or creek banks on stiff heavy soils, absent or very rare on the
sea face. Associated with low and light crowned species like Nypa
and young Rhizophora and Bruguiera." In his key he distinguishes
it from A. alba and A. marina as follows: "Small tree to 55 ft.
tall; bark grey to chocolate-brown, often lenticellate, leaves
spatulate or oblong obovate, not whitish below; inland or on firm
clay river banks." He calls it the "api-api sudu". Of particular
interest is his observation that salt is excreted from the upper
leaf-surface of this species, but from the lower surface in A.
marina.
  In Sri Lanka collectors describe A. officinalis as a shrub or
small tree, 3--7 meters tall, with scaly bark, the corollas yellow
or orange, and the petals and exserted stamens 4 in number. They
have found it growing at the borders of or in mangrove forests.
  The Fournier 81, distributed as A. officinalis and so filed in
some herbaria, actually is A. germinans var. guayaquilensis (H.B.K.)
Moldenke, while Sumithraarachchi DBS.696 is A. marina (Forsk.)
Vierh.
  Additional citations: SRI LANKA: Jayasuriya 1356 (W—2802165);
Tirvengadum & Waas 465 (W—2768090). MALAYA: Singapore: McCusker
308 (Ld).

AVICENNIA SCHAUERIANA Stapf & Leechman
  Additional synonymy: Avicennia tomentosa Schar ex Wangerin in
Just, Bot. Jahresber. 50 (1): 44. 1929 [not A. tomentosa Blanco,

1845, nor Blume, 1918, nor R. Br., 1851, nor Jack, 1945, nor Jacq.,
1760, nor L., 1821, nor L. & Jacq., 1783, nor Lam., 1918, nor G.
F. W. Mey., 1818, nor Millsp., 1930, nor Nutt., 1947, nor Nutt. &
Br., 1832, nor Roxb., 1835, nor Sieber, 1844, nor Sw., 1864, nor
Vahl, 1921, nor Wall., 1851, nor Weigelt, 1851, nor Willd., 1800].
Avicennia schauerana "Stapf et Leechman ex Moldenke" apud Townsend,
Excerpt. Bot. A.4: 242 & 243. 1962. Avicennia schaueriana "Stapf
& Leechman ex Moldenke" apud Angely, Fl. Anal. Paran., ed. 1, 581.
1965. Avicennia schaueeriana Lamberti, Univ. São Paulo Fac. Fil-
os. Bol. 317 [Bot. 23]: 40, sphalm. 1969. Avicennia tomentosa
sensu Marc. ex Moldenke, Fifth Summ. 1: 394, in syn. 1971. Avicennia schaueriana "Stapf et Leechmann ex Moldenke" apud Angely, Fl.
Anal. & Fitogeogr. Est. S. Paulo, ed. 1, 4: 841. 1971. Avicenia
tomentosa Duarte ex Moldenke, Phytologia 28: 453, in syn. 1974.
Avicennia schaueriana f. glabrescens Moldenke, Phytologia 34: 485,
1976. Avicennia schaeereana Stapf & Lehm. ex Moldenke, Phytologia
36: 41, in syn. 1977.

    Additional & emended bibliography: Schenck, Flora 72 [ser. 2,
44]: [83]—[89]. 1889; H. Hallier, Meded. Rijks Herb. Leid. 37:
87. 1918; Wangerin in Just, Bot. Jahresber. 50 (1): 44. 1929; Dan-
sereau, Biogeogr. 132—134. 1957; Bascope, Bernardi, Jorgensen,
Hueck, & Lamprecht, Inst. Forest. Latinoam. Invest. Capac. Descrip.
Arb. Forest. 5, imp. 1, 13, 16, & 51. 1959; Braga, Pl. Nordest.,
ed. 2, 348. 1960; Reitz, Sellowia 13: 44/46 & 109. 1961; Townsend,
Excerpt. Bot. A.41: 242 & 243. 1962; Melchior in Engl., Syllab.
Pflanzenfam., ed. 12, 2: 438. 1964; Angely, Fl. Anal. Paran., ed.
1, 581. 1965; Hocking, Excerpt. Bot. A.13: 570. 1968; Moldenke,
Phytologia 15: 478. 1968; Moldenke, Résumé Suppl. 16: 15 (1968)
and 17: 8. 1968; Lamberti, Univ. São Paulo Fac. Filos. Bol. 317
[Bot. 23]: 40—46, 49—52, 55, 58—60, 76, 89, 102—104, 118, 121—
124, 126—128, 136, 137, 143, 145, 147—149, 151, 153, 157—160,
167—175, 206, & 207, fig. 10—14, 17, 18, & 78—88. 1969; Bascope, Bernardi, Jorgensen, Hueck, & Lamprecht, Inst. Forest.
Latinoam. Invest. Capac. Descrip. Arb. Forest. 5, imp. 2, 13, 16,
& 51. 1970; Gibson, Fieldiana Bot. 24 (9): 177. 1970; Reitz, Sellowia 22: 17. 1970; Angely, Fl. Anal. Fitogeogr. Est. S. Paulo,
ed. 1, 4: 841 & ii, map 1396. 1971; Moldenke, Fifth Summ. 1: 109,
111, 112, 129, 131, 147, 188, 393, & 394 (1971) and 2: 774 &
839. 1971; Odum, Fundament. Ecol., ed. 3, 347. 1971; V. J. Chapm.,
Trop. Ecol. 11: 5, 7, & 8, fig. 3. 1972; Moldenke, Phytologia 28:
453 (1974), 33: 255, 256, 258, 262, & 266 (1976), 34: 72, 75, 84,
170, 172, 180, 193, 200, 271, & 485 (1976), 35: 13 (1976), and
36: 31, 32, 34, 41, & 410. 1977; Moldenke, Biol. Abstr. 63: 2452.
1977.

    Additional illustrations: Dansereau, Biogeogr. 134. 1957; Lam-
berti, Univ. São Paulo Fac. Filos. Bol. 317 [Bot. 23]: 49—52, 59,
60, & 168—175, fig. 10—14, 17, 18, & 78—88. 1969; Odum, Fundament. Ecol., ed. 3, 347. 1971.

    This typical form of the species differs from the following one
in having its leaf-blades completely glabrous on both surfaces.

Recent collectors describe the plant as a tree, 5—18 meters tall, the fragrant "flowers visited by flies", record the additional vernacular names, "black parwa", "mangle negro", and "siriuva", and have found it growing "in mangrove", on coastal banks, around lagoons, on salt flats, and "na areia", flowering from February to April and from June to October, fruiting in October. Many of the flowering specimens exhibit remarkably elongated floral axes with scattered flowers. Cowan & Forster report it a "locally frequent tree" in Trinidad.

The corollas are said to have been "white" on Araujo 1245, 1316a, & 1320 and on Cowan & Forster 1252, "whitish" or "alvacente" on Hatschbach 1920, 29144, & 38583, and "white with yellow interior" on Prance 21149.

The Angely (1971) work cited in the bibliography above is sometimes cited as "1970", the title-page date, but the work was not actually published until 1971.

Gooding (1965) cites Gooding 553 from Barbados , deposited in the British Museum herbarium. Broadway found the species on Grenada, while both Hahn and Plée found it on Martinique, so it seems apparent that it is not confined to northern and eastern South American coasts. This fact supports the possibility of A. germinans var. guayaquilensis having a similar distribution, since the methods of dispersal for both taxa are obviously identical.

Angely (1971) cites A. schaueriana from Maranhão, Ceará, Paraíba, Rio Grande do Norte, Pernambuco, Goiás, Bahia, Espirito Santo, Rio de Janeiro, São Paulo, Paraná, and Santa Catarina. Braga (1960) also lists it from Ceará.

Gibson (1970) makes the remarkable statement that A. bicolor Standl., of Central America, "closely resembles A. schaueriana Stapf & Leechman of the West Indies and South America, which differs only in having fewer, more congested, less complex inflorescences. If they should prove synonymous, the name A. bicolor Standley takes precedence". Actually, these two taxa a very distinct and almost impossible to confuse.

Schenck (1889) describes the plant now known as A. schaueriana as "A. tomentosa", referring to it only middle and south Brazilian specimens. His figures 1—6 on plate 3 of his work are sometimes cited as illustrative of this taxon, but actually (as he plainly states) they illustrate the unrelated Laguncularia racemosa.

Melchior (1964) erroneously implies that A. tomentosa Jacq. is a synonym of A. schaueriana. However, it seems most probable to me now that, in view of Stapf and Leechman's copious and careful field and herbarium notes (cfr. Phytologia 7: 283—286. 1960), Jacquin's A. tomentosa (1760) may actually be a synonym of what is now known as A. germinans var. cumanensis (H.B.K.) Moldenke, rather than of typical A. germinans (L.) L. as hitherto regarded by me and by many other authors.

It should be noted here that when I first proposed the name, A. schaueriana, based on the work of Stapf and Leechman, I gave as

synonyms "'A. tomentosa Jacq.' as used by Schauer in A. DC., Prodr.
11: 699—700 (1847) and many subsequent authors; A. nitida var.
trinitensis Moldenke". I cited Britton 2595 in part, Broadway
5817 in part, and Trin. Bot. Gard. Herb. 2402, 5221, 5405 in part,
7988, 8695, 9516, 10461, 12651, & 12656, all from Trinidad, but
none cited as the type.

Schauer (1847) describes what he calls A. tomentosa Jacq. as
having "foliis obovate-ellipticis obtusissimis in petiolum atten-
uatis supra demum subnitidis subtus candicantibus aetate interdum
glabrescentibus.....Folia 3 poll. longa, 15—18 lin. lata, facie
demum glabrata neque admodum nitida, subtus indumento demum aboles-
cente induta, siccando decolorescentia neque vero nigrescentia."
He cites from Rio de Janeiro: Gaudichaud s.n., Pohl s.n., Riedel
s.n., from Bahia: Blanchet 1427, Martius 108, Salzmann 430, from
Colombia: Humboldt & Bonpland s.n. [Cumana], from the West Indies:
Jacquin s.n., Perrottet s.n., from Mexico: Humboldt & Bonpland s.
n., and from Ecuador: Gaudichaud s.n., Humboldt & Bonpland s.n.
Obviously, both his description and his cited material include
representatives of several taxa. No type was designated, but it
is to be assumed that Jacquin's collection from the West Indies
is the type of the true A. tomentosa Jacq. and this is usually
regarded as conspecific with typical A. germinans (L.) L.
Schauer's description does not apply to Jacquin's plant but apparently applies to a mixture of the pubescent- and glabrous-
leaved forms of A. schaueriana.

My A. nitida var. trinitensis is validated by a designated
type: R. L. Brooks s.n. [Herb. Trin. Bot. Gard. 12656] from the
Caroni Swamp, Trinidad, collected on May 29, 1932, and deposited
in the Britton Herbarium at the New York Botanical Garden. This
collection can, therefore, also be regarded as the type of typical A. schaueriana.

Avicennia schaueriana f. glabrescens, based on R. C. Marshall
s.n. [Herb. Trin. Bot. Gard. Herb. 12651], was described before
it was realized that the actual type of this species is the
glabrous-leaved form. It is a straight and unequivocal synonym
of typical A. schaueriana. All material previously cited by me
in this series of notes as f. glabrescens should be shifted back
to typical A. schaueriana.

Material of A. schaueriana has been widely misidentified and
distributed in herbaria as A. germinans (L.) L., A. nitida Jacq.,
and A. tomentosa Jacq. Some of the collections cited below (e.g.,
N. L. Britton 2595, Hahn 759, and Othmer s.n. [17/XI/03] were
previously erroneously cited by me as A. germinans. Araujo 1316,
Dusén 202, and Riedel & Luschnath 1007, as originally distributed,
are mixtures of typical A. schaueriana and f. candicans, so it
seems probable that, at least in Rio de Janeiro, the two forms
grow in very close proximity to each other. Hatschbach 14076, dis-

tributed as typical A. schaueriana, seems better placed as f. candicans. The Sellow s.n. [Brasilia] in the Britton Herbarium actually is Vitex schaueriana Moldenke.

Additional & amended citations: TRINIDAD AND TOBAGO: Trinidad: N. L. Britton 2595 (W--119444); Cowan & Forster 1252 (N, W--2287738, Z); R. C. Marshall s.n. [Herb. Trin. Bot. Gard. 12651] (N, R); Othmer 157 (Mu). TRINIDAD OFFSHORE ISLANDS: Patos: Britton, Hazen, & Mendelson 541 (N, W--1047019). VENEZUELA: Delta Amacuro: Budowski 118-18 (Gz, Kh, N). BRAZIL: Guanabara: Duarte 5017 [Herb. Brad. 12117; Herb. Jard. Bot. Rio Jan. 110275] (Mu, N, W--2650229). Pará: Prance 21149 (Ld). Paraná: Hatschbach 1920 (N), 29144 (N, S), 38583 (Mu). Rio de Janeiro: Araujo 1207 [Herb. FEEMA 12144] (Ld), 1245 [Herb. FEEMA 12285] (Ld), 1316a (Ac), 1320 [Herb. FEEMA 12330] (Pf); Nadeaud s.n. [Victhersby, XI.1862] (N, W--2547120); Riedel & Luschnath 1007b (N). Gobernador Island: G. Pabst 7385 [Herb. Brad. 27668] (Mu, N, N). Havarata Island: J. Vidal s.n. [Herb. Rio Jan. 31546] (N).

AVICENNIA SCHAUERIANA f. CANDICANS Moldenke, Phytologia 35: 15. 1976.

Bibliography: Moldenke, Phytologia 35: 15 (1976) and 36: 31, 32, 34, & 410. 1977.

This form differs from the typical form of the species in having the lower surface of its leaf-blades more or less very densely canescent-puberulent or farinaceous in the manner of A. germinans (L.) L. and its varieties.

Collectors describe this plant as a shrub or small tree, 1—17 meters tall, the trunk to 12.5 cm. in diameter at breast height, with pneumatophores. The corollas are said to have been "white" on Araujo 1244 & 1316, Pabst 5421, Pickel 3210, and Reitz 5525 and merely "whitish" on Hatschbach 14096 and Hatschbach & Guimarães 21397. It has been found growing in the mangrove formation in coastal regions, in salt water, in litoral and mangrove swamps, restinga, coastal ridges and muddy beaches, on "ground sometimes covered by high tide", along creeks of partially salt water, and in "banhado salgado", at altitudes of 1—5 m., flowering from February to April and June to December, fruiting in May. Froes reports that it furnishes "good lumber and fuel". Vasconcelos Sobrinho 251 includes a photograph of the collection locality. Araujo reports the tree as frequent and the flowers as fragrant, noting that the "folhas cobertas de um indumento acinzentado na superfície inferior".

Vernacular names reported for this form are "blaka parwa", "ciriba preta", "ciriuba", "fromarina", "mangue", "mangue canoé", "mangue seriva", "mangue siriba", "siriuba", and "siriuva".

Araujo 1316, Dusén 202, Lützelburg 401, and Riedel & Luschnath 1007, as originally distributed, are mixtures of this and the typical form of the species.

Material of A. schaueriana f. candicans has been misidentified
and distributed in many herbaria as A. elliptica Thunb., A. nitida Jacq., typical A. schaueriana Stapf & Leechman, "A. schaueriana
Stapf & Lehm.", and A. tomentosa Jacq.

It would appear that this form is a highly inconstant one. The
underside of the leaves on Lützelburg 16059 and s.n. [Caji, XII.
1910] exhibit very obscure puberulence, a condition also seen on
Martius s.n. [in litt. oceani ad Soteropolisi et Ilheos, Jan.
1819]. On Lützelburg 12521 and s.n. [VIII.1912], Martius s.n. [in-
ter Rhizophoras ad litt. oceani prope Rio], and Othmer s.n. [17.
XI.03] some of the (mostly smaller) leaves are subglabrous, glab-
rate, or practically glabrous.

In a previous publication (1960) I erroneously cited specimens
of Blanchet 1427, Gaudichaud s.n. [Rio Janeiro], and Luschnath s.n.
[Martius 108] as "cotypes" of A. schaueriana merely because they
were cited by Schauer (1847) in his mis-application of Jacquin's
"A. tomentosa". However, as pointed out under A. schaueriana in
the present installment of these notes, A. schaueriana should be
typified by the same collection which typifies A. nitida var. trinitensis and not by any of the many collections cited by Schauer
which actually represent several different taxa.

Citations: TRINIDAD AND TOBAGO: Trinidad: Othmer s.n. [17/XI/
03] (Mu--4054). GUYANA: Leechman s.n. [near Georgetown, 1917] (N,
N). SURINAM: Geijskes s.n. [5-II-1943] (N). BRAZIL: Bahia: Lützelburg 401 in part (Mu, Mu), s.n. [VIII.1912] (Mu, N--photo, Z--
photo); Martius s.n. [in maritimis prov. Bahiensis, 1819] (Mu--
1690), s.n. [in litt. oceani ad Soteropolisi et Ilheos, Jan. 1819]
(Mu--1082). Ceará: Dias da Rocha 108 [Herb. Inst. Bot. S. Paulo
7933] (N, Sp--7933); Drouet 2548 (E--1110545, F--857467, F--
949348, N, S, W--1594887). Guanabara: Moldenke & Moldenke 19606
(F--isotype, Mg--isotype, Mr--isotype, N--type, No--isotype, Ot--
isotype, S--isotype, Sm--isotype, Ss--isotype). Maranhão: Fróes
1812 (B, Bm, Cb, Cb, E--1042134, F--707083, I, K, Mi, N, P, S, Ut,
W--1660151). Paraíba: Lützelburg 12521 (Mu). Paraná: Dusén s.n.
[Herb. Mus. Nac. Rio Jan. 5004] (N); Hatschbach 14096 (Ac, Mu, N,
W--2564872, Ws); Hatschbach & Guimarães 21397 (Mi, N); Stellfeld
799 [Herb. Mus. Paran. 1799] (N). Pernambuco: Vasconcelos Sobrinho 251 [photo 136] (It, N, Ug). Rio de Janeiro: Araujo 1141 [Herb.
Inst. Conserv. Nat. 11687] (Fe), 1244 [Herb. FEEMA 12286] (Ld),
1310 [Herb. FEEMA 12329] (Ac), 1316 (Ld); Dusén 202 in part (W--
1055656); Gaudichaud 464 (B, Br, Cb, Dc, N); Glaziou 1362 (Br, Br,
Br, Br, Cp, Cp, F--667195, It, K, N, P); Herb. Mus. Nac. 6 (C),
s.n. [Sebastianopolis] (C); Herb. Mus. Nac. Rio Jan. 31572 (N),
31775 (N); F. C. Hoehne s.n. [Herb. Inst. Bot. S. Paulo 24908]
(N, Sp--24908); Luschnath s.n. [Herb. Martius 108] (Br, Dc, K, M,
Mu--1081, P, V); Lützelburg 16059 (Mu), s.n. [Caju, XII.1910] (Mu,

N--photo, Z--photo); Martius s.n. [inter Rhizophoras ad litt.
oceani prope Rio] (Mu--1084); Riedel & Luschnath 1007a (N); Sellow
304 (N); Wilkes U. S. Exp. Exped. s.n. (T, W--59278, W--59279).
Santa Catarina: Reitz 5525 (N); Reitz & Klein 683 [Herb. Barb.
Rodr. 6961] (N, S, W--2123175, W--2281838), 1414 (W--2142574). São
Paulo: F. C. Hoehne s.n. [Herb. Inst. Bot. S. Paulo 30854] (It, K,
N, Sp); Lofgren s.n. [Herb. Com. Geogr. & Geol. 3062; Herb. Inst.
Bot. S. Paulo 15596] (N, Sp); Pickel 3210 (It, N, Sf); A. Saint-
Hilaire C2.7665 [1665] (N, P); Usteri s.n. [Herb. Inst. Bot. S.
Paulo 15598] (N, Sp). Florianopolis Island: Rambo 50320 (Lm, N,
S). Gobernador Island: Lanna Sobrinho 1905 [Herb. FEEMA 8469]
(Z); G. Pabst 5421 [Herb. Brad. 21211] (Mu, N). Pinheiros Island:
Azevedo 3 [Herb. Inst. Conserv. Nat. 10329] (Fe). Santa Catarina
Island: Reitz 5088 [Herb. Barb. Rodr. 6363] (N).

AVICENNIA TONDUZII Moldenke
    Additional bibliography: Bascope, Bernardi, Jorgensen, Hueck, &
Lamprecht, Inst. Forest. Latinoam. Invest. Capao. Descrip. Arb.
Forest. 5, imp. 1, 13, 17, & 51. 1959; Hocking, Excerpt. Bot. A.
12: 425. 1967; Moldenke, Biol. Abstr. 49: 4199. 1968; Bascope,
Bernardi, Jorgensen, Hueck, & Lamprecht, Inst. Forest. Latinoam.
Invest. Capao. Descrip. Arb. Forest. 5, imp. 2, 13, 17, & 51.
1970; V. J. Chapm., Trop. Ecol. 11: 5, fig. 3. 1970; Moldenke,
Fifth Summ. 1: 87, 115, & 394 (1971) and 2: 839. 1971; Moldenke in
Woodson, Schery, & al., Ann. Mo. Bot. Gard. 60: 150, 153, & 154.
1973; "H. R.", Biol. Abstr. 57: 1904. 1974; Moldenke, Phytologia
32: 438 (1975), 33: 269 (1976), and 34: 74. 1976; Hocking, Excerpt.
Bot. A.26: 422. 1976.
    The Killip & Cuatrecasas 38660 collection, cited below, was
previously incorrectly cited as A. germinans (L.) L.
    Collectors report that A. tonduzii is "a tree sometimes 20 m.
tall", the corollas white, and have found it growing in swamps,
flowering in March and June. The Kupper 1552, distributed as A.
tonduzii, actually is A. germinans var. guayaquilensis (H.B.K.)
Moldenke, a very closely related taxon. It is probable that some,
if not all, of the very long- and narrow-leaved collections cited
as the latter are, in fact, better placed as A. tonduzii. The two
taxa require much more intensive study, especially in the field.

ADDITIONAL NOTES ON THE GENUS VERBENA. XXVII

Harold N. Moldenke

VERBENA SUPINA L.

Emended synonymy: Verbena supina β hirsuta Ehrenb. ex Sweet, Hort. Brit., ed. 2, 419. 1830.

Additional & emended bibliography: H. Bock [Tragus], Stirp. Max. Germ. 102 & 211. 1552; Dodoens [L'Ecluse], Hist. Pl. 96 & 97. 1557; L., Hort. Cliff., imp. 1, 11. 1737; Strand in L., Amoen. Acad. 69: 449. 1756; R. A. Salisb., Prodr. 71. 1796; Sweet, Hort. Brit., ed. 1, 1: 325 (1826) and ed. 2, 419. 1830; G. Don in Loud., Hort. Brit., ed. 1, 247 (1830) and ed. 2, 247. 1832; Loud., Hort. Brit., ed. 2, 553. 1832; G. Don in Loud., Hort. Brit., ed. 3, 247. 1839; Sweet, Hort. Brit., ed. 3, 553. 1839; Schnitzlein, Iconogr. Fam. Nat. 2: 137 Verbenac. [3] & 137, fig. 4—22. 1856; Buek, Gen. Spec. Syn. Candoll. 3: 495 & 496. 1858; Strobl, Oesterr. Bot. Zeitschr. 33: 406. 1883; Dur. & Barr., Fl. Lib. Prodr. 193. 1910; Rolland, Fl. Populaire 8: 43. 1910; L., Hort. Cliff., imp. 2, 11. 1968; Bolkh., Grif, Matvej., & Zakhar., Chrom. Numb. Flow. Pl., imp. 1, 717. 1969; Scully, Treas. Am. Ind. Herbs 283. 1970; Bolkh., Grif, Matvej., & Zakhar., Chrom. Numb. Flow. Pl., imp. 2, 717. 1974; El—Gazzar, Egypt. Journ. Bot. 17: 75 & 78. 1974; Gilmour, Thom. Johnson 122. 1972; Täckholm, Stud. Fl. Egypt, ed. 2, [453], 454, 817, 830, & 876, pl. 156, fig. A. 1974; Kooiman, Act. Bot. Neerl. 24: 464. 1975; Moldenke, Phytologia 30: 159 & 191 (1975), 34: 268 (1976), and 36: 36, 40, 250, & 277. 1977.

Additional illustrations: Scully, Treas. Am. Ind. Herbs 283. 1970; Täckholm, Stud. Fl. Egypt, ed. 2, pl. 156, fig. A. 1974.

Linnaeus (1737) says of this species "Crescit in agro Salmanticensi". Rolland (1910) records the following additional vernacular names: "chamaedrys" ("par confusion, chez les apothicaires"), "columba supina", "licinia", "verbenaca supinaca", and "verveine basse". Don (1830) calls it the "supine vervain" and says that it was introduced into English gardens from Spain in 1640. Loudon (1832) calls it the "trailing vervain"; the var. hirsuta he calls "hairy vervain" and says that it was introduced from Egypt in 1829.

Täckholm (1974) comments that in Egypt "Sterile specimens" of this plant are "very similar to Ambrosia maritima in general appearance.

Scully (1970) reproduces what is obviously a pre-linnean herbal illustration of V. supina in her discussion of Amerind uses of V. hastata L. and V. stricta Vent., but fails to label it or state that it could not possibly have been involved in any Amerind usage since it does not occur in the New World.

It should also be recorded here that the Verbena supina mas of Dodoens is actually Veronica chamaedrys L. in the Scrophulariaceae.

The figure given by Schnitzlein (1856), previously cited as representing V. supina (as it is labeled), seems, rather, to be V. officinalis. Similarly, Bock (1552), although using the name, Verbena foemina, in his text, illustrates it with a woodcut which very obviously depicts V. officinalis.

The Sibthorpe & Smith (1809) reference is often dated "1806", but actually pages 219—442 of volume 1 were not issued until 1809.

Some specimens of V. supina (e.g., Sieber s.n. at Munich and Herb. Zuccarini s.n. [Hort. bot. Erlangensis circa 1819]) have leaves greatly resembling those of the North American V. bracteata Lag. & Rodr. Rauh 156 & 603 and Fischer s.n. [Argyptus] represent very much stunted plants with very small, deeply disseoted, canescent leaves. It is very possible that this species consists of several more or less distinct forms which may be deserving of nomenclatural recognition, like the forms and varieties of V. officinalis. Ehrenberg's var. hirsuta, from the deserts of northern Africa, may well be one of these. Much more study in the field is required to determine these matters. The Kunkel 10256, cited below, may represent f. erecta Moldenke, a form which is very difficult to be sure of from herbarium specimens alone.

The corollas are said to have been "pinkish-blue" on Kunkel 10256 when fresh.

The Kotschy s.n. [Aegyptus inferior], distributed as V. supina, is not verbenaceous — probably represents something in the Scrophulariaceae.

Additional citations: MACARONESIA: Gran Canaria: Kunkel 10256 (Mu). EIRE: Wiest 90 (Mu—384, Mu—388). FRANCE: Herb. Kummer s.n. (Mu—1270). SPAIN: Reverchon 81 (Mu). HUNGARY: Borbás 934 (Mu), s.n. [28/8/1879] (Mu—1581); Herb. Zuccarini s.n. (Mu—386); Janka 1853 (Mu), s.n. (Mu); Kovács 460 (Mu—4329). GREECE: Guicciardi s.n. [in m. Parnassi reg. 1855] (Mu); Herb. Zuccarini s.n. (Mu). MOROCCO: Rauh 156 (Mu), 603 (Mu). LIBYA: Laing s.n. [ad Tibisium rara] (Mu—387). EGYPT: Fischer s.n. [Aegyptus] (Mu—1269); Sieber s.n. (Mu—382). SUDAN: Nubia: Kotschy 323 (Mu—383, Mu—1580). IRAQ: Kotschy s.n. [pr. Mossul, D. 8 Sept. 1841] (Mu—385). CULTIVATED: Germany: Herb. Zuccarini s.n. [Hort. bot. Erlangensis circa 1819] (Mu—390). LOCALITY OF COLLECTION UNDETERMINED: Herb. Schmiedelian s.n. (Mu—378); Herb. Schreber s.n. (Mu—379, Mu—380); Schrank s.n. (Mu—381).

VERBENA SUPINA f. ERECTA Moldenke
Additional bibliography: Moldenke, Phytologia 28: 362, 392, 393, & 441 (1974), 34: 268 (1976), and 36: 36, 40, & 277. 1977.
Weber encountered this plant on sand dunes.
Additional citations: AUSTRALIA: South Australia: J. Z. Weber 2294 (Ac).

VERBENA TAMPENSIS Nash
    Additional bibliography: M. F. Baker, Fla. Wild Fls., ed. 2, .
imp. 1, 188. 1938; Ayensu, Rep. Endang. & Threat. Pl. Spp. 67 &
126. 1974; Moldenke, Phytologia 28: 393—394, 451, & 465. 1974; M.
F. Baker, Fla. Wild Fls., ed. 2, imp. 2, 188. 1976; Fleming, Genelle, & Long, Wild Fls. Fla. 15 & 67. 1976; Lakela, Long, Fleming,
& Genelle, Pl. Tampa Bay, ed. 3 [Bot. Lab. Univ. S. Fla. Contrib.
73:] 116 & 182. 1976; Long & Lakela, Fl. Trop. Fla., ed. 2, 741 &
961. 1976; Moldenke, Phytologia 36: 141 & 142. 1977.
    Illustrations: Fleming, Genelle, & Long, Wild Fls. Fla. 67 (in
color). 1976.
    Ayensu (1974) lists V. tampensis as one of the endangered or
threatened species of plants in the United States which need
conservation measures to ensure their survival. Lakela and her
associates (1976) assert that in the Tampa Bay [Florida] area it
grows in "hammocks, low ground, [and] coastal areas", flowering
in spring and fall. Fleming and his associates (1976) record
"vervain" as a common name for it.
    The J. A. Churchill s.n. [12 March 1956], distributed as V.
tampensis, actually is V. maritima Small.
    Additional citations: FLORIDA: Brevard Co.: Curtiss 1963 in
part (Mu—1544).

VERBENA TENERA Spreng.
    Additional & emended bibliography: Buek, Gen. Spec. Syn. Can-
doll. 3: 496. 1858; Voss in Vilm., Fl. Pleine Terr., ed. 1, 937
(1865), ed. 2, 2: 975 (1866), ed. 3, 1: 1198 (1870), and ed. 4,
1066. 1894; Bolkh., Grif, Matvej., & Zakhar., Chrom. Numb. Flow.
Pl., imp. 1, 717. 1969; Williamson, Sunset West. Gard. Book, imp.
11, 437. 1973; Bolkh., Grif, Matvej., & Zakhar., Chrom. Numb.
Flow. Pl., imp. 2, 717. 1974; Kooiman, Act. Bot. Neerl. 24: 464.
1975; López-Palacios, Revist. Fac. Farm. Univ. Los Andes 15: 89 &
94. 1975; Moldenke, Phytologia 30: 150, 152, 153, 166, & 171—172
(1975), 31: 412 (1975), and 36: 151. 1977.
    López-Palacios (1975) comments that "La V. tenera en Venezuela
aún no está muy difundida y hasta la fecha sólo existe una colec-
ción mía: (López-Palacios 2565 bis)". Araujo describes it as a
heliophilous herb very common along roadsides in Rio Grande do
Sul, Brazil, and refers to the flowers [corollas] as "red".
    The Princess Therese of Bavaria 282, distributed as V. tenera,
actually is V. microphylla H.B.K., while Hieronymus s.n. [14/XI/
1876] & s.n. [Montevideo], Kummer s.n. [Hort. Monac. 16.VIII.
1849], and Meebold 12838, 26825, & 27313 are V. temuisecta Briq.
    Additional citations: BRAZIL: Rio Grande do Sul: Araujo 1256
[Herb. FEEMA 12264] (Pf).

VERBENA TENERA var. MAONETTI Regel
    Additional bibliography: Vilm., Fl. Pleine Terr., ed. 1, 938
(1865), ed. 2, 2: 975 (1866), ed. 3, 1: 1198—1199 (1870), and ed.
4, 1066. 1894; Williamson, Sunset West. Gard. Book, imp. 11, 437.

1973; Moldenke, Phytologia 30: 172 (1975) and 31: 412. 1975.
    Additional illustrations: Voss in Vilm., Fl. Pleine Terr., ed.
4, 1066. 1894.
    Vilmorin (1863) describes this variety as having "Fleurs d'un
rose purpurin, à lobes alternativement marqués de raies blanches
disposées en étoile......La variété Mahoneti, introduit depuis
quelques années seulement dans les jardines, est remarquable par
ses nombreuses flours étoilées de blanc, d'un très joli effet;
elle paraît plus rustique et résiste mieux á la sécheresse que
l'espèce; aussi sa culture se généralise-t-elle de plus en plus.
On en forme des tapis d'une très grande élégance et de très
jolies potées. Il en existe aujourd'hui plusieurs sous-variétés
obtenues récemment de semis.....La variété Mahoneti et les
sous-variétés qui en sont issues produisent fort peu de graines,
aussi les multiplie-t-on d'ordinaire de boutures, faites à
l'automae en terrines et hivernées sous châssis, ou bien au prin-
temps sur couche avec des rameaux pris sur des piede conservés
sous verre pendant l'hiver."

VERBENA TENUISECTA Briq.
    Additional synonymy: Glandularia tenuisecta (Spreng.) López-
Palacios, Revist. Fac. Farm. Univ. Los Andes 15: 89. 1975.
Verbena utenisecta Briq. ex Molina R., Ceiba 19: 96, sphalm. 1975.
Verbena tenuisectum McReynolds ex Moldenke, Phytologia 34: 279, in
syn. 1976.
    Additional & emended bibliography: M. F. Baker, Fla. Wild Fls.,
ed. 2, imp. 1, 188. 1938; A. W. Anderson, How We Got Fls., imp. 1,
168 & 283 (1951) and imp. 2, 168 & 283. 1966; Ewan in Thieret,
Southwest. La. Journ. 7: 11. 1967; G. W. Thomas, Tex. Pl. Ecolog.
Summ. 78. 1969; Bolkh., Grif, Matvej., & Zakhar., Chrom. Numb.
Flow. Pl., imp. 1, 717 (1969) and imp. 2, 717. 1974; S. B. Jones,
Castanea 39: 137. 1974; Duncan & Foote, Wildfls. SE. U. S. 150,
[151], & 295. 1975; Kooiman, Act. Bot. Neerl. 24: 464. 1975;
López-Palacios, Revist. Fac. Farn. Univ. Los Andes 15: 89 & 93--94.
1975; Molina R., Ceiba 19: 96. 1975; Moldenke, Phytologia 30: 172--
173 (1975), 31: 374--376, 392, & 398 (1975), and 34: 250, 252,
260, 270, & 279. 1976; M. F. Baker, Fla. Wild Fls., ed. 2, imp. 2,
188. 1976; Fleming, Genelle, & Long, Wild Fls. Fla. 82. 1976; F. R.
Fosberg, Rhodora 38: 113. 1976; Lakela, Long, Fleming, & Genelle,
Pl. Tampa Bay, ed. 3 [Bot. Lab. Univ. S. Fla. Contrib. 73:] 116,
168, & 182. 1976; Long & Lakela, Fl. Trop. Fla., ed. 2, 741 & 961.
1976; Soukup, Biota 11: 19. 1976; E. H. Walker, Fl. Okin. & South.
Ryuk. 884. 1976; A. L. Moldenke, Phytologia 36: 87 & 88. 1977;
Moldenke, Phytologia 36: 29, 40, 126, 128, 131, 140, 141, 164,
216, 231, 288, & 291. 1977.
    Additional illustrations: Duncan & Foote, Wildfls. SE. U. S.
[151] (in color). 1975; Fleming, Genelle, & Long, Wild Fls. Fla. 8
[as "canadensis"] (in color). 1976.
    Molina (1975) records this species as cultivated in Honduras,
while Fosberg (1976) found it in gardens on St. Croix. Ryscroft
reports it "fairly common in patches along roadsides in Natal. Bay-

liss refers to it as semiprostrate on sandy roadsides, "introduced, now widespread", at 2000 feet altitude, in the Cape Province of South Africa, misidentifying it as V. bonariensis L. Thomas found it growing in sandy pinewoods in Louisiana. Lakela and her associates (1976) aver that in the Tampa Bay [Florida] area it inhabits "sandhills, lawns, [and] berms", flowering all year; they call it the "moss verbena" and mistakenly accredit the sceintific name to "Briz."

The corollas on Ryscroft 2574 are said to have been "mauve" in color when fresh.

Recent collectors refer to V. tenuisecta as a plant "with no distinctive odor" and have encountered it on pine hills, along roadsides at the edges of swamps, and in "sandy soil on neutral ground with common grasses predominant". Hester says that it is scarce in dry soil in Louisiana, but Brown, in the same state, refers to it as a "weed in vacant lots", a "common weed in longleaf pineland", and "very abundant on prairie area roadsides". He notes that the anthers are appendaged (which is to be expected since the species is a member of the section Glandularia). Moore reports it "very frequent in dry sandy soil".

The corollas are said to have been "violet" in color on Schinini 11600 and Schinini & Martinez Crovetto 12753, "purple" on B. Moore s.n., "dark-purple" on Jenevein s.n. and Killmer 25, "blue" on C. M. Allen 453, "pink, but many pure white clumps" on Bougere 2235, and "lavender-purple" on Webster & Wilbur 3282.

The Moldenke & Moldenke 26437, 26580, 26690, & 26724 specimens cited below are transfers from the L. H. Bailey Hortorium herbarium. Hatschbach 23884 was previously erroneously cited as V. aristigera S. Moore.

Material of V. tenuisecta has been misidentified and distributed in some herbaria as V. elegans H.B.K.

Additional & emended citations: NORTH CAROLINA: Cumberland Co.: Moldenke & Moldenke 30000 (Ac). Harnett Co.: Moldenke & Moldenke 30002 (Tu). Johnston Co.: Moldenke & Moldenke 30004 (Ld). Robeson Co.: Moldenke & Moldenke 29991 (Gz). SOUTH CAROLINA: Allendale Co.: Moldenke & Moldenke 29960 (Ld). Bamber Co.: Moldenke & Moldenke 29964 (Ld). Clarendon Co.: Moldenke & Moldenke 29973 (Kh). Dillon Co.: Moldenke & Moldenke 29990 (Ac). Florence Co.: Moldenke & Moldenke 29982 (Gz), 29985 (Tu). Orangeburg Co.: Moldenke & Moldenke 29969 (Ld). GEORGIA: Bryan Co.: Moldenke & Moldenke 29942 (Tu). Bulloch Co.: Moldenke & Moldenke 29947 (Ld). Camden Co.: Moldenke & Moldenke 29872 (Gz). Clinch Co.: Spindler 170 (Lv). Dougherty Co.: Moldenke & Moldenke 29362 (Ld). Glynn Co.: Moldenke & Moldenke 29876 (Tu). Grady Co.: Moldenke & Moldenke 29365 (Gz). Lee Co.: Moldenke & Moldenke 29344 (Kh), 29350 (Gz). Marion Co.: Moldenke & Moldenke 29324 (Ac). McIntosh Co.: Moldenke & Moldenke 29907 (Gz). Mitchell Co.: Moldenke & Moldenke 29364 (Ac). Schley Co.: Moldenke & Moldenke

29329 (Gz). Screven Co.: Moldenke & Moldenke 29951 (Ac). Sumter
Co.: Moldenke & Moldenke 29338 (Tu). Talbot Co.: Moldenke & Mol-
denke 29318 (Ld). Jekyll Island: Moldenke & Moldenke 29880 (Ld),
29890 (Ac). FLORIDA: Bay Co.: Moldenke & Moldenke 26690 (Gz).
Dixie Co.: Moldenke & Moldenke 29437 (Tu). Duval Co.: Moldenke &
Moldenke 26437 (Ln, Ws). Lafayette Co.: Moldenke & Moldenke 26580
(Ws). Levy Co.: Moldenke & Moldenke 29440 (Kh). Madison Co.: B.
Moore s.n. [8 Apr. 1961] (N). Taylor Co.: Moldenke & Moldenke
29423 (Ac). Wakulla Co.: Moldenke & Moldenke 29388 (Ld). ALABAMA:
Baldwin Co.: Moldenke & Moldenke 26724 (Sl). MISSISSIPPI: Harrison
Co.: Jenevein s.n. [13 Oct. 1973] (Lv). Lincoln Co.: Webster &
Wilbur 3282 (Mi). Pearl River Co.: J. A. Churchill s.n. [10 May
1955] (Ln—204152). LOUISIANA: Acadia Par.: Killmer 25 (Lv).
Beauregard Par.: Hester 517 (Lv). Bossier Par.: Robinette 148 (Lv).
Jefferson Davis Par.: C. A. Brown 17829 (Lv). Ouachita Par.: R. D.
Thomas 3912 (Kl—11437). Rapides Par.: C. A. Brown 17094 (Lv).
Saint Helena Par.: C. M. Allen 453 (Lv). Saint Tammany Par.: Bou-
gere 2235 (Lv). Vernon Par.: Hester 442 (Lv). Washington Par.: C.
A. Brown 5636 (Lv), 17749 (Lv), 18408 (Lv). TEXAS: Jefferson Co.:
C. A. Brown 18806 (Lv). Orange Co.: McReynolds 750353 (Lv). CALI-
FORNIA: Los Angeles Co.: Meebold 26825 (Mu), 27319 (Mu). BRAZIL:
Mato Grosso: Hatschbach 23884 (W—2705822). BOLIVIA: Sucre: Zöll-
ner 8098 (Gz). URUGUAY: Hieronymus s.n. [Montevideo] (Mu). AR-
GENTINA: Córdoba: Hieronymus s.n. [14.XI.1876] (Mu). Corrientes:
Schinini 11600 (Ld); Schinini & Martinez Crovetto 12753 (Ld).
Misiones: Montes 14663 (N). GERMANY: Brixle s.n. [Herb. Merxmül-
ler 14336] (Mu). SOUTH AFRICA: Cape Province: Bayliss BS.7344
(Mu); Ryscroft 2626 (Mu). Natal: Meebold 12838 (Mu); Ryscroft
2574 (Mu). CULTIVATED: France: Herb. Kummer s.n. [h. Paris.] (Mu—
1272). Germany: Herb. Hort. Monac. s.n. (Mu); Herb. Schwaegrichen
s.n. [Hort. Lips.] (Mu—1251); Herb. Zuccarini s.n. [h. Monac.
1836] (Mu—392); Hiendlmayr s.n. [Hort. Lipsiensis] (Mu—1271);
Kummer s.n. [Hort. Monac. 16.VIII.1849] (Mu—391); Kupper s.n.
[cult. h.b.M.] (Mu).

VERBENA TENUISECTA f. ALBA (Benary) Moldenke, Phytologia 36: 164.
   1977.
   Additional bibliography: Duncan & Foote, Wildfls. SE. U. S. 150
& [151]. 1975; Moldenke, Phytologia 30: 173 (1975), 31: 374—376
(1975), 34: 250 (1976), and 36: 128. 1977.
   Additional illustrations: Duncan & Foote, Wildfls. SE. U. S.
[151] (in color). 1975.
   In line with current thinking in taxonomic circles, the status
of this taxon, a mere color form, is reduced from the varietal
rank under which I have hitherto considered it to form rank.
   Bougere has misidentified this plant as V. bipinnatifida Nutt.

and notes that the corollas were "white when fresh, but press lavender"; his no. 2235, cited herein under typical V. tenuisecta.
may represent a mixture because its label states "flowers pink,
but many pure white clumps", collected in Saint Tammany Parish,
Louisiana.

Additional citations: SOUTH CAROLINA: Clarendon Co.: Moldenke
& Moldenke 29979 (Ac, Gz, Kh, Ld). GEORGIA: Grady Co.: Moldenke &
Moldenke 29369 (Gz). Lee Co.: Moldenke & Moldenke 29357 (Ld, Tu).
Marion Co.: Moldenke & Moldenke 29328 (Ac). FLORIDA: Dixie Co.:
Moldenke & Moldenke 29438 (Tu). LOUISIANA: Saint Tammany Par.:
Bougere 1996 (Lv).

VERBENA TENUISECTA f. RUBELLA Moldenke
    Additional bibliography: Moldenke, Phytologia 24: 239 & 241.
1972.
    Bougere 2235, cited herein under typical V. tenuisecta Briq.,
may actually represent this form because its accompanying label
says "flowers pink".

VERBENA TESSMANNII Moldenke
    Additional bibliography: Moldenke, Phytologia 24: 241. 1972;
Soukup, Biota 11: 19. 1976.

VERBENA TEUCRIIFOLIA Mart. & Gal.
    Additional & emended bibliography: Schau. in A. DC., Prodr. 11:
553--555. 1847; Buek, Gen. Spec. Syn. Candoll. 3: 494 & 496. 1858;
Bolkh., Grif, Matvej., & Zakhar., Chrom. Numb. Flow. Pl., imp. 1,
717 (1969) and imp. 2, 717. 1974; Moldenke, Phytologia 28: 398--399
& 432. 1974; Hinton & Rzedowski, Anal. Esc. Nac. Cienc. Biol. 21:
111. 1975; Moldenke, Phytologia 36: 149. 1977.
    Additional citations: MEXICO: México: Pringle 4180 (Mu--1778).
Nuevo León: Beaman 2667 (Ln--170700), 4460 (Ln--171523). San Luis
Potosí: Schaffner s.n. [San Luis Potosí, 1875-79] (Mu--1562).
Veracruz: Dodds 99 (Ln--199254); Troll 129 (Mu).

VERBENA TEUCRIIFOLIA var. COROLLULATA Perry
    Additional bibliography: Moldenke, Phytologia 28: 398--399 &
432. 1974.

VERBENA THYMOIDES Cham.
    Additional bibliography: Buek, Gen. Spec. Syn. Candoll. 3: 496.
1858; Moldenke, Phytologia 28: 399. 1974.
    Lindeman reports finding this plant along roadsides and the
corollas on his no. ICN.20891 are said by him to have been "blue"
when fresh, while those on Valls & al. 2127 were "red".
    Additional citations: BRAZIL: Rio Grande do Sul: Lindeman ICN.
20891 (Ut--320463); Valls & al. 2127 [Herb. ICN.10131] (Ut--320462).
State undetermined: Sellow s.n. [Brasilia] (Mu--397--cotype).

VERBENA TOWNSENDII Svenson

Additional synonymy: Verbena towsendii Svenson ex Balgooy, Pacif. Fl. Areas 3: 245, sphalm. 1975.

Additional bibliography: Moldenke, Phytologia 24: 243. 1972; Balgooy, Pacif. Pl. Areas 3: 245. 1975; Moldenke, Phytologia 36: 302. 1977; Van der Werff, Bot. Notiser 130: 96. 1977.

Van der Werff (1977) reduces V. galapagosensis Moldenke, V. stewartii Moldenke, and V. glabrata var. tenuispicata Moldenke to synonymy under what he refers to as a very variable V. townsendii, citing his nos. 1172, 1192, 1218, 1897, 2124, 2152, 2280, & 2286 as well as Adsersen 486, 512, 543, 572, 913, 920, 944, 969, 1148, & 1189, Hendrix s.n., Howell 9007, Stewart 3317, 3318, 3319, & 3320, and Vagvolgri s.n.

VERBENA TRIFIDA H.B.K.

Additional bibliography: G. Don in Loud., Hort. Brit., ed. 1, 247 (1830) and ed. 2, 247. 1832; Loud., Hort. Brit., ed. 2, 552. 1832; G. Don in Loud., Hort. Brit., ed. 3, 247. 1839; Sweet, Hort. Brit., ed. 3, 553. 1839; Buek, Gen. Spec. Syn. Candoll. 3: 496. 1858; Moldenke, Phytologia 28: 399. 1974; Soukup, Biota 11: 19. 1976.

Duque-Jaramillo describes this plant as a subshrub, 70 cm. tall, with the corollas "blanco-rosadas", and found it growing at 2625 m. altitude.

Don (1830) calls this the "trifid vervain" and claims that it was introduced into cultivation in English gardens from Mexico in 1818.

Additional citations: COLOMBIA: Cundinamarca: Duque-Jaramillo 2667 (N).

xVERBENA TRINITENSIS Moldenke

Additional bibliography: Moldenke, Phytologia 24: 237 & 244. 1972.

VERBENA TRISTACHYA Troncoso & Burkart

Additional & emended bibliography: Bolkh., Grif, Matvej., & Zakhar., Chrom. Numb. Flow. Pl., imp. 1, 715 (1969) and imp. 2, 715. 1974; Moldenke, Phytologia 30: 173. 1975.

VERBENA TUMIDULA Perry

Additional bibliography: G. W. Thomas, Tex. Pl. Ecolog. Summ. 78. 1969; Moldenke, Phytologia 28: 399. 1974.

Chiang and his associates encountered this rare plant in "calcarsous soil of izotal or encinar (almost chaparral) on steep limestone slopes and canyons", associated with Yucca carnerosana, Dasylirion, Nolina, and Quercus spp., at 1750—1775 m. altitude, flowering and fruiting in September.

Additional citations: MEXICO: Coahuila: Chiang C., Wendt, & Johnston 9223b (Ld).

VERBENA URTICIFOLIA L.

Additional & emended bibliography: L., Hort. Cliff., imp. 1, 11. 1737; R. A. Salisb., Prodr. 71. 1796; Stokes, Bot. Mat. Med. 1: 40.

1812; Sweet, Hort. Brit., ed. 1, 1: 325. 1826; G. Don in Loud., .
Hort. Brit., ed. 1, 246 (1830) and ed. 2, 246. 1832; Loud., Hort.
Brit., ed. 2, 552. 1832; G. Don in Loud., Hort. Brit., ed. 3, 246.
1839; Sweet, Hort. Brit., ed. 3, 553. 1839; Buek, Gen. Spec. Syn.
Candoll. 3: 494 & 496. 1858; Paine, Ann. Rep. Univ. N. Y. 18: [Pl.
Oneida Co.] 109. 1865; Kuntze, Rev. Gen. Pl. 2: 510. 1891; Conard,
Pl. Iowa 44. 1951; E. R. Spencer, Just Weeds, ed. 2, xii, 199—
201, & 332, fig. 64. 1957; R. A. Davidson, State Univ. Iowa Stud.
Nat. Hist. 20 (2): 77. 1959; Hall & Thompson, Cranbrook Inst. Sci.
Bull. 39: 74. 1959; Cooperrider, State Univ. Iowa Stud. Nat. Hist.
20 (5): 70. 1962; L., Hort. Cliff., imp. 2, 11. 1968; E. R. Spen-
cer, All About Weeds xii, 199—201, & 332, fig. 64. 1968; Barker,
Univ. Kans. Sci. Bull. 48: 571. 1969; Bolkh., Grif, Matvej., &
Zakhar., Chrom. Numb. Flow. Pl., imp. 1, 717. 1969; G. W. Thomas,
Tex. Pl. Ecolog. Summ. 78. 1969; Hathaway & Ramsey, Castanea 38:
77. 1973; Bolkh., Grif, Matvej., & Zakhar., Chrom. Numb. Flow. Pl.,
imp. 2, 717. 1974; E. T. Browne, Castanea 39: 183. 1974; R. D.
Gibbs, Chemotax. Flow. Pl. 3: 1753—1755 (1974) and 4: 2295. 1974;
S. B. Jones, Castanea 39: 137. 1974; León & Alain, Fl. Cuba, imp.
2, 2: 281. 1974; Rousseau, Géogr. Florist. Qué. [Trav. Doc. Cent.
Étud. Nord 7:] 377, 467, 479, 504, 516, 644, & 788, map 829. 1974;
Van Saun & Kemp, Bull. Torrey Bot. Club 101: 371. 1974; [Bard],
Bull. Torrey Bot. Club 102: 431. 1975; D. S. & H. B. Correll,
Aquat. & Wetland Pl. SW. U. S., imp. 2, 2: 1396, 1399, & 1775.
1975; Hocking, Excerpt. Bot. A.26: 6. 1975; Kooiman, Act. Bot.
Neerl. 24: 464. 1975; A. L. Moldenke, Phytologia 32: 375. 1975;
Perkins, Estes, & Thorp, Bull. Torrey Bot. Club 102: 194—198.
1975; H. D. Wils., Vasc. Pl. Holmes Co. Cat. 54. 1975; Moldenke,
Phytologia 30: 146, 168, 169, & 173—178 (1975), 31: 412 (1975),
and 34: 247, 249, & 250. 1976; Anon., Biol. Abstr. 61: AC1.732.
1976; Van Bruggen, Vasc. Pl. S. Dak. 369, 536, & 537. 1976;
[Voss], Mich. Bot. 15: 237. 1976; Ziegler & Sohmer, Contrib. Herb.
Univ. Wisc. LaCrosse 13: 16. 1976; Greller, Bull. Torrey Bot.
Club 104: 176. 1977; Moldenke, Phytologia 36: 28, 29, 36, 134,
135, 217, 221, 228, 229, 297, & 303—306. 1977; F. H. Montgomery,
Seeds & Fruits 202, fig. 3, & 230. 1977.

Additional illustrations: E. R. Spencer, Just Weeds [200], fig.
64. 1957; E. R. Spencer, All About Weeds [200], fig. 64. 1968; F.
H. Montgomery, Seeds & Fruits 202, fig. 3. 1977.

Don (1830) calls this species the "nettle-leaved vervain" and
says that it was introduced into cultivation in English gardens
from North America in 1683, while what he calls V. diffusa, the
"diffuse vervain", was introduced from North America in 1818.

Wilson (1975) encountered V. urticifolia in alluvial woods and
low fields in Ohio. Hathaway & Ramsey (1973) found it in Pittsylvania County, Virginia, and Browne (1974) in Stone County, Arkansas. Spencer (1957, 1968) avers that the species is native to
"Tropical America", but this is erroneous: it is a purely east-
temperate North American plant. Tans reports it "common in heavily grazed pastures in sunny sites on upland soil" and growing with
Cirsium canadense, C. vulgare, Ambrosia trifida, and A. artemisii-

folia in Wisconsin. Davidson (1959) declares that in Iowa it occurs "Usually in alluvial thickets and open places; frequent",
while Cooperrider (1962) refers to it as "Common. Alluvial thickets and woods; roadside thickets". In Kansas Barker (1969) reports
it as "Common, in floodplain woods and on wooded slopes...through-
out the area". Merxmüller found it "verwildert" in Germany.

Other recent collectors refer to this plant as "upright, with
small white flowers" and have encountered it in shaded waste places,
swamps, ditches in open pastures, and open fields, in sandy soil, a-
long roadsides, in open areas around fields, in open grassy sandy
soil along rivers, and in "low semi-wet spots in Pinus taeda forests", associated with Rubus, Ambrosia, Cassia, and Sesbania. Claycomb avers that it is "rare" in Lafayette Parish, Louisiana, but in
other parts of that state Allen records it as "locally frequent in
open areas associated with Polygonum" and Wurzlow says "common in
waste places and roadsides".

The corollas are said to have been "white", as usual, on C. A.
Allen 1161, C. A. Brown 3893, Curry, Martin, & Allen 305 & 390, and
H. R. Wilson 302.

Gibbs (1974) reports cyanogenesis and leucoanthocyanin absent
from the leaves of V. urticifolia, syringin absent from the stems,
and the Ehrlich test giving negative results in the leaves.

Perkins and his associates (1975) report that V. urticifolia is
highly autogamous, nectar is present in the flowers but not in sufficient quantity for measurement, "Wilting of the persistent corolla.....brings the pollen into contact with the stigma, as evidenced
by pollen tube growth on stigmas of bagged inflorescences", the
corollas are the smallest (2 mm.) of the 4 species studied. "Verbena urticifolia produces relatively taller plants (45—135 cm.)
and an inflorescence pattern similar to V. halei, but because it
has more branches (5—144, mean = 40.4), and the distance between
the flowers is only about 2.1 mm, the inflorescences appear denser".
They found the following insects visiting the flowers: Diptera:
Allograpta sp., Baccha sp. (with Verbena pollen on head), Dolicho-
podidae sp., Paragus sp., Systropus sp. (with pollen on head), Hymenoptera: Dialictus sp. (with pollen on head), Sphecodes sp., and
Lepitoptera: Leptotes marinus. They found 4 plants with 6219 potential seeds had a 66.5 percent seed-set when insect-visited,
while 8 bagged plants, with 10,653 potential seeds, had a 47.3 percent seed-set.

Montgomery (1977) describes the seeds as "Nutlets 1.8 x 0.7 x
0.6 mm, rounded dorsal surface obscurely veined and finely reticulate".

Material has been misidentified and distributed in some herbaria
as V. scabra Vahl. On the other hand, the J. A. Churchill s.n. and
Tans 1454-3, distributed as V. urticifolia, actually represent its
var. leiocarpa Perry & Fernald, M. F. Spencer 994 is V. lasiostachys
Link, Robinette 203 is V. halei Small, and C. A. Brown 4084a, Eggers
996, Lindheimer 1077, Sintenis 767' Stam 60, and Thomas & al. 10859

are V. scabra Vahl.

Additional citations: NEW JERSEY: Union Co.: Moldenke & Moldenke 30154 (W). PENNSYLVANIA: Dauphin Co.: Moldenke & Moldenke 31156 (Ut). Northampton Co.: Herb. Schreber s.n. [Bethlehem] (Mu—403), s.n. [Nazareth, 1787] (Mu—402). Schuylkill Co.: Moldenke & Moldenke 31160 (Ac, Ld, Mu). County undetermined: Herb. Zuccarini s.n. [e Pennsylvania] (Mu—407). VIRGINIA: Nelson Co.: Freer, Ramsey, & Ramsey 17334 (Lc). MISSISSIPPI: County undetermined: Herb. Kummer s.n. [Mississippi] (Mu—1270). OHIO: Hamilton Co.: Frank s.n. [1837 (Mu). ILLINOIS: Adams Co.: Purpus 132 (Mu—4291). INDIANA: Marion Co.: Frazee s.n. [July 8, 1885] (Lc). Tippecanoe Co.: Bresinsky s. n. [2.9.1967] (Mu). MICHIGAN: Ingham Co.: Flanders 110 (Ln—226391) W. D. Stevens 1599 (Ln—237018). Wayne Co.: G. Stewart s.n. [Aug. 1898] (Ln—142429). WISCONSIN: Dane Co.: Tans 1478-20 (Ts). Jefferson Co.: Tans 1475-1 (Ts). MISSOURI: Marion Co.: J. Davis s.n. [25-7-18] (Mu). Pulaski Co.: Meebold 25489 (Mu). Saint Louis: Goehring 427 (Lv); Mühlenbeck 1293 (Mu). County undetermined: Martens s.n. (Mu). LOUISIANA: Bossier Par.: Correll & Correll 10057 (Lv). East Baton Rouge Par.: C. A. Brown 1071 (Lv); Hunt 6 (Lv). Lafayette Par.: Claycomb s.n. [June 25, 1942] (Lv). Lincoln Par.: J. A. Moore 5370 (Lv). Pointe Coupee Par.: C. A. Brown 3893 (Lv); M. Chaney 42 (Lv). Saint Helena Par.: C. M. Allen 349 (Lv), 1161 (Lv); Kirkpatrick 16 (Lv). Saint Tammany Par.: Arsène 11083 (Lv), 12069 (Lv). Tangipahoa Par.: H. R. Wilson 302 (Lv), 419 (Lv). Terrebonne Par.: Arceneaux 383 (Lv); Wurzlow s.n. [June 12, 1912] (Lv, Lv). West Feliciana Par.: Curry, Martin, & Allen 305 (Lv), 390 (Lv). GERMANY: Merxmüller 14338 (Mu). CULTIVATED: Germany: Herb. Hort. Monac. s.n. [20.IX.04] (Mu); Herb. Zuccarini s.n. [Hort. Bot. Monac.] (Mu—408, Mu—410); Prince Paul of Wurtemberg s.n. [Hort. Mergentheim 1840] (Mu—1586). LOCALITY OF COLLECTION UNDETERMINED: Herb. Mus. Bot. Landishuth s.n. (Mu—400); Herb. Reg. Monac s.n. (Mu—398); Herb. Schreber 279 [e Carolina] (Mu—401), s. n. [ex America boreali] (Mu—404), s.n. (Mu—405); Herb. Schmiedelian s.n. (Mu—400); Herb. Zuccarini s.n. (Mu—409); Hiendlmayr s.n. (Mu—1274); Hooker s.n. [Un. States] (Mu—414); Prince Paul of Wurtemberg s.n. [Verein. St. 1832] (Mu—1585).

VERBENA URTICIFOLIA var. LEIOCARPA Perry & Fernald
Additional bibliography: Hall & Thompson, Cranbrook Inst. Sci. Bull. 39: 74. 1959; G. W. Thomas, Tex. Pl. Ecolog. Summ. 78. 1969; D. S. & H. B. Correll, Aquat. & Wetland Pl. SW. U. S., imp. 2, 2: 1396, 1399, & 175. 1975; Moldenke, Phytologia 30: 176 & 177 (1975), 34: 247 (1976), and 36: 29 & 297. 1977.
Tans found this variety "common in heavily grazed pastures on nearly level upland ground", growing in association with Carduus nutans, Verbena stricta, V. hastata, and Verbascum thapsus, in Wis-

consin, flowering as early as May, the corollas white. Churchill
encountered it in pastures and on shaley lake shores. Hall &
Thompson (1959) report it from Oakland County, Michigan, where it
grows "In open woods, along roadsides, and in meadows. Occasional."

Additional citations: NEW YORK: Schuyler Co.: J. A. Churchill
s.n. [23 August 1937] (Ln—203427). NEW JERSEY: Union Co.: Mol-
denke & Moldenke 30953 (Ac, Ld, Tu). VIRGINIA: Surrey Co.: J. A.
Churchill s.n. [22 August 1970] (Ln—230920). WISCONSIN: Walworth
Co.: Hansen & Tans 1422 (Ts); Tans 1454-3 (Ts). MISSOURI: Shannon
Co.: Meebold 25328 (Mu). LOCALITY OF COLLECTION UNDETERMINED:
Herb. Schmiedelian s.n. [America borealis] (Mu—399).

VERBENA VALERIANOIDES H.B.K.

Additional bibliography: Buek, Gen. Spec. Syn. Candoll. 3: 496.
1858; Robledo, Bot. Med. 392. 1924; Moldenke, Phytologia 30: 178.
1975; López-Palacios, Revist. Fac. Farm. Univ. Los Andes 17: 50.
1976.

Robledo (1924) records the popular name, "verbena", for this
plant. López-Palacios refers to it as "hierba de unos 60—80 cms.
Hojas mas o menos lineales. Cabezuelas reducidas y congestas"
and found it growing at 2650 m. altitude, flowering in August. He
is of the opinion that his no. 3639 represents a natural hybrid
with V. litoralis H.B.K. because the "Hojas basales con las de V.
litoralis y las superiores como las de V. valerianoides; cabezuelas como las de ésta última." In a letter to me, dated January 16,
1976, he says "3639 queda definitivamente en Verbena valerianoides
HBK. Habrá que completar la descripción de HBK haciendo notar la
variación de las hojas inferiores según sus observaciones. No vale
la pena hacer una separación, ni siguiera de forma." In his 1976
publication he says: "Schauer la separa de la V. litoralis por sus
hojas enteras y lineales, y de esa manera la describe Bonpland;
pero ello no es absolutamente cierto. encontrádose algunas hojas
basales anchas y de borde dentado, muy similares de las de V. litoralis; López-Palacios 3639 (COL), Bogotá."

Additional citations: COLOMBIA: Cundinamarca: López-Palacios
3637 (N, Z), 3639 (N, Z).

VERBENA VARIABILIS Moldenke

Additional bibliography: Moldenke, Phytologia 24: 252. 1972;
Soukup, Biota 11: 19. 1976.

VERBENA VILLIFOLIA Hayek

Additional bibliography: Moldenke, Phytologia 24: 252. 1972;
Soukup, Biota 11: 19. 1976.

VERBENA VIOLATA Rojas

Additional & emended bibliography: Krapovickas, Bol. Soc. Argent. Bot. 11, Supl. 269. 1970; Moldenke, Phytologia 30: 178 (1975)

and 31: 388. 1975.

VERBENA WEBERBAUERI Hayek
    Additional bibliography: Moldenke, Phytologia 24: 252—253.
1972; Soukup, Biota 11: 19. 1976.

VERBENA WRIGHTII A. Gray
    Additional & emended bibliography: G. W. Thomas, Tex. Pl. Eco-
log. Summ. 78. 1969; Bolkh., Grif, Matvej., & Zakhar., Chrom.
Numb. Flow. Pl., imp. 1, 717 (1969) and imp. 2, 717. 1974; E. H.
Jordan, Checklist Organ Pipe Natl. Mon. 7. 1975; Moldenke, Phytologia 30: 178—179 (1975), 34: 251 (1976), and 36: 124, 128, 147,
& 148. 1977.
    The corollas are described as having been "blue" on Semple &
Love 321 and these collectors encountered the plant in dry sandy
soil near roadsides. On Van Devender & Van Devender s.n. [28
March 1976] the corollas are said to have been "light-purple"
    Higgins has found V. wrightii in limestone soil of oak-juniper
communities and in sandy soil in short-grass prairie communities,
as well as in "rocky or gravelly limestone with sotol-Acacia association", while Lehto and his associates encountered it among
volcanic rocks. Weber & Livingston found it growing on steep
shale hillsides; Sperry refers to it as "frequent" in Pecos County,
Texas. Wendt and his associates report ot from calcareous gravel
"in matorral subdesértico con espinosas laterales y inerne on
steep-walled narrow limestone canyons", associated with Acacia
berlandieri and Eupatorium solidaginifolium. Stotz found it to be
"fairly common in rocky wash lined with Prosopis and Acacia".
Thomas (1969) calls it the "Wright verbena".
    The Dziekanowski, Dunn, & Bennett 2393 and S. stephens 75643,
distributed as V. wrightii, actually are V. ambrosifolia Rydb.,
while D. Howe s.n. [1 October 1968] is V. ambrosifolia f. eglandulosa Perry, and Higgins 8790 and L. M. Andrews 259 are V. gooddingii Briq.
    Additional citations: COLORADO: Archuleta Co.: Weber & Living-
ston 6258 (Mi). TEXAS: El Paso Co.: Meebold 24224 (Mu). Hudspeth
Co.: Higgins 8552 (N). Pecos Co.: Semple & Love 321 (W—2732739);
Sperry 3081 (Sd—70647). NEW MEXICO: Bernalillo Co.: Meebold
22491 (Mu). Dona Ana Co.: Meebold 26694 (Mu). Eddy Co.: Higgins,
Higgins, & Higgins 9849 (N). Grant Co.: O. B. Metcalfe 126 (Mu—
4104). San Miguel Co.: Higgins 8884 (N). ARIZONA: Apache Co.:
Lehto, McGill, Nash, & Pinkava 11566 (W—2734658). Graham Co.:
Stotz 16 (N). Santa Cruz Co.: Van Deventer & Van Deventer s.n.
[28 March 1976] (Ld). MEXICO: Coahuila: Wendt, Chiang, & Johnston 9279a (Ld).

VERBENA XUTHA Lehm.
    Additional synonymy: Verbena xatha Lehm. ex Moldenke, Phytolo-

gia 34: 279, in syn. 1976. Verbena virginianum L. ex Moldenke,
Phytologia 34: 279, in syn. 1976.

Additional & emended bibliography: Buek, Gen. Spec. Syn. Candoll. 3: 495 & 496. 1858; G. W. Thomas, Tex. Pl. Ecclog. Summ. 78.
1969; Bolkh., Grif, Matvej., & Zakhar., Chrom. Numb. Flow. Pl.,
imp. 1, 717 (1969) and imp. 2, 717. 1974; Moldenke, Phytologia 30:
179--180 (1975), 34: 250, 251, & 279 (1976), and 36: 29, 124, 277,
& 294. 1977.

Recent collectors refer to this plant as upright, with pubescent "flowers" and no fragrance. They have encountered it in sandy soil, fallow fields, open dry roadsides, on the river side of
levees, and on "spoils" with Bermuda and Dallas grass. In addition to the months previously reported, it has been collected in
fruit in April. Piehl refers to it as infrequent in pastures in
Louisiana, where, however, Spindler reports it as a "frequent
perennial in vacant lots" and Hester found it plentiful in woods
and also plentiful along roadsides. Demaree found it "common in
bottoms" at 320 feet altitude in Arkansas.

Thomas (1969) calls this species the "coarse verbena", while
Claycomb calls it "blue vervain".

The corollas are said to have been "lavender" on Piehl s.n.,
"violet" on Schroer 71, "light-purple" on Spindler 71, and "purple"
on Curry, Martin, & Allen 594.

The Montz 2485, distributed as V. xutha, actually is xV. alleni
Moldenke, while Meebold 27301 is V. canescens H.B.K. and D. E. Ellis 58 is V. halei Small.

Additional citations: MISSOURI: Saint Louis: Prince Paul of Würtemberg s.n. [St. Louis, 1832] (Mu--1587). ARKANSAS: Sevier Co.:
Demaree 70559 (Ld). LOUISIANA: Assumption Par.: Maisonneaux s.n.
[June 1917] (Lv). Cameron Par.: C. A. Brown 9272 (Lv); Schroer
71 (Lv); Spindler 71 (Lv). East Baton Rouge Par.: C. A. Brown
1063 (Lv), 1381 (Lv); T. S. Jones s.n. [May 18th 1899] (Lv); Ro-
bertson s.n. [June 5, 1899] (Lv); Simon 688 (Lv). Iberia Par.:
Hester 642 (Lv). Lafayette Par.: Claycomb s.n. [July 2, 1942]
(Lv). Madison Par.: Piehl s.n. [1 Oct. 1972] (Lv). Natchitoches
Par.: Hester 158 (Lv). Pointe Coupee Par.: M. Chaney 402 (Lv,
Lv). Red River Par.: Hester 355 (Lv). Saint Martin Par.: Correll
& Correll 9447 (Lv); Hester 657 (Lv). Saint Mary Par.: Hester 796
(Lv). Terrebonne Par.: Bynum, Ingram, & Jaynes s.n. [Apr. 22,
1933] (Lv); Wurzlow s.n. [May 18, 1912] (Lv). Choupique Island:
Montz 2294 (Lv). Grand Isle: C. A. Brown 1980 (Lv); Cangemi & An-
drus 85 (Lv). Turnbull Island: Curry, Martin, & Allen 594 (Lv).
TEXAS: Orange Co.: J. A. Churchill s.n. [1 May 1955] (Ln--204156).
CULTIVATED: Germany: Prince Paul of Würtemberg s.n. [Hort. Mergen-
theim] (Mu--1588).   [to be continued]

Observations on Hawaiian Panicum and Sapindus
Hawaiian Plant Studies 61

Harold St. John
Bishop Museum, Honolulu, Hawaii 96818

Gramineae

Panicum conjugens Skottsb., Göteb. Bot. Trädg.,
   Meddel. 15: 298, figs. 106, 125-135, 1944.
   Lectotype: Hawaiian Islands, Kauai Island,
Alakai, bog along trail from Lehua makanoe toward
Kilohana, 13/8/1938, O. H. Selling 2,886 (BISH),
here designated.
   The species was based on three collections,
without a positive designation of a holotype.
One specimen was mentioned as a smaller form,
a second one was said to be possibly the same.
The third, Selling 2,886, agrees with the
diagnosis, so it is here chosen as lectotype.

Panicum gracilius (Skottsb.) comb. nov.
   P. hillebrandianum Hitchc., var. gracilius
      Skottsb., Göteb. Bot. Trädg., Meddel. 15:
      296, figs. 118-124, 1944.
   P. Hillebrandianum is native to the bogs of
west Maui. It has the spikelets 2.7-3.3 mm long;
first glume 0.8-2.1 mm long; second glume 2.5-3.3
mm long; sterile lemma 2.5-3 mm long, and the
sterile palea 1.1-1.5 mm long. P. gracilius of
the Alakai bogs, Kauai, has the spikelets 2.3-2.8
mm long; first glume 1-1.7 mm long, second glume
2.4-2.6 mm long; sterile lemma 2.3-2.5 mm long;
and the sterile palea 1.5-1.6 mm long.
   It seems to be a clearly distinct species.

Sapindaceae

Sapindus Thurstonii Rock, Hawaii Board Agric. &
      For., Bull. 1: 6, fig. 2, pl. 3, 1911; Fedde,
      Repert. 10: 368, 1912.
   S. saponaria L., var. Thurstonii (Rock) Skottsb.,
      Göteb. Bot. Trädg., Meddel. 2: 244, 1926.
   S. Saponaria L., forma inaequalis sensu Radlk.,
      as to Hawaiian tree, Engler's Pflanzenreich
      IV, 165(3): 646, 650, 1932; Fagerlund &
      Mitchell, Hawaii Natl. Park, Nat. Hist. Bull.

9: 45, 1944; St. John, Pacif. Trop. Bot.
Gard., Mem. 1: 225, 1973; non (DC.) Radlk.,
in Engler's Pflanzenreich IV, 165(3): 646,
1932.

S. Saponaria L., forma microcarpus sensu Radlk.
as to Hawaiian tree, Engler's Pflanzenreich
IV, 165(3): 646,651, 1932, non (Jardin)
Radlk., l. c. 646, 1932.

S. Saponaria, as to Hawaiian tree, Rock, Ind.
Trees Hawaii, 271-273, pl. 104-106, 1913;
Degener, Ferns Fl. Pl. Hawaii Natl. Park
202, 204, pl. 55, (1930); Lamoureux,
Trailside Pl. Hawaii Natl. Park, 47, color
fig., (1977); non L., Sp. Pl. 367, 1753.

The Hawaiian native Sapindus tree, called
"manele" or "a'e," occuring on Hawaii and Maui,
in botanical classification has fluctuated from
species, to forma, to variety, to a synonym of
the tropical American S. Saponaria, or to two
of its formae. J. F. Rock first described it in
1911 as a new species, S. Thurstonii, but in 1913,
after reexamination of the material, he reduced
it and made it a synonym of the American
S. Saponaria. The mongrapher Radlkofer placed
it in two of his formae under S. Saponaria.
Skottsberg in 1926, on review of the problem, noted
that the Hawaiian trees had leaves with a narrower
leaf rhachis and smaller fruit. He then published
it as S. Saponaria L., var Thurstonii (Rock)
Skottsb.

During several visits to large herbaria, the
writer has studied extensive collections of the
tropical American S. Saponaria L. He found that
it had the petioles winged or wingless; rhachis
winged; leaflets (6) 8-14; sepals glabrous without,
or sparsely remotely pubuerulent at base, but the
margins ciliate; petals with a 1 mm claw, white
pilose, the limb 1.3-1.5 mm long, broadly elliptic,
minutely ciliolate, but otherwise glabrous;
filaments to 3 mm in length, the lower 2/3 villous.
The Hawaiian S. Thurstonii has the petioles
wingless; rhachis wingless; leaflets 6-10 (-12);
sepals densely pilosulous without, except in the up
per part covered in bud; the margins ciliolate;

petals with a 0.7 mm pilose claw and a limb 2.8
mm long, 1.5 mm wide, elliptic, pilose within and
without, except near the ciliate upper margin;
filaments to 2.5 mm long, the lower 2/3 densely
pilose.

In conclusion, the differences tabulate[d] above
seem adequate for the recognition of the Hawaiian
trees as an endemic species, properly named
Sapinus Thurstonii Rock.

ADDITIONAL NOTES ON THE ERIOCAULACEAE. LXXII

Harold N. Moldenke

ERIOCAULACEAE Lindl.
 Additional & emended bibliography: Rottb., Act. Lit. Univ. Hafn.
1: 272, pl. 1, fig. 1. 1778; Sweet, Hort. Brit., ed. 1, 1: 435
(1826) and ed. 2, 546 & 597. 1830; G. Don in Loud., Hort. Brit.
Suppl. 1: 633. 1832; Loud., Hort. Brit., ed. 2, 719. 1832; Sweet,
Hort. Brit., ed. 3, 719. 1839; G. Don in Loud., Hort. Brit. Suppl.
2: 547. 1850; Engl., Syllab. Pflanzenfam., ed. 2, 86 (1898), ed.
3, 92 (1903), ed. 5, 94 (1907), and ed. 6, 99. 1909; E. D. Merr.
in Merr. & Merritt, Philip. Journ. Sci. Bot. 5: 301. 1910; Praeger, Journ. Roy. Hortic. Soc. Lond. 36: 302—303, fig. 107. 1910;
Praeger, Irish Natur. 21: 26. 1912; Gilg in Engl., Syllab. Pflanzenfam., ed. 7, 138 & 139, fig. 140 (1912) and ed. 8, 140 & 141,
fig. 140. 1919; Fedde in Just, Bot. Jahresber. 44: 19 (1922) and
45 (1): 517, 520, & 549. 1923; Fedde & Schust. in Just, Bot. Jahresber. 45 (1): 20. 1923; Gilg in Engl., Syllab. Pflanzenfam., ed.
9 & 10, 152, fig. 144. 1924; Kräusel in Just, Bot. Jahresber. 44:
1161 & 1163. 1926; Fedde in Just, Bot. Jahresber. 44: 1415. 1927;
Diels in Engl., Syllab. Pflanzenfam., ed. 11, 154, fig. 144. 1936;
León, Fl. Cuba, imp. 1, 1: 278—284, 423, 426, 428, 435, & 436,
fig. 112 & 113. 1946; M. R. Henderson, Malay. Nat. Journ. 6: 212.
1950; Gaussen, Viart, Legris, & Labroue, Trav. Sect. Scient.
Techn. Inst. Franç. Pond. Hors Ser. 5: 61. 1965; Boivin & Cayouette, Naturaliste Canad. 94: 524. 1967; Naik, Journ. Indian Bot.
Soc. 52: 108—113. 1973; Lepage, Naturaliste Canad. 101: 928.
1974; Rousseau, Géogr. Florist. Qué. [Trav. Doc. Cent. Étud. Nord
7:] 120—121, 382, 470, 480, 498, 550, 586, 625, 705, & 762, map
221 & 222. 1974; Satake, Journ. Jap. Bot. 49: 180—183, 237—240,
& 313—314. 1974; Bole, Excerpt. Bot. A.26: 302. 1975; Cárdenas
de Guevara, Act. Bot. Venez. 10: 23, 35, 39, & [69]. 1975; Hinton
& Rzedowski, Anal. Esc. Nac. Cienc. Biol. 21: 61. 1975; Hocking,
Excerpt. Bot. A.26: 6, 89, & 90. 1975; Hurusawa, Excerpt. Bot. A.
26: 99. 1975; J. A. Steyerm., Act. Bot. Venez. 10: 220, 225, 226,
& 232. 1975; R. D. Wood, Hydrobot. Meth. 15. 1975; Anon., Biol.
Abstr. 61: AC1.298, AC1.667, & AC1.718. 1976; Duke, Phytologia
34: 24. 1976; Fleming, Genelle, & Long, Wild Fls. Fla. 25. 1976;
Follmann-Schrag, Excerpt. Bot. A.26: 508 & 513. 1976; Hocking,
Excerpt. Bot. A.28: 259. 1976; Keys, Chinese Herbs 290 & 374.
1976; Krug, Excerpt. Bot. A.26: 415. 1976; Lakela, Long, Fleming,
& Genelle, Pl. Tampa Bay, ed. 3 [Bot. Lab. Univ. S. Fla. Contrib.
73:] 38—39, 159, & 172. 1976; Long & Lakela, Fl. Trop. Fla., ed.
2, x, 17, 259—262, 938, 944, & 958. 1976; A. L. Moldenke, Phytologia 35: 62. 1976; Moldenke & Sm. in Reitz, Fl. Illus. Catar. I
Erio: [1]—103. 1976; Monteiro-Scanavacca & Mazzoni, Bol. Bot.
Univ. S. Paulo 4: 23—30 & [105]—111, fig. 1—6. 1976; Monteiro-
Scanavacca, Mazzoni, & Giulietti, Bol. Bot. Univ. S. Paulo 4:
[61]—72, fig. 1—15. 1976; Rogerson & Becker, Bull. Torrey Bot.

Club 103: 145. 1976; Soukup, Biota 11: 22. 1976; J. L. Thomas,
Bull. Ala. Mus. Nat. Hist. 2: 9. 1976; Moldenke, Phytologia 34:
247—249, 252—254, 256—260, 262—268, 271—278, 281, 390—406,
485—497, 499—503, 505—509, & 511 (1976), 35: 14—36 (1976), 35:
109—131, 252—264, 277—322, 332—364, 420—458, & 507—511
(1977), and 36: 28—32, 34—40, 42, 43, 45, 47, 49—51, 54—85, &
116. 1977; Moldenke, Biol. Abstr. 63: 2452 & 2461. 1977; A. L. Moldenke, Phytologia 36: 416. 1977; F. H. Montgomery, Seeds & Fruits
108 & 219, fig. 4. 1977; Rogerson, Becker, & Prince, Bull. Torrey
Bot. Club 104: 82. 1977.

BLASTOCAULON Ruhl.
   Additional bibliography: Rousseau, Géogr. Florist. Qué. [Trav.
Doc. Cent. Étud. Nord 7:] 120—121, 382, 470, 480, 498, 550, 625,
705, & 762, maps 221 & 222. 1974; Follmann-Schrag, Excerpt. Bot.
A.26: 503. 1976; Moldenke, Phytologia 34: 390—391 & 499 (1976),
35: 14 (1976), and 35: 287, 288, 309, & 407. 1977; F. H. Montgomery,
Seeds & Fruits 108 & 219, fig. 4. 1977.

CARPTOTEPALA Moldenke
   Additional bibliography: Cárdenas de Guevara, Act. Bot. Venez.
10: 36 & [69]. 1975; Follmann-Schrag, Excerpt. Bot. A.26: 504.
1976; Moldenke, Phytologia 34: 272, 391, & 500 (1976), 35: 31
(1976), and 35: 507. 1977.

COMANTHERA L. B. Sm.
   Additional bibliography: Cárdenas de Guevara, Act. Bot. Venez.
10: 36 & [69]. 1975; Follmann-Schrag, Excerpt. Bot. A.26: 508.
1976; Moldenke, Phytologia 34: 391—392 & 501 (1976), 35: 306,
359, & 507 (1977), and 36: 74 & 75. 1977.

COMANTHERA KEGELIANA (Körn.) Moldenke
   Additional bibliography: Moldenke, Phytologia 34: 391—392
(1976), 35: 306 & 359 (1977), and 36: 74 & 75. 1977.

ERIOCAULON Gron.
   Additional & emended bibliography: Sweet, Hort. Brit., ed. 1, 1:
435 (1826) and ed. 2, 546 & 597. 1830; Loud., Hort. Brit., ed. 2,
719. 1832; G. Don in Loud., Hort. Brit. Suppl. 1: 633. 1832;
Sweet, Hort. Brit., ed. 3, 719. 1839; G. Don in Loud., Hort. Brit.
Suppl. 2: 547. 1850; Engl., Syllab. Pflanzenfam., ed. 2, 86 (1898),
ed. 3, 92 (1903), ed. 5, 94 (1907), and ed. 6, 99. 1909; Praeger,
Journ. Roy. Hort. Soc. Lond. 36: 302—303, fig. 107. 1910; Praeger,
Irish Natur. 21: 26. 1912; Gilg in Engl., Syllab. Pflanzenfam., ed.
7, 138 (1912) and ed. 8, 140. 1919; Fedde & Schust. in Just, Bot.
Jahresber. 44: 19. 1922; Fedde in Just, Bot. Jahresber. 45 (1):
520. 1923; Fedde & Schust. in Just, Bot. Jahresber. 45 (1): 20.
1923; Gilg in Engl., Syllab. Pflanzenfam., ed. 9 & 10, 152. 1924;
Kräusel in Just, Bot. Jahresber. 44: 1161 & 1163. 1924; Fedde in
Just, Bot. Jahresber. 44: 1415. 1927; Diels in Engl., Syllab.
Pflanzenfam., ed. 11, 154. 1936; León, Fl. Cuba, imp. 1, 1: 278—

284 & 423, fig. 112. 1946; M. R. Henderson, Malay. Nat. Journ. 6:
202. 1950; K. Jones, Taxon 9: 183. 1960; Lam & Leenhouts, Blumea
10 (2): xvi. 1960; Gaussen, Viart, Legris, & Labroue, Trav. Sect.
Scient. Techn. Inst. Franç. Pond. Hors Ser. 5: 61. 1965; Boivin &
Cayouette, Naturaliste Canad. 94: 524. 1967; Naik, Journ. Indian
Bot. Soc. 52: 108—113, fig. 1—3. 1973; Lepage, Naturaliste Canad. 101: 928. 1974; Rousseau, Géogr. Florist. Qué. [Trav. Doc.
Cent. Étud. Nord 7:] 120, 382, 470, 480, 498, 509, 550, 625, 705,
& 762, maps 221 & 222. 1974; Satake, Journ. Jap. Bot. 49: 180—
183, 237—240, & 313—314. 1974; Cárdenas de Guevara, Act. Bot.
Venez. 10: 36—37 & [69]. 1975; Hinton & Rzedowski, Anal. Esc.
Nac. Cienc. Biol. 21: 61. 1975; Bole, Excerpt. Bot. A.26: 302.
1975; Hocking, Excerpt. Bot. A.26: 89 & 90. 1975; Hurusawa, Ex-
cerpt.Bot. A.26: 99. 1975; J. A. Steyerm., Act. Bot. Venez. 10:
220, 225, 226, & 232. 1975; R. D. Wood, Hydrobot. Meth. 15. 1975;
Duke, Phytologia 34: 24. 1976; Fleming, Genelle, & Long, Wild Fls.
Fla. 24. 1976; Follmann-Schrag, Excerpt. Bot. A.26: 508. 1976;
Keys, Chinese Herbs 290 & 374. 1976; Krug, Excerpt. Bot. A.26:
415. 1976; Lakela, Long, Fleming, & Genelle, Pl. Tampa Bay, ed. 3
[Bot. Lab. Univ. S. Fla. Contrib. 73:] 38 & 159. 1976; Long &
Lakela, Fl. Trop. Fla., ed. 2, x, 17, 259—261, & 938. 1976; A. L.
Moldenke, Phytologia 35: 62. 1976; Moldenke & Sm. in Reitz, Fl.
Illus. Catar. I Erio: 4—40, 43, 49, 55, 58, 67, 73, 89, 94, 95,
& 98—100, pl. 1—5. 1976; Monteiro-Scanavacca & Mazzoni, Bol.
Bot. Univ. S. Paulo 4: [23], 24, & 27. 1976; Monteiro-Scanavacca,
Mazzoni, & Giulietti, Bol. Bot. Univ. S. Paulo 4: 65 & 66. 1976;
J. L. Thomas, Bull. Ala. Mus. Nat. Hist. 2: 9. 1976; Moldenke,
Phytologia 34: 248, 249, 252, 254, 256, 260, 262—268, 273, 274,
277, 278, 392—406, 485—497, 502, & 503 (1976), 35: 35, 36, &
62 (1976), 35: 116, 117, 121—124, 128, 129, 131, 254, 256, 286,
288, 289, 292, 295, 303, 309, 310, 317, 318, 320—322, 341, 347,
350, 354, 359, 421—423, 425—427, 429, 454, 455, 457, & 508
(1977), and 36: 28, 30, 34, 37—40, 42, 56, 57, 68, 72, 80—82,
& 84. 1977; Moldenke, Biol. Abstr. 63: 2461. 1977; A. L. Moldenke,
Phytologia 36: 416. 1977.

Sweet (1839) classifies this genus in the Restiaceae, as was
done also by many other early authors.

The J. Kohlmeyer 2039, distributed as Eriocaulon sp., actually
is Lachnocaulon anceps (Walt.) Morong, while Kohlmeyer 2358
[Herb. Hamann 1243] is Mesanthemum radicans (Benth.) Körn., Hamann 2895 is Syngonanthus huberi Ruhl., and Hamann 2894 is Syngonanthus longipes Gleason.

ERIOCAULON ACHITON Körn.
    Additional bibliography: Moldenke, Phytologia 34: 264, 392, &
399. 1976.

ERIOCAULON ALPESTRE Hook. f. & Thoms.
    Additional synonymy: Eriocaulon alpestre var. alpestre [Hook.
f. & Thoms.] apud Van Royen, Blumea 10: 127. 1960.

Additional bibliography: Van Royen, Blumea 10: 127. 1960; Moldenke, Phytologia 34: 393, 488, & 491 (1976) and 36: 42. 1977.

ERIOCAULON ALTOGIBBOSUM Ruhl.
Additional bibliography: Moldenke, Phytologia 29: 88. 1974.
Hatschbach encountered this plant on "margens arenosas de corrego, zona de cerrado", flowering and fruiting in October.
Additional citations: BRAZIL: Goiás: Hatschbach 38920 (Z).

ERIOCAULON AQUATICUM (J. Hill) Druce
Additional synonymy: Eriocaulon septangulare var. septangulare [With.] apud Rousseau, Géogr. Florist. Qué. [Trav. Doc. Cent. Étud. Nord 7:] 120. 1974.
Additional bibliography: Sweet, Hort. Brit., ed. 1, 1: 435 (1826) and ed. 2, 546. 1830; Loud., Hort. Brit., ed. 2, 719. 1832; Sweet, Hort. Brit., ed. 3, 719. 1839; Engl., Syllab. Pflanzenfam., ed. 2, 86 (1898), ed. 3, 92 (1903), ed. 5, 94 (1907), and ed. 6, 99. 1909; Praeger, Journ. Hort. Soc. Lond. 36: 302—303, fig. 107, 1910; Praeger, Irish Natur. 21: 26. 1912; Gilg in Engl., Syllab. Pflanzenfam., ed. 7, 138 (1912), ed. 8, 140 (1919), and ed. 9 & 10, 152. 1924; Kräusel in Just, Bot. Jahresber. 44: 1161 & 1163. 1926; Fedde in Just, Bot. Jahresber. 44: 1415. 1927; Diels in Engl., Syllab. Pflanzenfam., ed. 11, 154. 1936; Monteiro-Scanavacca, Mazzoni, & Giulietti, Bol. Bot. Univ. S. Paulo 4: 65 & 66. 1976; Moldenke, Phytologia 34: 393 (1976) and 36: 57. 1977.
Additional & emended illustrations: Praeger, Journ. Hort. Soc. Lond. 36: fig. 107. 1910; Melchior in Engl., Syllab. Pflanzenfam., ed. 12, 2: 556, fig. 230 A—C. 1964.
Loudon (1832) lists this species as a garden plant in Great Britain, introduced from Scotland (where it is native), but it is not certain that he actually means that it is (or was) in cultivation in England. He calls it the "jointed pipewort".

ERIOCAULON ARECHAVALETAE Herter
Additional bibliography: Moldenke, Phytologia 24: 342—343 (1972) and 33: 153 & 183. 1976.

ERIOCAULON ARENICOLA Britton & Small
Additional & emended bibliography: Fedde & Schust. in Just, Bot. Jahresber. 45 (1): 20. 1923; León, Fl. Cuba, imp. 1, 1: 280 & 423. 1946; Moldenke, Phytologia 32: 464. 1975.

ERIOCAULON ARGENTINUM Castell.
Additional bibliography: Moldenke, Phytologia 29: 90. 1974; Moldenke & Sm. in Reitz, Fl. Ilus. Catar. I Erio: 7, 25—28, & 98, pl. 5, fig. 8 & 9. 1976; Moldenke, Phytologia 36: 72. 1977.
Additional illustrations: Moldenke & Sm. in Reitz, Fl. Ilus. Catar. I Erio: 25, pl. 5, fig. 8 & 9. 1976.
Pedersen encountered this species in wet ground, flowering and fruiting in October.
Additional citations: ARGENTINA: Corrientes: Pedersen 9266 (N).

ERIOCAULON ATABAPENSE Moldenke
   Additional bibliography: Moldenke, Phytologia 29: 91. 1974;
Cárdenas de Guevara, Act. Bot. Venez. 10: 37. 1975.

ERIOCAULON ATRATUM Körn.
   Additional bibliography: Moldenke, Phytologia 34: 393—394
(1976) and 35: 354. 1977.
   Recent collectors describe this plant as having the inflorescence-heads "white" or "white and fluffy" or "fluffy and grayish"
and have encountered it in wet places along roadsides in muddy
patana grasslands, "in tussock grass and on mountain tops", and
along the shady banks of streams near teafields, as well as in
small areas of marshy grassland in forest openings, at 1500—2250
meters altitude, flowering in November, and both flowering and
fruiting in February, March, and October. Jayasuriya & Sumithraarachchi refer to it as "common on rocky-sandy island in river".
   Material has been misidentified and distributed in some herbaria as E. longicuspe Hook. f.
   Additional citations: SRI LANKA: Davidse & Sumithraarachchi
8002 (W—2784452), 8035 (W—2784394); Jayasuriya & Sumithraarachchi 1567 (W—2768303); Sohmer & Waas 8722 (Lc, W—2784444);
Sumithraarachchi DBS.114 (W—2767926); Sumithraarachchi & Jayasuriya DBS.178 (W—2784402).

ERIOCAULON ATRATUM var. MAJOR Thwaites
   Additional bibliography: Moldenke, Phytologia 34: 394 (1976)
and 35: 350 & 354. 1977.
   Recent collectors describe the flower-heads of this plant as
"white" and have found it growing in open grass in marshy clearings of montane forests, at 2250 m. altitude, flowering in October.
   Additional citations: SRI LANKA: Davidse & Sumithraarachchi
7952 (W—2784398), 8003 (W—2784453).

ERIOCAULON AUSTRALE R. Br.
   Additional bibliography: Sweet, Hort. Brit., ed. 1, 1: 435
(1826) and ed. 2, 546. 1830; Loud., Hort. Brit., ed. 2, 719.
1832; Sweet, Hort. Brit., ed. 3, 719. 1839; Moldenke, Phytologia
34: 264, 265, 267, 394—395, 494, & 495 (1976) and 36: 38 & 39.
1977.
   Mrs. Clemens has collected what may be this species or E.
willdenovianum in highway ditches with bush margins, in swamps
near the sea, and in swamps with Restio and Pimelea. Her nos.
42258, 42404, & 44153, all cited below, exhibit the very narrow
and elongated basal leaves overtopping the inflorescences so
characteristic of E. willdenovianum, rather than the shorter
narrow ones characterizing most of the Asiatic plants referred
here. These two taxa, along with the very similar E. sexangulare L., require more careful study. Balgooy's collection, also
cited below, has the typical short narrow leaves and he describes

the plant as an herb with the flower-heads "gray-brown" and the anthers white. He found it growing in sandy places along streams in a water-catchment area in rainforest, at 250 m. altitude.

Both Sweet (1830) and Loudon (1832) list this plant as cultivated in England, introduced from New South Wales in 1822. They call it the "New Holland pipewort".

Additional citations: MALAYA: Penang: Balgooy 2406 (Ac). AUSTRALIA: Queensland: M. S. Clemens 42258 (Mi), 42361 (Mi), 42404 (Mi). GREAT BARRIER REEF: Stradbroke: M. S. Clemens 44153 (Mi).

ERIOCAULON AUSTRALE f. PROLIFERUM Moldenke
Additional bibliography: Moldenke, Phytologia 34: 265, 267, 395, & 494. 1976.
Additional citations: MALAYA: Johore: Khatijah & Bastiah KLU. 18541 (Ac--photo of type, N--photo of type, Z--photo of type).

ERIOCAULON BENTHAMI Kunth
Additional bibliography: Hinton & Rzedowski, Anal. Esc. Nac. Cienc. Biol. 21: 61. 1975; Moldenke, Phytologia 32: 465. 1975.

ERIOCAULON BREVIPEDUNCULATUM Merr., Philip. Journ. Sci. Bot. 2: 265. 1907 [not E. brevipedunculatum Suesseng. & Heine, 1960].
Additional bibliography: Moldenke, Phytologia 29: 95 (1974) and 36: 42. 1977.

ERIOCAULON BREVISCAPUM Körn.
Additional bibliography: Moldenke, Phytologia 32: 466 (1975) and 36: 37. 1977.
Additional citations: INDIA: Meghalaya: Myrthong 1496 (N).

ERIOCAULON BROWNIANUM Mart.
Additional bibliography: Moldenke, Phytologia 32: 466 & 468 (1975), 34: 395 (1976), and 36: 37. 1977.

Waas refers to this plant as an herb with "white" flower-heads and found it growing at the edge of a stream in secondary montane forest at 4400 feet altitude, flowering and fruiting in January. Bernardi refers to the leaves as "plicate" [probably meaning conduplicate] and the heads as "white". He encountered the plant at 1700 meters altitude, flowering in December.

The Comanor 980, R. W. Read 2040 & 2270, and Waas 840, distributed as E. brownianum actually are E. nilagirense Steud. Bernardi 16094 is a mixture with var. latifolium Moldenke.

Additional citations: INDIA: Meghalaya: Myrthong 1022 (N). SRI LANKA: Bernardi 15945 (N, W--2807708), 16094 in part (W—2807707); Waas 998 (N).

ERIOCAULON BROWNIANUM var. LATIFOLIUM Moldenke
Additional bibliography: Moldenke, Phytologia 34: 395 (1976) and 36: 37. 1977.
Recent collectors describe the flower-heads of this plant as

"white" or "white and fluffy" and encountered it along trails and
streams and in open marshy grassy areas at edges of streams, at
1700--1800 m. altitude, although Sohmer and his associates refer
to it as "rare" along roads, flowering in October. Bernardi describes the plant as "herba caespitosa, caule crasso in humo,
folia aequitantia sat lata!"

Additional citations: INDIA: Meghalaya: Myrthong 1325 (N). SRI
LANKA: Bernardi 16093 (Mu, N); Davidse & Sumithraarachchi 7993
(W--2784400); Sohmer, Jayasuriya, & Eliezer 8347 (Lc, W--2767934);
Sohmer & Waas 8699 (Lc); Sumithraarachchi & Jayasuriya DBS.190 (W-
2767929).

ERIOCAULON CEYLANICUM Körn.

Additional bibliography: Moldenke, Phytologia 34: 277 & 396.
1976.

Waas refers to this plant as an herb with a "blue flower" (apparently an error in observation) and found it growing by streams,
flowering and fruiting in October. Davidse refers to the flower-
heads as "white" and encountered the plant "in compact patches in
grassland and forest" at 7000 feet altitude.

The Read & Desautels 2272, distributed as E. ceylanicum, seems,
rather, to be E. dalzellii Körn.

Additional citations: SRI LANKA: Davidse 7607 (W--2784456);
Waas 117 (W--2767932).

ERIOCAULON CINEREUM R. Br.

Additional bibliography: Satake, Journ. Jap. Bot. 49: 237. 1974;
Hurusawa, Excerpt. Bot. A.26: 99. 1975; Keys, Chinese Herbs 290 &
374. 1976; Moldenke, Phytologia 34: 396--398, 403, 488, & 491. 1976;
A. L. Moldenke, Phytologia 36: 416. 1977.

Additional illustrations: Keys, Chinese Herbs 290. 1976.

Waas refers to this plant as an herb, to 15 cm. tall, and encountered it growing at the edge of a water-hole in "secondary scrub
near stream edge". Davidse & Sumithraarachchi found it "in full
rain-fed pool, the plants mostly submerged, on rock outcrops with
bare rocks, small grassy soil pockets, and scattered low trees, at
100 m. altitude, and describe it as having "white" flowers.

Keys (1976) reports that in China the entire plant is "Pungent,
sweet" and is used as an antiphlogistic and ophthalmic, the "Dose,
5--10 gm." Bernardi refers to the flower-heads as "fuscous".

Additional citations: INDIA: Meghalaya: Myrthong 1211 (N). SRI
LANKA: Bernardi 15946 (W--2807706); Davidse & Sumithraarachchi 8948
(Ld); R. W. Read 2176 (Ld); Waas 751 (W--2784493).

ERIOCAULON COLLETTII Hook. f.

Additional bibliography: Naik, Journ. Indian Bot. Soc. 52: 109--
111, fig. 1. 1973; Moldenke, Phytologia 29: 101. 1974.

Illustrations: Naik, Journ. Indian Bot. Soc. 52: 109, fig. 1.
1973.

ERIOCAULON COLLINUM Hook. f.
Additional bibliography: Moldenke, Phytologia 34: 263 & 398.
1976.
Recent collectors refer to the heads of this species as "black"
or "gray" and have encountered it in boggy ground, along streams,
and in roadside wet places among muddy patana grasses, flowering
and fruiting from February to April and in September.
The Waas 1000, distributed as E. collinum, actually is E.
thwaitesii Körn. instead.
Additional citations: SRI LANKA: Jayasuriya 2393 (Ld); Read &
Desautels 2274 (Ld); Sumithraarachchi DBS.113 (W—2767927); Sumithraarachchi & Jayasuriya DBS.187 (W—2767928), DBS.205 (W—
2768319); Sumithraarachchi & Waas DBS.261 (W—2768302), DBS.300
(W—276923).

ERIOCAULON COMPRESSUM Lam.
Additional bibliography: Lakela, Long, Fleming, & Genelle, Pl.
Tampa Bay, ed. 3 [Bot. Lab. Univ. S. Fla. Contrib. 73:] 38. 1976;
Long & Lakela, Fl. Trop. Fla., ed. 2, 259, 260, & 938. 1976; Mol-
denke, Phytologia 34: 398 (1976) and 36: 57. 1977.
LeBlanc encountered this plant in open pine woodlands and cut-
over marsh areas. Correll & Popenoe refer to it as growing in
the mud of savanna marshes, the heads "white".
Additional citations: FLORIDA: Martin Co.: Correll & Popenoe
48037 (N). ALABAMA: Mobile Co.: LeBlanc 270 (Ac).

ERIOCAULON CRASSISCAPUM Bong.
Additional synonymy: Eriocaulon molle "Mart. ex Körn." apud
Moldenke & Sm. in Reitz, Fl. Ilus. Catar. I Erio: 99, in syn. 1976.
Additional bibliography: Moldenke, Phytologia 32: 470. 1975;
Moldenke & Sm. in Reitz, Fl. Ilus. Catar. I Erio: 7, 25, 32—33,
& 99, pl. 5, fig. 18—23. 1976.
Additional illustrations: Moldenke & Sm. in Reitz, Fl. Ilus.
Catar. I Erio: 25, pl. 5, fig. 18—23. 1976.
This species is listed by Moldenke & Smith (1976) as probably
occurring in Santa Catarina, Brazil, although no actual specimens
of it from that state have as yet been seen by me. The vernacular
names, "capipoatinga" and "sempre-viva-do-campo", are recorded for
it and it is said to flower from June to January.

ERIOCAULON CRISTATUM Mart.
Additional bibliography: Moldenke, Phytologia 32: 470 (1975),
33: 14 (1976), and 34: 400. 1976.
Additional citations: INDIA: Meghalaya: Myrthong 1021 (N),
1203 (N).

ERIOCAULON CUBENSE Ruhl.
Additional & emended bibliography: León, Fl. Cuba, imp. 1, 1:
280 & 423. 1946; Moldenke, Phytologia 32: 470. 1975.

ERIOCAULON CUSPIDATUM Dalz.
    Additional bibliography: Moldenke, Phytologia 32: 470. 1975.
    Additional citations: INDIA: Union Territory: Vartak RD.3 (Ac).

ERIOCAULON DALZELLII Körn.
    Additional bibliography: Moldenke, Phytologia 32: 484. 1976.
    Recent collectors have found this plant growing in boggy
ground, flowering and fruiting in September, and have misidentified it as E. ceylanicum Körn.
    Additional citations: SRI LANKA: Read & Desautels 2272 (Ld).

ERIOCAULON DECANGULARE L.
    Additional bibliography: Sweet, Hort. Brit., ed. 2, 597. 1830;
Loud., Hort. Brit., ed. 2, 719. 1832; Sweet, Hort. Brit., ed. 3,
719. 1839; G. Don in Loud., Hort. Brit. Suppl. 2: 547. 1850;
Fleming, Genelle, & Long, Wild Fls. Fla. 25. 1976; Lakela, Long,
Fleming, & Genelle, Pl. Tampa Bay, ed. 3 [Bot. Lab. Univ. S. Fla.
Contrib. 73:] 38. 1976; Long & Lakela, Fl. Trop. Fla., ed. 2, 17,
259, 260, & 938. 1976; Moldenke & Sm. in Reitz, Fl. Ilus. Catar.
I Erio: 6 & 99. 1976; Monteiro-Scanavacca, Mazzoni, & Giulietti,
Bol. Bot. Univ. S. Paulo 4: 65. 1976; Moldenke, Phytologia 34:
249, 273, 274, 277, & 398 (1976) and 36: 28 & 39. 1977.
    Additional illustrations: Fleming, Genelle, & Long, Wild Fls.
Fla. 25 (in color). 1976.
    Lakela and her associates (1976) call this the "giant pipewort"
and state that in the Tampa Bay area of Florida it grows in wet
soil and swamps, flowering in summer. Correll & Popenoe refer to
it as growing in the mud of savanna marshes and describe the
flower-heads as "white".
    On July 7, 1977, I personally observed this species in cultiva-
tion (very successfully!) in a greenhouse at the New York Botanical Garden. Sweet (1830), Loudon (1832), and Don (1850) all list
it as among the plants cultivated in England, introduced from
North America in 1826. Fleming and his associates (1976) call it
by the vernacular name of "hatpins": a very appropriate name if
one is old enough to remember the hatpins of ladies' millinery.
    The Webster & Wilbur 3199, distributed as typical E. decangu-
lare, is better regarded as f. parviceps Moldenke.
    Additional citations: SOUTH CAROLINA: Georgetown Co.: Kohlmeyer
& Kohlmeyer 324 (Hm). FLORIDA: Martin Co.: Correll & Popenoe
48038 (N).

ERIOCAULON DECANGULARE f. PARVICEPS Moldenke
    Additional bibliography: Moldenke, Phytologia 32: 489. 1976.
    Additional citations: NORTH CAROLINA: J. Kohlmeyer 2036 (Hm).
Onslow Co.: J. Kohlmeyer 2033 (Hm). County undetermined: J. Kohl-
meyer 2035 [Hofmann National Forest] (Hm). TEXAS: Tyler Co.:
Webster & Wilbur 3199 (Mi).

ERIOCAULON DECEMFLORUM Maxim.
    Additional bibliography: Moldenke, Phytologia 29: 113. 1974; Sa-

take, Journ. Jap. Bot. 49: 314. 1974.

ERIOCAULON DEPAUPERATUM Merr. in Merr. & Merritt, Philip. Journ.
     Sci. Bot. 5: 336—337. 1910.
     Additional & emended bibliography: E. D. Merr. in Merr. & Mer-
ritt, Philip. Journ. Sci. Bot. 5: 336—337. 1910; Moldenke, Phytologia 34: 399. 1976.

ERIOCAULON DIANAE var. LONGIBRACTEATUM Ftson
     Additional bibliography: Moldenke, Phytologia 34: 392 & 399.
1976.
     Maxwell describes the bracts of this plant as "gray-tan" and
encountered it "in open wet sandy marsh zone near a savanna", at
75 m. altitude.
     Additional citations: THAILAND: Maxwell 75-1067 (Ac).

ERIOCAULON DICTYOPHYLLUM Körn.
     Additional synonymy: Eriocaulon dictyophyllum "Mart. ex Mol-
denke" apud Moldenke & Sm. in Reitz, Fl. Ilus. Catar. I Erio: 99,
in syn. 1976. Eriocaulon fluviatile "Bong. ex Moldenke" apud
Moldenke & Sm. in Reitz, Fl. Ilus. Catar. I Erio: 99, in syn.
1976. Paepalanthus dictyophyllus "Mart. ex Moldenke" apud Moldenke & Sm. in Reitz, Fl. Ilus. Catar. I Erio: 101, in syn. 1976.
     Additional bibliography: Moldenke, Phytologia 29: 194 (1974),
34: 399 (1976), and 35: 121. 1976; Moldenke & Sm. in Reitz, Fl.
Ilus. Catar. I Erio: 7, 25, 38—40, 99, & 101, pl. 5, fig. 32.
1976.
     Illustrations: Moldenke & Sm. in Reitz, Fl. Ilus. Catar. I
Erio: 25, pl. 5, fig. 32. 1976.
     The vernacular names, "capim-manso", "capipoatinga", "gravatá-
manso", and "sempre-viva-dos- campos", have been recorded for
this species (and many others) and it is said to flower in December and January.

ERIOCAULON DIMORPHOPETALUM Moldenke
     Additional bibliography: Cárdenas de Guevara, Act. Bot. Venez.
10: 37. 1975; Moldenke, Phytologia 34: 399. 1976.

ERIOCAULON DIOECUM Ruhl.
     Additional & emended bibliography: León, Fl. Cuba, imp. 1, 1:
280 & 423. 1946; Moldenke, Phytologia 32: 490. 1976.

ERIOCAULON ECHINOSPERMOIDEUM Ruhl.
     Additional & emended bibliography: León, Fl. Cuba, imp. 1, 1:
281 & 423. 1946; Moldenke, Phytologia 32: 490. 1976.

ERIOCAULON ECHINOSPERMUM C. Wright
     Additional & emended bibliography: León, Fl. Cuba, imp. 1, 1:
279--280 & 423. 1946; Moldenke, Phytologia 32: 490. 1976.

ERIOCAULON ECHINULATUM Mart.
    Additional bibliography: Moldenke, Phytologia 32: 263 & 399—
400. 1976.
    The Charoenphol, Larsen, & Warncke 4684 collection, cited below,
is a mixture of this species with (mostly) E. luzulaefolium Mart.
It was found growing in open wet grassland at 1100 m. altitude.
    Additional citations: THAILAND: Charoenphol, Larsen, & Warncke
4684 in part (Mu).

ERIOCAULON EHRENBERGIANUM Klotzsch
    Additional bibliography: Moldenke, Phytologia 32: 490—491 &
499 (1976) and 34: 273. 1976.
    The Schaffner 31, cited below, is a mixture with E. microceph-
alum H.B.K.
    Additional citations: MEXICO: México: Schaffner 31 in part
(Ut—3286108).

ERIOCAULON EKMANNII Ruhl.
    Additional & emended bibliography: León, Fl. Cuba, imp. 1, 1:
280—281 & 423. 1946; Moldenke, Phytologia 32: 491. 1976.

ERIOCAULON ELICHRYSOIDES Bong.
    Additional bibliography: Moldenke, Phytologia 34: 400. 1976;
Monteiro-Scanavacca & Mazzoni, Bol. Bot. Univ. S. Paulo 4: [23],
24, & 27. 1976.
    Monteiro-Scanavacca & Mazzoni (1976) report that there is no
vegetative reproduction from the apex of the inflorescence in
this species (as there is in so many other species of the genus).
They cite Semit 4449 from Minas Gerais, Brazil.

ERIOCAULON FULIGINOSUM C. Wright
    Additional & emended bibliography: León, Fl. Cuba, imp. 1, 1:
280 & 423. 1946; Moldenke, Phytologia 32: 492—493 & 505 (1976)
ans 33: 11 & 184. 1976.
    Liesner & Dwyer describe this plant as "with leaves submerged
in water and heads above water of pond" and found it growing at
sealevel, flowering and fruiting in January.
    Additional citations: BELIZE: Liesner & Dwyer 1665 (Bm, Ld).

ERIOCAULON FUSIFORME Britton & Small
    Additional & emended bibliography: Fedde & Schust. in Just, Bot.
Jahresber. 45 (1): 20. 1923; León, Fl. Cuba, imp. 1, 1: 280 & 423.
1946; Moldenke, Phytologia 32: 493. 1976.

ERIOCAULON GOMPHRENOIDES Kunth
    Additional bibliography: Moldenke, Phytologia 29: 198. 1974;
Moldenke & Sm. in Reitz, Fl. Ilus. Catar. I Erio: 6, 13, 15—19, &
99, pl. 2, fig. 7—12, & pl. 3 & 4. 1976.
    Illustrations: Moldenke & Sm. in Reitz, Fl. Ilus. Catar. I Erio:
13, 16, & 17, pl. 2, fig. 7—12, & pl. 3 & 4. 1976.
    The vernacular names, "capim-manso", "capipoatinga", "gravatá-

manso", and "sempre-viva-do-campo" are recorded for this species
(and many others) and it is said to flower from October to December. Hatschbach encountered it on "campo locaes brejosos
junta a afloramentos de arenito".

It seems most unlikely to me now that the Smith, Reitz, &
Klein 7683, previously cited by me as E. gomphrenoides, actually represents this taxon. On at least some plants of this collection, the sheath-apex is very plainly bilobed, which is not
as it should be in E. gomphrenoides according to the original
description. It seems likely to me now that this collection
represents the closely related E. megapotamicum Malme which is
supposed to have such sheaths.

Additional citations: BRAZIL: Paraná: Hatschbach 39220 (Ld).

ERIOCAULON GRAPHITINUM F. Muell. & Tate
Additional bibliography: Fedde & Schust. in Just, Bot. Jahresber. 45 (1): 20. 1923; Moldenke, Phytologia 32: 493. 1976.

ERIOCAULON GREGATUM Körn.
Additional bibliography: Moldenke, Phytologia 29: 198. 1974.
Additional citations: INDIA: Meghalaya: Myrthong 1051 (N).

ERIOCAULON HETEROLEPIS var. NIGRICANS Körn.
Additional bibliography: Moldenke, Phytologia 32: 494—495
(1976) and 36: 38. 1977.
Sinclair and his associates refer to this plant as "rare" and
found it "sparingly in flower in damp places by springs" at
10,580 feet altitude, flowering in June. Material has been misidentified and distributed in some herbaria as E. beccarii Suesseng. & Heine and E. hookerianum Stapf.
Additional citations: GREATER SUNDA ISLANDS: Sabah: Sinclair,
Kadim b. Tassim, & Kapis b. Sisiron 9137 (Mu).

ERIOCAULON HETEROPETALUM Ruhl.
Additional & emended bibliography: León, Fl. Cuba, imp. 1, 1:
281 & 423. 1946; Moldenke, Phytologia 32: 495. 1976.

ERIOCAULON HONDOENSE Satake
Additional bibliography: Moldenke, Phytologia 34: 400—401 &
404. 1976.
Additional citations: JAPAN: Honshu: Togashi MT.6849 [Fl. Jap.
Exsic. 67] (N), MT.6857 [Fl. Jap. Exsic. 68] (N).

ERIOCAULON HUMBOLDTII Kunth
Additional bibliography: Cárdenas de Guevara, Act. Bot. Venez.
10: 36. 1975; J. A. Steyerm., Act. Bot. Venez. 10: 220, 226, &
232. 1975; Moldenke, Phytologia 34: 401. 1976.
Hatschbach has encountered this plant "nas agua razas de corrego".
Additional citations: VENEZUELA: Bolívar: Hamann 2890 (Hm),
2891 (Hm); Hertel s.n. [Canaima, 3.4.1969] (Hm). BRAZIL: Bahia:

Hatschbach 39465 (Ld).

ERIOCAULON INFIRMUM Steud.
Additional bibliography: Moldenke, Phytologia 32: 495--496 (1976), 33: 9 (1976), 34: 266, 267, 401, & 494 (1976), and 36: 37 & 72. 1977.
Additional citations: INDIA: Meghalaya: Myrthong 1056 (N).

ERIOCAULON INSULARE Ruhl.
Additional & emended bibliography: León, Fl. Cuba, imp. 1, 1: 281 & 423. 1946; Moldenke, Phytologia 32: 496. 1976.

ERIOCAULON JAUENSE Moldenke
Additional bibliography: Moldenke, Phytologia 32: 496. 1976.
Additional citations: MOUNTED ILLUSTRATIONS: Mem. N. Y. Bot. Gard. 23: 849, fig. 4. 1972 (N--photo).

ERIOCAULON KLOTZSCHII Moldenke
Additional bibliography: Moldenke, Phytologia 29: 201. 1974.
Additional citations: GUYANA: R. Schomburgk 107 (Ut--325369B--isotype).

ERIOCAULON LACUSTRE Ruhl.
Additional & emended bibliography: León, Fl. Cuba, imp. 1, 1: 281 & 423. 1946; Moldenke, Phytologia 32: 497. 1976.

ERIOCAULON LANIGERUM H. Lecomte
Additional bibliography: Moldenke, Phytologia 26: 27. 1973; Satake, Journ. Jap. Bot. 49: 240. 1974.

ERIOCAULON LAOSENSE var. MAXWELLII Moldenke, Phytologia 35: 109--111. 1977.
Bibliography: Moldenke, Phytologia 35: 109--111 (1977) and 36: 38. 1977.
Illustrations: Moldenke, Phytologia 35: 110. 1977.
Citations: THAILAND: Maxwell 74-376 (Ac--type, Z--isotype, Z--drawings of type).

ERIOCAULON LEPTOPHYLLUM Kunth
Additional bibliography: Moldenke, Phytologia 26: 460. 1973; Moldenke & Sm. in Reitz, Fl. Ilus. Catar I Erio: 6, 23--26, & 99, pl. 5, fig. 1--7. 1976.
Additional illustrations: Moldenke & Sm. in Reitz, Fl. Ilus. Catar. I Erio: 25, pl. 5, fig. 1--7. 1976.
The Pedersen 9266, distributed as E. leptophyllum and so filed in some herbaria, actually is E. argentinum Castell.

ERIOCAULON LEUCOGENES Ridl.
Additional bibliography: Fedde & Schust. in Just, Bot. Jahresber. 44: 19. 1922; Moldenke, Phytologia 26: 27. 1973.

ERIOCAULON LIGULATUM (Vell.) L. B. Sm.

Additional synonymy: Eriocaulon elichrysoides "sensu Kunth" apud Moldenke & Sm. in Reitz, Fl. Ilus. Catar. I Erio: 99, in syn. 1976. Eriocaulon kunthii var. j "Körn. ex Alv. Silv." apud Moldenke & Sm. in Reitz, Fl. Ilus. Catar. I Erio: 99, in syn. 1976.

Additional bibliography: Moldenke, Phytologia 34: 402. 1976; Moldenke & Sm. in Reitz, Fl. Ilus. Catar. I Erio: 6—11, 18, & 98—100, pl. 1. 1976.

Additional illustrations: Moldenke & Sm. in Reitz, Fl. Ilus. Catar. I Erio: 9, pl. 1. 1976.

Dombrowski refers to this plant as "abundant in brejo" (sedge meadow). The vernacular names, "capim-manso", "capipoatinga", "gravatá-manso", and "sempre-viva-do-campo", have been recorded for this (and many other) species and it is said to flower from September to November.

Additional citations: BRAZIL: Paraná: Dombrowski 6446 (Ld).

ERIOCAULON LINEARE Small

Additional bibliography: J. L. Thomas, Bull. Ala. Mus. Nat. Hist. 2: 9. 1976; Moldenke, Phytologia 34: 402 (1976) and 36: 29 & 57. 1977.

Thomas (1976) lists this species as a "Species of special concern" from the conservation standpoint in Alabama, recording it from only Escambia,.Geneva, and Houston counties in that state; it is, however, also known from Covington and Baldwin counties.

ERIOCAULON LINEARIFOLIUM Körn.

Additional bibliography: Moldenke, Phytologia 32: 498. 1976.

The Anderson, Stieber, & Kirkbride "36510" cited by me in a previous installment of these notes (1973) is a typographic error for no. 36810.

ERIOCAULON LONGICUSPE Hook. f.

Additional bibliography: Moldenke, Phytologia 32: 498 (1976), 33: 10 & 14 (1976), and 34: 263. 1976.

The Jayasuriya & Sumithraarachchi 1567, distributed as E. longicuspe, actually is E. atratum Körn.

ERIOCAULON LUZULAEFOLIUM Mart.

Additional bibliography: Naik, Journ. Indian Bot. Soc. 52: 111-113, fig. 3. 1973; Bole, Excerpt. Bot. A.26: 302. 1975; Moldenke, Phytologia 32: 498. 1976.

Additional illustrations: Naik, Journ. Indian Bot. Soc. 52: 112, fig. 3. 1973.

Recent collectors have encountered this plant in open wet grasslands, at 1100—1700 m. altitude, flowering in September.

The Charoenphol, Larsen, & Warncke 4684 collection, cited below, is a mixture with (a little) E. echinulatum Mart.

Additional citations: THAILAND: Charoenphol, Larsen, & Warncke 4684 in part (Mu); Larsen & Larsen 34463 (Ac).

ERIOCAULON MAGNIFICUM Ruhl.
   Additional bibliography: Moldenke, Phytologia 34: 403. 1976;
Moldenke & Sm. in Reitz, Fl. Ilus. Catar. I Erio: 6, 13, 20—21, &
99, pl. 2, fig. 13—18. 1976.
   Additional illustrations: Moldenke & Sm. in Reitz, Fl. Ilus.
Catar. I Erio: 13, pl. 2, fig. 13—18. 1976.
   The vernacular names, "capim-manso", "capipoatinga", "gravatá-
manso", and "sempre-viva-do-campo", have been recorded for this
(and many other) species and it is said to flower from September
to February.

ERIOCAULON MAGNUM Abbiatti
   Additional bibliography: Moldenke, Phytologia 34: 403. 1976.
   Schinini and his associates have encountered this plant "en
embalsado".
   Additional citations: ARGENTINA: Corrientes: Schinini & al.
12085 (Ac).

ERIOCAULON MAJUSCULUM Ruhl.
   Additional bibliography: Moldenke, Phytologia 24: 475. 1972.
   Bogner found what appears to be this species growing in a
marsh, at 2200 meters altitude, flowering in March, but distribu-
ted it erroneously as Paepalanthus sp.
   Additional citations: BRAZIL: Minas Gerais: Bogner 1166 (Mu).

ERIOCAULON MEGAPOTAMICUM Malme
   Additional bibliography: Moldenke, Phytologia 34: 403 (1976)
and 36: 34. 1977.
   Dombrowski reports this plant "frequent" on sandy campos,
flowering and fruiting in October.
   Smith, Reitz, & Klein 7683, cited below, was previously cited
by me as E. gomphrenoides Kunth, but seems (at least in some
specimens) to have its peduncular sheath-apices definitely bilobed -- in E. gomphrenoides they are said in the original description to be truncate. It was found growing in bogs at 1650 m.
altitude, flowering in November.
   Additional citations: BRAZIL: Paraná: Dombrowski 6546 (Z). Santa Catarina: Smith, Reitz, & Klein 7683 (N, Z).

ERIOCAULON MELANOCEPHALUM Kunth
   Additional & emended bibliography: León, Fl. Cuba, imp. 1, 1:
281 & 423. 1946; Moldenke, Phytologia 34: 256 & 403. 1976.

ERIOCAULON MICROCEPHALUM H.B.K.
   Additional bibliography: Moldenke, Phytologia 32: 491 & 499—
500 (1976), 33: 47 (1976), 35: 117, 128, 129, & 347 (1977), and
36: 30. 1977.
   Weber and his associates encountered this species in a páramo
bog with Azorella, Plantago rigida, Valeriana, and Ciminalis at
3310 m. altitude. Other recent collectors in Ecuador found it

"submersed in [a] small pond in areas of dry scrub 1—3 m. tall,
in "xerophytic scrub 2—3 m. tall intermingled with meadows and
drier grasslands", and "by small wet spring in humid páramo vegetation with an abundance of Espeletia hartwegiana", at altitudes
of 2900--4350 m., flowering and fruiting in May. Humbles found
it "with Espeletia". In Mexico, McGill and his associates encountered it in meadow bogs in pine forests on rocky slopes, at
8450 feet altitude.
    The Schaffner 31, cited below, is a mixture with E. ehrenberg-
ianum Klotzsch.
    Additional citations: MEXICO: Durango: McGill, Reeves, Nash, &
Pinkava P.13392 (N). México: Schaffner 31 in part (Ut—328610B).
ECUADOR: Azuay: Holm-Nielsen, Jeppesen, Løjtnant, & Øllgaard 4991
(N), 5054 (N). Carchi: Holm-Nielsen, Jeppesen, Løjtnant, & Øllgaard 5277 (N); Humbles 6086 (Ld); Weber, Gradstein, & Lanier s.
n. [3, 4 April 1976] (Ld).

ERIOCAULON MINIMUM Lam.
    Additional bibliography: Moldenke, Phytologia 32: 500. 1976.
    Jayasuriya & Faden collected what may be this species in shal-
low soil near a stream, at 760 meters altitude, where they note
that it was "common", the heads "gray". It is classified here
tentatively because it differs strikingly from other material so
determined.
    Additional citations: SRI LANKA: Jayasuriya & Faden 2410 (Z).

ERIOCAULON MISERRIMUM Ruhl.
    Additional & emended bibliography: León, Fl. Cuba, imp. 1, 1:
280 & 423. 1946; Moldenke, Phytologia 32: 501. 1976.

ERIOCAULON MISERUM Körn.
    Additional bibliography: Moldenke, Phytologia 29: 209. 1974.
    Additional citations: INDIA: Meghalaya: Myrthong 1379 (Z),
1476 (N).

ERIOCAULON MITOPHYLUM Hook. f.
    Additional bibliography: Moldenke, Phytologia 29: 209. 1974.
    Additional citations: INDIA: Meghalaya: Myrthong 1191 (N).

ERIOCAULON MODESTUM Kunth
    Additional synonymy: Eriocaulon hygropilus Mart. ex Moldenke &
Sm. in Reitz, Fl. Ilus. Catar. I Erio: 99, in syn. 1976. Erio-
caulon modestum "Auct. ex Herter" apud Moldenke & Sm. in Reitz,
Fl. Ilus. Catar. I Erio: 99, in syn. 1976. Eriocaulon modestum f.
elatior "Ruhl. ex Moldenke" apud Moldenke & Sm. in Reitz, Fl.
Ilus. Catar. I Erio: 99, in syn. 1976. Eriocaulon modestum f.
modestum [Kunth] apud Moldenke & Sm. in Reitz, Fl. Ilus. Catar.
I Erio: 28. 1976.
    Additional bibliography: Moldenke, Phytologia 34: 404. 1976;

Moldenke & Sm. in Reitz, Fl. Ilus. Catar. I Erio: 7, 25, 28—31,
99, & 100, pl. 5, fig. 10--17. 1976.
Additional illustrations: Moldenke & Sm. in Reitz, Fl. Ilus.
Catar. I Erio: 25, pl. 5, fig. 10--17. 1976.
Vernacular names recorded for this species are "capim-manso",
"capipoatinga", "gravatá-manso", and "sempre-viva-do-campo" (also
applied to many other species) and it is said to flower from
December to April.

ERIOCAULON MODESTUM var. BREVIFOLIUM Moldenke
Additional bibliography: Moldenke, Phytologia 32: 501. 1976.
The Irwin, Harley, & Smith "32195" cited by me in a previous
(1972) installment of these notes is a typographic error for no.
32175.

ERIOCAULON MODESTUM f. VIVIPARUM Herzog
Additional synonymy: Eriocaulon dusenii "Diógo ex Moldenke &
Sm." apud Moldenke & Sm. in Reitz, Fl. Ilus. Catar. I Erio: 31 &
99, in syn. 1976. Eriocaulon modestum f. viviparum "Herzog ex
Moldenke & Smith" apud Moldenke & Sm. in Reitz, Fl. Ilus. Catar.
I Erio: 99, in syn. 1976.
Additional bibliography: Moldenke, Phytologia 29: 209. 1974;
Moldenke & Sm. in Reitz, Fl. Ilus. Catar. I Erio: 31 & 99. 1976.
The vernacular names, "capim-manso", "capipoatinga-de-broto",
"gravatá-manso", and "sempre-viva-do-campo", are recorded for
this plant (most of which are applied to many other taxa in this
genus) and it is said to flower in December.

ERIOCAULON NANELLUM Ohwi
Additional bibliography: Moldenke, Phytologia 26: 31. 1973;
Satake, Journ. Jap. Bot. 49: 313--314. 1974; Hurusawa, Excerpt.
Bot. A.26: 99. 1975.

ERIOCAULON NANELLUM var. PILIFERUM Satake, Journ. Jap. Bot. 49:
313--314. 1974.
Bibliography: Satake, Journ. Jap. Bot. 49: 313--314. 1974;
Hurusawa, Excerpt. Bot. A.26: 99. 1975.
This variety is distinguished by the pilose petals on the pistillate florets, the longer leaves and peduncles (10--20 cm.),
and the whitish acute involucral bracts. It is based on B. Kawa-
mura 316852 from Konuma moor, at an altitude of about 970 meters,
near Shiobara-machi, Tochigi prefecture, Honshu, Japan, collected
on August 7, 1973, and deposited in the herbarium of the National
Science Museum in Tokyo. The leaves are 5 cm. long.

ERIOCAULON NEO-CALEDONICUM Schlecht.
Additional bibliography: Moldenke, Phytologia 34: 405. 1976.
Additional citations: NEW CALEDONIA: Franc 266 in part (W—
1112352--cotype).

ERIOCAULON NEPALENSE Prescott

Additional bibliography: Satake, Journ. Jap. Bot. 49: 237—239. 1974; Hurusawa, Excerpt. Bot. A.26: 99. 1975; Moldenke, Phytologia 34: 405. 1976.

Additional citations: INDIA: Meghalaya: Myrthong 1561 (N).

ERIOCAULON NEPALENSE var. LAOSENSE Satake, Journ. Jap. Bot. 49: 237—239, fig. 1 & 2. 1974.

Bibliography: Satake, Journ. Jap. Bot. 49: 237—239, fig. 1 & 2. 1974; Hurusawa, Excerpt. Bot. A.26: 99. 1975.

Illustrations: Satake, Journ. Jap. Bot. 49: 238, fig. 1 & 2. 1974.

This variety is based on T. Tuyama L.57379 from 2 km. south of Ban Phu Phao, about 23 km. northeast of Phangsavanh, Laos, collected on January 2, 1958, and deposited in the Botanical Institute herbarium in Tokyo. It differs in its broader leaves and in other respects.

ERIOCAULON NILAGIRENSE Steud.

Additional bibliography: Moldenke, Phytologia 34: 405—406. 1976.

Recent collectors describe this plant as large, herbaceous, fleshy, 75 cm. tall, not deep-rooted, the leaves elongate, linear-oblong, the flower-heads "snowy-white, semiglobose, to 1.4 cm. in diameter", and the flowers white, although Waas comments that the "flower-cone [is] blackish when mature". They have found it growing in boggy ground and along footpaths by streams in secondary montane forests, at 1700—2500 m. altitude. Comanor refers to it as "frequent in running water", while Cramer found it to be quite "common".

The Bernardi 16094 is a mixture with E. brownianum Mart.

Additional citations: SRI LANKA: Bernardi 16094 in part (W— 2807707); Comanor 980 (W); Cramer 3259 (W—2615826); R. W. Read 2040 (Ld), 2270 (Ld); Waas 840 (W, W—2784502).

ERIOCAULON NILAGIRENSE f. PARVIFOLIUM Moldenke

Additional bibliography: Moldenke, Phytologia 34: 406. 1976.

Read & Desautels found this plant growing in boggy ground. Davidse refers to it as cespitose, with white flower-heads, and encountered it in marshy montane tussock grassland along streamlets, at 7000 feet altitude.

Additional citations: SRI LANKA: Davidse 7604 (W—2784405); Read & Desautels 2271 (Ld).

ERIOCAULON NOVOGUINEENSE Van Royen

Additional bibliography: Moldenke, Phytologia 32: 503. 1976.

Croft & Leland describe what appears to be this species as a "very hard cushion herb, the leaves semi-glossy dark-green, the flowers light-brown, the roots white".

Additional citations: PAPUA NEW GUINEA: Croft & Leland LAE. 65874 (Mu, Z).

ERIOCAULON ODORATUM Dalz.

Additional synonymy: Eriocaulon ordoratum Moldenke, Biol. Ab-. str. 63: 2461, sphalm. 1977.

Additional bibliography: Moldenke, Phytologia 34: 485—486. 1976; Moldenke, Biol. Abstr. 63: 2461. 1977.

ERIOCAULON OLIVACEUM Moldenke

Additional & emended bibliography: León, Fl. Cuba, imp. 1, 1: 280 & 423. 1946; Moldenke, Phytologia 32: 503. 1976.

ERIOCAULON ORYZETORUM Mart.

Additional bibliography: Moldenke, Phytologia 32: 503—504. 1976.

The Larsens encountered this plant in wet grassland at 600—700 meters altitude, flowering and fruiting in September.

Additional citations: INDIA: Meghalaya: Myrthong 1491 (N). THAILAND: Larsen & Larsen 34135 (Ac).

ERIOCAULON OVOIDEUM Britton & Small

Additional & emended bibliography: Fedde & Schust. in Just, Bot. Jahresber. 45 (1): 20. 1923; León, Fl. Cuba, imp. 1, 1: 280 & 423. 1946; Moldenke, Phytologia 34: 486 (1976) and 36: 42. 1977.

The Eriocaulon ovoideum var. ulei Knuth listed by Fedde & Schuster (1923) is Dioscorea amarantoides var. ulei Knuth in the Dioscoreaceae.

ERIOCAULON PARAGUAYENSE Körn.

Additional bibliography: Moldenke, Phytologia 34: 486. 1976.

Prance and his associates encountered this plant in buriti-grass swamps at 720 m. altitude, flowering in October.

Additional citations: BRAZIL: Mato Grosso: Prance, Lleras, & Coêlho 18982 (Z).

ERIOCAULON PARKERI B. L. Robinson

Additional synonymy: Eriocaulon septangulare var. parkeri (Robins.) Boivin & Cayouette, Naturaliste Canad. 94: 524. 1967.

Additional & emended bibliography: Boivin & Cayouette, Naturaliste Canad. 94: 524. 1967; Rousseau, Géogr. Florist. Qué. [Trav. Doc. Cent. Étud. Nord 7:] 120—121, 382, 480, 550, 625, & 762, map 222. 1974; Moldenke, Phytologia 34: 486 (1976) and 36: 57. 1977.

ERIOCAULON PELLUCIDUM Michx.

Additional & emended bibliography: Lepage, Naturaliste Canad. 101: 928. 1974; Krug, Excerpt. Bot. A,26: 415. 1976; Rousseau, Géogr. Florist. Qué. [Trav. Doc. Cent. Étud. Nord. 7:] 120—121, 470, 498. 625, 705, & 762, maps 221 & 222. 1974; A. L. Moldenke, Phytologia 35: 62. 1976; Monteiro-Scanavacca, Mazzoni, & Giulietti, Bol. Bot. Univ. S. Paulo 4: 66. 1976; Moldenke, Phytologia 34: 486—487 (1976) and 36: 57. 1977; F. H. Montgomery, Seeds & Fruits 108 & 219, fig. 4. 1977.

Additional illustrations: F. H. Montgomery, Seeds & Fruits 108, fig. 4. 1977.

Montgomery (1977) describes the seeds of this plant as "0.7 x 0.5 x 0.5 mm, elliptic 4—5 in l.s., elliptic 6 in c.s.; surface obscurely rugulose, light brown".

The J. P. Standley 33, distributed as what is now called E. pellucidum, actually is Lachnocaulon glabrum Körn., while W. Bennett 417/73 is Syngonanthus flavidulus (Michx.) Ruhl.

ERIOCAULON PINARENSE Ruhl.

Additional & emended bibliography: León, Fl. Cuba, imp. 1, 1: 281 & 423. 1946; Moldenke, Phytologia 32: 492 & 505. 1976.

ERIOCAULON PSEUDOCOMPRESSUM Ruhl.

Additional & emended bibliography: León, Fl. Cuba, imp. 1, 1: 279 & 423, fig. 112. 1946; León & Alain, Fl. Cuba, imp. 2, 1: 279 & 423, fig. 112. 1974; Moldenke, Phytologia 34: 487. 1976.

Additional & emended illustrations: León, Fl. Cuba, imp. 1, 1: 279, fig. 112. 1946; León & Alain, Fl. Cuba, imp. 2, 1: 279, fig. 112. 1974.

ERIOCAULON QUINQUANGULARE L.   .

Additional bibliography: Moldenke, Phytologia 34: 487—488. 1976; Monteiro-Scanavacca, Mazzoni, & Giulietti, Bol. Bot. Univ. S. Paulo 4: 65. 1976.

Read reports finding this plant growing "above water level when that is very low" in Sri Lanka.

The Comanor 880, distributed as E. quinquangulare, seems better placed as E. walkeri Hook. f.

Additional citations: SRI LANKA: R. W. Read 2177 (Ld); Sumithraarachchi DBS.666 (W—2806327).

ERIOCAULON RAVENELII Chapm.

Additional bibliography: Moldenke, Phytologia 29: 222. 1974; Lakela, Long, Fleming, & Genelle, Pl. Tampa Bay, ed. 3 [Bot. Lab. Univ. S. Fla. Contrib. 73:] 38. 1976; Long & Lakela, Fl. Trop. Fla., ed. 2, x, 259—[261], & 938. 1976.

Additional illustrations: Long & Lakela, Fl. Trop. Fla., ed. 2, [261]. 1976.

Lakela and her associates (1976) calls this species the "southern pipewort" and state that in the Tampa Bay area of Florida it inhabits pinelands and low ground, flowering in summer.

Additional citations: FLORIDA: Hillsborough Co.: A. P. Garber 37 (W—936873), s.n. [Tampa, Sept. 1877] (W—45320). Levy Co.: A. P. Garber s.n. [Levy Co., Nov. 1877] (W—45320).

ERIOCAULON REITZII Moldenke & Smith

Additional bibliography: Moldenke, Phytologia 29: 222. 1974; Moldenke & Sm. in Reitz, Fl. Ilus. Catar. I Erio: 6, 13, 22—24, & 100, pl. 2, fig. 19—24. 1976.

Illustrations: Moldenke & Sm. in Reitz, Fl. Ilus. Catar. I Erio: 13, pl. 2, fig. 19—24. 1976.

Although this binomial is marked as "spec. nov." in Moldenke &
Smith (1976) by editorial error, it was first actually validly
published in 1973 while the paper referred to was in press.

Vernacular names recorded for this species are "capim-manso",
"capipoatinga-de-reitz", "gravatá-manso", and "sempre-viva-do-
campo" and the plant is said to flower in October.

ERIOCAULON ROBUSTIUS (Maxim.) Mak.
Additional bibliography: Moldenke, Phytologia 34: 488 & 491.
1976; Moldenke, Biol. Abstr. 63: 2461. 1977.

ERIOCAULON ROBUSTO-BROWNIANUM Ruhl.
Additional bibliography: Naik, Journ. Indian Bot. Soc. 52: 109—
111, fig. 2. 1973; Bole, Excerpt. Bot. A.26: 302. 1975; Moldenke,
Phytologia 34: 488. 1976.
Additional illustrations: Naik, Journ. Indian Bot. Soc. 52: 110,
fig. 2. 1973.

ERIOCAULON ROLLANDII Rousseau
Synonymy: Eriocaulon septangulare f. rollandii (Rousseau) Le-
page, Naturaliste Canad. 101: 928. 1974.
Additional bibliography: Moldenke, Phytologia 25: 69. 1972;
Lepage, Naturaliste Canad. 101: 928. 1974; Krug, Excerpt. Bot. A.
26: 415. 1976.
This is a terrestrial species, not seen anywhere in water in
its range in the Rimouski and Témiscouata regions of Québec.

ERIOCAULON SATAKEANUM Tatew. & Itô
Additional bibliography: Moldenke, Phytologia 25: 70. 1972; Satake, Journ. Jap. Bot. 49: 180. 1974.

ERIOCAULON SCARIOSUM J. E. Sm.
Additional bibliography: Moldenke, Phytologia 34: 489 (1976) and
36: 39. 1977.
The Clemens collections cited below exhibit a remarkable diversity in size of plant and length of peduncle. On her no. 42443
some plants are only 1.5—2.5 cm. tall and are apparently in full
anthesis, while on no. 42099 some peduncles are up to 22 cm. in
length. The small plants closely resemble E. nanum R. Br., but
the heads are distinctly hairy under a handlens, while in E. nanum
they are smooth. Mrs. Clemens encountered these plants along
highway ditches, in swamps with Stylidium, and grassy places in
boglets, flowering in April, June, October, and November, fruiting
in April.
Additional citations: AUSTRALIA: Queensland: M. S. Clemens 42099
(Mi), 42360 (Mi), 42370 (Mi), 42443 (Mi), 44371 (Mi). GREAT BARRI-
ER REEF: Bribie: M. S. Clemens 44113 (Mi).

ERIOCAULON SCLEROCEPHALUM Ruhl.
Additional & emended bibliography: León, Fl. Cuba, imp. 1, 1: 281

& 423. 1946; Moldenke, Phytologia 33: 12 & 15. 1976.

ERIOCAULON SEEMANNII Moldenke
 Additional bibliography: Moldenke, Phytologia 34: 489. 1976.
Seymour has encountered this plant in pinebarrens.
 Additional citations: NICARAGUA: Cabo Gracias a Díos: F. C.
<u>Seymour 3611</u> (Ft).

ERIOCAULON SELLOWIANUM Kunth
 Additional synonymy: <u>Eriocaulon</u> <u>selloviana</u> "Kunth ex Moldenke"
apud Moldenke & Sm. in Reitz, Fl. Ilus. Catar. I Erio: 100, in
syn. 1976. <u>Eriocaulon</u> <u>sellowianus</u> [Kunth apud Angely" ex Moldan-
ke & Sm. in Reitz, Fl. Ilust. Catar. I Erio: 100, in syn. 1976.
<u>Eriocaulon</u> <u>sellowianum</u> var. <u>sellowianum</u> [Kunth] apud Moldenke &
Sm. in Reitz, Fl. Ilust. Catar. I Erio: 34. 1976.
 Additional bibliography: Moldenke, Phytologia 34: 489 (1976),
35: 121 (1976), and 36: 34. 1977; Moldenke & Sm. in Reitz, Fl.
Ilust. Catar. I Erio: 7, 25, 33—38, 98, & 100, pl. 5, fig. 30 &
31. 1976.
 Additional illustrations: Moldenke & Sm. in Reitz, Fl. Ilust.
Catar. I Erio: 25, pl. 5, fig. 30 & 31. 1976.
 Hatschbach encountered this plant in "brejo, base de chapada",
flowering and fruiting in October. Dombrowski reports it "fre-
quente em beira de banhado". Vernacular names recorded for it are
"caá-guaço", "capim-manso", "capipoatinga-de-sellow", "gravatá-
manso", and "sempre-viva-do-campu" and it is said to flower from
December to February.
 Additional citations: BRAZIL: Goiás: <u>hatschbach 39022</u> (Ld).
Paraná: <u>Dombrowski 5848</u> (Ld), <u>6712</u> (Ld).

ERIOCAULON SELLOWIANUM var. LONGIFOLIUM Moldenke
 Additional bibliography: Moldenke & Sm. in Reitz, Fl. Ilust.
Catar. I Erio: 34 & 36—37. 1976; Moldenke, Phytologia 34: 489
(1976) and 36: 34. 1977.
 Hatschbach encountered this plant in "brejo" (sedge meadow),
flowering in October, and refers to the flowers as "white".
 Vernacular names recorded for this variety are "capim-manso",
"capipoatinga-de-sellow-de-folha-grande", "gravatá-manso", and
"sempre-viva-do-campo" and it is said to flower in January and
February. It is included in the Santa Catarina (Brazil) flora
on the supposition that it occurs there, although no material
from that state has yet been seen by me.
 Additional citations: BRAZIL: Mato Grosso: <u>Hatschbach 25254</u>
(Ld).

ERIOCAULON SELLOWIANUM var. PARANENSE (Moldenke) Moldenke & Smith
 Additional bibliography: Moldenke, Phytologia 34: 489—490
(1976) and 35: 121. 1976; Moldenke & Sm. in Reitz, Fl. Ilust. Ca-
tar. I Erio: 25, 34, 37—38, 99, & 100, pl. 5, fig. 24—29. 1976.
 Illustrations: Moldenke & Sm. in Reitz, Fl. Ilust. Catar. I
Erio: 25, pl. 5, fig. 24—29. 1976.

Dombrowski reports this plant as "abundant" or "frequent" in "banhado" and found it in flower there in October and November. The inflorescence-heads on her no. 6256 are globose and very hard and firm, not at all compressed in drying, much like those of typical E. decangulare L. It may well be incorrectly placed here.

Vernacular names recorded for this variety are "capim-manso", "capipoatinga-de-sellow-do-parana", "gravatá-manso", and "sempreviva-do-campo" and it is said to flower in December in Santa Catarina.

Additional citations: BRAZIL: Paraná: Dombrowski 6256 (Ia), 6549 (Ld), 6678 (Ld).

ERIOCAULON SETACEUM L.
  Additional bibliography: Moldenke, Phytologia 34: 490 & 494 (1976) and 36: 57. 1977; Moldenke, Biol. Abstr. 63: 2461. 1977.
  Additional citations: INDIA: Meghalaya: Myrthong 1091 (N).

ERIOCAULON SETACEUM var. CAPILLUS-NAIADIS (Hook. f.) Moldenke
  Additional bibliography: Moldenke, Phytologia 34: 490. 1976.
  Fosberg describes this plant as having its flower-heads "grayish, emergent, lower filiform submerged leaves green" and refers to it as "very common in fallow rice field, growing in 10—20 cm. of water". Sumithraarachchi reports the "surface roots green", while Waas comments "not very deep-rooted in the mud in waterhole near paddy field" and found it in both flower and fruit in August.
  Additional citations: SRI LANKA: F. R. Fosberg 51799 (W); Sumithraarachchi DBS.670 (N), 678 (W—2806328); Waas 735 (W—2784485).

ERIOCAULON SEXANGULARE L.
  Additional bibliography: G. Don in Loud., Hort. Brit. Suppl. 1: 633 (1832) and 2: 547. 1850; M. R. Henderson, Malay. Nat. Journ. 6: 212. 1950; Moldenke, Phytologia 34: 490—496 (1976) and 36: 38 & 40. 1977; Moldenke, Biol. Abstr. 63: 2461. 1977.
  Waas describes this species as an "herb to 30 cm., flower buds white" and found it growing in secondary forest close to a stream; Balakrishnan encountered it in "paddy marsh, growing with Xyris", flowering and fruiting in January. Sumithraarachchi reports finding it "bordering paddy fields". Balgooy encountered it in sandy places "along a stream in rainforest" and refers to it as an herb with gray-brown flower-heads and white anthers.
  The Waas 923, distributed as E. sexangulare, actually is E. willdenovianum Moldenke.
  Additional citations: SRI LANKA: Balakrishnan NBK.1157 (W); Davidse 7826 (W—2784425); Sumithraarachchi DBS.669 (N, W—2806326); Waas 887 (W—2784431). MALAYA: Penang: Balgooy 2406 (N).

ERIOCAULON SIGMOIDEUM C. Wright
  Additional & emended bibliography: León, Fl. Cuba, imp. 1, 1: 280 & 423. 1946; Moldenke, Phytologia 33: 12 & 15. 1976.

ERIOCAULON SOLLYANUM Royle
    Additional bibliography: Moldenke, Phytologia 34: 492—493
(1976) and 36: 38. 1977.
    Additional citations: INDIA: Meghalaya: Myrthong 1239 (N).

ERIOCAULON SPRUCEANUM f. VIVIPARUM Moldenke
    Additional bibliography: Moldenke, Phytologia 33: 15—16 (1976)
and 36: 34. 1977.
    Recent collectors have found this plant growing "in dry catin-
ga" vegetation.
    Additional citations: BRAZIL: Roraima: Murça Pires, Cavalcante,
Magnago, & Silva 13980 (Ld).

ERIOCAULON STAINTONII Satake, Journ. Jap. Bot. 49: 314—317. 1974.
    Bibliography: Satake, Journ. Jap. Bot. 49: 314—317. 1974; Hu-
rusawa, Excerpt. Bot. A.26: 99. 1975.
    This new species is described from Nepal.

ERIOCAULON STEYERMARKII Moldenke
    Additional bibliography: Cárdenas de Guevara, Act. Bot. Venez.
10: 36. 1975; Moldenke, Phytologia 33: 16. 1976.

ERIOCAULON STRICTUM Milne-Redhead
    Additional bibliography: Moldenke, Phytologia 26: 464. 1973.
    A letter from Robert Wingfield to me, dated April 19, 1977,
lists this species from Mafia island and cites Vesey-Fitzgerald
5213/3 at Kew and at the East Africa Herbarium in Nairobi.

ERIOCAULON TENUIFOLIUM Klotzsch
    Additional bibliography: Moldenke, Phytologia 34: 493. 1976.
    Recent collectors speak of this plant as "common", describe the
flowering-heads as white, and found it in anthesis in June.
    Additional citations: GUYANA: Goodland 515 (N). BRAZIL: Rorai-
ma: Murça Pires & Leite 273 [Herb. IPEAN 14797] (Ld).

ERIOCAULON TENUISSIMUM Nakai
    Additional bibliography: Fedde & Schust. in Just, Bot. Jahres-
ber. 45 (1): 20. 1923; Moldenke, Phytologia 29: 233. 1974.

ERIOCAULON TEXENSE Körn.
    Additional bibliography: J. L. Thomas, Bull. Ala. Mus. Nat.
Hist. 2: 9. 1976; Moldenke, Phytologia 34: 493 (1976) and 36: 29.
1977.
    Correll and his associates have encountered this plant "in seep-
age in scrub oak pinelands", flowering in May. Thomas (1976) includes it among his "Species of special concern" from the conservation standpoint in Alabama and records it from pitcherplant bogs
in only Escambia, Mobile, and Washington counties in that state. As
yet I have seen no material confirming these records for that state.
    Additional citations: TEXAS: Tyler Co.: Correll, Johnston, &
Edwin 22333 (N).

ERIOCAULON THWAITESII Körn.

Additional bibliography: Moldenke, Phytologia 34: 494 & 495. 1976.

Waas refers to this plant as an herb, to 15 cm. tall, the "flower-heads white" [but they are virtually black on his dried specimens], and found the plant "in secondary forest edge near [a] stream". Others have found it in grassy roadside ditches through tea plantations.

Material of this species has been misidentified and distributed in some herbaria as E. collinum Hook. f.

Additional citations: SRI LANKA: Davidse 7822 (W—2784424); Davidse & Sumithraarachchi 7923 (W—2784393); Waas 1000 (W).

ERIOCAULON TRUNCATUM Hamilt.

Additional bibliography: M. R. Henderson, Malay. Nat. Journ. 6: 212. 1950; Moldenke, Phytologia 34: 491 & 493—495. 1976; Moldenke, Biol. Abstr, 63: 2461. 1977.

The Bernardi 15946, distributed as E. truncatum, actually is E. cinereum R. Br.

Additional citations: INDIA: Meghalaya: Myrthong 1466 (N).

ERIOCAULON TUTIDAE Satake, Journ. Jap. Bot. 49: 180—183, fig. 1 & 2. 1974.

Bibliography: Satake, Journ. Jap. Bot. 49: 180—183, fig. 1 & 2. 1974; Hurusawa, Excerpt. Bot. A.26: 99. 1975.

Illustrations: Satake, Journ. Jap. Bot. 49: 181 & 182, fig. 1 & 2. 1974.

This species is based on Z. Tutida s.n. from the Garimegi moor near Tashiro, Ugo-machi, Akita prefecture, Honshu, Japan, collected on September 14, 1973, and deposited in the herbarium of the National Science Museum in Tokyo. It is named in honor of the collector.

ERIOCAULON TUYAMAE Satake, Journ. Jap. Bot. 49: 237, 239, & 240, fig. 3 & 4. 1974.

Bibliography: Satake, Journ. Jap. Bot. 49: 237, 239, & 240, fig. 3 & 4. 1974.

Illustrations: Satake, Journ. Jap. Bot. 49: 239, fig. 3 & 4. 1974.

The type of this species was collected by T. Tuyama (no. L.57378) -- in whose honor it is named — 2 km. south of Ban Phu Phao, about 23 km. northeast of Phangsavanh, Laos, on January 2, 1958, and is deposited in the herbarium of the Botanical Institute in Tokyo. Satake (1974) claims that "This species is near to E. lanigerum Lecomte in having ciliolate leaves, peduncles, vaginas, involucral and floral bracts, however the female sepals are glabrous and carinate or narrowly alate on the back side".

ERIOCAULON ULAEI Ruhl.

Additional synonymy: Eriocaulon ulaei var. ulaei [Ruhl.] apud

Moldenke & Sm. in Reitz, Fl. Ilus. Catar. I Erio: 12. 1976.
    Additional bibliography: Moldenke, Phytologia 29: 285. 1974;
Moldenke & Sm. in Reitz, Fl. Ilus. Catar. I Erio: 6, 11—15, &
100, pl. 2, fig. 1—6. 1976.
    Illustrations: Moldenke & Sm. in Reitz, Fl. Ilus. Catar. I Er-
io: 13, pl. 2, fig. 1—6. 1976.
    Vernacular names recorded for this species are "capim-manso",
"capipoatinga-de-ule", "gravatá-manso", and "sempre-viva-do-campo"
and it is said to flower in January in Santa Catarina, Brazil.

ERIOCAULON ULAEI var. RADIOSUM Ruhl.
    Additional bibliography: Moldenke, Phytologia 29: 236. 1974;
Moldenke & Sm. in Reitz, Fl. Ilus. Catar. I Erio: 12, 14, 15, &
100. 1976.

ERIOCAULON WALKERI Hook. f.
    Additional bibliography: Moldenke, Phytologia 34: 495. 1976.
Comanor encountered this plant "in open scrub community on sand,
in herbaceous community".
    Additional citations: SRI LANKA: Comanor 880 (W).

ERIOCAULON WILLDENOVIANUM Moldenke
    Additional bibliography: Moldenke, Phytologia 34: 491 & 495—
496. 1976.
    Read and Jayaweera encountered this plant in swampy ground
"under Hevea trees". It has been found in flower and fruit in
December. Material has been misidentified and distributed in some
herbaria as E. brownianum Mart. Poilane 8068 was previously erroneously reported as E. sexangulare L.
    Additional citations: SRI LANKA: Read & Jayaweera 2319 (W);
Waas 923 (W). INDOCHINA: Tonkin: Poilane 8068 (N).

LACHNOCAULON Kunth
    Additional & emended bibliography: León, Fl. Cuba, imp. 1, 1:
279, 284, & 426. 1946; Follmann-Schrag, Excerpt. Bot. A.26: 510.
1976; Lakela, Long, Fleming, & Genelle, Pl. Tampa Bay, ed. 3 [Bot.
Lab. Univ. S. Fla. Contrib. 73:] 38—39, 151, & 165. 1976; Long &
Lakela, Fl. Trop. Fla., ed. 2, 259, 260, 262, & 944. 1976; Moldenke, Phytologia 34: 497 (1976), 35: 14 (1976), 35: 111, 288, 309, &
509 (1977), and 36: 28—31, 43, & 57. 1977; Moldenke, Biol. Abstr.
63: 2461. 1977.
    Lakela and her associates (1976) gives "bog buttons" as the
common name for members of this genus.
    The Pollard s.n. [Tampa, March 7, 1898], distributed as Lachno-
caulon, actually is Syngonanthus flavidulus (Michx.) Ruhl.

LACHNOCAULON ANCEPS (Walt.) Morong
    Additional synonymy: Lachnocaulon anceps (Walt.) DC. ex Molden-
ke, Phytologia 36: 43, in syn. 1977.
    Additional & emended bibliography: León, Fl. Cuba, imp. 1, 1:
284 & 426. 1946; Long & Lakela, Fl. Trop. Fla., ed. 2, 260, 944, &

962. 1976; Moldenke, Phytologia 35: 14 (1976) and 36: 28--31, 43,. & 57. 1977.

Hebert found this plant growing on open wooded hillsides and wooded marshes.

Long & Lakela (1976) regard L. floridanum Small and L. glabrum Körn. as synonyms of L. anceps. It seems to me that the latter, at least, is abundantly distinct, while the former is much more closely allied to L. engleri Ruhl. or L. minus (Chapm.) Small. Specimens intermediate between L. anceps and L. glabrum probably represent L. anceps f. glabrescens Moldenke.

Recent collectors have encountered L. anceps on moist slopes of sand hills. Arsène refers to it as "rare" in Saint Tammany Parish, Louisiana, while Mohr reports it "very common" in Mobile County, Alabama. Kral describes it as "forming large circular clumps in moist sandy peat of sedge bogs" and forming "large clumps, not as large nor as narrow-leaved as nearby L. beyrichianum in rather dry sand of longleaf pine - saw-palmetto forests".

L. C. Anderson, in a personal communication to me, lists this species (sens. lat.?) from the following counties of Florida: Bay, Bradford, Brevard, Broward, Charlotte, Dixie, Franklin, Gulf, Hillsboro, Holmes, Indian River, Jackson, Lee, Levy, Liberty, Madison, Manatee, Martin, Nassau, Okaloosa, Okeechobee, Orange, Osceola, Pinellas, St. Johns, St. Lucie, Santa Rosa, Seminole, Sumter, Taylor, Union, Volusia, and Wakulla.

Material of this species has been misidentified and distributed in some herbaria as Eriocaulon sp. On the other hand, the Brass 15148, Ekman 12410, and R. Kral 20039 & 20420, distributed as (and in the case of the Ekman collection previously cited by me as) typical L. anceps, are better placed as f. glabrescens Moldenke, while Combs & Baker 1113, Coville 70 & 101, Fox & Boyce 3780, McCarthy s.n. [Julio 1885] & s.n. [N. Carolina 1885], C. Mohr s.n. [Aug. 18, 1879], and Reynolds s.n. [St. Augustine, June '75] are L. beyrichianum Sporleder, R. Kral 18012, 18048, 18288, & 20424, J. P. Standley 33, and P. C. Standley 52589 are L. glabrum Körn., Drushel 9642, E. S. Ford 4644, R. Kral 15637, Meislahn 158a, C. Mohr s.n. [April 20, 1868], and O'Neill 7786 (in part) are L. minus (Chapm.) Small, and Milligan s.n. [May 1890] is Syngonanthus flavidulus (Michx.) Ruhl. Kral & Ricks 16991, at least in the United States National Herbarium, is a mixture with fragments of Panicum sp.

Additional citations: VIRGINIA: Dinwiddie Co.: Fernald & Long 6120 (W--1682737). Greensville Co.: Smith & Hodgdon Pl. Exsicc. Gray. 1028 (W--1828401). Nansemond Co.: Fernald, Long, & Clement 15239 (W--2003605). Prince George Co.: Fernald, Long, & Smart 5698 (W--1682490). NORTH CAROLINA: Beaufort Co.: Godfrey 4403 (W--1767462). Bladen Co.: Biltmore Herb. 2755 (W--331153). Brunswick Co.: Godfrey & Shunk 4118 (W--1767244). Columbus Co.: Godfrey &

Shunk 4190 (W—1767299). Craven Co.: Kearney 1940 (W—356593).
Greene Co.: Godfrey 4316 (W—1767396). Harnett Co.: Godfrey 4253
(W—1767344). Martin Co.: Drushel 10141 (W—1688981). New Hanover Co.: M. A. Chase 3153 (W—594230); Coville 261 (W—45321).
Onslow Co.: Godfrey 4488 (W—1767535). Pender Co.: Hyams s.n.
[Burgaw, Aug. 1879] (W—152112). Sampson Co.: Godfrey 4511 (W—
1767552). SOUTH CAROLINA: Charleston Co.: Stewart s.n. [Charles-
ton] (W—202807). Darlington Co.: Norton s.n. [Hartsville, July
8, 1920] (W—1070504). Georgetown Co.: Godfrey & Tryon 51 (W—
1836996); J. Kohlmeyer 2038 (Hm). Kershaw Co.: House 2644 (W—
514155). Lexington Co.: Godfrey & Tryon 1210 (W—1837924).
GEORGIA: Calhoun Co.: Thorne 3589 (W—2005898). Clinch Co.: R.
Kral 24289 (W—2470347). Douglas Co.: Cronquist 5424 (W—1928743).
Emanuel Co.: R. M. Harper 804 (W—400280). Macon Co.: Pyron & Mc-
Vaugh 498 (W—1811260). Miller Co.: Thorne 4426 (W—2005929).
Pierce Co.: R. Kral 24150 (W—2470416). Screven Co.: R. Kral
24052 (W—2470432). Sumter Co.: R. M. Harper 443 (W—384447).
Worth Co.: Pollard & Maxon 562 (W—443076). FLORIDA: Alachua Co.:
O'Neill 633 (W—1241628), 746 (W—1241627). Calhoun Co.: Canby s.
n. [Magnolia] (W—202501). Clay Co.: Canby s.n. [Hibernia, 1869]
(W—45323, W—204860). Duval Co.: Curtiss 3021 (W—45324, W—
936868), 4139 (W—218358), 4861 (W—224477), s.n. [Jacksonville]
(W—152110); Fredholm 104 (W—264414). Lake Co.: Nash 1942 (W—
252693, W—309072). Lee Co.: Francis 61 (W—1036539). Levy Co.:
Kral & Kral 6593 (W—2308330). Nassau Co.: Godfrey & Lindsey
56895 (W—2329838). Okeechobee Co.: R. Kral 20478 (W—2470304).
Orange Co.: O'Neill s.n. [Bithlo, June 17, 1929] (W—1488705).
Osceola Co.: R. Kral 20468 (W—2470368); Swallen 300 (W—1631128).
Volusia Co.: R. Kral 20443 (W—2470337), 20453 (W—2470404). Coun-
ty undetermined: Biltmore Herb. 2755a (W—335142); Chapman s.n.
[Florida] (W—45326); Herb. Chapman 554 (W—936769), s.n. [Flor-
ida] (W—957067). ALABAMA: DeKalb Co.: A. Ruth 125 (W—345309).
Mobile Co.: Hebert 238 (Ac); Mackenzie 4059 (W—648847); Mohr s.n.
[June 1879] (W—784497), s.n. [June-Aug.] (W—784498), s.n. [July]
(W—152111). MISSISSIPPI: Covington Co.: Webster & Wilbur 3337
(Mi, W—2068060). Jackson Co.: Skehan s.n. [Ocean Springs, 5/7/
1895] (W—309071); Tracy 5031 (W—341108). Pearl River Co.: R.
Kral 17332 (W—2470308). LOUISIANA: Beauregard Par.: R. Kral
20156 (W—2470389), 20197 (W—2460390); Kral & Ricks 16991 (W—
2470367). Rapides Par.: R. Kral 20069 (W—2470369). Saint Tam-
many Par.: Arsène 12142 (W—1033047), 12315 (W—1033056). TEXAS:
Newton Co.: Correll, Johnston, & Edwin 22288 (N).

LACHNOCAULON ANCEPS f. GLABRESCENS Moldenke
    Additional bibliography: Moldenke, Phytologia 33: 20 & 21 (1976)

and 36: 28—31. 1977.

Brass describes this plant as having "whitish" flower-heads and forming cushions 20—35 cm. in diameter in dry low pinelands and on low sand-scrub ridges where he found it "abundant". Kral found it "usually at upper edge of bog with Pinguicula pumila on sandy clay peat of bog in longleaf pine country". On the Isla de Pinos, Cuba, where it has previously been regarded (incorrectly) as typical L. anceps (Walt.) Morong, it inhabits moist sandy pinelands.

It is worth pointing out that even the type collection of this form exhibits the variability ascribed to it -- some of the peduncles are densely pilose with more or less appressed antrorse hairs, others partially or completely glabrous, apparently not always connected with age. It has been found in flower in May and November and both in flower and fruit in July.

Material of this form has usually been identified and distributed in herbaria as typical L. anceps (Walt.) Morong or as L. minus (Chapm.) Small.

Additional citations: SOUTH CAROLINA: Kershaw Co.: House 2685 (W--514159). FLORIDA: Highlands Co.: Brass 15148 (W--2065381). Manatee Co.: S. M. Tracy 7585 (W--4422332--isotype). Martin Co.: R. Kral 20420 (W--2470305). LOUISIANA: Vernon Par.: R. Kral 20039 (W--2470388). ISLA DE PINOS: Ekman 12410 (N, S, W--1302554).

LACHNOCAULON BEYRICHIANUM Sporleder

Additional bibliography: Moldenke, Phytologia 33: 21 (1976) and 36: 29. 1977.

Combs and Baker describe this plant as a "common perennial in bunches in open sandy scrub"; Fox and Boyce found it "in turkey oak community on lakeshores"; Harper encountered it "in rather dry pine barrens in Neocene geologic formation overlaid by Lafayette and Columbia". It has been found in flower and fruit in June, July, and September. Coville encountered it on savannas in North Carolina.

Usually the flowering- and fruiting-heads of this species are rather soft, flattening out under pressure in drying [e.g., R. M. Harper 1491), but in Godfrey 4680 they are quite round and hard as in L. minus (Chapm.) Small; the leaves, however, are narrow and thin as in typical L. beyrichianum. Perhaps it represents a hybrid. Kral 20378 exhibits leaves that are somewhat broader than usual and similar to those of L. minus.

McCarthy s.n. [N. Carolina], Reynolds s.n. [St. Augustine, June '75], and Coville 101 comprise exceptionally small 1- or 2-peduncled plants with especially small heads, perhaps because of immaturity and poor growing conditions. The United States National Herbarium specimen of Mohr s.n. [Aug. 18, 1879] from Alabama includes 2 small scapes of L. engleri Ruhl., possibly accidentally included during mounting and actually belonging to Herb. Chapman s.n. [St. Andrew's Bay] from Florida.

Material of L. beyrichianum has been misidentified and distributed in some herbaria as L. digynum Körn. On the other hand, the Curtiss 3022, distributed as L. beyrichianum, actually is the type collection of L. eciliatum Small, while Nash 1184 is the type collection of L. engleri Ruhl. and R. M. Harper 1607, Mohr s.n. [April 20, 1868], and Nash 148, 1295, & 1855 are L. minus (Chapm.) Small. R. Kral 20418 appears to be a mixture with L. glabrum Körn.

Additional citations: NORTH CAROLINA: Bladen Co.: Fox & Boyce 3780 (W--2265510). New Hanover Co.: Coville 70 (W--45322), 101 (W--45318); Godfrey 4680 (W--1767654). County undetermined: Mc Carthy s.n. [Julio 1885] (W--152109), s.n. [N. Carolina] (W--45325). GEORGIA: Charlton Co.: R. M. Harper 1491 (W--431796). FLORIDA: Martin Co.: R. Kral 20418 in part (W--2470379). Orange Co.: Combs & Baker 1113 (W--592685). Saint Johns Co.: Reynolds s. n. [St. Augustine, June '75] (W--936867). Saint Lucie Co.: R. Kral 20378 (W--2470381). Volusia Co.: R. Kral 20441 (W--2470338). ALABAMA: Mobile Co.: Mohr s.n. [Aug. 18, 1879] (W--784496).

LACHNOCAULON CUBENSE Ruhl.
Additional & emended bibliography: León, Fl. Cuba, imp. 1, 1: 284 & 426. 1946; Moldenke, Phytologia 33: 21. 1976.

LACHNOCAULON DIGYNUM Körn.
Additional synonymy: Lachnocaulon digynum Sporl. ex Moldenke, Phytologia 36: 43, in syn. 1977.
Additional bibliography: Moldenke, Phytologia 33: 21 (1976) and 36: 43. 1977.
Additional citations: FLORIDA: Bay Co.: R. Kral 15667 (W--2470435). Escambia Co.: R. Kral 17634 (W--2470436), 23169 (W--2470424).

LACHNOCAULON ECILIATUM Small
Additional bibliography: Moldenke, Phytologia 26: 184. 1973; Lakela, Long, Fleming, & Genelle, Pl. Tampa Bay, ed. 3 [Bot. Lab. Univ. S. Fla. Contrib. 73:] 39 & 165. 1976; Long & Lakela, Fl. Trop. Fla., ed. 2, 262 & 944. 1976; Moldenke, Phytologia 36: 29. 1977.
Lakela and her associates (1976) reduce L. eciliatum to the synonymy of L. minus (Chapm.) Small. It seems to me, however, that the flowering- and fruiting-heads of L. eciliatum are usually softer and more apt to crush and become flattened under the pressure of the plant-press.
Brass describes L. eciliatum as gregarious and very abundant on open sandy lakeshores, flowering and fruiting in January. Material has been misidentified and distributed in some herbaria as L. beyrichianum Sporleder, L. engleri Ruhl., and L. minus (Chapm.) Small. In L. minus the flowering- and fruiting-heads are normally quite hard, and do not flatten out in pressing unless very immature.

BOOK REVIEWS

Alma L. Moldenke

"ANNUAL REVIEW OF PLANT PHYSIOLOGY" Volume 26 edited by Winslow R.
   Briggs with Paul B. Green & Russell L. Jones, 531 pp., illus.,
   Annual Reviews Inc., Palo Alto, California 94306. 1975.
   $15.00 in U.S.A., $15.50 foreign.

The prefatory chapter in this volume recognizes Dr. Robert Hill,
whose survey article is entitled "Days of Visual Spectroscopy and
records the "scientific developments.....emanating from the colors
of blood and grass". The following 23 papers cover recent work
and ideas such as: Plasmodesmata as channels for symplastic trans-
port even for viral size; Biochemistry of pathogenesis with the
hypothesis that gene-for-gene pathogens have on their surfaces
molecules that differ from race to race by one glycosyl and prove
virulent to a host if not "recognized" by it; Tree photosynthesis
reporting that all trees except mangroves (what genera?) so far
classify as C_3 plants; Chlorophyll-proteins; Stomatal action with
an excellent operative diagram; and Incompatibility and pollen-
stigma interaction especially in Oenothera, Lilium, Brassica,
Petunia and Ambrosia in which the principal hay fever allergin --
antigen E -- has been located in the interbacular cavities of the
exine and also in the intine.
   These and other pertinent topics are well presented. On p. 103
a species name of Acer and on p. 110 one of Tilia are misspelled.
Each paper is provided, as is characteristic of all "Annual Reviews", with important bibliographic material.

"ANNUAL REVIEW OF ECOLOGY AND SYSTEMATICS" Volume 6 edited by Richard F. Johnston with Peter W. Frank & Charles D. Michener,
   viii & 422 pp., illus., Annual Reviews Inc., Palo Alto, California 94306. 1975. $15.00 U.S.A. & $15.50 foreign.

This volume continues in the fine tradition of this series with
15 timely, well documented papers. Among others, they include
Fretwell's appreciation of Robert MacArthur's impact on ecology
through methodology and his enthusiasm, rumen microbial ecosystems
whose total products tend to be "remarkably constant", tropical
rain forests whose productivity varies seasonally with the trans-
piration rate, cannibalism in certain insects, rotifers, etc. as a
"normal response to many environmental factors", demographic consequences of human infanticide, energetics of pollination affect-
ing divergent and convergent evolution of flowers, late Quaternary
climatic change in Africa, the first survey of butterfly ecology by
Lawrence Gilbert and Michael Singer, and Paul Ehrlich's population
498

biology of coral reef fishes with comparisons and contrasts with butterfly populations.

"KNOW YOUR LILIES" by Richmond E. Harrison, 84 pp., illus., A. H. & A. W. Reed, Wellington, Auckland, Sydney & Melbourne, in U.S.A. imported for sale by Charles E. Tuttle Co., Inc., Rutland, Vermont 05701. 1971. $17.50.

My consternation at the fat price for this thin book — even though the author is a world-famous horticulturist and lily speci_ alist — was completely obliterated by my wonder at its many glorious color plates — produced mostly by the author's photographer son Charles and also by Herman v. Wall. From Agapanthus to Zante_deschia there are 32 sketches and descriptions of over 100 plants bearing the common name "lily" but which are not members of the genus Lilium.

The cultivated species of Lilium are described as well as their horticultural history and growing conditions. Asiatic, Martagon, Candidum, American, Longiflorum, Trumpet, Oriental and other new hybrids are similarly treated. Pollen can be refrigerated and often kept viable for about two months, increasing breeding possibilities.

"THE INDIGENOUS TREES OF THE HAWAIIAN ISLANDS" by Joseph F. Rock, 2nd edition introduced by Sherwin Carlquist, xx & 548 pp., illus., Pacific Tropical Botanical Garden, Lawaii on Kauai & Hana on Maui, Hawaii. Charles H. Tuttle Company Inc., Publishers, Tokyo, Japan, & Rutland, Vermont 05701. 1974. $22.50 slip-cased.

For several reasons this new edition is and will be welcomed: first and foremost, for the needed increased availability of this long recognized botanical classic;   secondly for the neat and complete copy of the original 1913 privately published text with its regional descriptions, keys, plant descriptions, common names and uses; thirdly for the fine reproduction of the 215 photographic plates of great technical excellence (Rock was originally trained as a professional photographer); fourth for its enrichment by Sherwin Carlquist's introduction to this new edition tracing the probable origin of and need to protect the remnants of the original forests; and fifth for the helpful addenda to this new edition supplied by Derral Herbst listing all text names with any revisions in modern nomenclature and geographic distribution among these islands.

Index to authors in Volume Thirty-six

- - - - - - - - - - - - - - - -

Index to supraspecific scientific names in Volume Thirty-six

- - - - - - - - - - - - - - - -

Publication dates

Vol. 35, No. 5 — April 2, 1972

Vol. 35, No. 6 — May 2, 1977

Vol. 36, No. 1 — June 18, 1977

Vol. 36, No. 2 — June 23, 1977

Vol. 36, No. 3 — July 15, 1977

Note to Librarians: Through an unfortunate printer's error the front cover of most copies of Volume 35, Number 5, bear the printed statement "No. 1". The correct "No. 5" appears on the runninghead of all the even-numbered pages from page 330 to page 400.

Lightning Source UK Ltd.
Milton Keynes UK
UKHW011848140219
337178UK00015B/429/P